**Advances
in Electrochemical
Science and Engineering**

Volume 10
Electrochemical
Surface Modification

Advances in Electrochemical Science and Engineering

Advisory Board

Prof. Elton Cairns, University of California, Berkeley, California, USA
Prof. Adam Heller, University of Texas, Austin, Texas, USA
Prof. Dieter Landolt, Ecole Polytechnique Fédérale, Lausanne, Switzerland
Prof. Roger Parsons, University of Southampton, Southampton, UK
Prof. Laurie Peter, University of Bath, Bath, UK
Prof. Sergio Trasatti, Università di Milano, Milano, Italy
Prof. Lubomyr Romankiw, IBM Watson Research Center, Yorktown Heights, USA

In collaboration with the International Society of Electrochemistry

Advances in Electrochemical Science and Engineering

Volume 10
Electrochemical Surface Modification
Thin Films, Functionalization and Characterization

Edited by
Richard C. Alkire, Dieter M. Kolb, Jacek Lipkowski,
and Philip N. Ross

WILEY-VCH

WILEY-VCH Verlag GmbH & Co. KGaA

The Editors

Prof. Richard C. Alkire
University of Illinois
600 South Mathews Avenue
Urbana, IL 61801
USA

Prof. Dieter M. Kolb
University of Ulm
Institute of Electrochemistry
Albert-Einstein-Allee 47
89081 Ulm
Germany

Prof. Jacek Lipkowski
University of Guelph
Department of Chemistry
N1G 2W1 Guelph, Ontario
Canada

Prof. Philip N. Ross
Lawrence Berkeley National Laboratory
Materials Science Department
1 Cyclotron Road MS 2-100
Berkeley, CA 94720-0001
USA

All books published by Wiley-VCH are carefully produced. Nevertheless, authors, editors, and publisher do not warrant the information contained in these books, including this book, to be free of errors. Readers are advised to keep in mind that statements, data, illustrations, procedural details or other items may inadvertently be inaccurate.

Library of Congress Card No.: applied for

British Library Cataloguing-in-Publication Data
A catalogue record for this book is available from the British Library.

Bibliographic information published by Die Deutsche Bibliothek
Die Deutsche Bibliothek lists this publication in the Deutsche Nationalbibliografie; detailed bibliographic data are available in the Internet at http://dnb.d-nb.de.

© 2008 WILEY-VCH Verlag GmbH & Co. KGaA, Weinheim Germany

All rights reserved (including those of translation into other languages). No part of this book may be reproduced in any form – by photoprinting, microfilm, or any other means – nor transmitted or translated into a machine language without written permission from the publishers. Registered names, trademarks, etc. used in this book, even when not specifically marked as such, are not to be considered unprotected by law.

Typesetting Thomson Digital, Noida, India
Printing Strauss GmbH, Mörlenbach
Binding Litges & Dopf GmbH, Heppenheim
Cover Design Grafik-Design Schulz, Fußgöheim

Printed in the Federal Republic of Germany
Printed on acid-free paper

ISBN: 978-3-527-31419-5
ISSN: 0938-5193

Contents

Series Preface IX
Volume Preface XI
List of Contributors XIII

1	**Valve Metal, Si and Ceramic Oxides as Dielectric Films for Passive and Active Electronic Devices** 1	
	Alexander Michaelis	
1.1	Introduction 1	
1.1.1	Experimental Approaches 2	
1.2	Fundamentals and Experimental Details 5	
1.2.1	Electrochemical Oxide Layer Formation on Valve Metals 5	
1.2.2	The $C(U)$ Curve of a Valve Metal Electrode 7	
1.2.3	Application of Lasers in Electrochemistry 8	
1.2.3.1	Thermal Effects 9	
1.2.4	Electrochemical Photocurrent Measurements (Optical/Electrical Method Class), Introduction of a New Model 10	
1.2.4.1	Photocurrent Model for Ultra-thin, Amorphous Films With TiO_2 as an Example 11	
1.3	Valve Metal Systems 15	
1.3.1	Ti/TiO_2 System 15	
1.3.1.1	Experimental Details 15	
1.3.1.2	Determination of Ti Substrate Grain Orientation by SAME 17	
1.3.1.3	Photocurrent Spectra and $i_{ph}(U)$ Measurements on Single Ti/TiO_2 Grains 18	
1.3.1.4	Microscopic Modification of the TiO_2 Films by Means of Laser Scanning 19	
1.3.1.5	Characterization of the Modified TiO_2 Films 21	
1.3.1.6	Photoresist Microelectrochemistry (Nanoliter Droplet Method) 25	
1.3.1.7	Applications of Photoresist Microelectrodes 28	
1.3.1.8	Summary and Conclusions for the Ti/TiO_2 System 36	

1.3.2	Zr/ZrO_2 and Hf/HfO_2 Systems 37	
1.3.2.1	Zr/ZrO_2 37	
1.3.2.2	Hf/HfO_2 46	
1.3.3	Systems: Nb/Nb_2O_5, Ta/Ta_2O_5 and Al/Al_2O_3 48	
1.3.3.1	Nb/Nb_2O_5 System 49	
1.3.3.2	Al/Al_2O_3 System 53	
1.3.3.3	Ta/Ta_2O_5 System 54	
1.3.4	Application of Valve Metals in Electrolytic Capacitor Manufacturing 57	
1.3.4.1	Capacitor Fundamentals 57	
1.3.4.2	Capacitor Device Types and Production of Ta Capacitors 62	
1.3.4.3	Current Development Trends for Ta Capacitors and Research Issues Involved 65	
1.3.4.4	Effect of Oxygen Content and Sinter Conditions on Dislocation Formation 67	
1.3.4.5	Thermal Runaway 70	
1.4	Si/SiO_2 System 74	
1.4.1	Application of the Si/SiO_2 System 77	
1.4.1.1	Si/SiO_2 in MOSFETs 77	
1.4.1.2	Si/SiO_2 in DRAMs 80	
1.4.1.3	DRAM Storage Capacitor (Deep Trench) 82	
1.4.2	Alternative Dielectric Materials 90	
1.4.2.1	Ta_2O_5 92	
1.4.2.2	Ti/TiO_2 95	
1.5	Summary and Conclusions 96	
	References 99	
2	**Superconformal Film Growth** 107	
	Thomas P. Moffat, Daniel Wheeler, and Daniel Josell	
2.1	Introduction 107	
2.2	Destabilizing Influences 108	
2.3	Stabilization and Smoothing Mechanisms 110	
2.3.1	Geometric Leveling 110	
2.3.2	Inhibitor-based Leveling 110	
2.3.3	Brightening by Grain Refinement 111	
2.3.4	Catalyst-derived Brightening 112	
2.3.5	Stabilization Across Length Scales 112	
2.4	Additive Processes 113	
2.4.1	Adsorption Kinetics 117	
2.4.2	Surface Segregation versus Consumption Processes 117	
2.4.2.1	Adsorbates Segregated onto Growing Surface 118	
2.4.2.2	Adsorbates Incorporated into Growing Deposit 119	
2.4.2.3	Deactivation of Adsorbate 121	
2.4.3	Adsorbate Evolution 121	
2.4.4	Impact on Microstructure 122	
2.4.5	Quantifying Adsorbate Inhibition of Metal Deposition 125	

2.4.6	Co-adsorption Effects *130*	
2.4.7	Catalysis of Metal Deposition *134*	
2.4.8	Activation of Blocked Electrodes by Competitive Adsorption of a Catalyst *135*	
2.4.9	Catalyst Function and Consumption *138*	
2.4.10	Quantifying the Effects of Competitive Adsorption on Metal Deposition *141*	
2.4.10.1	Site Dependence of Charge Transfer Kinetics *142*	
2.4.10.2	Catalyst Evolution *143*	
2.4.10.3	SPS Adsorption from the Electrolyte *143*	
2.5	Interface Motion and Morphological Evolution *146*	
2.5.1	Shape Change Simulations *146*	
2.5.2	Geometric Leveling *150*	
2.5.3	Inhibitor-based Leveling *153*	
2.5.3.1	Feature Filling *153*	
2.5.3.2	Stability Analysis *160*	
2.5.4	Catalyst-derived Brightening *161*	
2.5.4.1	Feature Filling *161*	
2.5.4.2	Stability Analysis *173*	
2.5.5	Bridging the Length Scales *176*	
2.6	Conclusions and Outlook *179*	
	References *179*	
3	**Transition Metal Macrocycles as Electrocatalysts for Dioxygen Reduction** *191*	
	Daniel A. Scherson, Attila Palencsár, Yuriy Tolmachev, and Ionel Stefan	
3.1	Introduction *191*	
3.1.1	Electrocatalysis *192*	
3.1.2	Dioxygen Reduction in Aqueous Electrolytes: General Aspects *193*	
3.1.3	Transition Metal Macrocycles *199*	
3.1.3.1	General Characteristics *199*	
3.1.3.2	Electrocatalytic Properties Toward Oxygen Reduction *201*	
3.2	Homogeneous Electrocatalysis *204*	
3.2.1	Intrinsic Properties of Solution Phase Transition Metal Macrocycles *204*	
3.2.1.1	Formal Redox Potentials and Diffusion Coefficients *204*	
3.2.1.2	Molecular Speciation *209*	
3.2.1.3	Rates of Heterogeneous Electron Transfer Reactions *211*	
3.2.2	Macrocyclic-Mediated Reduction of Dioxygen in Aqueous Electrolytes *212*	
3.2.2.1	Model Systems *212*	
3.3	Heterogeneous Electrocatalysis *219*	
3.3.1	Adsorption Isotherms *220*	
3.3.2	Chemically Modified Electrodes *221*	
3.3.2.1	Preparation and Electrochemical Characterization *221*	

3.3.2.2	*In situ* Spectroscopic Characterization 226
3.3.3	Redox Active Chemically Modified Electrodes 232
3.3.3.1	Thermodynamic Aspects 232
3.3.3.2	Redox Speciation 235
3.3.3.3	Redox Dynamics 238
3.3.4	Electrocatalytic Aspects of Dioxygen Reduction 241
3.3.4.1	Theoretical Considerations 241
3.3.4.2	Model Systems 244
3.4	Thermal Activation of Transition Metal Macrocycles 269
3.4.1	Brief Introduction 269
3.4.2	Electrochemical Characterization 269
3.4.2.1	Cyclic Voltammetry 270
3.4.2.2	Oxygen Reduction Polarization Curves 271
3.4.3	Spectroscopic and Structural Characterization 273
3.4.3.1	Pyrolysis-Mass Spectrometry 273
3.4.3.2	Mossbauer Effect Spectroscopy 277
3.4.3.3	X-ray Absorption Fine Structure 278
3.4.3.4	X-ray Photoelectron Spectroscopy 281
3.4.4	*In Situ* and Quasi *In Situ* Spectroscopic Characterization 281
3.4.5	Concluding Remarks 283
	References 285

4	**Multiscale Modeling and Design of Electrochemical Systems** 289
	Richard D. Braatz, Edmund G. Seebauer, and Richard C. Alkire
4.1	Introduction 289
4.2	Background and Motivation 291
4.2.1	Multiscale Simulation 291
4.2.2	Electrochemical Systems 293
4.2.3	Microelectronic Applications 295
4.2.4	Nanoscale Science and Technology 296
4.2.5	Other Electrochemical Applications 297
4.3	Trend Toward Atomistic/Molecular Simulation 298
4.3.1	Integrated Circuit Example 298
4.3.2	Continuum Methods 300
4.3.3	Molecular Simulation Methods 300
4.3.4	Coarse-grained Simulation Methods 303
4.4	Multiscale Simulation 304
4.5	Challenges and Requirements of Multiscale Modeling 310
4.6	Addressing the Challenges in Multiscale Modeling 311
4.7	Design Based on Multiscale Models 315
4.8	Concluding Remarks 322
	References 324

Index 335

Series Preface

This tenth volume continues the merger of two well-established monograph series. One point of origin began in 1961 with the editorial collaboration of Paul Delahay and Charles Tobias, and which continued in 1976 with Heinz Gerischer and Charles Tobias. Their efforts led in 1987 to the series "Advances in Electrochemical Science and Engineering", which was continued in 1997 by Richard Alkire and Dieter Kolb. The second point of origin is the series "Frontiers in Electrochemistry" established in 1992 by Jacek Lipkowski and Philip N. Ross. With this volume the concept of topical volumes is introduced to the new series of Advances that resulted from the merger. The favourable reception of the Frontiers in Electrochemistry and the first eight volumes of Advances, and the steady increase of interest in electrochemical science and technology provide good reasons for the continuation of these editions with the same high standards. The purpose of the series is to provide high quality advanced reviews of topics of both fundamental and practical importance for the experienced reader.

Richard Alkire
Dieter Kolb
Jacek Lipkowski
Philip Ross

Volume Preface

The purpose of the series is to provide high quality advanced reviews of topics of both fundamental and practical importance for the experienced reader. This volume focuses on spectacular recent developments in electrochemical surface modification for the purpose of forming thin films and surface regions that have unique functional properties and that can be well-characterized by multi-scale mathematical models.

Michaelis reviews the application of valve metals in electronics based on the dielectric properties of ultra-thin films. Following presentation of fundamental principles and experimental details, the discussion of valve metal systems includes thin film oxide behavior of Ti, Zr, Hf, Nb, Ta, and Al. The application of these valve metal systems in electrolytic capacitor manufacturing is discussed with emphasis on current development trends and research issues. In addition, special emphasis on SiO_2 dielectric films is provided for integrated circuit applications associated with dynamic random access memory chip fabrication.

Moffat, Wheeler and Josell review electrodeposition systems in which superconformal film growth results in progressive reduction of surface roughness. Special emphasis is given to quantitative studies on the relation between electrochemical reaction phenomena and morphological evolution during filling of micro- and nanometer scale features. The complex interplay between electrolyte additives and the metal electrodeposition process is covered in detail. Quantification of additive behavior in order to predict interface motion and morphology evolution is treated with special emphasis on fabrication of on-chip copper interconnects.

Scherson, Palenscar, Tolmachev and Stefan provide a critical review of transition metal macrocycles, in both intact and thermally activated forms, as electrocatalysts for dioxygen reduction in aqueous electrolytes. An introduction is provided to fundamental aspects of electrocatalysis, oxygen reduction, and transition metal macrocycles. Since the theoretical and experimental tools used for investigation of homogeneous and heterogeneous electrocatalysis are considerably different, these topics are given separate discussion. The influence of the electrode surface on adsorbed macrocycles, and their influence on mechanism and rates of O_2 reduction is treated in detail. Issues related to pyrolyzed macrocycles are also described.

Advances in Electrochemical Science and Engineering, Vol. 10.
Edited by Richard C. Alkire, Dieter M. Kolb, Jacek Lipkowski and Philip N. Ross
Copyright © 2008 WILEY-VCH Verlag GmbH & Co. KGaA, Weinheim
ISBN: 978-3-527-31419-5

Braatz, Seebauer and Alkire discuss recent developments in molecular simulation, multiscale simulation, and multiscale systems engineering, and how these developments enable the targeted design of processes and products at the molecular scale. They point to the need to model electrochemical phenomena at the molecular scale, and to link such models to the macroscopic level where engineering control and optimization strategies are implemented. Following a summary of continuum and molecular-scale simulation approaches, they outline the key challenges of multiscale modeling and its integration with experimental data. Also covered is the use of multiscale models for engineering design and optimization of processes and products where quality is determined by events at the molecular scale.

Urbana, Illinois, May 2008   *Richard Alkire*

List of Contributors

Richard C. Alkire
Department of Chemical and
Biomolecular Engineering
University of Illinois
Urbana, IL 61801
USA

Richard D. Braatz
Department of Chemical and
Biomolecular Engineering
University of Illinois
Urbana, IL 61801
USA

Daniel Josell
Materials Science and Engineering
Laboratory
National Institute of Science and
Technology
Gaithersburg, MD 20899
USA

Alexander Michaelis
Fraunhofer Institute for Ceramic
Technologies and Systems
Winterbergstraße 28
01277 Dresden
Germany

Thomas P. Moffat
Materials Science and Engineering
Laboratory
National Institute of Science and
Technology
Gaithersburg, MD 20899
USA

Attila Palencsár
Ernest B. Yeager Center for
Electrochemical Sciences
Department of Chemistry
Case Western Reserve University
Cleveland, OH 44106-7078
USA

Edmund G. Seebauer
Department of Chemical and
Biomolecular Engineering
University of Illinois
Urbana, IL 61801
USA

Daniel A. Scherson
Ernest B. Yeager Center for
Electrochemical Sciences
Department of Chemistry
Case Western Reserve University
Cleveland, OH 44106-7078
USA

Ionel Stefan
Ernest B. Yeager Center for
Electrochemical Sciences
Department of Chemistry
Case Western Reserve University
Cleveland, OH 44106-7078
USA

Yuriy Tolmachev
Kent State University
Kent, OH 44242
USA

Daniel Wheeler
Materials Science and Engineering
Laboratory
National Institute of Science and
Technology
Gaithersburg, MD 20899
USA

1
Valve Metal, Si and Ceramic Oxides as Dielectric Films for Passive and Active Electronic Devices

Alexander Michaelis

1.1
Introduction

Owing to their low electrochemical potential the group IVB and VB valve metals Ti, Zr, Hf, V, Nb and Ta readily react with water or oxygen to form a dense, protecting passive layer. This also holds for Al. Because of these protecting oxide films, the valve metals show an exceptional resistance towards corrosion in many aggressive environments. This explains why valve metals are widely used in the construction of chemical apparatus. An extensive number of papers and reviews exist that focus on the corrosion resistance of valve metal oxides [e.g. 1, 2].

In this treatise however another field is covered, namely the application of valve metals in electronics. The valve metal oxides show outstanding dielectric properties, which make them a key component in many passive and active devices such as capacitors, resistors and ICs (integrated circuits). Nevertheless, the most important oxide system in electronics is still Si/SiO_2, which can be considered the bench mark system. Consequently, one chapter of this book is dedicated to Si/SiO_2 with special emphasis on its use in DRAM (dynamic random access memory) microchip fabrication, which can be considered the most advanced and demanding technology in microelectronics, as will be explained in Chapter 4. Electrochemically, p-Si/SiO_2 also behaves like a valve metal. Regarding their electronic properties, the valve metal oxides are related to another class of oxides, namely the perovskite structured ferroelectrics: $BaTiO_3$, $(Ba,Sr)TiO_3$ (BST), $Pb(Zr_xTi_{1-x})O_3$ (PZT), $SrBi_2Ta_2O_9$ (SBT). These advanced electroceramics outperform the simple valve metal oxides in many respects, such as, by their exceptional high relative dielectric constant (ε_r) (sometimes also called relative permittivity). Therefore, both the valve metal and the perovskite oxides compete in many applications. For example, in passive component capacitor manufacturing Al/Al_2O_3 and Ta/Ta_2O_5 electrolyte capacitors share the market with multilayer ceramic capacitors (MLCC) that are $BaTiO_3$ based. However, owing to their more complex chemistry, the integration of perovskite layers into devices is difficult when ultra-thin nano-dielectric films are needed.

Therefore, this treatise focuses on the less complex valve metal oxides as ultra-thin nano-dielectrics (<100 nm). Some relevant oxide systems are summarized in Table 1.1. In anticipation of the results presented later, some parameters that are important for electronic applications are also given. For application as a capacitor dielectric material, for example, a high relative dielectric constant ε_r is important but not sufficient. As indicated in Table 1.1 the anodically formed valve metal oxides differ in structure (some are amorphous, some crystalline), electronic behavior (some oxides behave like an n-type semiconductor), and some oxides show a pronounced texture dependence of oxide growth, that is the oxide properties vary with the crystallographic orientation of the substrate surface. Depending on the desired application of the oxides, an optimum combination of these properties has to be found. As will be discussed in the following chapters, Table 1.1 has to be used carefully, that is, it has to be kept in mind that the given parameters hold for certain 'typical' experimental conditions only. For instance, in the case of Ti/TiO$_2$ the texture dependence can be reduced by switching from potentiodynamic to potentiostatic oxide formation conditions. Therefore, Table 1.1 only gives a first, rough orientation of the electronic properties and behavior of the oxides. Depending on the experimental formation conditions a wide range of properties can often be obtained. This also explains why literature data often show a huge variation for almost all relevant parameters. Consequently, if data are to be compared, the experimental conditions have to be carefully considered.

1.1.1
Experimental Approaches

For the study of the oxide systems, electrochemical and optical methods can be applied. Valve metal oxides are formed conventionally electrochemically, with the important exception of SiO$_2$, which is usually formed by thermal processing or chemical/physical vapor deposition (CVD/PVD). Conventional electrochemical methods show an excellent vertical resolution for the fundamental study of ultra-thin nano-oxide films on solid surfaces, but their application to micro- and nano-systems, which are of special importance in electronics, is limited because of the poor lateral resolution, with typical electrode areas in the mm^2 to cm^2 range. In order to gain deeper insight into local electrode reactions, microscopic *in situ* methods must be developed and employed that feature both high lateral and vertical resolution. Here an important new approach to address and solve this issue, that is the combination of conventional electrochemistry with optical and new electrochemical micro-methods, is presented. This enables one to characterize or/and modify solid surfaces at high lateral resolution, which is only limited by refraction of light in the μm^2 range. This is especially important for electronic applications as the critical dimensions of relevant structures are extremely small (sub-micron range) and the effect of structural heterogeneities on the film properties is essential for the reliability of the devices. Some fundamental aspects of optical methods employing focused light and their application to electrochemistry are discussed with emphasis on laser light sources. To allow for a high lateral resolution, even by conventional

Table 1.1 Relative dielectric constant ε_r, anodic formation factor m, density ρ, band gap energy E_g, cation transfer coefficient t_+, bias dependence and electronic behavior (SC = semiconductor), structure (a = amorphous, c = crystalline), texture dependence for some of the oxide systems described in this treatise (see also [20]).

System	ε_r	m (nm V^{-1})	ρ (g cm^{-3})	E_g (eV)	t_+	Bias dep. Electronic behavior	Structure	Texture dep.
Ti/TiO$_2$	56 (anodic) [3]	1.4–2.2	3.9	3.4	0.35	Yes, n-type SC	a	Yes
Rutile	110 powder 90 ⊥, 170 ∥	[6, 7]	4.25	3.0	[3]		[8]	
Anatase	48 (powder)		3.99	3.6				
Brookite	[4, 5]		4.13	3.0				
Zr/ZrO$_2$ [17]	27 [9]	2.6–3.3	5.7	5	0 [10]	No, dielectric	c [8]	Yes
Ta/Ta$_2$O$_5$	25 [11, 12]	1.8 [13]	8.5	5.2	0.5 [14]	No, dielectric	a [15, 16]	No
Nb/Nb$_2$O$_5$ [18]	42 [19]	2.9 [20]	4.4	3.4	0.5 [10]	Yes, n-type SC	a [8]	No
Al/Al$_2$O$_3$ [21]	9 [22–24]	1.3 [25]	3.1	7	0.5 [26]	No, dielectric	a [27]	No
Hf/HfO$_2$ [28]	16 [29]	2.0 [31]	10	5.1	0 [30]	No, dielectric	c [8]	**Yes**
p-Si/SiO$_2$ [32, 33]	3.8 [34, 35]	0.7	2.2	9	0	No, dielectric	a [36]	Yes
BaTiO$_3$	≈2000		6			Ferroelectric		
WO$_3$	23 [37]	1.8 [38]	6.5	3.1	0.5 [10]		a [8]	

electrochemistry itself, the nl-droplet method first introduced by Kudelka [6] is applied. Depending on the input/output signals, the *micro-methods* used throughout this text can be divided into three different classes:

1. Optical/optical methods (exciting and measured signal = light beam). Examples are the application of well established optical methods, such as ellipsometry, reflection spectroscopy, beam deflection, interferometry, Raman-spectroscopy, etc., along with electrochemistry [40–45]. The incident light beam is focused on a microscopic portion (e.g. a single substrate grain) of a macroscopic electrode and the optical information in the reflected or deflected light beam is analyzed and correlated with the simultaneously recorded electrochemical signals. For instance, micro-ellipsometry can be performed during cyclovoltammetric formation of a passive layer to record film thickness growth on a particular grain of a polycrystalline macroscopic electrode. Examples of this important application will be discussed in full detail later (Section 1.3). In order to allow for an accurate correlation of the signals to the deliberately chosen electrode areas, microscopic control is essential. Figure 1.1 shows a typical experimental setup for this purpose. *SAME* (spectroscopic anisotropy micro-ellipsometry) is introduced as the most important spectroscopic method. This newly developed method allows for the determination of structural properties such as the crystal orientation of the surface film along with conventional parameters, for example, the film thickness. In

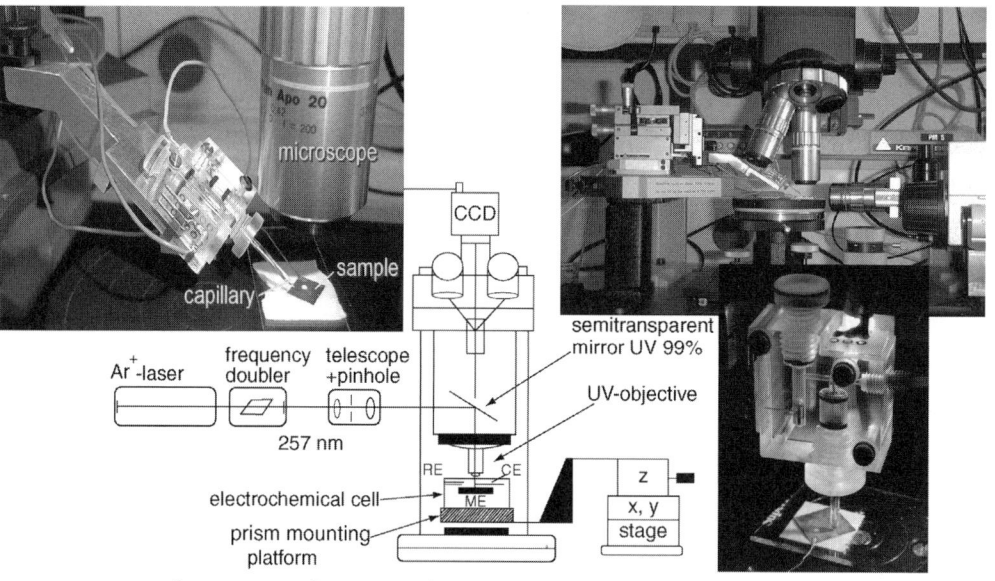

Figure 1.1 Typical experimental setup used for the combined application of micro-optical and electrochemical methods along with *in situ* microscopic sample monitoring. The electrochemical cell is mounted on an x,y,z-stage and can be scanned with sub-micron resolution. Besides the nl-photoresist droplet, the 'contact cell' shown can also be used [39].

the case of electroceramics such as YBCO even the chemical composition can be determined by SAME [308].

2. Optical/electrical or optical/chemical methods [measured signals such as electric current (photocurrent i_{ph}), voltage, reaction rates, etc.]. A characteristic for this class is that the light beam is used to locally induce a reaction [46–51]. The type of reaction depends on thermal and electronic properties of the solid, the neighboring phase and energy or power density of the light beam. In metals, energy dissipation is very fast. Consequently, most of the energy is transformed into heat resulting in melting or even evaporation of the substrate. In dielectric materials bond breaking with subsequent reactions dominates. In semiconductors dissociation and ionization, reactions are often observed if high power densities are used. However, at low power densities electron/hole pair formation with an associated change of the potential distribution at the interface dominates. Both electron transfer (ETR) in addition to ion transfer (ITR) reactions can be induced. For instance, hole accumulation can cause redox reactions and film growth or removal (corrosion, etching) as will be discussed later. Laser-induced reactions under electrochemical conditions at low power densities, that is at quasi-isothermal conditions will be emphasized.

3. Electrical/electrical microelectrode methods under optical (microscopic) control. This class covers the conventional electrochemical methods, that is, application of an electrical signal (e.g. potential) to the electrode to induce a reaction. The role of optics here is twofold:

   - Samples are prepared by means of photolithography, that is, the local resolution of the electrical measurement is determined by focused laser radiation. In this context it has to be kept in mind that the high lateral resolution achieved in modern device and micro-system technology and research is mainly due to optical processing, that is, photolithography. Detailed examples for application of photoresist microelectrochemistry, namely the new photoresist nl-droplet method, will be given.
   - In order to measure on defined areas the electrode must be microscopically controlled (see Figure 1.1).

1.2
Fundamentals and Experimental Details

In this section some fundamentals of the methods used later are summarized. In particular, electrochemistry and photocurrent measurements (optical/electrical) are treated here.

1.2.1
Electrochemical Oxide Layer Formation on Valve Metals

Valve metals bear this name due to the fact that they pass current in only one direction (rectifying effect), that is, the anodic current is low for potentials lower than the formation potential but a strong cathodic evolution of hydrogen is possible. Experimentally the following behavior is found for the anodic oxide formation [52, 295]:

$$i = i_0 e^{(\beta E)} \tag{1.1}$$

where i is the oxide formation current, i_0 and β are material-dependent constants and E is the electric field strength in the oxide. This equation is called the *high field equation*. The corresponding high field model assumes an E-field supported, thermally activated ion movement (hopping) along the distance a, with a typically being the lattice parameter. There are many papers on this subject where the above equation is derived assuming specific mechanisms [e.g. 52–61]. Depending on the details in the used models, the rate determining step may be located within the oxide (Verwey, Güntherschulze and Betz), the oxide/electrolyte (Sato) or metal/oxide (Cabrera–Mott) interface. A detailed discussion and derivation of the high field equation can be found in Ref. [20]. An equation of the following form can be derived:

$$i = \nu \rho a \cdot e^{-\frac{W}{RT}} e^{\frac{\alpha a z F E}{RT}} \tag{1.2}$$

with ν = hopping frequency, α = transfer coefficient, ρ = concentration of mobile charge, a = hopping distance, W = activation energy for hopping, R = gas constant, T = temperature and F = Faraday's number.

This equation is identical to the experimental Equation 1.1 using the equalities:

$$\beta = \frac{\alpha a z F}{RT} \quad \text{and} \quad i_0 = \nu \rho a \cdot e^{-\frac{W}{RT}} \tag{1.3}$$

The parameters occurring in these equations are illustrated in Figure 1.2.

For homogeneous oxide layers the field strength E can be determined from the potential drop $\Delta\varphi_{ox}$ across the oxide and the layer thickness d_f.

$$i = i_0 e^{(\beta E)} = i_0 e^{\left(\beta \frac{\Delta\varphi_{ox}}{d_f}\right)} \tag{1.4}$$

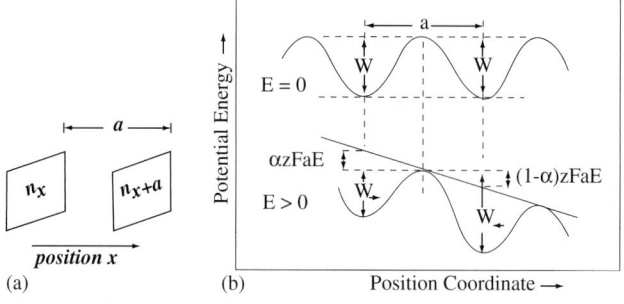

Figure 1.2 Illustration of the high-field model: (a) adjacent lattice planes in the oxide and (b) effect of the E-field on the activation energy W (source Ref. [20]).

For *galvanostatic oxide growth* (log i = const) the E field has to be constant resulting immediately in the proportionality:

$$d_f \propto \Delta\varphi_{ox} \Rightarrow d_f = m \cdot \Delta\varphi_{ox} \quad (1.5)$$

This equation defines the oxide formation factor: m (nm V^{-1}), that is, the layer thickness increases linearly with the applied potential.

Equation 1.4 directly implies the time behavior (rate) of oxide growth. An inverse logarithmic growth law according to the following expression can be derived to a good approximation (Cabrera and Mott [55]):

$$\frac{1}{d_f} = A - \frac{1}{\beta \cdot \Delta\varphi_{ox}} \cdot \ln t \quad (1.6)$$

with A being a constant.

1.2.2
The C(U) Curve of a Valve Metal Electrode

The total potential drop $\Delta\varphi_{tot}$ across an electrode is made up of four parts: the potential drop at the metal/oxide interface, the potential drop in the oxide, the potential drop in the Helmholtz layer and the potential drop in the electrolyte (diffuse double layer):

$$\Delta\varphi_{tot} = \Delta\varphi_{Me/ox} + \Delta\varphi_{ox} + \Delta\varphi_H + \Delta\varphi_{el} \quad (1.7)$$

The potential drop $\Delta\varphi_{tot}$ correlates with the applied electrode voltage ΔU by: $\Delta\varphi_{tot} = \Delta U - $ const. For high electrolyte concentrations (>0.1 M), the last part $\Delta\varphi_{el}$ can be neglected. In addition, the first part $\Delta\varphi_{Me/ox}$ is normally small. The potential drop in the Helmholtz layer is usually determined by the reaction of adsorption of protons and hydroxyl groups. The corresponding reaction rate is high, that is, equilibrium is reached almost instantaneously. Therefore, the corresponding potential drop is constant resulting in a constant double layer capacitance of about 20 µF cm^{-2}. The potential drop $\Delta\varphi_{ox}$ can be calculated using the Poisson equation $\nabla E = \rho_c/(\varepsilon_r \varepsilon_0)$ with ρ_c being the charge density given by the defect state concentration N_D. This is:

$$\Delta\varphi_{ox} = \Delta\varphi_{ox,scl} + \Delta\varphi_{ox,isol} = -\frac{eN_D d_{scl}^2}{2\varepsilon_r \varepsilon_0} + \Delta\varphi_{id,isol} \quad (1.8)$$

where $\Delta\varphi_{id,isol}$ describes the potential drop in the ideal insulator part, e is the electronic charge, d_{scl} is the extension of the space charge layer scl. Inversion of the $\Delta\varphi_{ox,scl}$ semiconductor part of this equation yields the following expression for the extension of the space charge layer:

$$d_{scl} = \sqrt{\frac{2\varepsilon_r \varepsilon_0 \Delta\varphi_{ox,scl}}{eN_D}} \quad (1.9)$$

Hence a potential (bias) dependent capacitance C_{scl} can be defined with use of the simple capacitor equation $C_{scl}/A = \varepsilon_r \varepsilon_0 / d_{scl}$. (As the E field is completely shielded

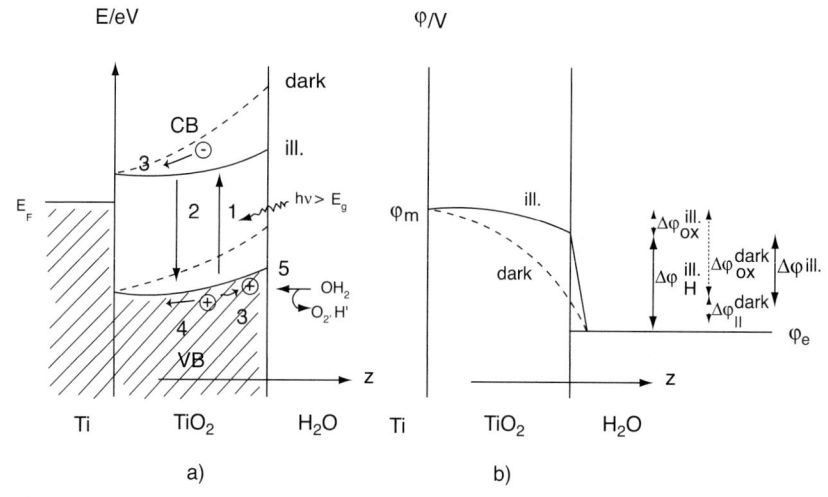

Figure 1.3 (a) Band scheme and (b) potential drops for the dark and illuminated situations. The Ti/TiO$_2$ system is chosen here for illustration.

inside the scl region, the potential drop $\Delta\varphi_{id,isol}$ vanishes for small applied voltages.) Linearized with respect to the potential drop this yields the so called *Schottky–Mott relationship* [62, 310] (for $T > 0$, a $(K_B T/e)$ part has to be subtracted, which describes the thermal discharging of defect states into the conduction band):

$$\frac{1}{C_{scl}^2} = \frac{2}{\varepsilon_r \varepsilon_0 e N_D} \cdot \left(\Delta\varphi_{ox, scl} - \frac{K_B T}{e} \right) \tag{1.10}$$

This equation holds for the case of small applied voltages when the extension of the scl is smaller than the oxide thickness. Evaluation of the slope of the $1/C_{scl}^2$ ($\Delta\varphi_{ox,scl}$) curve allows for the determination of the product $\varepsilon_r N_D$. Once the scl reaches completely through the oxide film, the capacitance is determined by d_f, again yielding the conventional potential independent capacitance $C/A = \varepsilon_r \varepsilon_0 / d_f$. A schematic representation of the relevant potential drops as well as the band structure is shown in Figure 1.3 for the case of the Ti/TiO$_2$ system. An example of an illuminated surface (induced photocurrent) is also shown here, which is required later.

1.2.3
Application of Lasers in Electrochemistry

As laser radiation plays an important role in some of the experimental methods employed later, some fundamental aspects of lasers are discussed briefly here. The fact that lasers are more and more replacing conventional light sources in all types of optical applications can be understood by considering their fundamental advantages. Perhaps the most striking feature of laser radiation is its directionality due to a very low angle of divergence, typically in the order of 1 rad. This has basically two important implications:

1. Laser radiation can be focused down to dimensions of the wavelength of light.
2. A large amount of energy can be concentrated into small spots, yielding high power densities.

In fact, the focusability of laser radiation is the most applied feature. Optimum focusing of laser radiation demands a high mode quality of the laser beam. The best suited mode is the TEM$_{00}$, which shows a Gaussian spatial intensity distribution. This fundamental mode can always be obtained by using intra-cavity apertures, which are small enough to suppress the higher modes. For lasers operated in the TEM$_{00}$ mode, the minimum obtainable spot diameter s (diffraction limit) is given by Airy's formula:

$$s = \lambda \frac{f}{D\pi} \tag{1.11}$$

where f is the focal length of the focusing system used, λ is the wavelength of the laser and D is the maximum effective beam diameter (aperture) of the focusing system (beam expanders are an advantage).

However, it is not only the spot size s that governs the final spatial resolution of the electrode reaction, but illumination of the sample is just the first step in a whole series of induced processes resulting in the final state of the system. For instance, in semiconductor interfaces the lateral hole drift causes an extension of the reaction zone. For the thermally activated processes, the heat conductivity of the sample determines the reaction zone. Therefore, both optical and physical effects determine the final spatial resolution of optical/electrical laser methods.

Other important properties of laser radiation are its narrow spectral bandwidth, coherence and polarization. These features are used in numerous spectroscopic and interferometric applications, such as Raman, polarization, interference spectroscopy, spectroscopic ellipsometry, holography. Moreover, the possibility to create very short light pulses by Q-switching or mode-locking should be mentioned. Pulses below fs-duration are possible and allow for the elucidation of fast transient phenomena such as short spontaneous lifetimes or fast relaxation processes. Laser illumination results in thermal heating, chemical reactions and/or potential changes. In the following experiments laser-induced reactions under electrochemical conditions at low power densities, that is, at quasi-isothermal conditions, are emphasized. Thermal processes are discussed briefly below.

1.2.3.1 Thermal Effects

With the assumption of a Gaussian temperature distribution (TEM$_{00}$-mode) on the laser affected zone, the temperature increase ΔT can be estimated by the following equation [63]:

$$\Delta T = \frac{A}{2\kappa\sqrt{\pi/2}} \cdot P \cdot r \cdot \arctan\sqrt{\frac{8\kappa t}{\rho c r^2}} \tag{1.12}$$

where A is absorption, ρ is density, c is specific heat, κ is heat conductivity, r is spot radius $= s/2$ and P is power density.

For times $t \gg \pi r^2 \rho \chi/[8\kappa\,(t > 10^{-2}\,\text{s})]$ the temperature becomes time independent, a continuous wave (cw) condition. The local temperature is then given by:

$$\Delta T = \frac{APr}{2\kappa} \cdot \sqrt{\frac{\pi}{2}} \qquad (1.13)$$

The temperature effect often disturbs precise measurements if isothermal conditions cannot be maintained and if it leads to damage of the sample. Equation 1.12 shows that at constant power density the temperature effect decreases with decreasing pulse times. Therefore, the application of short pulses may be of advantage to avoid damages. If, however, the modification of the surface requires a large amount of total energy, it should be delivered with low power density. On the other hand, there are numerous applications of the thermal heating. It can be used to evaporate or to dissociate the substrate (LAMMA) [64], to enhance reaction rates at the surface or the convection of the electrolyte [65–67]. Finally, it can be employed to study electrode reaction rate constants and the dynamics of the double layer [68].

1.2.4
Electrochemical Photocurrent Measurements (Optical/Electrical Method Class), Introduction of a New Model

The theoretical formalisms for the interpretation of the pure optical (optical/optical) and the pure electrochemical methods are well established and can be found in numerous publications. It is far beyond the scope of this treatise to review the corresponding models. However, the quantitative interpretation of optical/electrical measurements (photoelectrochemsitry) is still incomplete and will be discussed here. For instance, multiple internal reflections within the thin films must be considered, causing i_{ph} to depend on the layer thickness. This experimental fact has already been described by Schultze et al. [69] and later by Di-Quarto and coworkers [70, 71]. However, no quantitative analysis or model were given. The quantitative optical model was independently derived by Smyrl and coworkers [72, 73], Gärtner [75] and Michaelis and Schultze [74]. This optical model has to be combined with the Butler–Gärtner model [75, 76], which itself holds only for crystalline bulk materials. The extended model describes i_{ph} as a function of the layer thickness d_f, the optical constants of the substrate and layer, the extension of the space charge layer d_{scl}, and such electronic properties as the defect state concentration N_D and the relative dielectric constant ε_r. However, this model is still not accurate for an exact description of the i_{ph} behavior of amorphous oxide layers, but another, E-field depending term, taking the recombination efficiency $r(E)$ into account, must be coupled to the model. This yields a further dependence of i_{ph} on the film thickness and on the applied anodic potential. This complete photocurrent model consisting of three different parts, namely the Gärtner-, optics- and $r(E)$-parts, was first introduced by Michaelis and Schultze [74] and allows for a complete description of the experimental results presented later. Consequently, both electronic and optical film properties can principally be determined allowing for a mutual confirmation of the pure optical and electrical methods. We will discuss in which potential and film thickness ranges the different parts of the model dominate and whether a quantitative evaluation of optical and electronic properties is really possible.

1.2.4.1 Photocurrent Model for Ultra-thin, Amorphous Films With TiO$_2$ as an Example

Illumination of anodically polarized n-type semiconducting oxide films such as TiO$_2$ with photon energies above the band gap energy ($E_g \approx 3.5$ eV for TiO$_2$) results in electron/hole pair formation. These charges are separated within the space charge layer (scl) causing a photocurrent i_{ph}. According to the Butler–Gärtner model, the main contribution to i_{ph} can be attributed to electron/hole pairs, which are generated directly within the space charge layer, namely i_{scl}. Another contribution is due to charges diffusing into the scl that is, $i_{ph} = i_{scl} + i_{diff}$. The generation of i_{ph} is in competition with recombination processes throughout the entire layer. These recombination processes are not considered explicitly in the Gärtner approach. The electronic layer parameters enter into the Gärtner model only because of their effect on the extension of the scl (d_{scl}) (see Equation 1.9):

$$d_{scl} = \sqrt{\frac{2\varepsilon_r \varepsilon_0 (\Delta\varphi_{ox})}{eN_D}} \tag{1.14}$$

However, the optical film properties describing the absorption of light, that is, the number of photons g generating the electron/hole pairs have not yet been considered.

This generation number g is plainly given by the change in the light intensity I per unit length z, that is:

$$g(z) = -\frac{d}{dz} I \tag{1.15}$$

In the classical Butler–Gärtner model, light absorption is simply described in terms of Lambert–Beer's law, containing just one optical constant, the extinction coefficient k. However, this approach holds for bulk semiconductors only. In the case of ultra-thin semiconducting layers dealt with here, multiple internal reflections must be considered [74–76] if the penetration depth δ of the light is larger than the film thickness d_f. This is illustrated in Figure 1.4. The light energy flux I into the volume now has to be described by the Poynting vector $\vec{I}$. A detailed description of the derivation of this quantity can be found in Ref. [74]. In this case, n and k of both the layer and substrate enter into the expression for i_{ph} and, even more importantly, the layer thickness d_f is now considered quantitatively. In principle, this optical model can be coupled to any electronic model. Here, the Butler–Gärtner approach will be kept, yielding

$$i_{ph} = -e \int_0^{d_{scl}} \left(\frac{d\vec{I}(z)}{dz}\right) dz \tag{1.16}$$

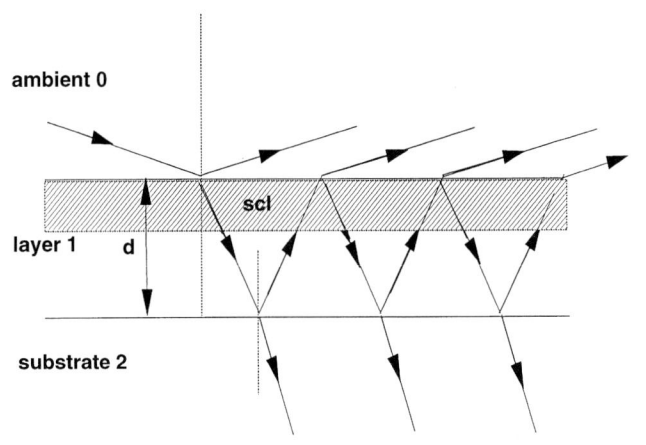

Figure 1.4 Multiple internal reflections within ultra-thin semiconducting films (penetration depth of light $\delta \gg d_f$). The extension of the space charge layer scl depends on the applied potential U and on the electronic layer properties (defect density N_D, dielectric constant ε). It is assumed that only electron/hole pairs generated within the scl contribute to the photocurrent [74].

if only the contribution of electron/hole pairs generated within the scl is considered. It can be shown that the neglection of the diffusion contribution is small [75, 80]. More severe and decisive is the complete neglection of recombination processes in Equation 1.16. Recombination processes are of particular interest for amorphous semiconductors because of the large number of recombination centers. The recombination, that is the efficiency of the electron/hole pair separation, must clearly depend on the E-field gradient, which itself is a function of the applied potential and the layer thickness. Therefore, a recombination factor $r(E)$ must be coupled to the above equation. For $r(E)$ the following approach obeying the boundary conditions of $r(E) = 0$ for vanishing E-fields and $r(E) = 1$ for high fields can be chosen:

$$r(E) = \left(1 - \frac{1}{\text{const} \cdot E + 1}\right) \tag{1.17}$$

yielding

$$\dot{i}_{ph} = -e\left(1 - \frac{1}{\text{const} \cdot \frac{\Delta\varphi_{ox}}{d_{scl}} + 1}\right) \cdot \int_0^{d_{scl}} \left(\frac{d\vec{I}(z)}{dz}\right) dz \tag{1.18}$$

with a linear E-field approximation. The unknown constant in this equation can be determined by fitting of the experimental $i_{ph}(U_{tot})$ curves. The function used for $r(E)$ is a good approximation of the expression for the recombination efficiency derived by Pai and Enck [77] for amorphous materials, which has already been discussed in this

Table 1.2 Contributions of the different parts [d_{scl} change, recombination efficiency $r(E)$, optics] of the presented photocurrent model to the potential and thickness dependence of the photocurrent [$i_{ph}(U)$ and $i_{ph}(d)$]. The 'up arrow', 'down arrow' symbols refer to an increase and decrease in the corresponding parameters, respectively.

		Varied parameter			Effect on i_{ph} divided into the different parts of the model		
Case		d_{lay}	U	d_{scl}	$r(E)$	Multiple reflections	Lambert–Beer
1	$d_{scl} < d_{lay}$	↑	const	const	const	↑↓	not valid
2	and	const	↑	↑	↑	const	n.v.
3	$d_{lay} \ll \delta$	↑	↑	↑	↑	↑↓	n.v.
4	$d_{scl} < d_{lay}$	↑	const	const	const	n.v.	↓
5	and	const	↑	↑	↑	n.v.	const
6	$d_{lay} > \delta$	↑	↑	↑	↑	n.v.	↓
7	$d_{scl} = d_{lay}$	↑	const	↑	↓	↑↓	n.v.
8	and	const	↑	↑	↑	const	n.v.
9	$d_{lay} \ll \delta$	↑	↑	↑	↑	↑↓	n.v.
10	$d_{scl} = d_{lay}$	↑	const	↑	↓	n.v.	↓
11	and	const	↑	↑	↑	n.v.	const
12	$d_{lay} > \delta$	↑	↑	↑	↑	n.v.	↓

context in Refs. [71, 76]. The comparison of both equations shows that the constant has to be a function of the thermalization length, which itself depends on the mobility of the charge carriers and the photon energy of the exciting light.

In Table 1.2 the contributions of the three different parts of the model to the experimental output signals are summarized for various cases. The three parts are:

- The classical Gärtner part describing the effect of the extension of the space charge layer d_{scl}, which itself depends on the electronic properties ε_r and N_D.

- The exact optical model describing multiple internal reflections, which are important for thin films where the penetration depth of the light δ is larger than the layer thickness. Because of this part, i_{ph} depends on the layer thickness and the optical constants of both substrate and layer. This optical part mainly explains the observed variation in $i_{ph}(d_f)$ found in the photocurrent spectra.

- The third part describes the electron/hole pair recombination efficiency $r(E)$ which causes a further dependence of i_{ph} on the layer thickness for very thin films, high potentials, respectively ($d_{scl} = d_f$), and superimposes on the thickness dependence due to the optical part.

The 'up arrow', 'down arrow' symbols in the table denote an increase or decrease in the applied potential U_{tot} and d_f or i_{ph}, respectively. For example, the i_{ph} thickness

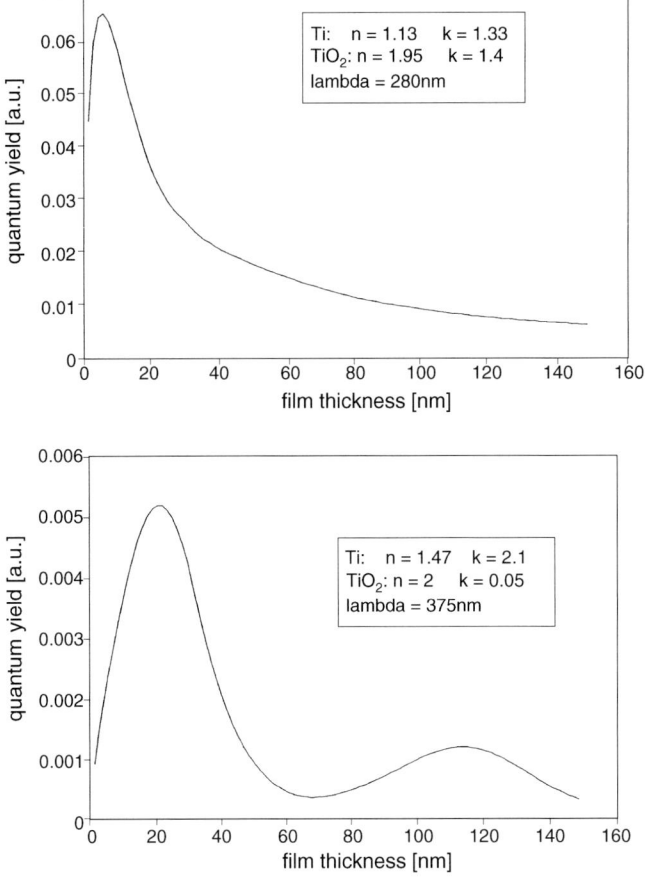

Figure 1.5 Simulations according to the photocurrent model for two different wavelengths. The optical constants used are given in the insets. It was assumed that the E-field gradient reaches through the entire film (dielectric behavior), applied potential 2 V.

dependence in cases 1 and 4 ($d_{scl} < d_f$) is given solely by the effect of multiple internal reflections or the Lambert–Beer law, depending on whether the penetration depth δ of the light is larger or smaller than d_f. Therefore, these cases can be used to judge the optical model and to determine the optical parameters. In the case of the ultra-thin films only $d_f \ll \delta$ applies. The experiments for the valve metal oxides presented later can be understood by referring to particular cases of this table.

Figure 1.5 shows $i_{ph}(d_f)$ simulations according to the model for two different wavelengths (280 and 375 nm) under potentiostatic conditions (applied potential 2 V). The optical constants used for these calculations were determined by ellipsometry (isotropic approximation) and are given in the insets of the figures. Owing to the strong light absorption at 280 nm ($k = 1.4$) no interference is observed at this wavelength. The decrease in the photocurrent at high film thicknesses is due to the dominating E-field effect $r(E)$ corresponding to cases 7 and 10 of Table 1.2. For this, an insulating TiO$_2$ film behavior was assumed, that is, the E-field gradient reaches through the entire film ($d_{scl} = d_f$). At 375 nm (3.3 eV) case 7 clearly dominates ($\delta \ll d_f$), that is, the interference effect is obvious. Note that in this case the photon energy is below the direct bandgap energy of TiO$_2$. In this spectral range the photocurrent is caused by indirect or intraband excitations. As this intraband mechanism differs from the direct interband one, the electronic part of the model is not fully applicable and slight deviations between measurement and simulation have to be expected (see the experiments below).

1.3 Valve Metal Systems

1.3.1 Ti/TiO$_2$ System

As a first example for the combined application of the aforementioned methods for characterization and modification of local areas on technically relevant polycrystalline materials, the Ti/TiO$_2$ system is used. All subsequent experiments were performed under microscopic control on well-defined and deliberately chosen single substrate grains. For this the photoresist microelectrode method was applied, which allows all types of electrochemical measurements at high lateral resolution to be performed in a droplet of nl-volume (Kudelka [78]). As the optical/optical method, micro-ellipsometry, in particular SAME, will be emphasized as it allows for a complete determination of both the optical parameters of the system and the film thickness d_f [81–83]. SAME uses a generalized ellipsometry approach and allows for the determination of crystallographic properties such as the orientation of substrate grains, which than can be correlated with the film properties [308]. This allows for quantification of the effect of the substrate/film interface on film properties. For instance, recrystallization of semiconducting films or epitaxial film growth can be studied. The electronic film parameters are correlated with the substrate texture yielding an insight into the growth mechanisms of this system. Figure 1.6 illustrates the combined application of ellipsometry and the nl-photoresist microelectrode method on technically relevant polycrystalline electrode materials. Additionally, photocurrent spectroscopy and UV-laser scanning measurements are presented to give an example for the aforementioned optical/electrical methods.

1.3.1.1 Experimental Details
Owing to its pronounced heterogeneity, the Ti/TiO$_2$ system is particularly suitable for demonstrating the advantages of micro-methods. The heterogeneity of the Ti/TiO$_2$ system can immediately be observed visually, by looking at a Ti/TiO$_2$ surface with open grain boundaries through a microscope. The TiO$_2$ passive film properties significantly depend on the substrate texture resulting in a variation of the TiO$_2$ interference colors from grain to grain. The objective of the presented experiments is

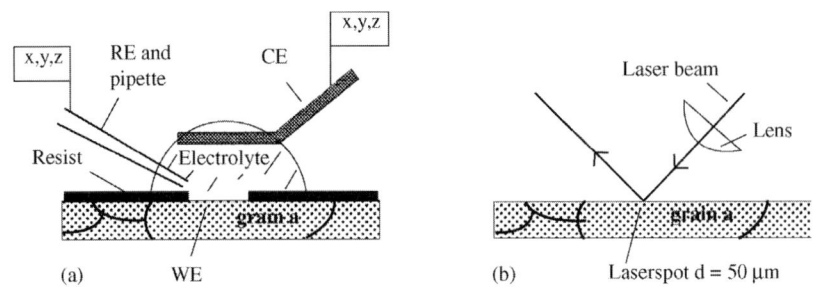

Figure 1.6 Electrochemical and optical micro-methods under microscopic control: (a) photoresist nl-droplet method according to Kudelka and (b) optical micro-methods can be applied on the same portion of the heterogeneous sample. Measurements on a single grain of technically relevant polycrystalline samples can be performed.

the quantification of this phenomenon. For all experiments Ti high-purity (99.98%) samples with a coarse grain texture were used (grain diameter >600 µm). This allows several independent experiments to be performed on each single grain. The crystallographic orientation of the substrate grains was determined *in situ* by SAME. In order to prepare samples with open grain boundaries, the surfaces were electropolished. As the electrolyte 0.5 M H_2SO_4 was chosen. Potentials are given with respect to the standard hydrogen electrode (SHE) throughout this treatise.

For ellipsometry, a rotating analyzer configuration with an HeNe laser (632.8 nm) as the light source was employed. SAME requires rotation of the sample around the surface normal (denoted by an angle α). In order to allow for this sample rotation, a particular *in situ* cell was constructed. For this, the cell windows were mounted on the ellipsometer arms using an alignable connector and simply dipped into the electrolyte. Therefore, rotation of the entire cell containing the sample in three-electrode configuration is possible without affecting the window alignment [81, 84]. For the optical/electrical (photoelectrochemical) experiments at high lateral resolution, the setup shown in Figure 1.1 was used (UV laser scanning setup) [79, 80]. The light of a frequency doubled Ar^+-laser ($\lambda = 257$ nm, i.e. 4.8 eV) is focused onto the sample surface by means of the same microscope that is used to control the optical/optical and electrical/electrical methods, that is the measurements can be carried out simultaneously. A minimum spot radius of about 1 µm was achieved yielding power densities up to some kW cm^{-2}. The sample/cell configuration can be scanned normal to the light beam by using a computerized x,y-stage (step size = 1 µm). For the detection of photocurrent spectra, the laser is substituted by a standard 1 kW Xe high pressure arc lamp and a monochromator. In this instance the lateral resolution is about 50 µm. The whole set-up is based on a modular concept, that is all measurements can be performed simultaneously under microscopic control and further tools such as reflection-spectroscopy or video-microscopy can be added. Details of the nl-droplet photoresist method (electrical/electrical) are given below.

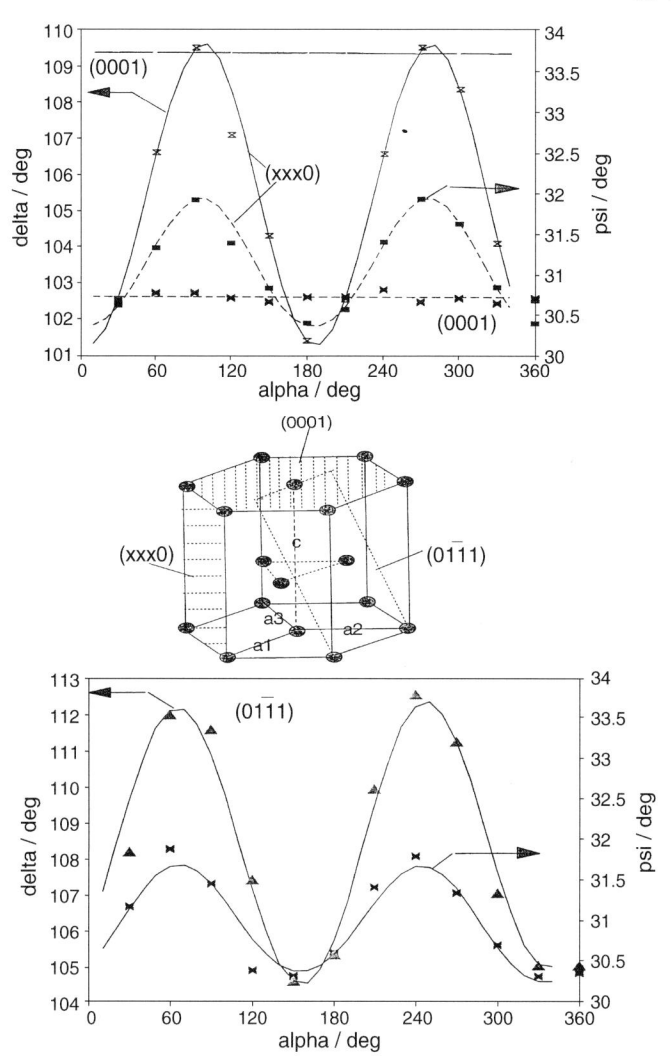

Figure 1.7 SAME on three different Ti grains with (0001), (01̄11) and (xxx0) orientation. The crystallographic orientations are illustrated in the displayed Ti lattice.

1.3.1.2 Determination of Ti Substrate Grain Orientation by SAME

In order to determine the crystallographic orientation of single Ti grains, SAME was applied. SAME yields the angle φ between the surface normal and the optical axis (c-axis). For this, the ellipsometric measurables Δ and Ψ are measured as a function of the angle α, which describes rotation of the sample about the surface normal. The corresponding experimental curves are shown in Figure 1.7 for the example of two different Ti grains. The grains are covered by anodically formed 4 V TiO$_2$ layers, which themselves are amorphous and therefore do not contribute to the $\Delta(\alpha)$ and

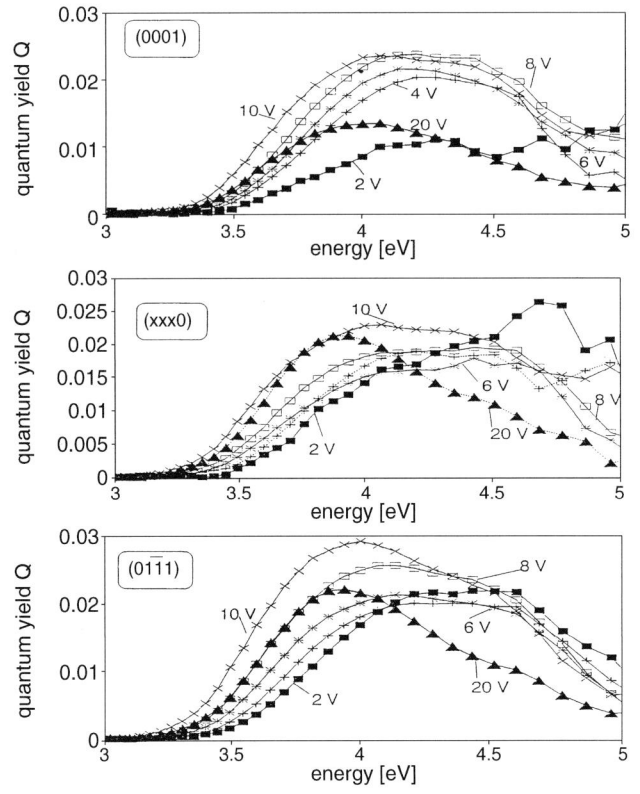

Figure 1.8 Photocurrent spectra (quantum yield Q) of TiO_2 passive films for different film thicknesses (formation potential, respectively). The measurements were carried out on three differently oriented Ti single grains. Electrolyte 0.5 M H_2SO_4, anodic potential 2 V.

$\Psi(\alpha)$ variation. The quantitative analysis of these curves allows the determination of the crystallographic orientation [82, 83, 308]. The planes determined from the curves of Figure 1.7 are illustrated in the Ti crystal structure shown. The (0001) surface ($\varphi = 0°$) is the closest packed surface with 1.15 a^{-2} (lattice parameters: $a = 0.2951$ nm, $c = 0.4679$ nm), (xxx0) refers to any plane perpendicular to the c-axis ($\varphi = 90°$).

1.3.1.3 Photocurrent Spectra and $i_{ph}(U)$ Measurements on Single Ti/TiO$_2$ Grains

Micro-photocurrent spectra, (quantum yield Q) taken on different oriented single Ti grains in 0.5 M H_2SO_4 are shown in Figure 1.8. The TiO_2 films were formed potentiodynamically in the same solution at the indicated formation potentials (2 V up to 20 V) before the spectra were recorded at a constant anodic potential of 1 V. Consequently, the data show the dependence of the photocurrent spectra on both the TiO_2 film thickness and the grain orientation. The formation factor was measured for each of the different orientations by ellipsometry (value averaged over all orientations: 2 nm V^{-1}). This allows for a quantitative correlation of the spectra with the film thickness. Evaluation of the Q^2 and $Q^{0.5}$ curves allows for the determination of the direct and indirect bandgap energies. E_{dir} varies between 3.7 and 3.4 eV for the different orientations and shows a slight decrease with increasing film thickness. The same holds for E_{indir}, which varies between 3.3 and 3 eV. For energies below the bandgap, an Urbach tail was found with a slope of 120 meV for the disorder energy. This indicates the amorphizity of the TiO_2 passive films, that is, there is no sharp bandgap but a mobility gap. The mobility gap energy is close to the bandgap energies of the TiO_2 modification anatase, therefore indicating a relationship between this modification and the amorphous TiO_2 layers found here (near order). The variation of i_{ph} with the Ti grain orientation confirms that in spite of the amorphicity of the TiO_2 films, their properties depend sensitively on the substrate texture. It has to be emphasized that the dependence of i_{ph} on d_f caused by internal reflections and the recombination processes mentioned in the above photocurrent model is evident across the entire spectral range: up to 10 V i_{ph} increases, thereafter i_{ph} decreases. This interference pattern is in excellent agreement with the simulation curves shown in Figure 1.5, therefore confirming this model. In order to allow for a more quantitative check of the model, the photocurrent was measured as a function of the formation potential, film thickness, respectively. Four experimental curves covering the spectral range from 280 up to 375 nm are shown in Figure 1.9 for an 0001 oriented grain. During illumination, an anodic potential of 2 V was applied. For the wavelength of $\lambda = 280$ nm, clearly case 10 from Table 1.2 applies, that is, the $r(E)$ term dominates over the entire thickness range. As the penetration depth of the light increases with increasing wavelength, the interference effect becomes more and more prominent at longer wavelengths, confirming the contribution of the internal reflections to i_{ph}. The simulations shown in Figure 1.9 agree well with the experimental curves. The minor deviations between model and experiment can be attributed to the different mechanisms for direct and sub-band excitations.

1.3.1.4 Microscopic Modification of the TiO_2 Films by Means of Laser Scanning

The migration of holes to the oxide/electrolyte interface leads to an accumulation of positive charge at the surface, which causes a change to the potential distribution within the oxide, namely the photopotential. Under potentiostatic conditions, the potential drop in the Helmholtz layer will be increased accordingly, enhancing the charge transfer of oxygen through the oxide/electrolyte interface. This affects the film formation which itself is a simple anodic ion transfer reaction without direct contribution of holes as:

$$Ti + 2H_2O \rightarrow TiO_2 + 4H^+ + 4e^- \tag{1.19}$$

On the other hand, the holes are directly involved in two reactions, oxygen formation and photo-corrosion, as:

$$4h^+ + 2H_2O \rightarrow O_2 + 4H^+ \tag{1.20}$$

and

$$4h^+ + TiO_2 \rightarrow Ti^{4+} + O_2 \tag{1.21}$$

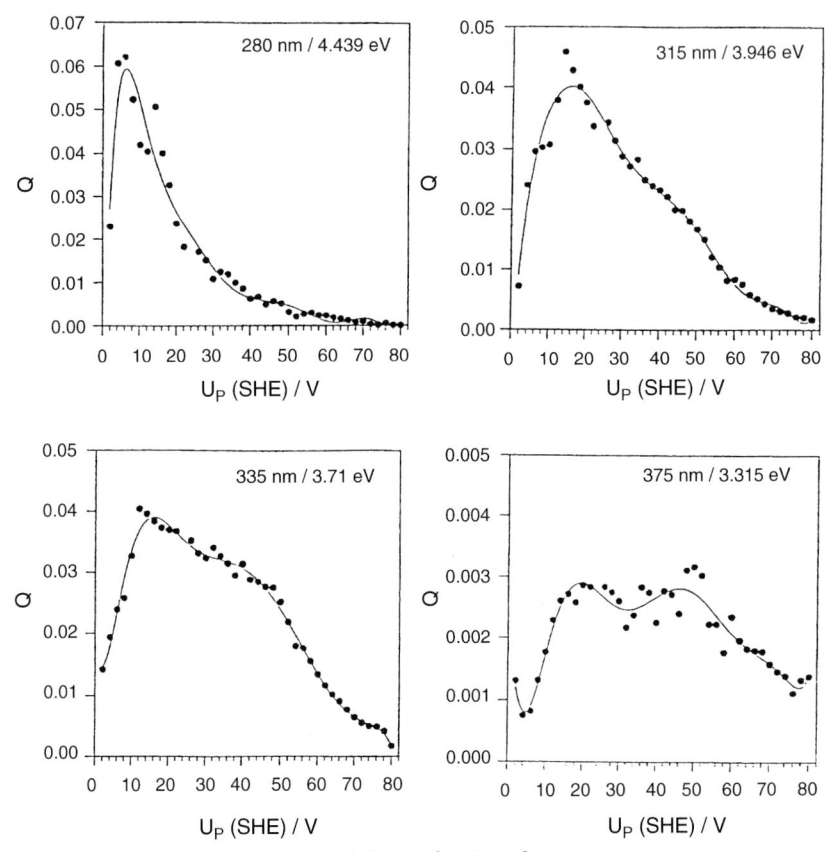

Figure 1.9 Photocurrent (quantum yield Q) as a function of the TiO_2 formation potential U_P, film thickness, respectively. During illumination a potential of 2 V was applied to the electrode.

As will be shown below, at high anodic potentials reaction 19 dominates allowing for a light induced modification of the surfaces, that is, laser induced film growth. This effect was first discovered by Leitner and Schultze [49]. An example from Michaelis [7] is demonstrated in Figure 1.10, where the results of so-called locodynamic experiments across grain boundaries between single Ti grains of known orientation are shown. For this, the illuminated samples were scanned in μm-steps normal to the laser beam while the electrode potential was increased, as in a conventional cyclic voltammogram. The advantage of this procedure is that the laser beam always hits an almost fresh surface area, allowing the potential dependence of the photo-reactions to be investigated without previous modification of the initial semiconducting films. Figure 1.10a shows micrographs of the resulting modified Ti surfaces on four substrate grains. The surface orientations and the grain boundaries between the single grains are indicated. The locodynamic experiments were performed on anodic TiO_2 films formed potentiodynamically at 20 V. The UV-laser power was 50 kW cm^{-2}, the applied potential U_{loco} during illumination was cycled

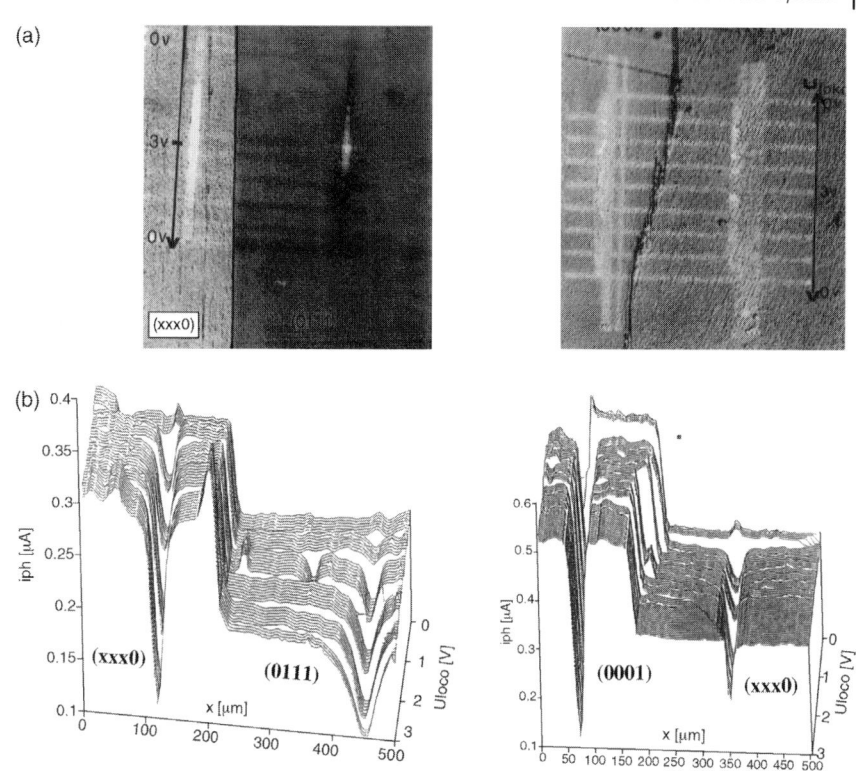

Figure 1.10 (a) Micrographs of the (0001)/(xxx0) and (0111)/(0001) grains (anodic 20 V TiO$_2$ films). The applied electrode potential U_{loco} was cycled between 0 and 3 V, laser power = 50 kW cm^{-2}. The visible laser traces indicate formation of a gradient in the TiO$_2$ film thickness. (b) Laser scanning analysis across the laser traces and the grain boundaries (imaging of the film properties).

between 0 and 3 V. The lateral dimensions of the laser modified line (parallel to the grain boundaries) significantly exceed the illuminated area (spot radius 2 μm), indicating lateral transport mechanisms such as hole diffusion beyond the laser focus. The color of the laser traces varies with the applied anodic potential U_{loco} indicating formation of a gradient in the film thickness.

1.3.1.5 Characterization of the Modified TiO$_2$ Films

1.3.1.5.1 UV-Laser Scanning Analysis

For a quantitative analysis of the laser-modified films, laser-scanning experiments normal to the laser traces and across the grain boundaries were carried out using reduced laser power below the threshold of modification (50 μW). The results [$i_{ph}(x,y)$] are shown in Figure 1.10b. The scale of the x-axis holds for both the micrograph and the corresponding $i_{ph}(x)$-curve. These curves give a sectional profile of the laser-modified site corresponding to the potential

U_{loco} applied in the locodynamic experiment previously. From the x-scan the lateral dimensions of the laser-induced modification can be evaluated. Scanning of the surface $i_{\text{ph}} = f(x,y)$ in both dimensions allows for an imaging of the entire potential dependence $i_{\text{ph}} = f(x, y = U_{\text{loco}})$ of the laser induced modification. The photocurrent fluctuations on the unmodified crystal surfaces are below 1%. Crossing from one crystal surface to the other (scanning across the grain boundary) yields an immediate i_{ph} change of about a factor of 2 due to the dependence of i_{ph} on the texture, as discussed previously. From the decay of this sudden change, the lateral resolution of the method can be estimated to be about 2 μm, which agrees well with the laser focus. Hence, it can be concluded that laser scanning allows imaging of passive layer properties at heterogeneities such as grain boundaries. For instance, as in Figure 1.10a, the fringed structure of the grain boundary between the (0001)/(xxx0) surfaces, visible in the micrograph can be imaged accurately.

1.3.1.5.2 Micro-Reflection Spectroscopy

Figure 1.11 shows the film thickness measured on top of the modified region by means of micro-reflection spectroscopy. For the evaluation of the film thickness, averaged optical constants were used, which were determined by ellipsometry. At locodynamic potentials exceeding 0.25 V an increase of the film thickness is observed, proving reaction 19 to be dominant in this potential regime. However, below 0.25 V a thinning of the TiO_2 film took place, which can be attributed to photocorrosion, i.e. reaction 21. This leads to the important conclusion that UV-laser illumination can be used for both a thickening (writing) or thinning (erasing) of semiconducting films. The applied potential (U_{loco}) determines which reaction dominates and therefore can be used to control the layer thickness. One important application of this method is the generation of semiconducting films with thickness gradients. Such gradients can be realized by means of locodynamic illumination of the sample surface.

Additionally, d_f and the lateral extension of the laser modified films depend on the substrate orientation, which therefore must be known. To do this, SAME can be

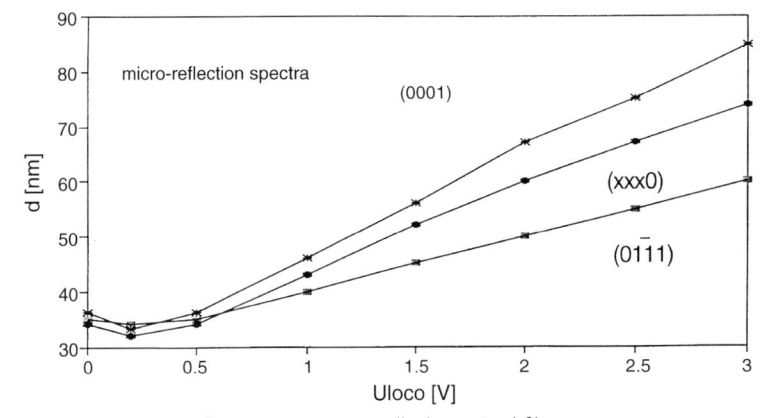

Figure 1.11 Micro-reflection spectroscopically determined film thickness of the UV-laser generated TiO_2 gradients as a function of the applied electrode potential U_{loco}.

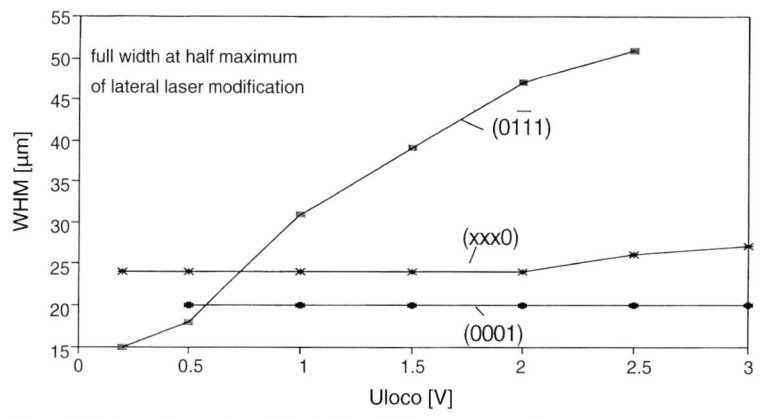

Figure 1.12 Lateral extension (full width at half maximum) of the TiO$_2$ gradients as a function of the applied electrode potential U_{loco}. The focus spot diameter during illumination was 2 μm.

applied *in situ*. For all substrate orientations it was found that the lateral extension of the modified regions exceeds the illuminated site significantly. In the case of the (0001) and (xxx0) orientation, the lateral extension of the modified site is almost independent of U_{loco}. In contrast, the (0$\bar{1}$11) surface shows a significant dependence of the laser line extension on U_{loco}. This is quantified in Figure 1.12, showing the full width at half maximum of the laser scans shown in Figure 1.10b as a function of U_{loco}. These curves correlate well with the reflection spectra results shown in Figure 1.11, that is the laser induced oxides are vertically thicker but less extended for the (0001) and (xxx0) than for the (0$\bar{1}$11) orientation. This suggests that the amount of the laser-induced oxide is determined by the number of electron/hole pairs generated and is almost constant for the different orientations. The difference in the lateral and vertical extension of the laser-modified region for the different grains is due to the different electronic properties of the original semiconducting films determining the lateral hole diffusion.

1.3.1.5.3 SAME Analysis of the Modified Films

In order to analyze the structure of the UV laser modified anodic TiO$_2$ films (formation potential: 20 V), SAME was carried out ($U_{loco} = 2$ V). Figure 1.13 shows the resulting SAME-curves $\Delta(\alpha)$ and $\Psi(\alpha)$ for the (0$\bar{1}$11) grain as an example. Three SAME experiments at different sites of the sample were performed, that is:

1. at large a distance from the laser modified region (characterization of unmodified passive film)
2. close to the illuminated area (effect of lateral hole diffusion on passive film)
3. on top of the illuminated area.

At a large distance from the laser illumination, the typical SAME curve for an amorphous TiO$_2$ with 35 nm thickness was measured. The film properties completely correspond to the films measured previously in Figure 1.7. More interesting were the SAME results from experiment 2. The corresponding curves show a significant

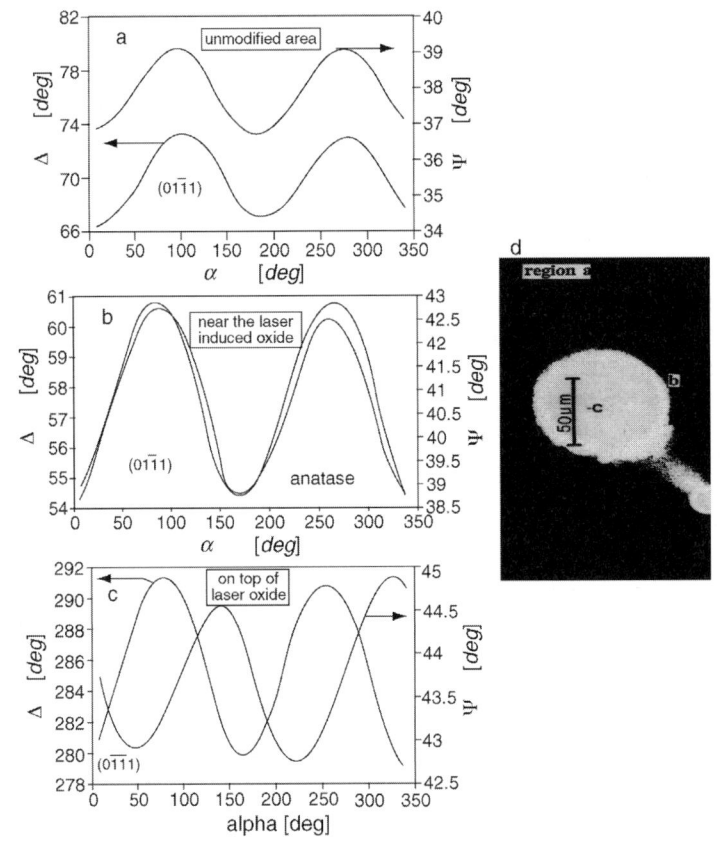

Figure 1.13 SAME curves recorded on potentiodynamically formed 20 V TiO$_2$ films on a (0111) oriented Ti substrate grain. (a) A large distance from a laser modified spot; (b) close to the modified region; (c) on top of the laser modified area; (d) photograph of laser modified surface.

change in amplitude, phase and average value in comparison with the SAME curve of experiment 1. The optical parameters of crystalline anatase were then derived, suggesting a recrystallization of the TiO$_2$ films. However, this phenomenon did not occur on the other two grain orientations. Lateral hole diffusion is most prominent on the (0111) orientation. This agrees with the laser scanning measurements of Figure 1.10 where the largest extension of the laser traces appears on the (0111) surface.

Experiment 3 again shows a clear change in amplitude, phase and average value of the $\Delta(\alpha)$ and $\Psi(\alpha)$ curves. The increase in the $\Delta(\alpha)$ and $\Psi(\alpha)$ average values can be attributed to an increased film thickness (reaction 19, i.e. film growth). The alteration of both phase and amplitude again indicates a structure change of the TiO$_2$ films. Quantitative SAME evaluation reveals that a two-layer model now has to be applied. It was found that the initial films (metal/oxide interface) recrystallized into anatase analogues in experiment 2 (optical parameters $n_{ao} = 2.48$, $n_o = 2.55$, ordinary and extraordinary refractive index, respectively). On top of this crystallized film an amorphous layer was found exhibiting a low density of about $3\,\mathrm{g\,cm^{-3}}$ (estimated by analysis of the determined optical constants [85, 86]. Therefore, these oxides can clearly be distinguished from each other and from the original anodic films. These experiments illustrate well that even on homogeneous electrode areas, in this case one single (0111)-oriented grain, optical micro-methods can be applied to characterize and elucidate local electrode reactions.

1.3.1.6 Photoresist Microelectrochemistry (Nanoliter Droplet Method)

As described in the preceding sections, the application of focused laser light is a powerful instrument for the detection of local thin film properties such as morphology, thickness and photocurrent. In order to complement this information in terms of electrochemical reactivity, such as electron and ion transfer rates, capacitance and potential, a flexible photolithographic technique in combination with electrochemical measurements in small droplets under optical control has been applied to scale down classical electrochemical experiments [6]. In comparison with the well established scanning electrochemical microscope (SECM) [87], the technique employed here uses a patterned fixed ultra-microelectrode with a well defined surface area and a three-electrode configuration. Hence, the local resolution is limited by the lithographic resolution in the sub-micrometer range and by the electronic detection limit of the current amplifier.

1.3.1.6.1 Preparation and Measurement Technique
In order to allow the preparation of the photoresist microelectrode at the point of interest, the preparation as well as the measurement was carried out under microscopic control as described in the flow chart in Figure 1.14. Following the resist-coating (positive tone, d_{resist} 1 µm) and prebake, the microelectrode is exposed under proximity conditions at the desired spot of the sample. A preconditioning of a highly reflective metal surface with HMDS (hexamethyldisilane) and an antireflective coating was avoided due to the possible impact on the electrochemical reactivity of the surface, though limiting the performance of the lithography. Pinhole masks are positioned by an x,y,z-stage and exposed by UV light, which is fed through the microscope. The distance between two adjacent microelectrodes can be as low as 50 µm. A postbake and the development completes the preparation.

A schematic drawing and the microscopic view of the experimental setup are shown in Figures 1.14 and 1.15. A glass capillary is filled with electrolyte and is connected to the reference electrode (e.g. Hg/Hg_2SO_4, $U_0 = +0.68$ V) and an electrolyte reservoir with a nanopipette, so that nanoliter droplets can be dispensed on the photoresist microelectrode. The diameter of the droplet is usually set to twice the diameter of the microelectrode. Micromanipulators are used to position the electrodes. A platinum wire is dipped into the droplet as a counter electrode, completing the experimental setup. Microelectrodes with a diameter of 5 µm can be realized using proximity exposure with reproducible surface areas. Figure 1.15 shows an example of a microelectrode preparation on single grains of a polycrystalline Ti sample as viewed

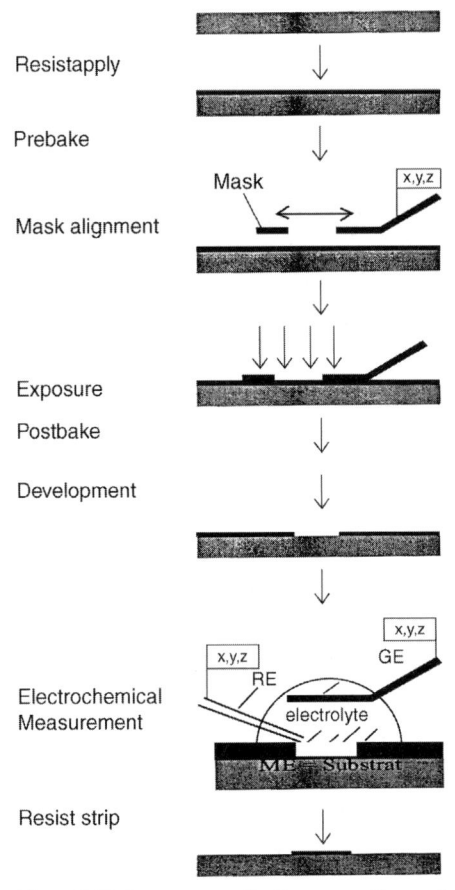

Figure 1.14 Process steps for the preparation and electrochemical measurement of photoresist microelectrodes. All steps are carried out under microscopic control with micromanipulators to handle the mask and the microelectrodes: GE, counter electrode; RE, reference electrode; ME = working electrode. (Method introduced by Kudelka and Schultze 1993 [6].).

with the microscope. Individual grains are identified on the electropolished surface when the resist is applied. The localization of the electrode reaction is demonstrated by the different colors of the anodic oxide layers. Each electrode can be addressed separately. A more detailed description, including the effects of the photoresist preparation on the electrochemical measurements, is given in Ref. [88].

1.3.1.6.2 **General Aspects** It is beyond the scope of this treatment to discuss the current and potential distribution for photoresist microelectrodes in detail. Therefore, only a few comments will be made on the influence of the geometry and the resist barrier on electrochemical measurements. Photoresist microelectrodes can be treated as a recessed microdisk electrode. The diffusion field and the current distribution are

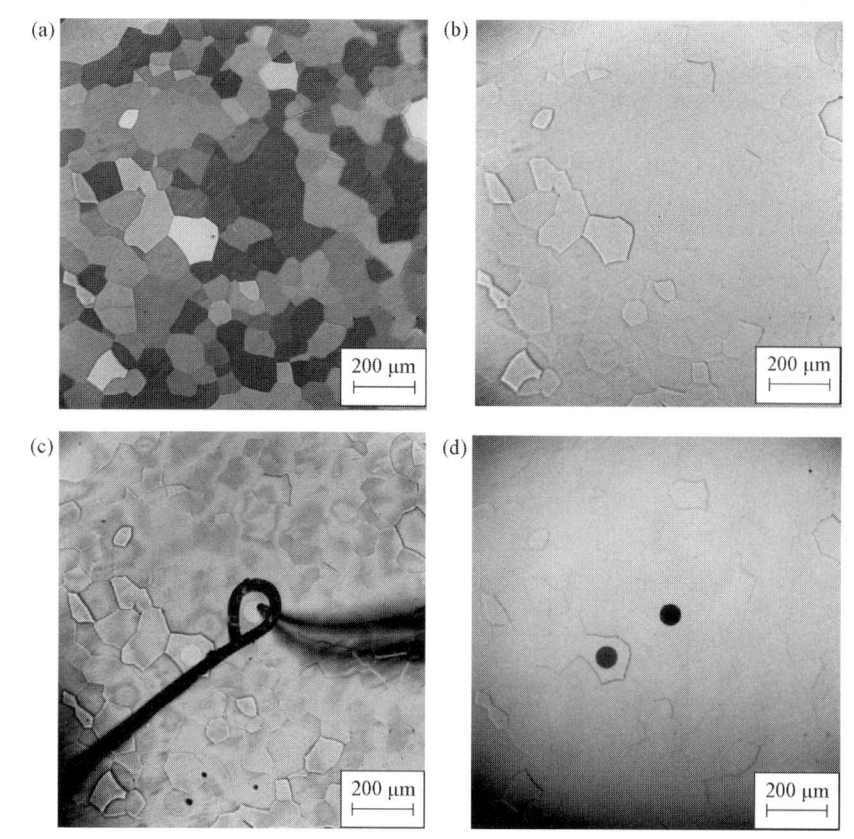

Figure 1.15 Microscopic view of sample preparation and measurement as expressed in Figure 1.14. The same area of a polycrystalline Ti sample is shown after different treatments: (a) with 20 V anodic oxide layer formed in 0.5 M H_2SO_4 potentiodynamically with 60 mV s^{-1}; (b) after electropolishing; (c) during the electrochemical measurement on a 100 μm photoresistmicroelectrode; and (d) after growing anodic oxide layers and stripping of the photoresist.

not hemispherical, as the aspect ratio increases. Model calculations [87, 88] have shown that for a microelectrode with an aspect ratio of 0.1 (diameter of the electrode $d_\mu = 10\,\mu m$, $d_{resist} = 1\,\mu m$) the resistivity increases by 15% in comparison with a planar microdisk electrode. For an aspect ratio of 1 the value increases to 50%. These limitations are negligible unless very high current densities in the order of $A\,cm^{-2}$ occur. For potential step measurements, high currents occur due to the capacitance charging of the double layer, microelectrodes with $d_\mu = 50\,\mu m$ have been used in this study. Here, the assumptions of planar microelectrodes are still applicable. The time constant for charging a capacitance of $C_H = 20\,\mu F\,cm^{-2}$ in 0.5 M H_2SO_4 (3.7 Ωcm) at a 50 μm microelectrode is less than 1 μs, thus allowing kinetic studies in the ms range, as will be demonstrated in the next section. In addition, the distance between the microelectrode and the reference and counter electrode was adjusted to some μm, which helps to control the potential of the electrode.

Owing to the finite electrolyte volume of the nanoliter droplet, the diffusion boundary layer reaches the dimensions of the droplet within seconds, assuming a diffusion coefficient of for example $D = 1\,10^{-5}\,\text{cm}^2\,\text{s}^{-1}$. A constant diffusion gradient can not be established because the bulk concentration is reduced. Thus, measurements of diffusion limited currents are not stable over time. The electron transfer reactions discussed below were carried out on anodic oxide layers with maximum current densities 3 to 4 orders of magnitude lower than the diffusion limited current densities, thus ensuring a stable support of consumables over the time of the measurement. In order to prevent evaporation of electrolyte, the ambient was saturated with water vapor.

The use of a small electrolyte covered resist area around the microelectrode is essential when capacitance measurements are performed, as the resist capacitance is parallel to the electrode capacitance. With a specific resistance of $\sigma = 10^{12}\,\Omega\,\text{cm}$ and a dielectric constant of $\varepsilon = 1.5$ for the resist, and assuming a typical electrode capacitance of $10\,\mu\text{F}\,\text{cm}^{-2}$ with a 50 µm electrode, an error of 5% is obtained, if the electrolyte covered surface is $10^{-3}\,\text{cm}^2$ ($f = 1000\,\text{Hz}$) [88]. Thus, for capacitance measurements, the use of nanoliter droplets is essential.

1.3.1.7 Applications of Photoresist Microelectrodes

In the following sections the electrochemical reactivity of single grains of polycrystalline Ti is explored by using the nl-droplet method. The results from electrochemical measurements and the optical laser techniques from the previous section are combined to yield a band structure model for anodically grown anodic oxide layers. Other applications of this method to study laser induced corrosion, texture dependent photocurrent and corrosion of anodic oxide films are described in Refs. [89, 90 and 91].

1.3.1.7.1 Cyclovoltammetry

Anodic Passive Layer Formation on Single Ti Grains   The following measurements were performed on single grains of polycrystalline Ti. In some instances the experimental results are discussed for two deliberately selected grains (a) and (b), because the aspects of the different behavior are represented sufficiently. Prior to the electrochemical measurements, the surface orientation of these grains and the thickness of the anodic oxide layers was determined by SAME. In Figure 1.16 several cyclic voltammograms are shown, which were obtained on a freshly electropolished Ti sample. Obviously, the texture has a significant influence on the anodic current density. The total anodic current density i is composed of four contributions:

$$i = i_{ox} + i_{O_2} + i_{corr} + i_C \tag{1.22}$$

where i_{ox} is oxide formation, i_{O_2} oxygen evolution, i_{corr} corrosion current and i_C is due to capacitive charging. The last two contributions can be neglected as i_C does not exceed $2\,\mu\text{A}\,\text{cm}^{-2}$ ($dU/dt = 50\,\text{mV}\,\text{s}^{-1}$, $C_H = 20\,\mu\text{F}\,\text{cm}^{-2}$) and i_{corr} was found to be less than 8% in the potentiodynamic preparation mode [92].

In the potential region from 0 to 3 V SHE the oxygen evolution can be excluded and the anodic current density is purely the oxide formation i_{ox}. Thus, the anodic charge of the oxide layer formation depends on the texture of the underlying Ti. For grain (a),

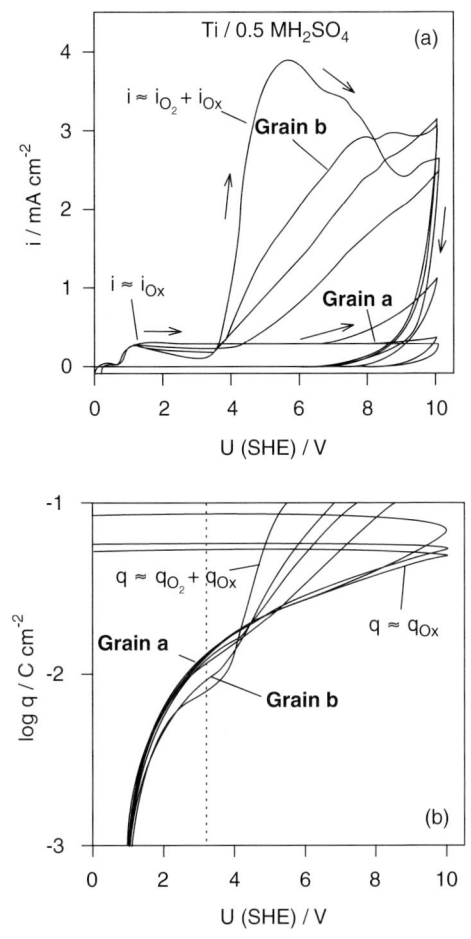

Figure 1.16 Current density (a) and charge (b) during the anodic oxidation on different grains of a polycrystalline Ti sample. Diameter of the microelectrode 20 µm, $dU/dt = 50\,mV\,s^{-1}$ in 0.5 M H_2SO_4 [6].

with an orientation angle $\varphi = 85°$, the current density is independent of the potential, corresponding to a proportional growth of the oxide thickness with the potential U. For grain (b), with an orientation angle $\varphi = 37°$, i decreases with U, thus the oxide formation rate is potential dependent. The relationship between surface orientation, expressed by the orientation angle φ, and anodic oxide charge q_{ox} is systematic, as plotted in Figure 1.17. At orientation angles <30°, [near to the (0001) orientation] the anodic charge is 30% less than at orientation angles >50%. A sharp transition is observed at 40°. The values for the (0001) and (xxx0) orientation are taken from measurements at macroscopic single crystal electrodes. These experiments prove that standard electrochemical measurements can be extended into the µm range.

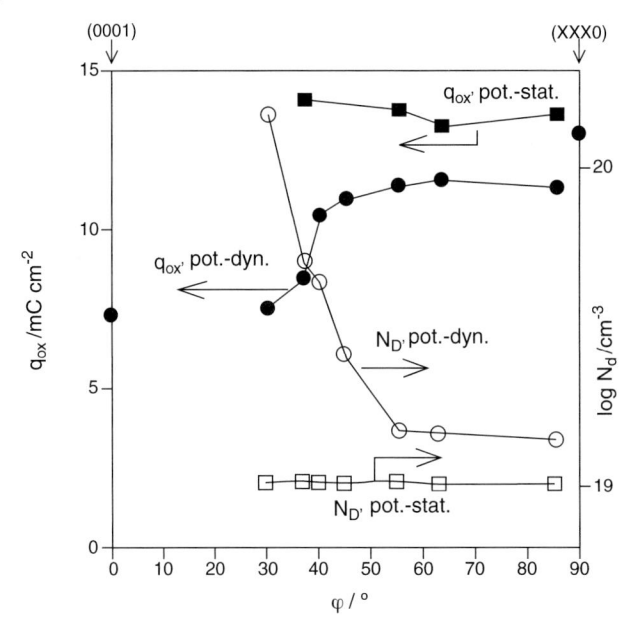

Figure 1.17 Anodic charge q_{ox} and donor density N_D as a function of the orientation angle φ for potentiodynamically (slow $dU/dt = 50\,mV\,s^{-1}$) and potentiostatically (fast) grown oxide layers. For the potentiodynamically grown oxides, the oxide charge was integrated from 0.5 to 3 V (SHE), for the potentiostatic grown oxides from 0 to 4 V. Values for (xxx0) and (0001) orientation were taken from measurements on macroscopic single crystal electrodes [6].

In the potential region above 3 V, additional oxygen evolution may contribute, which also depends on the surface orientation. On grain (a) [close to (xxx0)] only a slight increase in the current density can be observed at $U > 8\,V$, whereas on grain (b) [close to (0001)] the current rises up to 10-times the oxide formation current. A correlation with the oxide formation current can again be found. The oxygen evolution increases with a decrease in the anodic charge between 0.5 and 3 V. As demonstrated below, this is due to an increased conductivity of the oxide layers.

1.3.1.7.2 Capacitance Measurements As outlined earlier, the resist capacitance can be neglected, if the electrolyte covered resist surface is reduced, for example, $3\,10^{-4}\,cm^2$ for a 50 μm electrode. Typical capacitance values of a 50 μm diameter electrode at anodic potentials of $U > 5\,V$ are >100 pF. Figure 1.18a and b shows the capacitance measurements and the corresponding Schottky–Mott charts for four different grains, including again the grains a and b. Two different measurement conditions namely potentiodynamic and potentiostatic are compared (analogues with the conditions chosen for Figure 1.17). In addition, in Figure 1.18c the reciprocal capacitance is shown as a function of the formation potential for grains a and b under potentiodynamic conditions. Experimental values are represented by a straight line for grain (a) (xxx0) orientation. A dielectric constant of $\varepsilon = 34$ can be calculated by a simple capacitor model. In contrast, for example, for grain (b) (0001) a nonlinear behavior reflects the nonlinear growth. An average dielectric constant of $\varepsilon = 40$ can be

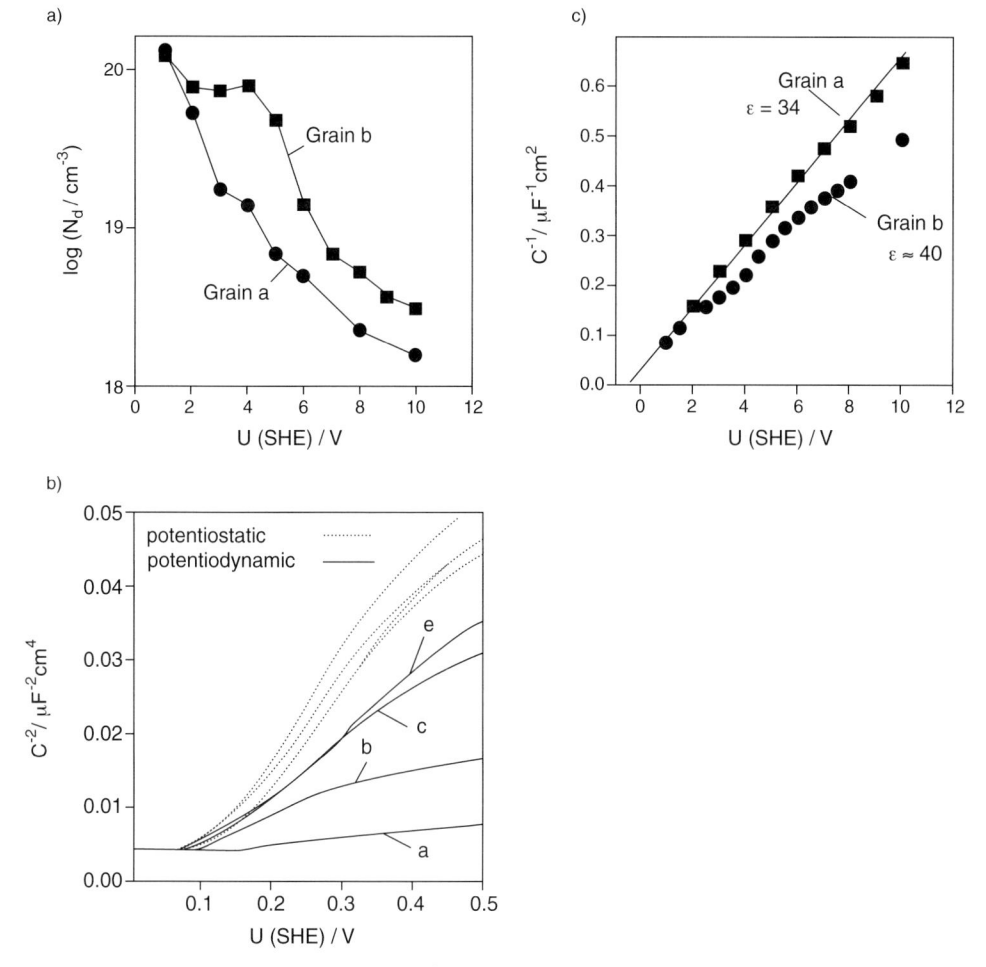

Figure 1.18 (a) Capacitance measurements and formation potential U for grains a and b (b) Schottky–Mott plot for four different Ti grains and two different experimental formation conditions (potentiodynamic and potentiostatic). (c) Reciprocal capacitance C plotted against the formation potential (potentiodynamic formation, 0.5 M H_2SO_4, 20 mV s^{-1}.) The capacitance was measured at the upper formation potential with $f = 1030$ Hz. The film thickness was measured ellipsometrically [6].

estimated, which still lies significantly below the values given in the literature (see e.g. Table 1.1). The corresponding evaluation for the potentiostatic oxide formation gives ε_r values much closer to the literature of about 60. This again confirms the importance of the experimental formation conditions on the electronic properties of the oxides. At lower potentials than the formation potential, for films formed at 2, 4, 6 and 10 V the capacitance values are plotted according to the Schottky–Mott equation in Figure 1.19 (only the potentiodynamic case is shown). The slope of the curves is a function of the relative dielectric constant ε_r and the donor density N_D as pointed out

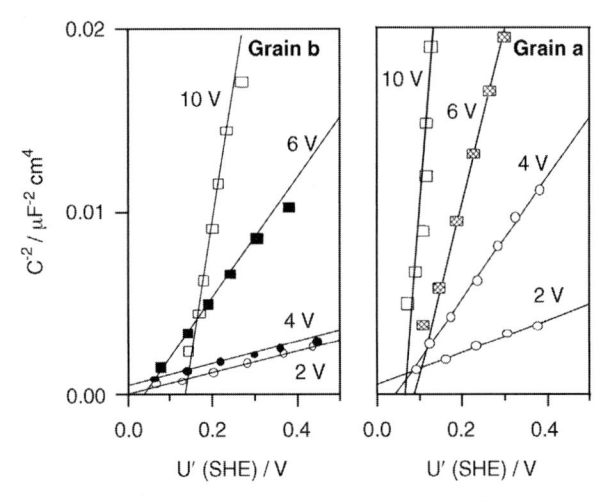

Figure 1.19 Evaluation of the capacitance according to the Schottky–Mott equation for grains a and b for oxide layers grown potentiodynamically at different formation potentials U (100 μm electrodes, $f = 1030$ Hz, $dU/dt = 20$ mV s^{-1}) [6].

in Section 1.2.2. Below 0.5 virtually straight lines are observed. Here only the slopes of the curves were evaluated. In the case of grain (a) (xxx0), the slope increases with the formation potential. In contrast, grain (b) (0001) shows an almost constant slope up to a formation potential of 4 V. According to the Schottky–Mott equation, donor densities are calculated and plotted as a function of the formation potential, shown in Figure 1.20. In the potential region between 3 and 5 V (SHE) donor densities differ

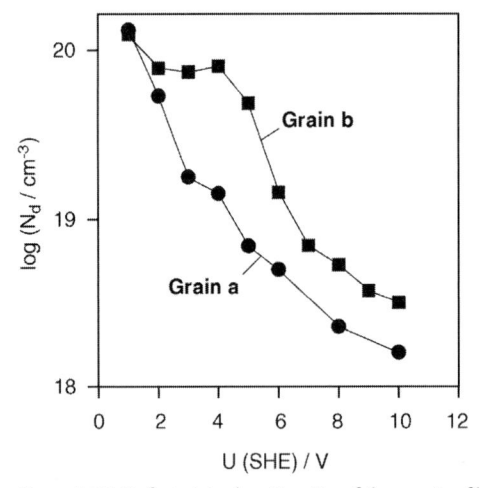

Figure 1.20 Defect state densities N_D of the passive films on grains a (xxx0) and b (0001) as a function of formation potential derived from the Schottky–Mott plots [6].

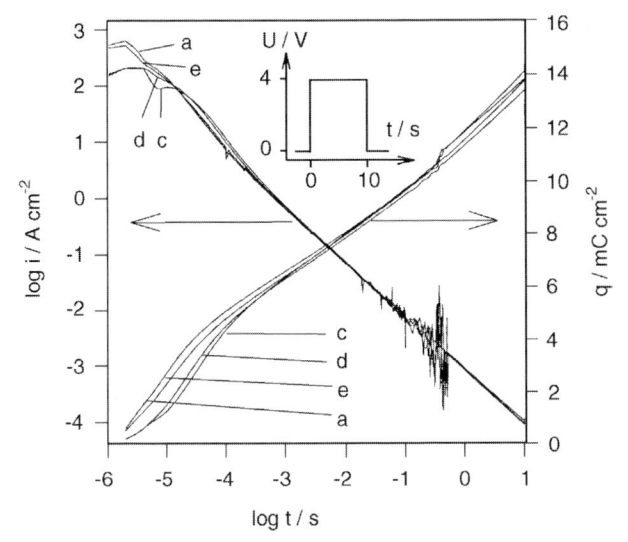

Figure 1.21 Double logarithmic plot of current density and time t for a potentiostatic pulse of from 0 to 4 V (SHE) on four different grains with different orientations. Microelectrode diameter was 100 μm.

by almost one order of magnitude, underlining the texture dependent electronic properties of the passive layers. Again, the change in the donor density is a systematic function of the crystal orientation, as already indicated in Figure 1.17 for potentiodynamically formed oxides.

1.3.1.7.3 Transient Measurements One of the main features of microelectrodes is their suitability for the study of fast electrode reactions. Owing to a low potential drop in the electrolyte and low absolute currents, the potential of the microelectrode can be controlled very accurately. In this instance, the formation of anodic oxide layers was investigated as a current response to a potential-step experiment (potentiostatic oxide formation). The potential of the oxide free electrode (natural oxide) is jumped abruptly to the formation potential. Figure 1.21 shows time and current in a double logarithmic plot for an anodic pulse from $U = 0$ to 4 V (SHE). The potential control of the electrode was achieved within a few microseconds. High current densities of more than 100 A cm^{-2} could be realized, with the absolute values of only some mA cm^{-2}. Time and current could be followed over almost seven orders of magnitude by using a fast transient recorder in combination with automatic amplifier sense switching. In contrast to the slowly grown potentiodynamically formed oxides, for all grain orientations, charge and current is almost identical. Again, prior to the measurement the orientation of each grain was determined by microellipsometry. Thus, for fast grown oxides the influence of the texture is decreased drastically. The slope $d\log i/d\log t = -1$ is in agreement with an oxide growth according to a high field mechanism. Additional evidence for the grain independent properties of the fast grown oxides has already been indicated in the capacitance measurements of

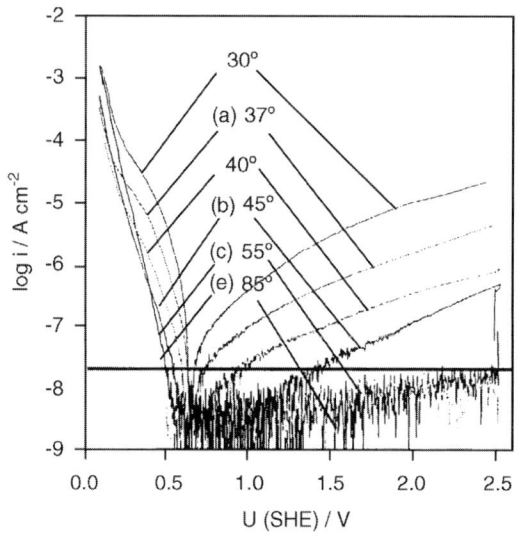

Figure 1.22 Tafel plots of ETR with 0.5 M Fe^{2+}/Fe^{3+} redox system in 0.5 M H_2SO_4 on different grains of a polycrystalline sample. The orientation of the grains is given as the angle φ. The current without the redox system is subtracted. The current detection limit is indicated by the horizontal line ($d = 200\,\mu m$, $dU/dt = 5\,mV\,s^{-1}$) [6].

Figure 1.18 and will be further confirmed by the electron transfer reaction (ETR) measurements below. The donor densities of the fast grown oxide layers are lower than for the potentiodynamically formed oxides, indicating a more isolating character (see Figure 1.17) [258].

1.3.1.7.4 **Electron Transfer Measurements** ETR measurements were carried out in a 0.5 M Fe^{2+}/Fe^{3+} redoxsystem in 0.5 M H_2SO_4 after the oxide layers were formed in a slow growth mode, also in 0.5 M H_2SO_4 with a formation potential of 4 V. Capacitive charging and ionic currents are subtracted, so that the Tafel plots in Figure 1.22 represent only the contribution of ETR. Again, each curve represents a grain with a different crystal orientation. As it is not the subject of this text to elaborate on the various mechanisms of ETR, only a few comments will be made here. A more detailed description is available in Ref. [42]. A direct electron exchange with the metal can be excluded due to the oxide thickness of 6.5–10 nm. On the cathodic and on the anodic site, log i decreases with increasing orientation angle φ, which means from (0001) to (xxx0) orientation. In the same direction, the donor density decreases, which results in a increasing extension of the space charge layer. As the extension of the space charge layer is proportional to $1/C$ at a given potential, the plot of log i versus C^{-1} in Figure 1.23 suggests that a tunneling mechanism is the rate-determining step. Oxide layers with the highest donor concentration show the highest anodic current densities because of the reduced tunneling distance to the conduction band

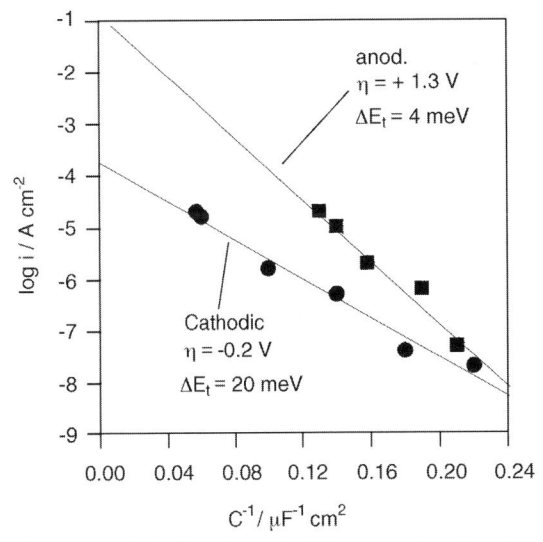

Figure 1.23 Cathodic and anodic ETR current densities at a constant overvoltage of $\eta = -0.2$ V and $\eta = 1.3$ V taken from Figure 1.22, plotted versus the corresponding reciprocal capacitance C^{-1} at the same η.

(reduced Debye length). For the (xxx0) orientation (grain e), the anodic reaction is totally blocked. On the cathodic branch at high overvoltages ($\eta < -500$ mV), the ETR is independent of texture, which is typical for a direct elastic tunneling process (see process 1 in Figure 1.24). At lower overvoltages a texture dependent shoulder in the current density is observed, which can be explained by a resonance tunneling process (see process 2 in Figure 1.24). Oxide layers formed by a potentistatic pulse experiment (see the previous section) show the behavior of grain (e), independent of the crystallographic orientation, underlining the insulating character.

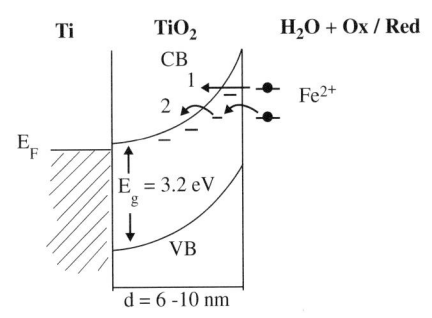

Figure 1.24 Schematic representation of electron transfer reaction to the conduction band for direct (1) and resonance tunneling (2).

1.3.1.8 Summary and Conclusions for the Ti/TiO$_2$ System

The complete employment of the micro-optical and micro-electrochemical methods, which were introduced in Section 1.2, has been demonstrated for the Ti/TiO$_2$ system. Owing to its pronounced heterogeneity Ti/TiO$_2$ is ideally suited for this purpose. For the other valve metal systems that will be discussed in the following sections only results with methods relevant for the specific system will be further presented. In order to elucidate local electrode reactions it was essential that all measurements were performed under microscopic control. This allows precise measurement on deliberately chosen electrode areas. For instance, all measurements presented in this section were carried out on the same single grains of only one electroploished Ti sample.

The properties of the semiconducting TiO$_2$ films depend sensitively on the substrate texture. The new SAME method was used for *in situ* determination of the substrate grain orientation. The electronic and optical TiO$_2$ film parameters were determined by means of photoresist micro-electrochemistry and micro-ellipsometry and then correlated with the Ti substrate grain orientation. The results using photoresist microelectrodes show that conventional electrochemical measurement techniques provide valuable information on the electronic properties, even in μm areas. The passive layer formation on Ti depends on the grain structure in terms of anodic charge, growth factor, oxygen evolution rate, electron transfer rate and donor density. The most important results can be summarized in a geometric film model shown schematically in Figure 1.25. Because of the semiconducting character of the passive film, the potential drop across the metal electrolyte interface is mainly located in the space charge layer. The extension of this space charge layer and the corresponding distribution and availability of energetic states determine the rate of electron transfer. The electron exchange takes place over the conduction band. For the low density packed Ti surface (xxx0), the oxide growth is a linear function of the potential. The donor density is comparably low, 10^{19} cm^{-3}. Anodic reactions such as the oxygen evolution reaction are blocked. On the close packed (0001) surface, the films grow more slowly above 2 V, are highly doped, 10^{20} cm^{-3}, and allow anodic reactions due to a thinner space charge layer and reduced tunnel barrier. High growth rates reduce the effect of texture and show a more isolating character. This explains the systematic variation of the film properties such as layer thickness and defect state concentration with the substrate orientation. The micro-electrochemically determined electronic and optical film parameters were confirmed by photocurrent measurements. The anodic TiO$_2$ films were found to be amorphous with mobility gap energies close to those of the TiO$_2$ modification anatase. The TiO$_2$ films were modified at high local resolution by means of focused UV-laser illumination. Both a thickening (writing) or thinning (erasing) of the films could be realized depending on the applied potential, which therefore could be used to control the layer thickness. This allows preparation of unique structures such as semiconducting films with thickness gradients. Pitting could be also induced.

Besides the pronounced dependence of the electronic oxide properties on the substrate orientation, an especially important result is that the film properties also depend sensitively on the formation conditions. This important result is illustrated in Figure 1.17 where the cases of potentiodynamic and potentiostatic formation are

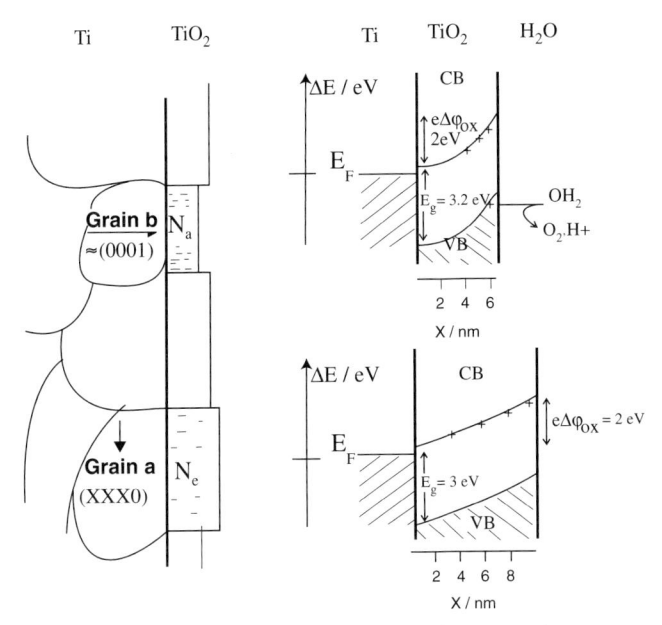

Figure 1.25 Calculated band structure of anodically grown oxides on (xxx0) and (0001) grains at a potential of $U=+2\,\text{V}$ (SHE). The oxides are formed at the same upper potential of $U_{ox}=4\,\text{V}$ with $d_{ox}=10$ and 6.5 nm. Dielectric constant, donor density and flat band potential were taken from measurements using photoresist.

compared. The texture dependence is suppressed significantly if a potentiostatic formation mode is used and the defect state concentration N_D is decreased. This is of paramount importance for electronic applications where a homogeneous oxide is desired, to ensure maximum reliability. Consequently, a potentiostatic formation mode should be chosen for Ti if an electrolyte capacitor similar to the well established Ta/Ta$_2$O$_5$ capacitors is to be developed. However, even then the defect state concentration of TiO$_2$ is still significantly higher than for Ta$_2$O$_5$, as is confirmed directly by capacitance measurements where Ti/TiO$_2$ always shows a pronounced bias dependence (Schottky–Mott behavior) whereas Ta/Ta$_2$O$_5$ shows only dielectric behavior with no texture dependence at all. Results for the Ta/Ta$_2$O$_5$ system will be presented in Section together with Nb/Nb$_2$O$_5$ and Al/Al$_2$O$_3$. Also, the Zr/ZrO$_2$ system shows dielectric behavior only, but here again a pronounced texture dependence can be found, as will be discussed in the following section.

1.3.2
Zr/ZrO$_2$ and Hf/HfO$_2$ Systems

1.3.2.1 Zr/ZrO$_2$

For Zr/ZrO$_2$ a strong effect of the substrate texture on the oxide properties can be expected. This expectation is based on the fact that the transport number of the

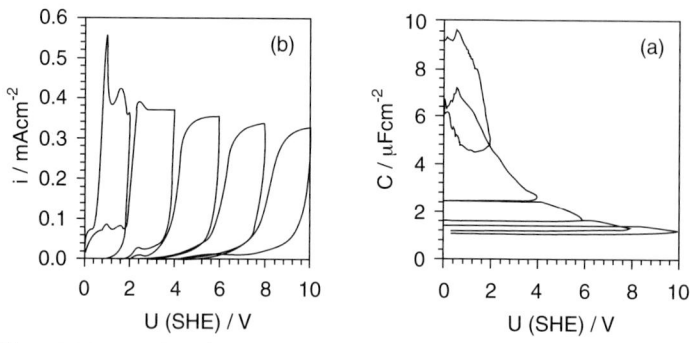

Figure 1.26 (a) Cyclic voltammogram and (b) simultaneous capacitance measurement (lock-in amplifier) during potentiodynamic ZrO_2 formation on a mechanically polished Zr sample: 0.5 M H_2SO_4, $dU/dt = 20$ mV/s, $f = 1013$ Hz, amplitude 1.13 mV [17].

cations during oxide growth is approximately zero (see Table 1.1). Thus, the role of the metal/oxide phase boundary can be expected to exert a pronounced influence on film formation and will therefore cause a strong correlation between substrate crystallography and passive film properties. Unlike the usually amorphous TiO_2 films, anodic ZrO_2 films are predominantly crystalline [93–96]. As all ZrO_2 modifications are optically anisotropic, they are amenable to SAME investigations, which should therefore reveal the potential grain-dependent differences in crystallinity. A first important elucidation of the electronic behavior of the Zr/ZrO_2 system can be drawn from capacitance measurements recorded during potentiodynamic oxide formation in 0.5 M H_2SO_4 at a sweep rate of 20 mV s^{-1}. This experiment, which was performed with a macroscopic, mechanically polished (homogenized) Zr electrode, is shown in Figure 1.26a. The cyclic voltammograms show the typical valve metal behavior similar to the Ti/TiO_2 measurements shown in Figure 1.16. However, different to with Ti/TiO_2, no oxygen evolution occurs at potentials higher than 3 V. This is a first hint at a more insulating behavior of ZrO_2. This finding is confirmed by the corresponding capacitance measurements, which show a complete bias independence. The defect state density N_D therefore has to be well below 10^{19} cm^{-3}, which is several orders of magnitude lower than the values for TiO_2. The important conclusion therefore is that ZrO_2 is an excellent dielectric. This finding is confirmed by Figure 1.26b, where a similar measurement is shown. Here an electropolished (open grain boundaries) sample was used and the sweep rate was 100 mV s^{-1}.

1.3.2.1.1 SAME

The SAME parameters relevant to the Zr/ZrO_2 system are shown schematically in Figure 1.27. The angle φ_{ox} describes the orientation of the oxide, the angle φ_{met} describes the orientation of the bare Zr surface, which has to be known before the derivation of φ_{ox}. Therefore, first SAME measurements on the bare Zr surfaces are presented. The results are used subsequently for the quantitative determination of the crystalline oxide parameters.

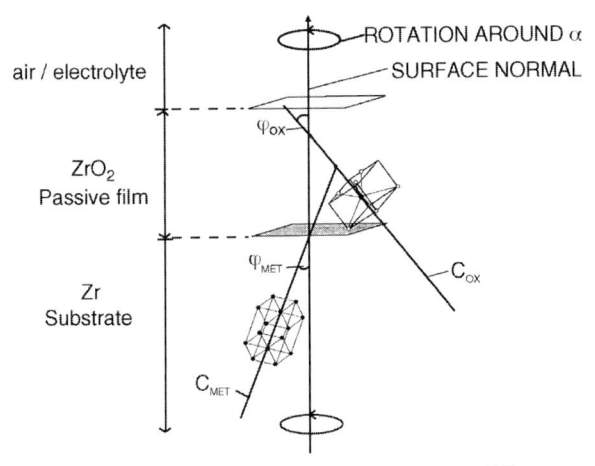

Figure 1.27 SAME parameters for the Zr/ZrO$_2$ system [17].

SAME on Bare Zr Surfaces As with Ti, Zr also crystallises in an hcp (hexagonal closed package) structure and is therefore also optically birefringent. In order to determine the ordinary (n_o) and extraordinary (n_{ao}) complex refractive indexes of the bare Zr substrate, SAME measurements on four differently oriented, freshly electropolished grains were carried out. These experimental curves were fitted simultaneously for each substrate by a least square fit procedure as explained in Ref. [308]. The resulting $\Delta(\alpha)$ and $\Psi(\alpha)$ simulation curves are shown as a function of the orientation angle φ in Figure 1.28. The experimental curves, which were recorded on the aforementioned four different Zr grains, are also indicated in this figure. Consequently, the corresponding grain orientation angles φ_{met} for these grains can be determined and are

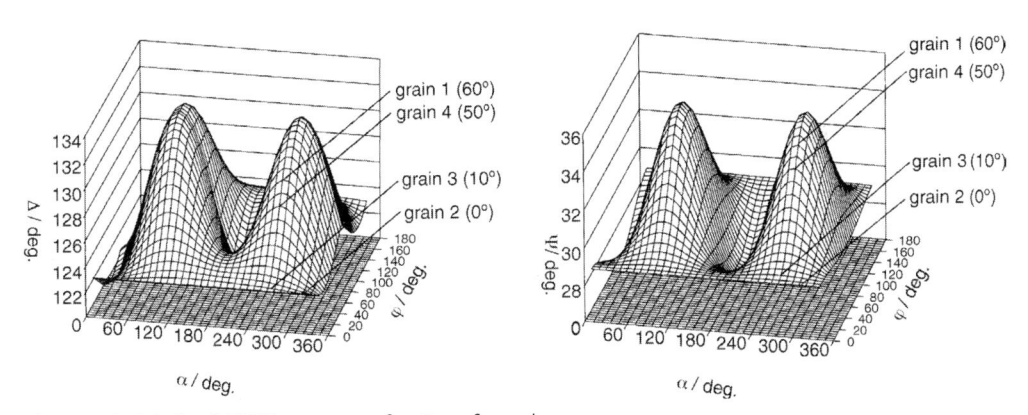

Figure 1.28 Calculated SAME curves as a function of sample rotation about angle α and as a function of the orientation angle φ_{met}. The experimental curves for four different Zr grains are indicated together with the orientation angles φ_{met} determined for these grains [84, 98].

also given in Figure 1.28. The substrate parameters calculated from the simulation for a wavelength of 632.8 nm are:

$$n_o(Zr) = 2.18 - i3.41 \text{ and } n_{ao} = 2.23 - i2.54$$

These parameters can be checked independently by their consistency with macroscopically determined ones on mechanically polished samples. Mechanical polishing yields an amorphous and therefore isotropic Beilby layer resulting in an averaged isotropic complex refractive index n_{iso}. The correlation between n_{iso} and the anisotropic parameters is given by the equation:

$$n_{iso} = 2/3 n_{ao} + 1/3 n_o$$

A value of $n_{iso}(Zr) = 2.21 - i2.83$ was measured showing excellent agreement with the calculated value, according to the above equation, therefore confirming the determined anisotropic parameters.

SAME on Zr/ZrO$_2$ The most important ZrO$_2$ modification is baddeleyeite, which has a monoclinic unit cell ($a = 0.5169$ nm, $b = 0.5232$ nm, $c = 0.5341$ nm, $\beta = 99\ 15''$). In addition, tetragonal and hexagonal ZrO$_2$ structures are discussed in the literature [94, 95]. Figure 1.29a–d shows SAME measurements taken on an (0111) oriented Zr grain for four different formation potentials (potentiodynamic, $20\,\mathrm{mV\,s^{-1}}$) between 10 and 40 V. Clearly, a strong dependence of the amplitudes and phase relationship between the $\Delta/\Psi(\alpha)$ curves on the formation potential is observed. These amplitude changes and phase shifts are clearly due to the ZrO$_2$ films that contribute to the anisotropy of the system. This specific anisotropy effect of the anodic oxide constitutes clear evidence for the formation of an ordered crystalline passive film with a well defined epitaxial relationship to the substrate grain. In contrast, no such effect [phase shifts in the $\Delta/\Psi(\alpha)$ curves] was found for TiO$_2$, indicating the amorphous status of the TiO$_2$ oxides. For completeness, the corresponding TiO$_2$ curves are also shown in Figure 1.29, e and f. The measured curves for ZrO$_2$ can be fitted assuming a film orientation angle ρ of 22.5° and the following optical constants

$$n_o(ZrO_2) = 2.13 - i0, \quad n_{ao}(ZrO_2) = 2.20 - i0$$

These values are consistent with literature data for monoclinic crystalline ZrO$_2$ [97]. The excellent fits of experimental data (dots) and simulations (solid lines) in fig. 1.29(a–d) is strong evidence for a *texture-dependence of crystallization processes* in anodic ZrO$_2$ films.

In spite of the biaxial nature of this material, a uniaxial approximation can be used because the two extraordinary indices are almost equal and can therefore be treated as being identical. Figure 1.30 shows the calculated $\Delta(\alpha)$ and $\Psi(\alpha)$ curves as a function of the film thickness using the determined parameters. Similar results were obtained on grains with orientation angles φ between 10 and 60°. No grains with higher orientation angles could be found on the investigated polycrystalline sample. The ZrO$_2$ films exhibited two major film axis orientations, namely $\varphi_f = 22.5°$ for $\varphi_{sub} > 30°$ and $\varphi_f = 45°$ for $\varphi_{sub} < 30°$. Apparently, in all these instances an anisotropic/anisotropic system exists and SAME reveals its full power for the study of

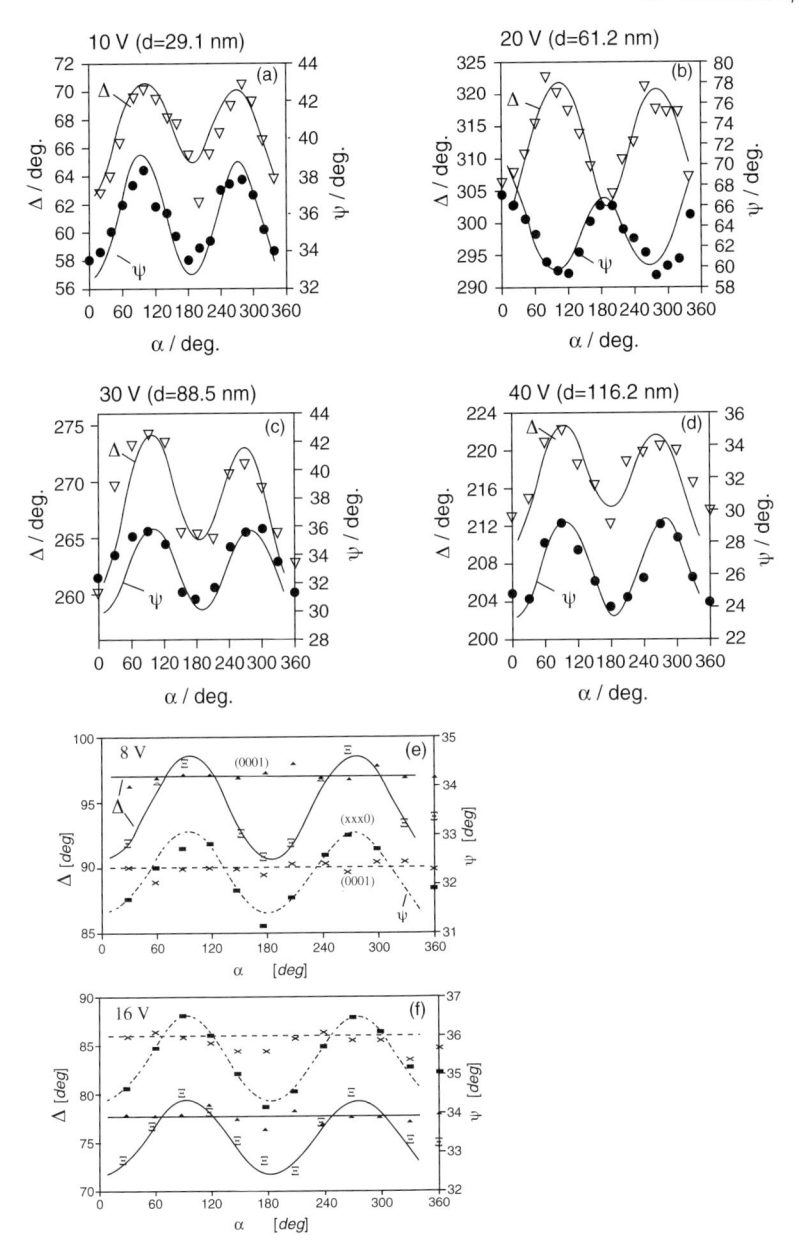

Figure 1.29 (a–d) SAME curves on an (0111) oriented Zr single grain after potentiodynamic formation to the indicated formation potentials (10–40 V). The symbols refer to the experimental data, the lines represent the fit results (SAME determination of the orientation angles). Significant amplitude variations and phase shift of the Δ curves with respect to the Ψ curves are observed, indicating an epitaxial, crystalline growth of the ZrO_2 [84]. (d and f) Similar curves for three different Ti grains (orientation indicated) at two different formation potentials. No amplitude and phase shifts occur constituting the amorphous character of the TiO_2 films [7].

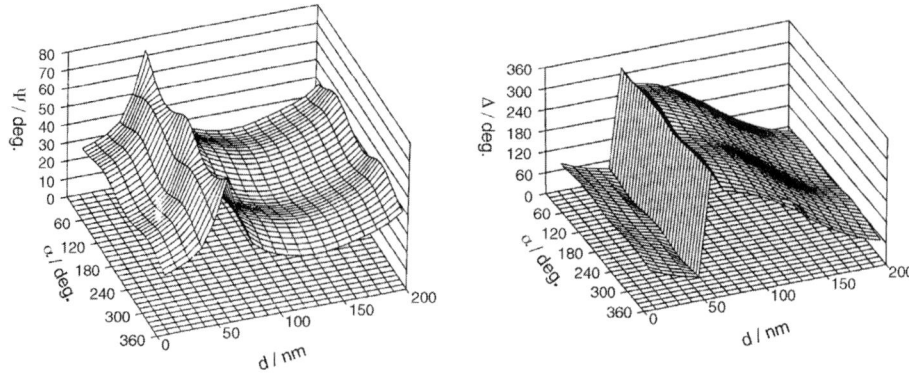

Figure 1.30 Calculated SAME curves. The dependence of the ZrO$_2$ thickness on the ellipsometric measureables is shown for the (0111) oriented Zr grain [17].

epitaxy relationships for this system class. However, on (0001) oriented grains (closest-packed surfaces) no anisotropy contribution of the films were observed indicating an amorphous or nano-crystalline character of the layers formed on this particular orientation. A similar behavior (amorphous film formation) is also found on mechanically polished samples. The epitaxial relationships found between substrate and films are summarized schematically in Figure 1.31, where the ellipsometrically determined film formation factors are also given.

1.3.2.1.2 **Photocurrent Spectra** Further correlation with the electronic structure of the films can be obtained by photocurrent measurements. The results are shown in Figure 1.32, with grains 1 and 2 as examples. As indicated from the photocurrent spectra (Figure 1.32a), the photocurrent quantum yield above the bandgap of 5.0 eV is

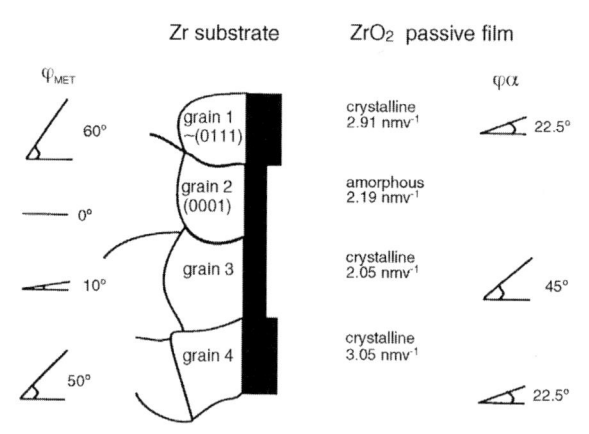

Figure 1.31 Model for the texture dependent oxide growth on Zr surfaces. The crystallographic orientation angles of substrate Zr and oxide film ZrO$_2$ are given in addition to the electrochemically (photoresist method) determined formation factors [17].

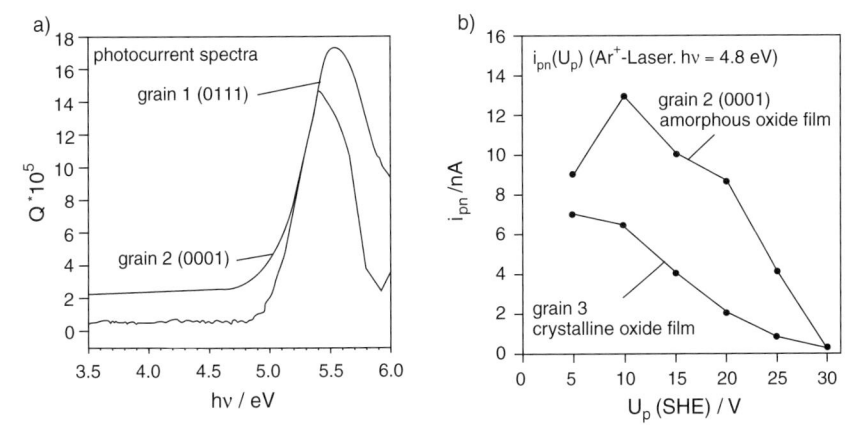

Figure 1.32 Photocurrent spectra: (a) as a function of photon energy for grains 1 and 2 and (b) as a function of the formation potential; 0.5 M H_2SO_4, 23 Hz, $U = 3$ V. The amorphous film on the grains exhibits a higher photocurrent in the sub-band regime due to the states within the mobility gap [17].

higher for the crystalline oxide films on grain 1. In contrast, the sub-bandgap photocurrent is much higher for the amorphous films on (0001) oriented grains (Figure 1.32b). This clearly points towards a higher density of states in the bandgap for the latter case, again corroborating the grain-dependent crystalline versus amorphous behavior found in the SAME experiments.

1.3.2.1.3 Photoresist Microelectrodes

The local cyclic voltammograms on photoresist-microelectrode prepared on grains 1 and 2 are shown in Figure 1.33a. Clearly, the shape of the curves is different for each case. On grain 2, the measured anodic

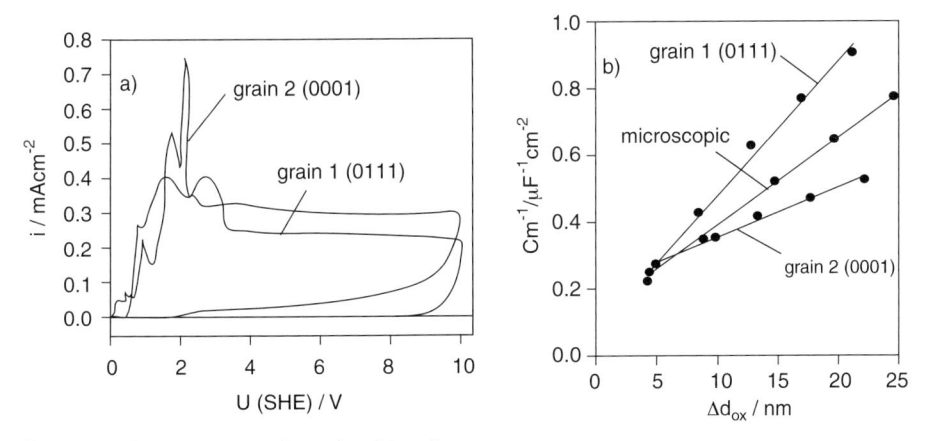

Figure 1.33 Photoresist microelectrodes: (a) cyclic voltammograms and (b) capacitance analysis for grains 1 and 2 [17].

current density is higher over the entire potential range, although the layer formation factor is lower (2.2 versus 2.9 nm V^{-1} for grain 1). This is probably due to extended side reactions in this instance, which are supported by the higher defect density. Different electronic properties can be deduced from analysis of local capacitance curves (Figure 1.33b). The capacitance is potential-independent for potentials below the formation potential, constituting the insulating behavior of ZrO_2. This insulating behavior is found on all Zr grains regardless of substrate orientation. However, the absolute capacitance values show a strong dependence on the substrate. Assuming a simple capacitor model this points towards a variation in the relative permittivities for the ZrO_2 films formed on different substrate grains. The crystalline films on grain 1 exhibit the lowest apparent dielectric constant whereas the amorphous films on (0001) faces represent the other extreme case. These relationships are shown in Figure 1.34 for a whole variety of grains, together with the layer formation charge.

1.3.2.1.4 **Conclusions for the Zr/ZrO$_2$ System** Over the entire ZrO_2 film thickness range investigated here (1 up to 100 nm) and under all experimental formation conditions (potentiostatic, potentiodynamic) no bias dependence of capacitance on the applied potential was found. Apart from passive film growth, no anodic reaction could be found in the cyclic voltammograms, indicating an ideal valve metal behavior. Therefore, in contrast to the n-type semiconducting TiO_2, passive ZrO_2 shows an ideal insulating behavior with a defect state concentration well below 10^{19} cm^{-3}.

SAME measurements performed on single grains of polycrystalline Zr reveal the presence of well ordered, single crystalline oxide films on all substrate orientations except (0001). The crystal structure of the oxides depends on the grain orientation in a clearly defined way. This finding is summarized in Figure 1.35, where the ZrO_2 film orientation angle and the oxide formation factor is shown as a function of the

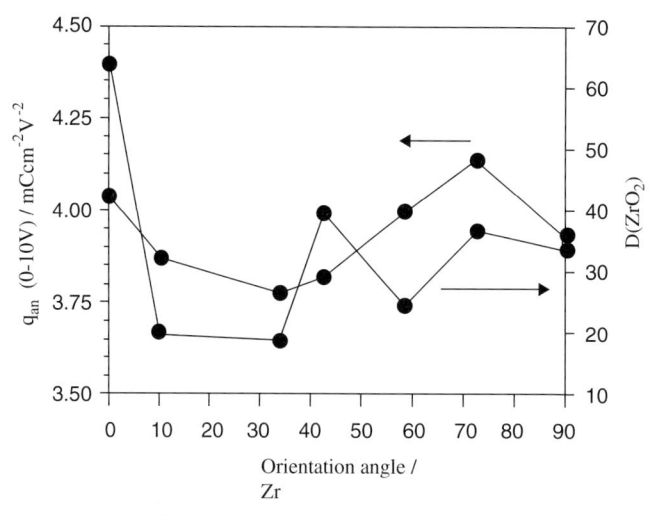

Figure 1.34 ZrO_2 film formation charge and relative dielectric constant ($\varepsilon_r = D$) as a function of the orientation angle of the Zr substrate [17].

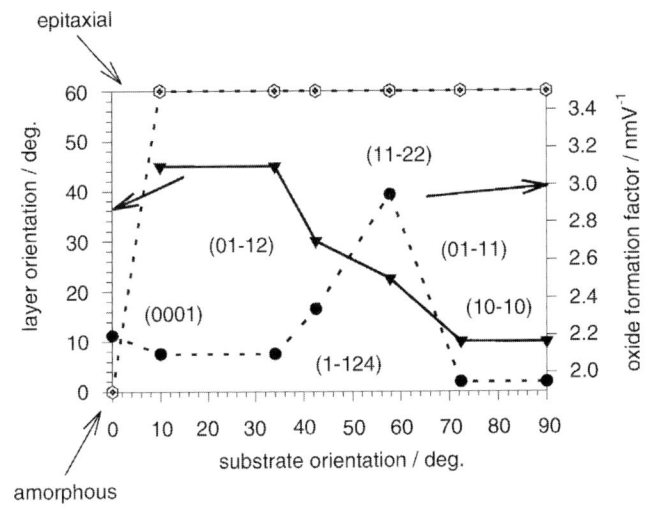

Figure 1.35 ZrO$_2$ film orientation angle and oxide formation factor as a function of the Zr substrate orientation [17].

substrate orientation. Therefore, as with the Ti/TiO$_2$ system, the results for Zr/ZrO$_2$ again emphasize the strong grain-dependence of passive film growth but unlike the usually amorphous TiO$_2$ films, anodic ZrO$_2$ films are predominantly crystalline. The corresponding SAME data show excellent agreement with model calculations within the framework of SAME theory. The crystal orientations of the substrate and layer are the only adjustable parameters, and they can be determined separately by SAME measurements on the bare substrate and anodized surfaces, respectively, thus strongly supporting the consistency of the model presented here. The anisotropic/anisotropic Zr/ZrO$_2$ system is therefore ideally suited to demonstrating the full power of SAME for analysis of crystallographic properties. Both the orientation of the Zr substrate and the orientation of the epitaxial growing ZrO$_2$ films could be determined quantitatively. The pronounced texture dependence of the ZrO$_2$ film properties confirms the decisive role of the metal/oxide phase boundary on anodic oxide formation. Consistently, it was found that the cation transport number for ZrO$_2$ formation is almost zero.

The measurements presented here illustrate impressively that it is essential to consider the effect of substrate heterogeneities on passive film growth. Measurements of passive films on macroscopic surfaces can, in principle, not be quantitatively understood in terms of classical models without taking texture effects into account, as macroscopic measurements can yield only averaged values.

1.3.2.1.5 Application of ZrO$_2$ to Electronics For potential electronic applications of ZrO$_2$ as a dielectric or insulating film the following facts are important:

1. ZrO$_2$ shows no bias dependence of capacitance therefore proving its excellent insulating behavior.

2. The relative dielectric constant ε_r of ZrO_2 depends strongly on the crystallographic orientation of the Zr substrate. The average value is about 27.
3. Zr/ZrO_2 shows a strong texture dependence of film formation.
4. The ZrO_2 films are predominately crystalline.

The last two points in particular are critical for potential applications in electronics as a thin dielectric film ($d_f < 300$ mn). Owing to grain boundaries in the crystalline film, which act as leakage paths, and due to electronic avalanche processes, which are much more likely in crystalline than in amorphous films (because of the higher trap density in amorphous films), the breakdown voltage will be low. Additionally, the oxide thickness variation due to the texture dependence of oxide growth may cause reliability problems in the electronic devices. Nevertheless, ZrO_2 has been discussed for electronic thin film applications as an additive for Ta_2O_5 dielectric films or even more interestingly as a nano-laminate component in successively deposited films [99–102]. As will be discussed next, a very similar behavior is found for the Hf/HfO_2 system.

1.3.2.2 Hf/HfO_2

Owing to their close similarity, it makes sense to treat the Hf/HfO_2 system in the same context as the $Zr/ZrO2$ system. The relationship between the two systems begins with their natural occurrence. Zr and Hf are typically found together ores, although Zr is much more frequent (megaton production) than Hf (about 100 ton per year). With the exception of nuclear shielding applications (reactor grade Zr), Zr materials always contain significant amounts of Hf as an impurity, which usually does no harm. Nuclear shielding is an exception because of the higher absorption efficiency of Hf (much higher density) for neutron radiation. Shielding of neutron radiation is one of the major applications of bulk quantities of Hf. However, as will be shown here, HfO_2 is also an excellent dielectric material (similar to ZrO_2), which makes it an interesting material for electronics at least in thick film applications ($d_f > 500$ nm). In recent publications HfO_2 has been discussed as a promising dopant material for advanced CVD electroceramic thin film dielectrics in IC-MOSFET (metal oxide semiconductor field effect transistor) [99–102, 104].

The first evidence of the excellent dielectric properties of HfO_2 films can be seen immediately in the anodic film formation cyclic voltammograms and the simultaneously recorded capacitance measurements shown in Figure 1.36. The experiments were performed in 0.5 M HNO_3 as the electrolyte with a sweep rate of $25\,\mathrm{mV\,s^{-1}}$ using a mechanically polished pure (99.9%) Hf electrode. The capacitance curves were recorded by means of a lock-in amplifier set to a frequency of 1013 Hz with a signal amplitude of 2 mV. Similar results were obtained in NaOH or borate buffer as the electrolytes indicating the outstanding stability of the oxides to aggressive environments. The results were taken from Ref. [103]. As for the Zr/ZrO_2 system, no bias dependence of capacitance is found for potentials below the formation potential confirming the insulating character of the oxides. Again an ideal valve metal behavior is found with almost zero current density for anodic potentials below the formation potential. The coulometrical analysis $[d_f = q_{ox}M/(zF\rho r)]$ of the cyclic

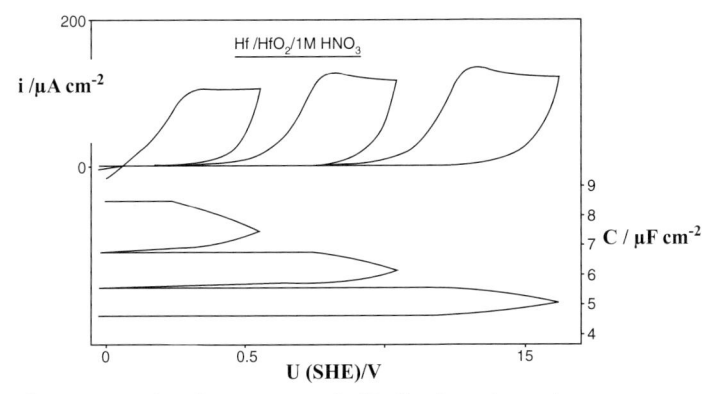

Figure 1.36 Cyclic voltammograms of HfO$_2$ film formation and simultaneously recorded capacitance measurements in 0.5 M HNO$_3$ [103].

voltammograms yields an HfO$_2$ formation factor of about 2 nm V^{-1}. For this, a surface roughness factor r of 1.5, $M = 210.5$ g cm^{-3} and density $\rho = 9.7$ g cm^{-3} were assumed. The corresponding curves for all three mentioned electrolytes are shown in Figure 1.37. Evaluation of the reciprocal capacitance as a function of the formation charge or film thickness, respectively, allows for determination of the dielectric constant (see Figure 1.37b). A dielectric constant ε_r of 16 could be determined, which is significantly lower than the corresponding values for ZrO$_2$ ($\varepsilon_r = 27$) and TiO$_2$ ($\varepsilon_r = 56$).

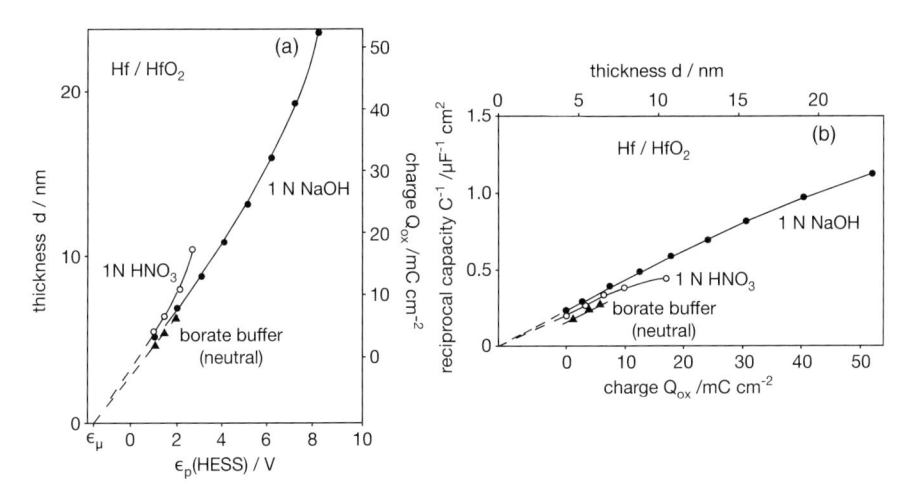

Figure 1.37 (a) Coulometrically determined film thickness and film formation charge as a function of the formation potential. Determination of formation factor, ca. 2 nm V^{-1}. (b) Reciprocal capacitance and film thickness as a function of formation charge for determination of the relative dielectric constant $\varepsilon_r = 16$ [103].

1.3.2.2.1 Texture Dependence of HfO_2 Film Formation

Hf also crystallises in an hcp structure and the HfO_2 oxides show the same monoclinic structure as the ZrO_2 oxides (coordination number = 7). Owing to the zero cation transfer number, again a pronounced texture dependence of the oxide formation with crystalline oxide films is to be expected. This can immediately be confirmed by looking at the interference colors of the anodically formed layers on an electropolished (open grain boundaries) electrode surface. As for the Zr/ZrO_2 system, the interference colors of the formed HfO_2 oxides vary from grain to grain. This means that all the measurements presented for Zr/ZrO_2 in the previous section could be repeated. In particular, SAME could be successfully applied again. This is a task for a forthcoming study and has not yet been carried out.

However, some important facts that are crucial for potential electronic applications have already been established at this stage. Basically, the same conclusions as for the Zr/ZrO_2 system hold, that is:

1. Anodic HfO_2 shows an insulating behavior without any bias dependence of capacitance.
2. The dielectric constant on mechanically polished surfaces is 16.
3. Hf/HfO_2 shows a strong texture dependence of film formation.
4. The oxide films are predominantly crystalline.

Therefore, as for ZrO_2, a thin film application of pure HfO_2 in electronics as a functional dielectric film is doubtful, since it tends to crystallization. However, it crystallization and grain boundary formation could be suppressed, then both HfO_2 and ZrO_2 could be of interest.

1.3.3
Systems: Nb/Nb_2O_5, Ta/Ta_2O_5 and Al/Al_2O_3

In contrast to the aforementioned systems Ti/TiO_2, Zr/ZrO_2 and Hf/HfO_2, the valve metal systems Nb/Nb_2O_5, Ta/Ta_2O_5 and Al/Al_2O_3 show no texture dependence on oxide growth at all. Consistently the cation transfer number of these systems is >0.5, indicating a less pronounced influence of the metal/oxide interface on electrochemical oxide formation. The anodically formed oxides are amorphous with no tendency to any crystallization even if they are exposed to high temperatures. These properties are very desirable for electronics, explaining why these materials dominate in passive component and IC applications where ultra-thin dielectrics are needed. Only the Si/SiO_2 system, which also exhibits amorphous oxides but with a pronounced texture dependence, can keep up with them.

The substrate atomic structure of all of these systems shows an isotropic, cubic symmetry. Therefore, SAME can not be applied here. Owing to the texture independence of oxide growth, already indicated by the homogeneous interference colors on electropolished surfaces (no grain to grain variations), application of SAME would make no sense anyway. Only a potential crystallization of the oxides that are predominantly optical anisotropic could possibly be detected. However, no indication of any crystallization under the experimental conditions used here was found. Therefore, no SAME measurements are presented in this section.

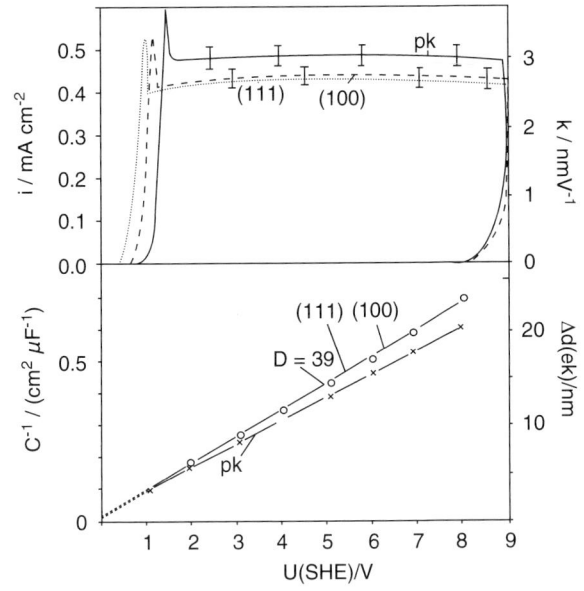

Figure 1.38 Potentiodynamic oxide formation cyclic voltammograms for a polycrystalline Nb electrode and two different single crystal Nb surfaces in addition to the corresponding reciprocal capacitance curves 0.5 M H_2SO_4, 100 mV s^{-1} [18].

1.3.3.1 Nb/Nb$_2$O$_5$ System

The texture independence of film formation for the Nb/Nb$_2$O$_5$ system is confirmed in Figure 1.38, where potentiodynamic film formation cyclic voltammograms for two differently oriented Nb surfaces, (111) and (100), and an electropolished polycrystalline surface (pk) are shown together with the corresponding reciprocal capacitance curves. The cyclic voltammograms show the typical valve metal behavior. The minor differences in the current densities are due to slight differences in the surface roughness of the samples. The higher surface roughness of the polycrystalline sample can be attributed to its open grain structure. This also explains the slight difference in the slopes of the reciprocal capacitance curves. There is a small difference in the oxide formation onset potential, which might indicate a minor texture dependence of native (thermal) oxide formation. In anticipation of the results presented later, it has to be pointed out that there is no indication of any texture dependence of anodic oxide formation. The results were taken from Ref. [18]. Evaluation of the reciprocal capacitance slope yields a dielectric constant ε_r of 39 in good agreement with literature values of 42. The determination of the formation factor by coulometric analysis of the formation charge and film thickness curves, respectively, is shown in Figure 1.39. A constant formation factor m of 2.8 nm V^{-1} was found, independent of the surface orientation. The measurements were performed in 0.5 M H_2SO_4, the potentiodynamic sweep rate was 100 mV s^{-1}. The

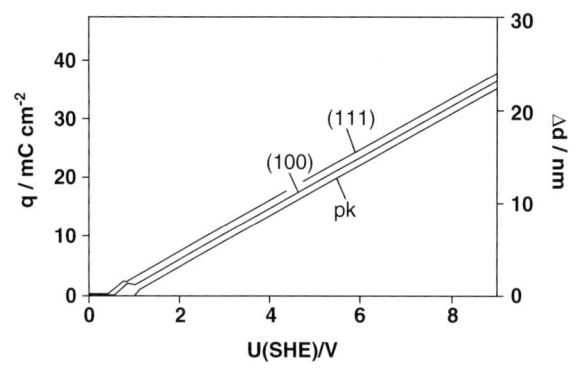

Figure 1.39 Oxide formation charge and coulometrically determined film thickness as a function of the formation potential for three different electrode surfaces [18].

capacitance was measured in the usual way during the oxide formation using a lock-in amplifier at 1013 Hz with a signal amplitude of 2 mV.

In Figure 1.40 a comparison of $C(V)$ curves for Nb, Ta, vanadium-doped Nb and Ta/Nb alloy electrodes is shown. Prior to these capacitance measurements, the oxides were formed at a formation potential of 30 V. In contrast to Ta, Nb and the Nb/Ta alloy show a clear bias dependence (Schottky–Mott behavior) of the capacitance indicating an n-type semiconducting behavior of the Nb_2O_5. Therefore, similar to TiO_2, Nb_2O_5 is not an ideal dielectric but exhibits a high defect state concentration N_D, which can be attributed to oxygen vacancies. This is an important finding with respect to a potential application of Nb oxide as a capacitor dielectric. In comparison with Ta_2O_5 or Al_2O_3, a higher dielectric leakage and less thermal stability with respect to

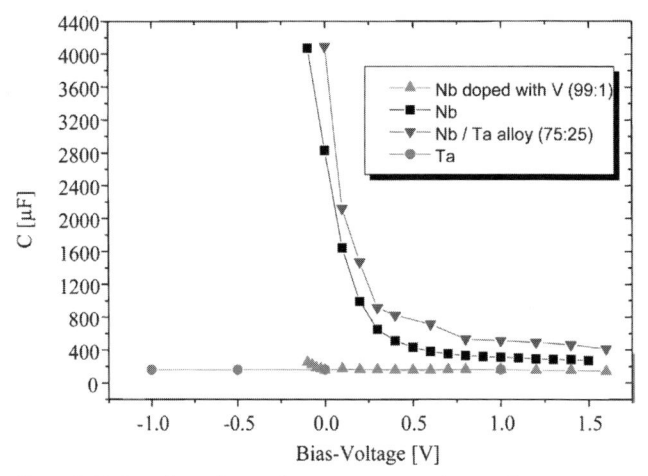

Figure 1.40 Bias dependence of capacitance on Nb, Ta, Nb/Ta alloy and V-doped Nb. Prior to these measurements an oxide was formed at a formation potential of 30 V.

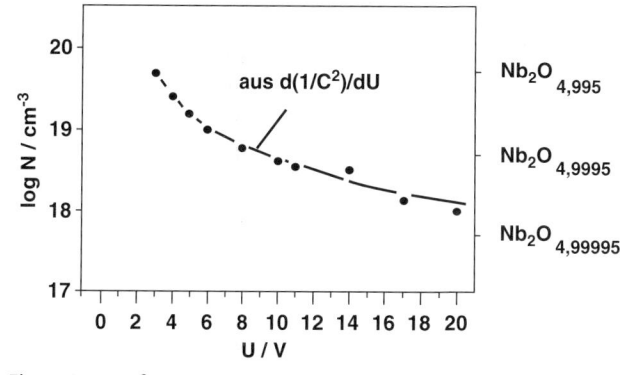

Figure 1.41 Defect state concentration N_D determined by Schottky–Mott analysis of capacitance curves and corresponding Nb oxide stoichiometry as a function of the formation potential [18].

dielectric breakdown can be expected for Nb oxides. Interestingly, the bias dependence can be completely suppressed by a V-doping. V seems to stabilize the Nb_2O_5 structure and decrease the oxygen vacancies [104]. The defect state, that is, the oxygen vacancy concentration of pure Nb, however, turns out to be a function of the formation potential and film thickness, respectively. This finding is shown in Figure 1.41. The N_D data were obtained by Schottky–Mott analysis of capacitance curves for various formation potentials. The defect state concentration decreases with increasing film thickness and seems to reach saturation at a formation potential above 20 V. The defect state concentration can be correlated with the stoichiometry of the Nb oxides, that is, the number x of oxygen vacancies in Nb_2O_{5-x} can be determined quantitatively. The result is also shown in Figure 1.41. For the 20 V film, a defect state concentration x in the ppm regime is found. Consistently, it was found that a V-doping in the ppm regime is sufficient to suppress the bias dependence.

For completeness, it has to be mentioned that through XP-spectra (X-ray photoelectron spectroscopy) an indication of the formation of an Nb suboxide film NbO_y between the Nb metal and the Nb_2O_5 was found [18]. Electrochemically this would point towards a two-step oxide formation. In the first step a thin metal-like, conducting suboxide forms at the metal/oxide interface according to the reaction:

$$Nb + xH_2O \rightarrow NbO_x + 2xH^+ + 2xe^- \qquad (1.23)$$

In the second step this oxide is further oxidized to the dielectric Nb_2O_5 form:

$$2NbO_x + (5-x)H_2O \rightarrow Nb_2O_5 + (10-2x)H^+ + (10-2x)e^- \qquad (1.24)$$

The oxide formation according to this two-step process is consistent with the thermodynamics. In the corresponding Pourbaix diagram, stable NbO and NbO_2 modifications exist between Nb and Nb_2O_5. The formation of the suboxide in the first step can not be seen in the capacitance measurements due to its conductivity. Therefore, a more anodic oxide formation potential U_{ox} is determined from the reciprocal capacitance curve. However, the slope of the curve used for determination

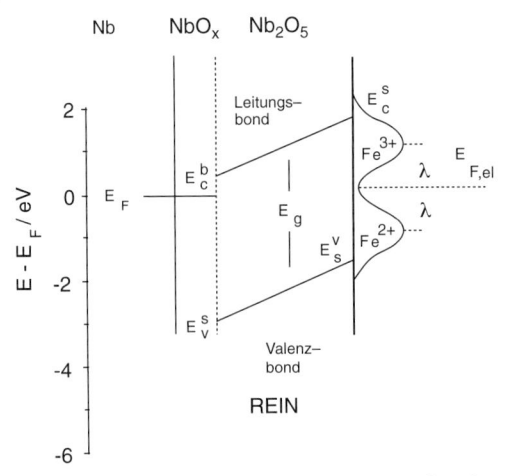

Figure 1.42 Resulting model for band structure for Nb and Nb_2O_5.

of the dielectric constant is not affected. The resulting model for the band structure is shown in Figure 1.42.

A good direct confirmation of the texture independence of the Nb oxide formation on Nb substrate is shown in Figure 1.43. Here SEM micrographs of sintered Nb-particles are shown after anodic formation of a Nb_2O_5 layer. As will be discussed in full detail later (chapter 3.4), such sponge-like structures are used for fabrication of ultra-high capacitance electrolyte capacitors. The Nb works as the anode with Nb_2O_5 as the capacitor dielectric. This structure is impregnated with MnO_2 which works as the cathode counter electrode. The cross-sections in Figure 1.43 clearly show the Nb cores, which are uniformly covered by the Nb_2O_5 film and a thin flake-like MnO_2 layer on top of the surface. The particles are crystalline, that is, a huge variation of crystal surfaces are exposed. The Nb_2O_5 film shows no thickness variation at all, which is

Figure 1.43 SEM cross-sections through $Nb/Nb_2O_5/MnO_2$ capacitor structures proving the uniform, texture independent anodic film formation on Nb, d_f ca. 90 nm [117].

beneficial behavior for such a type of capacitor application. Additional details about the electrochemical behavior of Nb/NbO$_2$ can be found in Refs. [105–152, 319, 235]. Nb/Nb$_2$O$_5$ for electronic applications, as a capacitor material in particular, is discussed in Refs. [153–164].

1.3.3.1.1 Electronic Application of Nb/Nb$_2$O$_5$ in Capacitors

The following facts are important for potential applications of thin Nb$_2$O$_5$ films as dielectrics in electronic devices:

1. Anodic Nb$_2$O$_5$ is amorphous with a dielectric constant ε_r of about 42.

2. There is no or only negligible texture dependence of film growth in the Nb/Nb$_2$O$_5$ system.

3. Nb$_2$O$_5$ shows a pronounced bias dependence of capacitance, that is, Nb$_2$O$_5$ is not a perfect insulator but shows n-type semiconducting behavior.

4. The anodic oxide formation factor m is about 2.9 nm V^{-1}. The higher dielectric constant for Nb$_2$O$_5$ (42) in comparison with Ta$_2$O$_5$ (25) is, therefore, compensated by the thicker anodic oxides for a constant formation voltage, that is, capacitance-wise no advantage (capacitance gain) for Nb$_2$O$_5$ with respect to Ta$_2$O$_5$ is to be expected. Conversely, the capacitance volume efficiency will degrade by a factor of about 2 due to the different densities of the metal substrates (8.6–16.6 g cm^{-3} for the metals Nb to Ta).

The first two points are beneficial for an electronic application. However, because of the last two points, no advantage with respect to tantalum, which will be discussed below, can be expected but the leakage current will be higher (point 3) and the volume efficiency for a specific charge powder will be lower. However, the cost of Nb is significantly below that of Ta.

1.3.3.2 Al/Al$_2$O$_3$ System

The Al/Al$_2$O$_3$ system is treated only briefly here, as the electrochemical film formation behavior corresponds to that of the previously discussed valve metal system. The oxide formation cyclic voltammograms and the corresponding capacitance measurements are shown in Figure 1.44. The experiments were performed in acetate buffer (pH 5.9) with a potentiodynamic sweep rate of 20 mV s^{-1}. The cyclic voltammograms show the typical valve metal behavior and the reverse capacitance curves show no dependence on the applied bias at all, indicating a perfect dielectric behavior of the oxides. The measurements were taken from Ref. [21]. The pronounced 'overshoot' in the cyclic voltammogram at the onset of oxide formation is discussed in full detail in Refs. [20, 21, 165]. Evaluation of the reciprocal capacitance in the usual way yields a dielectric constant of 9, in good agreement with literature values. The coulometric analysis of the oxide formation cyclic voltammograms gives an oxide formation factor m of 2.2 nm V^{-1}. There is no variation of the oxide interference colors on electropolished surfaces with open grain boundaries, confirming the texture independence of oxide growth. X-ray defraction analyses give no

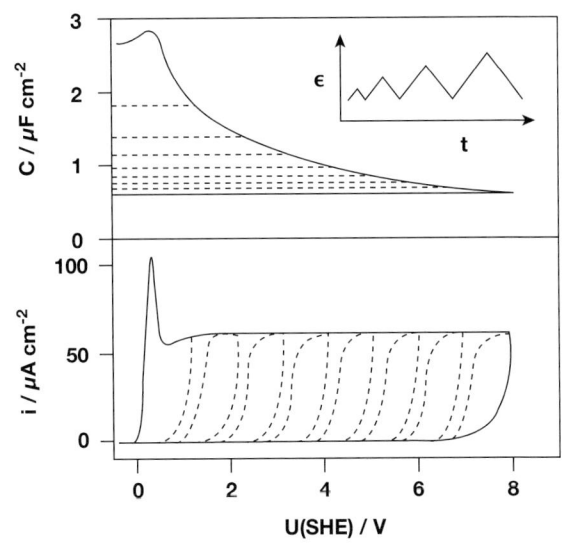

Figure 1.44 Cyclic voltammogram of oxide formation for Al/Al$_2$O$_3$ and corresponding capacitance measurements recorded simultaneously with the voltammograms. Electrolyte acetate buffer, $dU/dt = 20\,\text{mV s}^{-1}$ [21].

indication of any crystallinity, but the oxides are strictly amorphous. The Al$_2$O$_3$ properties can be summarized as follows:

1. Al$_2$O$_3$ shows dielectric behavior without any bias dependence of capacitance, $\varepsilon_r = 9$.
2. The Al$_2$O$_3$ films are amorphous.
3. There is no texture dependence of Al$_2$O$_3$ film growth.

Thus, these oxides exhibit perfect properties for application as dielectric materials in electronics. Consistently, Al$_2$O$_3$ is one of the most commonly applied electronic dielectrics.

A similar conclusion can be drawn for the Ta/Ta$_2$O$_5$ system, which is treated next. Here again the electrochemical behavior will be dealt with only briefly but an extensive account of the application of this system to passive component manufacturing and the research issues involved will be given. One of the major differences between the Al/Al$_2$O$_3$ and Ta/Ta$_2$O$_5$ systems shows up in the way devices such as electrolyte capacitors are fabricated. For Al an etched foil technology has to be employed, whereas for Ta a sintered powder technology is possible (the difference in the metal melting point between Ta/Nb/Al is 2996/2468/660 °C). The latter technology (sintering) provides a much higher surface area, therefore allowing higher capacitance volume efficiencies to be achieved.

1.3.3.3 Ta/Ta$_2$O$_5$ System

The typical electrochemical oxide formation on Ta is shown in Figure 1.45. Again simultaneously taken potentiodynamic cyclic voltammograms (upper part) and

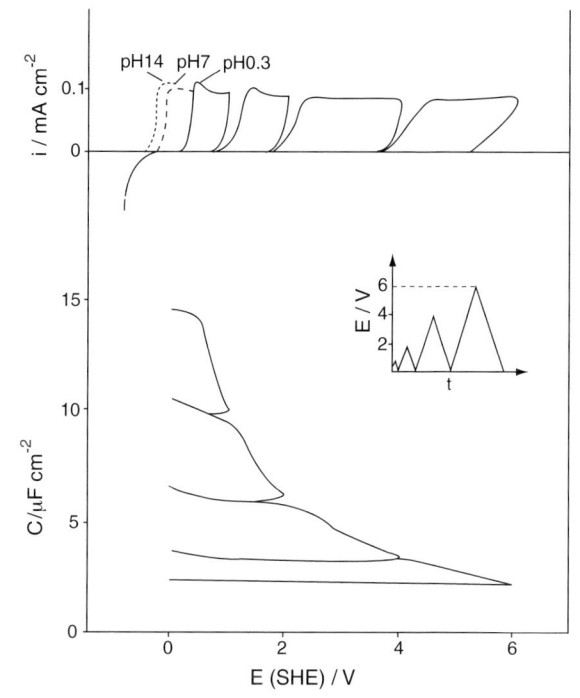

Figure 1.45 Potentiodynamic oxide formation on Ta and simultaneously recorded capacitance. 0.5 M H_2SO_4, with neutral (borate buffer) and alkaline (NaOH) electrolyte as dashed curves. $dU/dt = 25$ mV s^{-1} (source Refs. [11, 13]).

capacitance C (lower part) measurements are shown, with 0.5 M H_2SO_4 as the electrolyte. The potentiodynamic sweep started at 0 V (SHE) and ended at increasing potentials of 1, 2, 4 and 6 V, as shown in the small potential–time program (inset of Figure 1.45). In the upper potential curve, the result of a negative run down to −1 V is also shown for a qualitative demonstration of hydrogen evolution. The measurements were taken from Refs. [11, 13]. As for all of the aforementioned systems, the anodic oxide formation is an irreversible process, that is, a stable oxide is formed, which is not reduced in the reverse sweeps. According to literature [167, 168], the reaction

$$2Ta + 5H_2O \rightarrow Ta_2O_5 + 10H^+ + 10e^- \tag{1.25}$$

takes place with an equilibrium potential of about $U_o = 0.75$ V (HESS) [167] (−0.81 V given in Ref. [17]). In the negative run, the oxide formation stops almost immediately and the current density becomes 0. The oxide growth strictly follows the 'high field mechanism' with a reciprocal logarithmic film growth for potentiostatic [169, 170] and linear growth for potentiodynamic or galvanostatic conditions [166, 171], as discussed in Section 1.2.1. From the negligible current densities during repeated cycling between 0 V and the formation potential, it can be concluded that side reactions such as anodic oxygen evolution, anodic corrosion and anodic or cathodic reactions of electrolyte compounds can be excluded. Consequently, the anodic charge Q can again be taken for a coulometric thickness determination of the films formed. From this analysis a *formation factor*, m, of $1.9\,\text{nm}\,\text{V}^{-1}$ results. If the electrode is polarized in NaOH or borate buffer solution, the anodic oxide formation starts at lower potentials, referred to the SHE. This can be seen from the dashed (pH 7) and dotted lines (pH 14) in Figure 1.45 close to 0 V (see Refs. [11, 13]). The shift in the onset of oxide formation is the only difference between the acidic and neutral or alkaline solutions. The capacitance in Figure 1.45 shows the typical capacitor behavior, that is, a thickness proportional decrease during the oxide formation. Evaluation of the corresponding reciprocal capacitance as a function of film thickness allows determination of the dielectric constant. An ε_r of 25.3 was determined, in excellent agreement with literature values [172].

However, the reverse capacitance scans reveal an interesting effect. For extremely thin films, formed at formation potentials below 6 V, a significant bias dependence of capacitance exists. This bias dependence decreases with increasing formation potential, or film thickness, accordingly. In Figure 1.46 a closer look is taken at this phenomenon. Here, the Schottky–Mott curves for a whole variety of Ta_2O_5 film

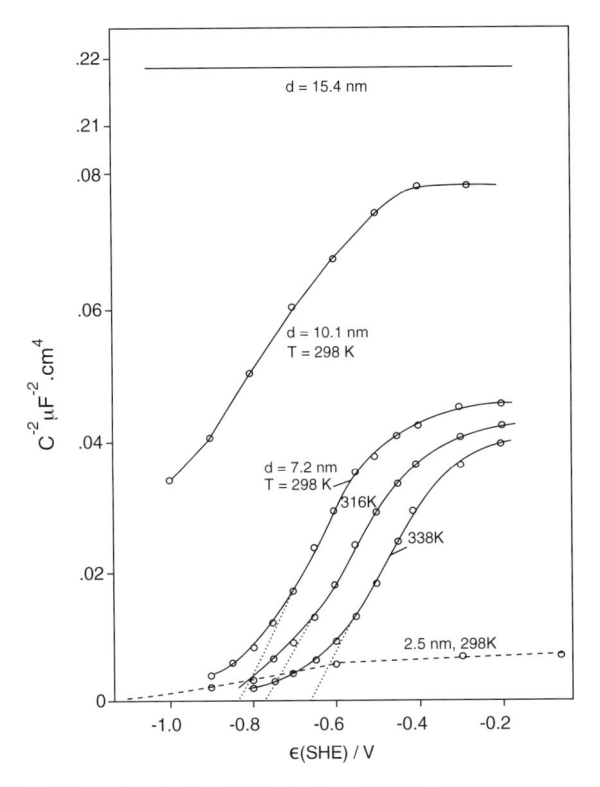

Figure 1.46 Schotky–Mott analysis of the capacitance measurements for a variety of Ta_2O_5 film thicknesses up to 15 nm (source Refs. [11, 13]).

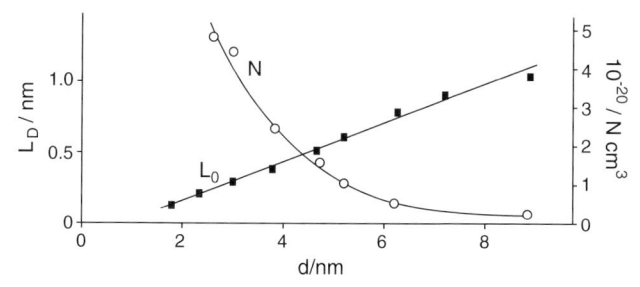

Figure 1.47 Defect state density N_D and Debye length L_D as a function of Ta_2O_5 film thickness (source Refs. [11, 13]).

thicknesses are shown. For the thickest film, $d_f = 15.4$ nm, the capacity is independent of the electrode potential, which means a clear insulating behavior. For thinner films, the electrode capacity increases and the slope of the Schottky–Mott curves decrease with decreasing potentials, which means an increasing n-type semiconducting behavior (increasing number of defect states N_D). In the region between -0.8 and -0.4 V almost straight lines are observed for film thickness values from 3 to 9 nm allowing the determination of the defect state concentration N_D. The corresponding result, that is, the decrease in the defect states with increasing film thickness, is shown in Figure 1.47 together with the Debye length L_D, which correlates to the space charge layer extension d_{scl} defined in Equation 1.14 previously by:

$$d_{scl} = L_D \sqrt{2e/kT \cdot \Delta\varphi_{ox,scl}} \tag{1.26}$$

This result corresponds to the results discussed earlier for Ti (see also [28]) and Nb (see also [173]). With increasing film thickness, the film is exhausted of defects N_D (decrease of oxygen vacancies) and behaves as a true insulator. For capacitor applications, usually a film thickness well above 20 nm is applied. Therefore, Ta/Ta_2O_5 is a perfect system for fabrication of dielectric layers in electronics. The Ta_2O_5 properties are very similar to those of Al_2O_3 and can be summarized as follows:

1. Ta_2O_5 shows perfect dielectric behavior without any bias dependence of capacitance for a film thickness >15 nm, $\varepsilon_r = 25$.
2. The Ta_2O_5 films are amorphous.
3. There is no texture dependence of film growth in the Ta/Ta_2O_5 system.

Further details about the electrochemical behavior of the Ta/Ta_2O_5 can be found in Refs. [174–234]. In the next section, the application of valve metal oxides in passive component fabrication with particular focus on the Ta/Ta_2O_5 system is discussed.

1.3.4
Application of Valve Metals in Electrolytic Capacitor Manufacturing

1.3.4.1 Capacitor Fundamentals

The basic function of a capacitor is the storage of electrical charge Q. This task is similar to that of a battery but in contrast to a battery, the charge in a capacitor has to be stored electrostatically and not by Faradaic processes. The capacitance of a capacitor is simply defined by the equation:

$$Q = CU$$

that is, the capacitance C is the factor of proportionality between the applied potential and the charge. The *capacitance* C of a capacitor is not to be confused with the *capacity* C_{bat} of a battery, which is defined in a completely different manner (the theoretical specific capacity C_{bat} of a battery is defined by $C_{bat} = nF/M$, where n is the number of electrons involved and M the mass of the active electrode material). Charging and discharging of a capacitor are fast electrostatic processes, and are only limited by the unavoidable series resistance R in the equivalent circuit of a real capacitor. In contrast, for a battery a constant potential is desirable, that is, as much charge as possible should be drawn from a battery at a constant rated voltage. A capacitor is optimized with respect to high power densities (high frequencies) whereas a battery is optimized for high energy densities.

These facts are illustrated in Figure 1.48 where the typical Q versus U curves for batteries and capacitors are compared. Two typical cyclic voltammograms for an

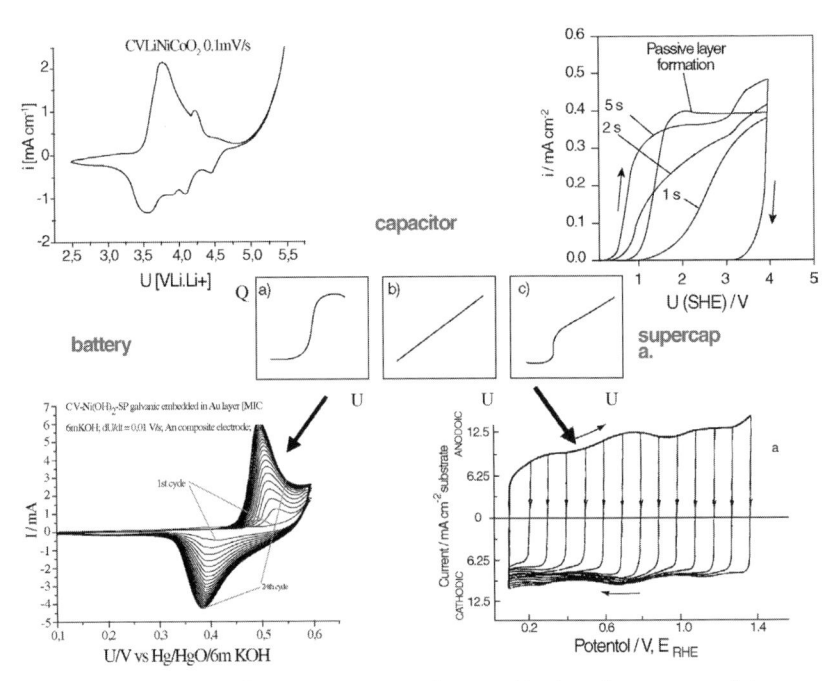

Figure 1.48 Comparison of capacitor, supercapacitor and battery behavior. On the left side, two typical cyclic voltammograms for battery electrodes [LiNiCo oxide, top, and Ni(OH)$_2$, bottom] are shown. Integration of the voltammograms yields the S-shape $Q(U)$-curve shown, which is typical for a battery. On the right bottom side, the voltammogram of the supercapacitor material RuO$_2$ is shown with an almost rectangular profile, yielding an almost capacitor-like $Q(U)$-curve. For comparison the cyclic voltammogram of Ti oxide formation is also shown on top. This CV is completely irreversible whereas the other CVs are reversible.

Ni(OH)$_2$ cathode (used in Ni/MH batteries) and an LiNiCo oxide electrode (used in Li-ion batteries) are shown to illustrate the behavior of a typical battery. The corresponding Q(U)-curve can be derived from the potentiodynamic voltammograms by a simple mathematical integration and shows an S-shaped curve, confirming that most of the charge can be drawn from the battery without significantly changing U. In this figure, the cyclic voltammogram of RuO$_2$, which behaves as a so-called supercapacitor is also shown. In this instance a combination of double layer capacitance (electrostatic) and Faradaic processes occurs, resulting in an almost capacitor-type Q(U) curve. The frequency response of such a supercapacitor (1 Hz regime) is much faster than for a conventional battery, whereas it is still many orders of magnitude slower than a typical capacitor. Supercapacitors are typically used to back-up batteries when high power peaks are needed, for example, for the ignition of a car or for short-term UPS (uninterrupted power supply) circuits. In contrast to the cyclic voltammograms of valve metal oxide formation, which were discussed in great detail in the previous sections, charging and discharging of batteries and capacitors are highly reversible processes.

The capacitance C of a simple parallel plate capacitor is given by:

$$C = \varepsilon_r \varepsilon_o A/d \tag{1.27}$$

where A is the plate surface, d the distance between the plates and $\varepsilon_o = 8.86\,\mathrm{pF\,m^{-1}}$ (permittivity of a vacuum). The complex dielectric constant ε_r describes the dielectric material between the capacitor plates:

$$\varepsilon_r = \varepsilon_r - i\,\varepsilon_{im} \tag{1.28}$$

The important part of this equation is the dielectric constant, which was used as such in the previous sections. It describes the increase in capacitance C in comparison with the case without a dielectric material (ε_r of vacuum = 1) C_0, that is $\varepsilon_r = C/C_0$. The physical reason behind the existence of the complex dielectric constant ε_r is due to the induction and orientation of electrical dipoles within the dielectric material in an applied external field. This phenomenon is called 'polarization'. The imaginary part ε_{im} describes the unavoidable polarization losses, which are caused by the energy dissipation (friction) during the orientation of the dipoles. Both the real and imaginary part depend on the frequency [$\varepsilon_r = \varepsilon_r(\omega)$] of the AC signals. Four types of polarization can be distinguished:

1. *Electron polarization*, which is due to the deformation of the electron hull of the atoms constituting the dielectric material. Owing to the small mass of the particles involved (electrons), the rate of dipole induction is very fast (highest frequency part).

2. *Ionic polarization*, which is due to an induced shift of cations with respect to anions in the dielectric material.

3. *Orientation polarization*, which is due to the orientation of molecules that exhibit a permanent dipole momentum.

4. *Space charge polarization*, which is due to accumulation of charge carriers at isolating grain boundaries.

By means of the real (Re) and imaginary (Im) parts of ε_r the so-called dissipation factor (DF) = $\tan \delta_\varepsilon$ can be defined as:

$$DF = \tan \delta_\varepsilon = \frac{\operatorname{Im} \varepsilon}{\operatorname{Re} \varepsilon} \tag{1.29}$$

The angle δ_ε describes the deviation of the phase shift between current and applied potential from the ideal value of 90°. Its projection on the real part of the total impedance Z therefore describes the ohmic part of the capacitor characteristic corresponding to a series resistance in the equivalent circuit. In practice, this number is used to specify the polarization losses of a capacitor. Analogously, the total loss factor $\tan \delta_C$ of a capacitor is defined by the ratio of the real and imaginary part of the total impedance Z. In addition to the polarization loss, this number considers unavoidable resistances due to lead wires and losses due to an imperfect insulation, plus the capacitive inductance L:

$$\tan \delta_C = \frac{\operatorname{Re} Z}{\operatorname{Im} Z} = \frac{1}{Q} \tag{1.30}$$

Through this equation the Q-factor is also defined, which, in addition, can be used to specify the losses or quality (Q) of the capacitor. In Figure 1.49 the parallel and series equivalent circuits of a capacitor are shown, along with the frequency dependence of the impedance Z. Both the parallel and series circuit can be used synonymously, that is they describe an identical behavior. (The series circuit is normally preferred for high frequencies.) ESR stands for 'equivalent series resistance'. The total impedance Z for the parallel circuit is:

$$Z = R_s + i\omega L_s + \frac{R_p}{1 + i\omega R_p C_p} \tag{1.31}$$

with

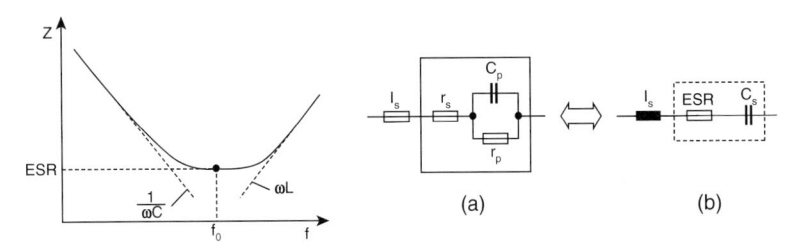

$$\operatorname{Re} Z = R_s + \frac{R_p}{1 + \omega^2 R_p^2 C_p^2}$$

Figure 1.49 Left: impedance Z of an electrolyte capacitor as a function of the signal frequency, where f_o is the resonance frequency. Right: identical equivalent circuits of a capacitor: (a) parallel circuit; (b) series circuit. The dashed boxes are identical according to the equations given in the text [236].

and

$$\text{Im } Z = -\frac{1}{\omega C_p}\left[\frac{1}{1+(\omega R_p C_p)^{-2}} - \omega^2 L_s C_p\right] \tag{1.32}$$

In order to describe an identical behavior, the dashed boxes of the equivalent circuit in Figure 1.49 have to be equal, resulting in the following relationship:

$$R_s + \frac{R_p}{1+i\omega R_p C_p} = ESR + \frac{1}{i\omega C_s} \tag{1.33}$$

From this equation it follows that:

$$C_s = C_p + \frac{1}{\omega^2 R_p^2 C_p} \tag{1.34}$$

and

$$ESR = R_s + \frac{R_p}{1+\omega^2 R_p^2 C_p^2} \tag{1.35}$$

Note that the relationship for ESR is identical to ReZ in the Equation 1.32. Using the definition

$$\frac{1}{R_p} = \omega C_p \tan \delta_\varepsilon + \frac{1}{R_{is}} \tag{1.36}$$

and neglecting the isolation resistance R_{is} of the dielectric yields, the following relationship between the ESR and the DF, $\tan \delta_\varepsilon$, respectively, is obtained:

$$ESR = R_s + \frac{1}{2\pi C_p}\frac{\tan \delta_\varepsilon}{1+\tan^2 \delta_\varepsilon}\cdot\frac{1}{f} \approx R_s + \frac{1}{2\pi C_N}\frac{\tan \delta_\varepsilon}{f} \tag{1.37}$$

For the last approximation it was assumed that C_p corresponds to the nominal capacitance C_N and that there are only small polarization losses, that is, $1 + \tan^2\delta_\varepsilon$ 1. (Each capacitor is specified by the manufacturer with a nominal capacitance C_N and nominal or rated voltage V_R.) The quantity f_o is the 'resonance frequency'. At f_0 a transition from a capacitive to an inductive behavior takes place. Further characteristic quantities for capacitors are:

- *Temperature coefficient* α describing the variation of the nominal capacitance C_N (specified for 20 °C) with the temperature: $C(T) = C_N [1 + \alpha(\text{temp} - 20\,°C)]$.
- *Moisture coefficient* β describing the capacitance change with the relative change of air moisture F in % humidity: $\beta = 1/C \, dC/dF$.
- *Retention time* (self-discharge time) constant τ describing the time needed for a potential drop of 37% after the capacitor was charged to the rated voltage V_R: $\tau = R_{is}C$, where R_{is} the isolation resistance. The associated power loss P is: $P = U^2/R_{is} = 2W/\tau$, where W is the energy stored in the capacitor.

Table 1.3 Comparison of the three capacitor types, polymer foil, ceramic and electrolyte capacitor, with some examples of typical applications [263].

Polymer foil capacitors:	Ceramic capacitors:	Electrolyte capacitors:
Dielectric material: Polymer (PP, PS, PET, PC), paper	Dielectric material: Ceramic (low K e.g. mica, high K BaTiO$_3$)	Dielectric material: Anodic valve metal oxides (Al$_2$O$_3$, Ta$_2$O$_5$)
• $\varepsilon_r = 2$–4 • $d = 1$–3 μm • C: 10 pF–10 μF • tan δ $= 0.2 \times 10^{-3}$–$5 \cdot 10^{-3}$	• $\varepsilon_r = 10$ – $>10^4$ • $d = 10$–30 μm • C: 1 pF–10 μF • Tan δ $= 1 \cdot 10^{-3}$–$50 \cdot 10^{-3}$	• $\varepsilon_r = 2$–27 • $d = 20$–500 nm • C: 10 μF–F • tan δ $= 40 \cdot 10^{-3}$
+: low cost	+: excellent multi-purpose capacitor	+: highest volume efficiency
−: temp. and moisture sensitive	−: aging, high temp. coeff.	−: high leakage current, low frequency, polar devices
Applications: Filter, high frequency, low loss, timing and tuning circuits, blocking	Applications: Low K: blocking, filtering/smoothing High K: coupling/decoupling, filtering/smoothing, energy storage	Typical applications: Blocking, coupling, decoupling, interference suppression, energy storage, filtering smoothing, RC and timing circuits, RFI

- *Ripple voltage* V_{rip} describes the maximum AC voltage that can be applied to a DC capacitor.

- *Break down voltage* (*BDV*) describes the highest voltage that can be applied to a capacitor without destroying the dielectric by a catastrophic breakdown.

1.3.4.2 Capacitor Device Types and Production of Ta Capacitors

Three major types of capacitor devices are used as passive components for electrical circuits, that is, polymer foil capacitors, ceramic capacitors, for example, multi-layer ceramic capacitors (MLCC), and electrolytic capacitors. Each type has its merits and disadvantages. An overview of the basic features of these capacitor types is given in Table 1.3. Electrolyte capacitors are used especially for high volume efficiency applications at relatively low voltages and frequencies. In the following discussion, the electrolytic type of capacitors are dealt with specifically. Although all valve metals mentioned so far have attracted some interest, only Ta and Al have entered the market place so far. Ta capacitors show the highest volume efficiency for capacitance and have excellent reliability properties, for example, there is no aging effect. However, Ta is a rare material and thus the cost of Ta capacitors is relatively high. Therefore, increasing of the specific charge (CV product) per unit of volume or weight has been the main goal in research and development of Ta electrolytic capacitors since their manufacturing began in the late 1950s. Higher CV provides smaller capacitor size and less consumption of the Ta powder for the same rating, which is attractive for both capacitor producers and customers. During this period of time the CV has increased more than ten times. This was mainly achieved by reducing the efficient size of the particles and by increasing of the porosity of the Ta powder used for sintering of the Ta anodes. The efficient particle size was reduced from tens of microns initially to about one micron at the present time. The reduction in the powder particle size and the adjustment of the porosity is achieved by a sophisticated chemical process. Two routes are possible. Both routes start with a 'Ta strip' solution, which is produced directly from ore concentrates by first dissolving the Ta and Nb contents of the ore in 70% HF solution and subsequently performing a solvent extraction in methyl isobutyl ketone (MIBK). The Ta remains in the MIBK solution, which is the Ta strip, whereas the Nb stays in the aqueous solution. The Ta strip can be processed in two ways, providing the two routes for Ta metal powder production:

1. A *K_2TaF_7 crystallization* process is performed by adding KCl and HF to the Ta strip with subsequent drying and filtering. This K_2TaF_7 salt is than reduced by Na in a hot salt melt (950 °C). This salt melt also contains deliberately chosen amounts of inert salts to achieve a certain dilution of the active compounds. The careful adjustment of the components in this salt mixture along with temperature and agitation speed allows the Ta nucleation process to be controlled and in turn the Ta particle morphology can be tailored. This primary powder morphology is than stabilized and further modified by downstream processing, including leaching, agglomeration, doping, consolidation and deoxidization steps.

2. A *Ta_2O_5 precipitation* process is performed by adding NH_3 to a Ta strip hydroxide solution. After subsequent calcination, a stable Ta_2O_5 powder is obtained. The details of the precipitation process determine the morphology of the oxide powder. Subsequently an Mg–steam reduction of the oxide powder is performed resulting in the Ta metal powder. In contrast to the first route, here both the parameters of the precipitation and the conditions of the reduction determine the final morphology of the metal powder. Consequently, there is one additional free parameter allowing for an even better adjustment of the metal powder properties. The downstream process steps after reduction are similar to the first route.

There are many further important process details for Ta metal powder (TaMP) production that can not be disclosed here, and which are the object of continuing R&D work. It should be mentioned that Nb capacitor grade metal powder can be produced in a method similar to the second route using the aqueous Nb solution mentioned above. Some further information can be found in Ref. [237–247]. Once the TaMP is finished it can be used by the capacitor manufacturer to produce the capacitor devices. Briefly, the following process steps are used in a conventional production line:

1. Insertion of a Ta wire into the TaMP and dry pressing of the green body (anode).

2. Sintering of the anode for formation of a porous sponge-like Ta structure.

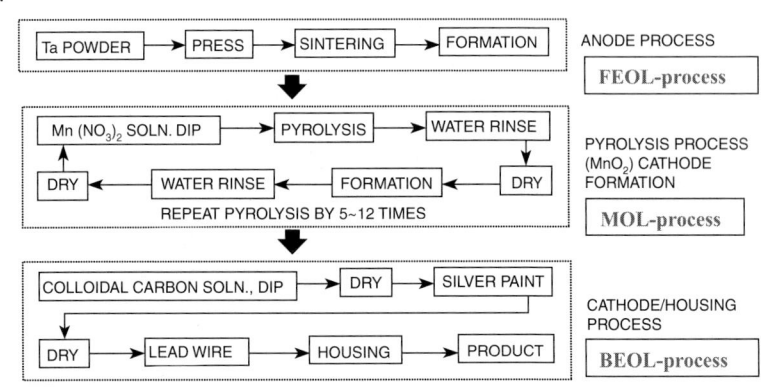

Figure 1.50 Ta capacitor production scheme.

3. Anodizing of the anode in H_3PO_4 for Ta_2O_5 dielectric formation.

4. Impregnation of the porous Ta anode with $Mn(NO_3)_2$ solution and subsequent pyrolysis for formation of the conducting MnO_2 counter electrode. This step is repeated several times in $Mn(NO_3)_2$ solutions with increasing concentrations to accomplish a thick MnO_2 layer.

5. Applying of a graphite and Ag layer to improve electrical contacts.

6. Encapsulation.

A detailed process flow is given in Figure 1.50. The resulting capacitor devices and the sponge-like capacitor structure are shown in Figure 1.51. In Figure 1.51d a scanning electron micrograph (SEM) of a cross-section through a broken particle is shown (analogous to Figure 1.43, which showed an Nb capacitor structure). The Ta metal core is covered by a homogeneous 100 nm thick Ta_2O_5 film. The excellent lateral uniformity of the oxide confirms the texture independence of anodic oxide formation for the Ta/Ta_2O_5 system. The oxide was formed in 0.1% H_2PO_3 as the electrolyte.

For Al electrolyte capacitors a different process must be applied that uses an etched Al foil instead of a powder. In contrast to the sintered powder, which constitutes a three-dimensional structure, a foil principally is only two-dimensional. Consequently, the absolute surface area, that is the volume efficiency of an Al capacitor, is much lower than for the Ta types, yielding much larger device sizes for comparable capacitances. New processes for Al capacitor manufacturing make use of a solid electrolyte, which is a combination of MnO_2 and a conductive polymer. However, in order to reduce the cost, most Al capacitors (98%) are still produced employing a wet electrolyte. These devices show a rather short lifetime (<5 years) as they are prone to drying out. Conversely, the lifetime of a solid Ta capacitor is literally unlimited, justifying their higher cost if better reliability is essential. The following treatment of current research issues in electrolyte capacitor development is limited to Ta exclusively.

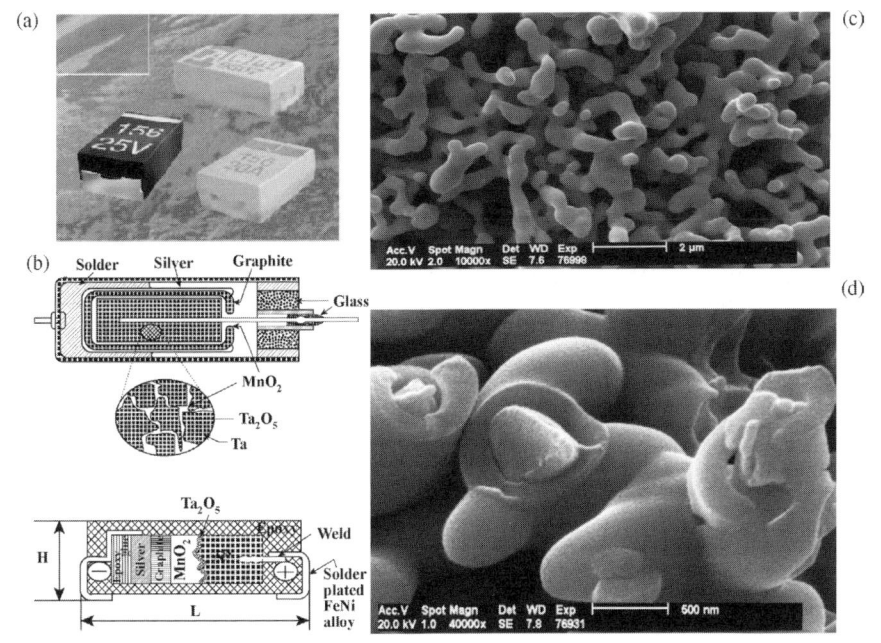

Figure 1.51 (a) Ta capacitors as surface mounted devices (SMD); (b) corresponding schemes; (c) SEM of sponge-like anode after oxide formation; (d) SEM with cross-section through a Ta/Ta$_2$O$_5$ particle.

1.3.4.3 Current Development Trends for Ta Capacitors and Research Issues Involved

As reasoned above, the main development trend for electrolyte capacitors is the increase in capacitance volume efficiency to reduce the size and cost of the components. This goal is achieved mainly by reducing the particle size of the TaMP. With decreasing particle size of the TaMP, the sintering temperature has to be reduced from 1900–2000 °C in the past to about 1200–1350 °C at the present time. While the two physical parameters (particle size and sintering temperature) were reduced gradually, the physical situation in the Ta capacitors changed, fundamentally causing several important problems. There are two major effects:

1. *Dislocation formation:* The oxygen content in the bulk of the Ta anodes reaches the solubility limit resulting in an increased dislocation formation (crystallization) within the amorphous oxide layer. These dislocations act as leakage paths in the oxide increasing the direct leakage current (DCL) and causing defects in the dielectric. Normally the defects are not fatal but are isolated by a self-healing mechanism, that is, the reaction $2MnO_2 \rightarrow Mn_2O_3 + \frac{1}{2}O_2$ takes place with the insulating Mn_2O_3 covering the defect (an illustration of the self-healing mechanism is shown in Figure 1.52). However, when the temperatures involved are too high thermal runaway can occur.

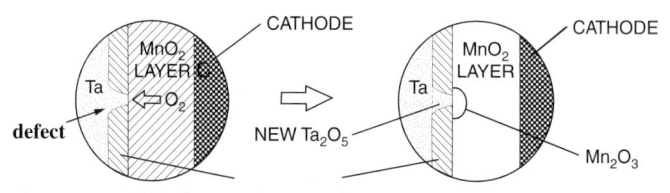

Figure 1.52 Self-healing mechanism of Ta capacitors.

2. *Thermal runaway:* With increasing volume efficiency more and more energy is stored at a constant volume, together with the increasing surface area and the presence of an oxygen source, that is MnO_2 (see reaction above), a fatal combination is generated that can lead to catastrophic failures, that is, thermal runaway (explosion) of the devices. This problem is aggravated by the weaker links within the sinter body (reduced sinter temperature) causing a reduction of the thermal conductivity. Therefore, a critical amount of heat can accumulate at the defects generated by dislocation formation.

Both effects enhance each other, that is, the increased defect density due to the increased number of dislocations increases the chance for a fatal defect (thermal runaway). Moreover, the higher energy density leads to an increased thermal stress (heat accumulation) expediting the dislocation formation. Therefore, a vicious circle is created, as is illustrated in Figure 1.53. In the following both effects are described in more detail and several possible ways out of this vicious circle are presented. For completeness, a third problem, that is the increasing difficulty for impregnation of the finer and finer particles with a conductive counter electrode (e.g. MnO_2) has to be mentioned here. This issue constitutes an important engineering problem but is only of minor scientific interest and will not be dealt with here.

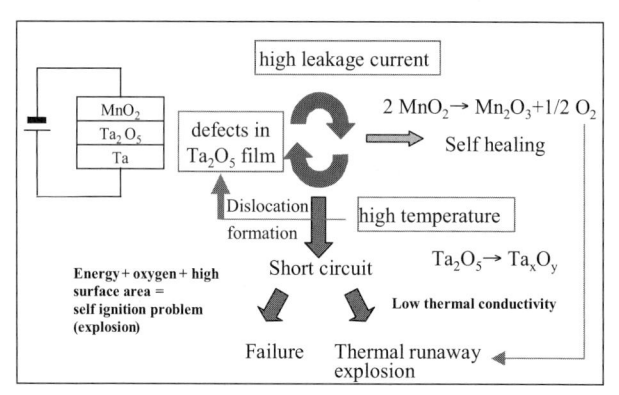

Figure 1.53 Thermal runaway vicious circle. Shrinking of TaMP leads to an increasing defect density due to a higher bulk oxygen concentration. The high associated leakage currents result in high temperatures, which induce further dislocations. Additionally, thermal runaway can occur due to the reduced thermal conductivity (lower sinter temperature leads to degradation of the interconnect structure, which reduces the thermal conductivity).

1.3.4.4 Effect of Oxygen Content and Sinter Conditions on Dislocation Formation

The dielectric Ta_2O_5 film thickness in the Ta capacitor typically varies between 30 and 300 nm depending on the formation voltage (ratio of rated- to formation-voltage is about 1:4). As a result of the low film thickness, the electrical field in the film at the rated voltage is about $10^6\,V\,cm^{-1}$. To withstand the associated high electric fields for a long time the dielectric should have an *amorphous structure*. Amorphous dielectrics have a high density of electronic traps, that is, there is a continuum of localized states within the band gap, which now becomes a mobility gap. Owing to these traps, mobility of the charge carriers stays low in the high electrical field, preventing their accumulation and therefore reducing the chance of an electrical breakdown (electronic avalanche effects). *This is why the anodic Ta_2O_5 film in a Ta capacitor is formed with conditions providing an amorphous structure. In fact, all successfully employed ultrathin film dielectrics for electronic applications are amorphous* (e.g. Si/SiO_2). Amorphous dielectrics are in principle not thermodynamically stable. They tend towards a spontaneous ordering and crystallization to reduce their internal energy. The crystallization process in amorphous Ta_2O_5 film begins at nucleation sites located at the Ta/Ta_2O_5 interface. These nucleation sites are related to defects on the Ta surface (impurity precipitates, crystalline grain boundaries, etc.) [248–251]. Growth of the crystalline inclusions induces mechanical stress in the amorphous matrix of the anodic oxide film and eventually causes damage to the dielectric. Figure 1.54 shows a crystalline 'dislocation' passing through the amorphous matrix of the Ta_2O_5 film, resulting in its complete disruption (picture taken from Ref. [248]). Besides mechanical stress, the dislocations also act as a leakage path through the dielectric, which is one of the major reasons for catastrophic failures of Ta capacitors [252]. Consequently, the crystallization rate is a principal factor governing capacitor performance and reliability.

The bulk oxygen content in the Ta anode strongly affects this crystallization process in the dielectric oxide film [249, 250]. If the oxygen concentration in the bulk approaches the solubility limit (about 3.200 wt. ppm for Ta at room

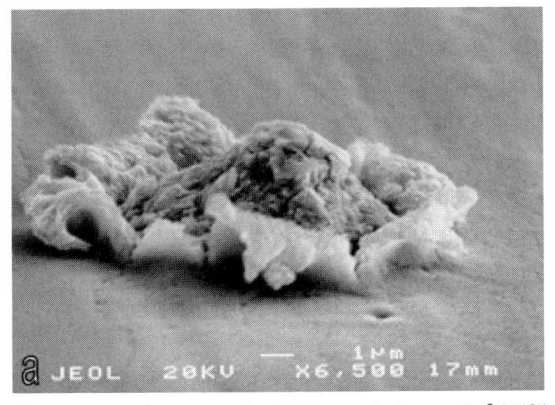

Figure 1.54 Dislocation in the Ta_2O_5 matrix (source Ref. [248]).

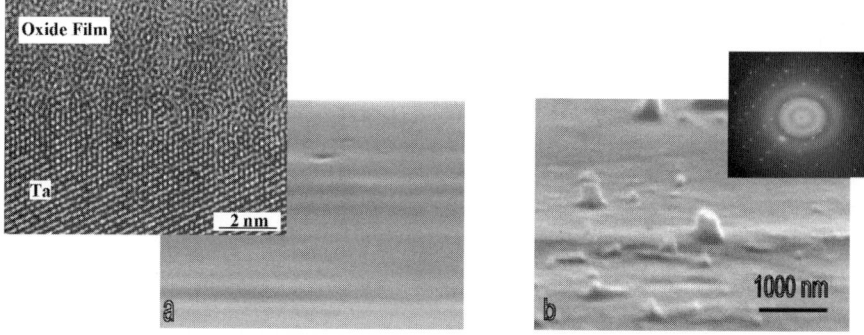

Figure 1.55 (a) Top view of the Ta_2O_5 dielectric; (b) bottom view showing that the crystallization nuclei originate at the Ta/oxide interface. The inset in the left corner shows a high resolution TEM shot of the Ta/oxide interface without nuclei. The Ta shows a crystalline lattice whereas the Ta_2O_5 is completely amorphous [248].

temperature), crystalline Ta_2O_5 particles precipitate from the solid solution of oxygen in Ta. In the vicinity of the anode surface these crystalline particles serve as effective crystallization nuclei forcing crystallization of the anodic oxide film. As these crystalline dislocations serve as leakage paths in the dielectric film, an associated significant increase in the direct current leakage (DCL) is observed. Figure 1.55 shows the top (a) and bottom (b) of the anodic oxide film after formation of the crystalline nuclei. As one can see, the top surface of the film is smooth with some scattered pits or pores. The bottom of the Ta_2O_5 film looks significantly different from the top. Peaks of various size and shape pass from the anode surface into the amorphous matrix. The diffraction pattern in the right corner of Figure 1.55b (TEM) shows that these peaks conform to the crystalline Ta_2O_5 phase. This is the initial stage of the crystallization process. Both density and size of these nuclei increase gradually with increasing oxygen content in the anode and rise sharply when the oxygen content reaches the solubility limit. Regardless of the particle size, all Ta anodes exhibit the same behavior with respect to crystallization once they reach the critical oxygen content [255]. However, the smaller the particle size, the sooner this critical oxygen content is reached. This is simply due to the increasing surface to volume ratio of the smaller particles. Therefore, at a constant rate of oxygen diffusion from the native surface oxide into the bulk, higher oxygen concentrations build up in the smaller particles. The higher the bulk oxygen content, the greater is the rate of thermal oxide growth at elevated temperatures. Therefore, thicker thermal surface oxides are formed on smaller particles [256, 257]. This is because the oxygen atoms absorbed by the Ta surface cannot be dissolved into the Ta volume when it is already filled with oxygen. They stay on top of the surface building up a thick thermal oxide layer. The thickness of this thermal surface oxide prior to anodization plays an important role in crystallization. The reason for this is that at elevated temperatures micro-crystals can be formed in the thermal oxide, which serve as nuclei for a further crystallization in the subsequently formed anodic oxides.

1.3.4.4.1 Impact of Sintering on Oxygen Concentration
For sintering temperatures >1750 °C, oxygen is evaporated out of the Ta anodes during the sintering process in vacuum. Additionally, most of other impurities, such as Fe, Ni and Cr, are also evaporated, providing a high purity of the final anodes. In this instance, crystallization of the anodic oxide film may only occur as a result of an eventual misprocessing. A few Ta capacitors, mostly for military applications, are still produced under such conditions. In contrast, at temperatures below this threshold temperature oxygen absorption by Ta takes place. The major source for the oxygen pick up during sintering is the native oxide on the top surface of the Ta particles [253], that is, heating in vacuum results in dissolving of the surface oxygen, which diffuses into the bulk of the Ta particles. As pointed out above, the finer the particles are (higher CV), the higher the absolute surface oxygen content, resulting in high oxygen contents of the sintered tantalum anodes. In fact, for very fine particles with a CV in excess of 70–80 kμF V g^{-1}, the dissolving of the natural surface oxide into the bulk during powder sintering results in the bulk oxygen content already approaching the solubility limit. This is a major obstacle for capacitor manufacturers. For low voltages the anodic oxide film is thinner than in the high voltage regime, and therefore, it is less susceptible to crystallization. Several measures that can be taken to reduce the dislocation formation are considered next. These measures are of fundamental importance for the implementation of ultra-high CV powder into Ta capacitors. The results can also be transferred to other valve metal systems (e.g. Nb).

1. *Fixing the Oxygen Problem by Deoxidization after Sintering* To minimize the oxygen content in the Ta powder, a deoxidizing step using magnesium after sintering can be applied. As the magnesium–oxygen bonding energy is higher than that for tantalum–oxygen, the magnesium reacts with oxygen in the tantalum, forming magnesium oxide molecules, which in turn evaporate out of the powder [259]. However, this deoxidizing results in mechanically very weak junctions between the porous anode and the incorporated lead wire (the sintering between the lead wire and the anode body is triggered by the oxide). In [248] a new type of sintering was mentioned, which ultimately seems to fix this problem. No details were given on this very promising process (patent pending). By means of this new, so-called 'Y-sintering' (Yuri Freeman) process, the application of ultra-high CV powder seems to be possible.

2. *Prevention of Crystallization by Thermal Annealing after Anodizing* Random crystalline dislocations can begin to grow even in a completely amorphous matrix of the anodic oxide film. In this instance, crystallization can be minimized by a thermal annealing process subsequent to the anodizing process. The following physical model was suggested in Ref. [260] to explain this phenomenon. During thermal annealing the amorphous Ta$_2$O$_5$ film undergoes high mechanical stress due to a large difference in the coefficient of thermal expansion between the Ta and Ta$_2$O$_5$ phases (8×10^{-7} and 6.5×10^{-5} K^{-1}, respectively) [261]. These thermal stresses can relax by sliding (shear) of the Ta$_2$O$_5$ film along the Ta$_2$O$_5$/Ta interface. Because the Ta$_2$O$_5$ crystals are embedded into the Ta anode and sit inside the

Ta_2O_5 film, this sliding will cause mechanical cutting of the crystals away from the metal surface. Loosing intimate contact with the metal surface undermines their further ability to grow. They are passively buried in the amorphous matrix of the anodic oxide film.

3. *Reduction of Thermal Budget after Anodizing* Heat has to be applied carefully as the opposite effect is also possible, that is, due to the increased film stress, nuclei, which were so far inactive, can be activated, initiating the growth of dislocations. Moreover, the activation energy for dislocation formation follows an Arrhenius law. Therefore, elevated temperatures can even induce dislocation formation. In this context, the conducting polymers should be mentioned. As will be pointed out below, conducting polymers have many advantages and can replace the MnO_2. For impregnation with conducting polymers no pyrolysis is needed, therefore significantly reducing the thermal budget of the downstream process.

4. *Reduction of Crystallization by Purity Improvement* One root cause for crystallization even with a reduced oxygen content can be attributed to further impurities at the particle surface. The trivial fix for this would be an improvement in purity, which is actually difficult to achieve with the conventional K_2-salt Na-reduction process described above (see p. 63). However, the second route, that is the Ta metal powder production by Mg reduction of the oxides, delivers purer metal powder grades (at least with respect to alkali elements). Therefore, this route may be superior for ultra-high CV powders. Moreover, higher purity is beneficial to achieve a higher BDV (break down voltage) as is known from E-beam melted powder.

5. *Reduction of Crystallization Rate by Adjustment of the Electrochemical Formation Conditions* As a last possibility for the reduction of dislocations in the amorphous oxide, the electrochemical formation conditions have to be mentioned. As was shown for the Ti/TiO_2 system in Figure 1.17, the formation conditions have a significant impact on the properties of the oxide films. The application of potentiostatic instead of potentiodynamic formation conditions allowed for a suppression of the texture dependence of oxide growth.

    Moreover, from semiconductor manufacturing processes (e.g. the DRAM process as will be described later) it is known that doping of the film can also stabilize an amorphous matrix. Such a doping could potentially also be achieved electrochemically by a corresponding addition of the dopants to the electrolyte.

1.3.4.5 Thermal Runaway

The root cause for the thermal runaway problem was identified to be due to the combination of high energy (high volume efficiency), large surface area and the presence of an oxygen source, that is the MnO_2 cathode. Initially the runaway is initiated by a defect in the dielectric layer causing high leakage and an associated temperature increase. Normally the defect is fixed inherently by the MnO_2 self-healing mechanism mentioned previously. When the defect density is too high, the self-healing fails and a permanent leakage path is generated leading to a short circuit in the component (device failure). This failure is fatal for the device and can destroy the whole circuit (electrical malfunction). However, it is not catastrophic. For high CV powder, which has to be sintered at a relatively low temperature, the much more severe failure mode of thermal runaway occurs. In contrast to the normal short circuit failure mode described above, the heat now accumulates due to an insufficient thermal conductivity in the sinter body [weak sinter necks (bridges)]. As a result, the threshold temperature for ignition of the Ta powder is reached and a sudden, catastrophic ignition of the whole surface takes place, which is fed by the oxygen provided by the MnO_2. The device literally explodes constituting a severe safety problem for the surrounding environment. An incident such as this has to be prevented by any means. Therefore, expensive and time-consuming 'burn in' tests are necessary in order to make sure that poor devices are screened out before they are sold. This significantly reduces the process yield. It also prevents larger capacitor case sizes from being produced with high-CV powder (the energy content increases with increasing case size). Thus, fixing the thermal runaway problem is of paramount importance. In the following, several measures that can be taken to reduce the thermal runaway issue are presented. This includes one ultimate solution to the problem, namely the use of conductive polymers as a replacement of the MnO_2 counter electrode that acts as the oxygen source.

1.3.4.5.1 **Reduction of Defects** Firstly, the defects in the dielectric film are responsible for the occurrence of a leakage path leading to a temperature increase. Therefore, reduction of defect number will reduce the chance for a thermal runaway significantly. For this, the exact measures already explained above can be taken (e.g. higher purity, adjusted formation conditions and doping). However, these measures can not eliminate the thermal runaway completely as some defects will always remain.

1.3.4.5.2 **Improvement of Thermal Conductivity of the Anode Sinter Body** For the initiation of a thermal runaway the threshold temperature T_{crit} for Ta powder ignition must be exceeded. T_{crit} itself is a function of the surface area, that is the larger the surface area the lower T_{crit} will be. In order to reach T_{crit} a minimum energy E_{crit} is required, which is of the order of some mJ for a 50 kμ FVg^{-1} Ta powder. Whether this critical energy can be accumulated depends on two major factors:

1. *The total energy stored in the device.* Therefore, the case size is critical. The larger the device, the more energy it contains. As an aside, the energy content also depends on the powder morphology: the larger the surface area the more energy can be stored per unit volume.

2. *Thermal conductivity of the sintered body.* If the thermal conductivity is high enough the critical temperature can not be reached locally but the energy is dissipated over the whole volume of the device.

The first factor leads to a simple measure, that is the use of smaller and smaller case sizes with increasing CV of the powder. It is fairly clear that this measure is not desirable as it strongly limits the design capabilities for capacitor devices. The second factor, that is improvement of the thermal conductivity, is much more interesting as in principle it allows for a fundamental solution to the problem. However, it has to be kept in mind that the processing of increasingly finer powders requires the sinter temperature to be decreased in order to preserve the powder morphology (the high surface). Consequently, the interconnections between the particles become weaker. This even tends to reduce the thermal conductivity. Moreover, as worse comes to worst, the energy density also increases with increasing CV product. Therefore, both parameters aggravate the thermal runaway problem under standard conditions. Nevertheless, two promising approaches to improving the thermal conductivity of the sinter body are discussed below:

1. *Improvement of the green body uniformity* by using optimized binders. Owing to the range in the particle size distribution, the contact within and between the agglomerates can vary, that is there are density fluctuations within the green body that are transferred into the final sinter body (accumulation of fines). The thermal conductivity in low density areas is especially poor and thermal runaway is therefore likely to originate in such weak spots. These density fluctuations can be suppressed by using optimized binders (organic compounds), which provide for a uniform distribution of fines throughout the sinter body.

2. *Addition of a sinter aid* that strengthens the sinter bridges between the particles. These sinter bridges are the weakest part in the sinter body and therefore determine the thermal conductivity. An improvement of the sinter bridges can be accomplished by adding even finer particles to the mix, which fit into the spaces in between the particles.

1.3.4.5.3 Conductive Polymers as the Cathode Material

The thermal runaway can ultimately be prevented by replacing the MnO_2 cathode with conductive polymers (CP). This eliminates the oxygen source, which is essential to feed the reaction. Various conducting polymers are considered for application as a cathode material. Among them, polyaniline (PANI) and polypyrole (PPY) were the first to attract interest. However, the long time and temperature stability of these polymers are significantly worse than the ones for a newer conducting polymer *poly-3,4-ethylene-dioxythiophene* (PEDT, tradename BAYTRON). For the application of PEDT two different process schemes are possible:

1. *Single step dipping* with a mixture of EDT (Baytron M)/polymerizer/organic solvent. Iron(III) p-toluene sulfonate (tradename Baytron C) can be used as the polymerizer.
2. *Multiple step process* with successive dipping of the anode into the EDT (Baytron M) and polymerizer (Baytron C) solution with intermediate drying steps.

Both processes have their advantages and disadvantages with respect to the process yield (e.g. different pot life). Further fine tuning of the processes is required to identify the optimum sequence. In Figure 1.56 a direct comparison of a Ta anode impregnated with MnO_2 and one impregnated with PEDT is shown. The thickness of the PEDT film is about 20 nm. Besides suppression of thermal runaway, the major advantage of CP is its better conductivity with respect to MnO_2. This significantly

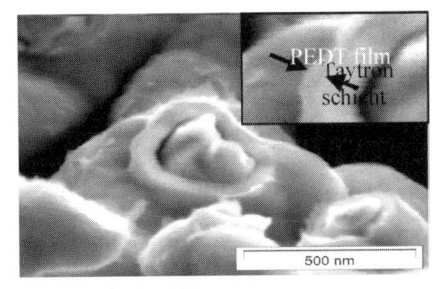

Figure 1.56 Left: Ta anode with MnO_2 cathode; Right: PEDT cathode [262].

reduces the ESR and therefore improves the performance of the capacitors. However, the cost of the CPs is much higher than that for MnO_2 and the integration process is less established, that is the yield is still low. Therefore, currently this new technology is only applied where an excellent ESR is needed and/or the thermal runaway problem is inherent. In the future the CPs that also can be applied to Al capacitors will significantly gain importance. Advantages and disadvantages of CPs are summarized in Figure 1.57 along with a cyclic voltammogram of a PEDT electrode. As is typical for most CPs the polymer is p-type semiconducting in the oxidized and insulating in the reduced state. PEDT shows an excellent reversibility with an almost rectangular profile over a potential range of 1.4 V. This is very similar to the corresponding voltammogram of RuO_2 that was shown in Figure 1.48, therefore indicating another capacitor related application of PEDT, namely as a supercapacitor material.

1.3.4.5.4 Alternative or Modified Anode Materials

In order to prevent or minimize thermal runaway, modified anode materials can also be considered. The minimum energy for a dust ignition in the case of pure Ta or Nb powder is below 3 mJ. Nitration of Ta and Nb (e.g. TaN and Ta_2N) increases the ignition energy to >10 J. Consequently, these materials are also less likely to be prone to ignition in solid capacitors. These

Advantages of Conducting Polymers, PEDT

Advantages of CP with respect to MnO_2
- Lower ESR
- No thermal runaway
- Failure mechanism (open for PEDT / short MnO_2)
- No pyrolysis needed --> lower thermal budget (Nb !!)

Advantage of PEDT with respect to PANI and PPY
- life time stability
- temperature stability up to 270 °C (lead-free soldering)
- toxicity

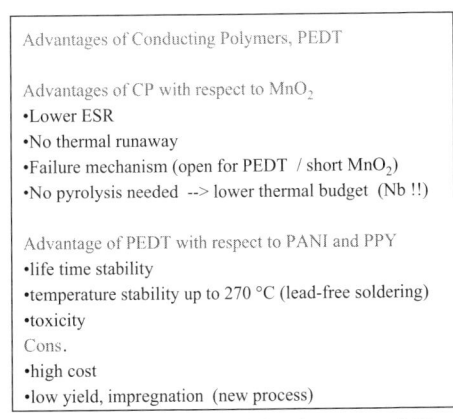

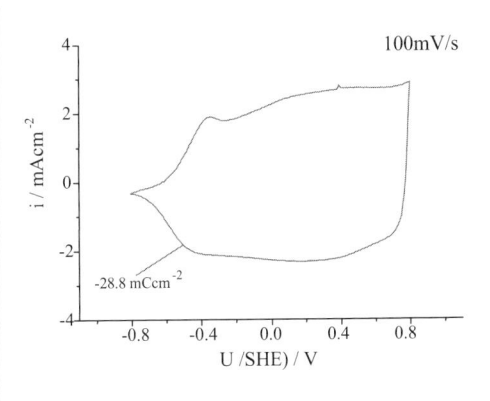

Cons.
- high cost
- low yield, impregnation (new process)

Figure 1.57 Advantages and disadvantages of CPs, and a cyclic voltammogram of PEDT in aqueous solution.

nitrides are still metallic conductors, which can be anodized to form a good dielectric oxide layer. However, the dielectric constant of the corresponding oxides is significantly smaller by about a factor of two, reducing the possible capacitance volume efficiency. Even more severe are the changes in the mechanical properties of these materials. Nitrides are conventionally used for abrasive-proof tools. They are exceptionally hard, which results in severe handling problems of the powder. It is therefore difficult to form a green body without destroying the tools that are used for pressing. Moreover, the sinter activity of the nitrides is poor, resulting in the need for a high sinter temperature in excess of 1600 °C. This in turn further reduces the remaining CV efficiency. Therefore, currently no promising route for the production of a high CV nitrided powder exists. For Nb a more promising modification, namely the suboxide NbO, exists. NbO is a good metallic conductor, which is just a little harder than pure Nb. Therefore, handling of this powder is easier than for the nitrides. This material is currently been investigated together with Nb as a potential complement material for supplementation of both Ta and Al capacitors.

In the next section another important oxide system for electronic applications, namely the Si/SiO$_2$ system, will be discussed.

1.4
Si/SiO$_2$ System

Without any exaggeration, Si/SiO$_2$ can be considered by far to be the most important system in electronics. This is true from a technological as well as from a scientific point of view. For no other system are so many publications available. Since the invention of the transistor in 1947 by Bardeen and Brattain [264] and Shockley [265], there has been a tremendous growth in semiconductor microelectronics. According to recent reports, Si still makes up about 98% of the world market share in microelectronics, with Ge, GaAs, InP, etc. covering the remaining 2%. (This does not include passive components as discussed in the previous sections.) However, the question is: why Si? Si is one of the most abundant elements, which can therefore be produced at low cost and the technology to grow large (currently 300 mm diameter) single crystals is well established. However, as shown in Table 1.4, the intrinsic carrier

Table 1.4 Selected properties of common semiconductors.

Properties	Si	Ge	GaAs	InP
Energy gap	1.1 (indir.)	0.7 (direct)	1.4 (direct)	1.4 (direct)
Density of states (cm^{-3})				
Conduction band	$3 \cdot 10^{19}$	$1 \cdot 10^{19}$	$5 \cdot 10^{17}$	
Valence band	$1 \cdot 10^{19}$	$6 \cdot 10^{18}$	$7 \cdot 10^{18}$	
Intrinsic carrier mobility at 300 K (cm^2 V^{-1})				
Electrons	1500	3800	8500	4000
Holes	450	1800	400	150

mobility of Si is lower than for all the other important semiconductors. Consequently, in terms of carrier mobilities, Si is inferior to the others, resulting in a lower signal speed of the corresponding devices. In addition, Si is an indirect semiconductor, which limits its use for optical applications. Therefore, performance wise Si seems not to be the optimum choice. However, there is one crucial advantage of Si that comes into play when the passivated Si surface, that is the Si/SiO$_2$ interface, is considered. Passivation of the Si surface significantly improves the electrical characteristics of the Si device. This is due to the fact that at the termination of any crystalline lattice, there are non-saturated chemical bonds. For Si, there are about 10^{15} cm^{-2} of these 'dangling bonds' at the surface, which is in the order of the number of surface atoms. These dangling bonds give rise to energy states within the band gap. According to Deal [266] theses defects are termed the interface trapped charge, Q_{it} (C cm^{-2}), and are often reported as the interface trap density D_{it} (number cm^{-2} eV^{-1}). These states can affect and even prevent electronic conduction because under device operation, the electrons that are flowing in response to an external electric field will exchange charge with the surface states instead of exchanging charge with the conduction and valence bands of Si. This disastrous effect is termed *Fermi level pinning* as the Fermi level of the semiconductor cannot move in response to the applied electric field but the charge carriers are trapped in these interface energy states.

The passivation of the Si surface results in formation of an *amorphous SiO$_2$ film*, tying up the dangling bonds and reducing their number by the enormous amount of up to five orders of magnitude to about 10^{10} cm^{-2} eV^{-1}. No other semiconductor with a surface film has been able to achieve electronic surface passivation to the same extent that Si can with SiO$_2$. This is the main reason why the Si/SiO$_2$ system dominates in microelectronics, that is Si and SiO$_2$ are the 'ideal couple' with SiO$_2$ being an excellent dielectric film, as will be pointed out below.

The passivation can be achieved in several manners. The most frequently used one is a thermal passivation [33–36], by the simple exposure of the Si surface to an O$_2$ ambience at elevated temperature. In contrast, an electrochemical passivation can be performed at room temperature, which is beneficial when a low thermal budget is desired. Both the thermal and the electrochemical treatment result in a consumption of Si from the surface, that is the SiO$_2$ grows into the surface (about 1/3). This can be critical for very large scale integration (VLSI) structures. If consumption of Si is not acceptable, a deposition of SiO$_2$ by chemical or physical vapor deposition (PVD/CVD) techniques is necessary.

A good overview on the various passivation and deposition processes can be found in Refs. [267–269]. In Table 1.5 the resulting D_{it} trap densities for the various possible passivation techniques are shown. Thermal passivation yields the highest interface quality, that is the lowest D_{it} can be achieved. Quality wise the electrochemical passivation is next. However, electrochemical reactions at a semiconductor surface are only possible in the accumulation mode. Therefore, anodic reactions only take place at p-type doped Si electrodes (accumulation of majority charge carriers, i.e. holes), whereas on n-Si only reduction reactions are possible. Consequently, only p-type doped Si can be anodically passivated. This can be changed by an illumination

Table 1.5 Comparison of surface states D_{it} for various SiO_2 preparation techniques.

SiO_2 formation	Approximate number of surface states D_{it} (cm^{-2})
Thermal oxidation	10^{10}
Anodic oxidation of p-Si	10^{10}–10^{11}
CVD	10^{10}–10^{12}
PVD	10^{10}–10^{13}
No SiO_2 on Si	10^{15}

of the surface with photon energies above the band gap (1.1 eV for Si). Illumination generates a minority carrier even in the depletion mode (holes accumulate at the surface), that is under illumination n-type Si basically behaves as p-type Si and can be anodically oxidized. The corresponding formation cyclic voltammograms are shown in Figure 1.58 (from Ref. [32]).

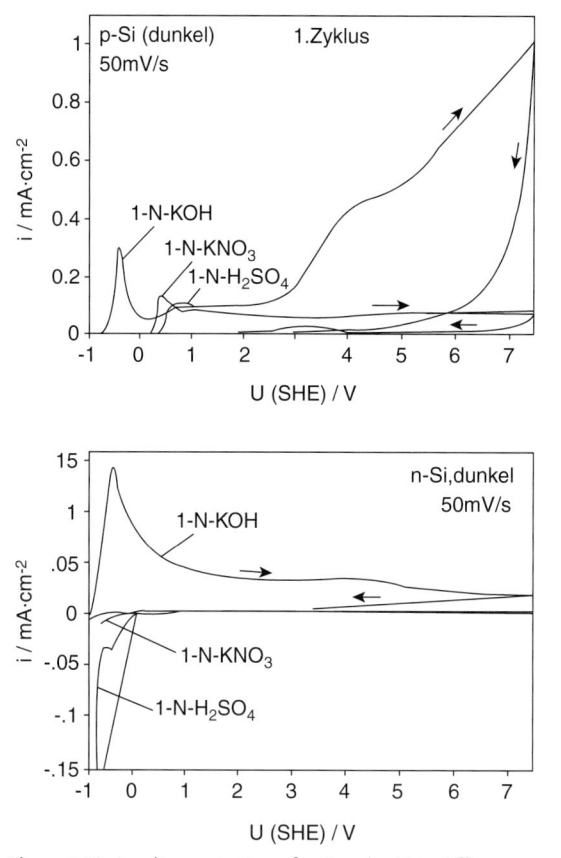

Figure 1.58 Anodic passivation of p-Si and n-Si in different electrolytes without illumination. In KOH corrosion takes place, 50 mV s^{-1} (source Ref. [32]).

1.4 Si/SiO$_2$ System

Regarding oxide formation, this behavior relates to that of the valve metals treated earlier. Analogous to the valve metal oxides in the previous sections we now focus on the use of SiO$_2$ as a dielectric film for capacitors. In contrast to the prior sections, it is not the passive components but integrated circuits (IC) that are considered. There are two main applications of functional dielectrics in microelectronics, namely for MOSFET gate insulation and for storage capacitor nodes. Besides these applications, SiO$_2$ is used for many more purposes in integrated circuits, such as an ILD (inter layer dielectric between metal lines of a chip), STI (shallow trench isolation) fill, or LOCOS (local oxidation of silicon) isolation between active areas (also called field oxides), and for sacrificial hard-masks in lithography [267–269].

1.4.1
Application of the Si/SiO$_2$ System

1.4.1.1 Si/SiO$_2$ in MOSFETs

In the same way that Si is at the heart of microelectronics, the metal oxide semiconductor field effect transistor, or MOSFET, is at the core of most IC devices. The MOSFET is a four-terminal device, which includes a metal gate electrode, heavily doped source and drain regions and a substrate electrode. The metal and Si substrate are separated by a dielectric, such as SiO$_2$. For an n-channel or N-MOSFET, current conduction is via electron flow and not hole conduction. The N-MOSFET devices are fabricated on p-Si substrates and the source and drain regions are heavily doped (low resistance) with an n-type material, such as arsenic or phosphorus. When a negative voltage is applied to the metal gate electrode, the majority carriers, or holes, respond to the applied electric field and accumulate at the Si/SiO$_2$ interface. Consequently, the transistor is in the 'off' state, as current cannot flow from source to drain. Conversely, when a positive bias is applied to the metal gate, the minority carriers or electrons invert the Si/SiO$_2$ interface, generating a conducting channel between source and drain. The device is switched on. In Figure 1.59 a C-MOSFET, or complementary-MOSFET structure is shown for illustration. The CMOS design consists of consecutive MOSFETs of different polarities (N-MOSFET and P-MOSFET), which are integrated in series on one chip. This design minimizes leakage paths due to consecutive pn-junctions in reverse bias, resulting in low power consumption and high speed [272].

The requirements for higher density integrated circuits and faster switching and communication between the devices on a circuit have driven the microelectronics industry to *scaling*, which means a miniaturization of devices. The definition of scaling is: to reduce all parts of a device without affecting the electrical characteristics of the device. In current ULSI (ultra-large scale integration), a minimum lateral feature size well below 100 nm is targeted. Microelectronics have already therefore entered the realms of *'nanotechnology'*. The minimum feature size is defined by the gate length L (distance between source and drain, i.e. channel length) of the MOSFET. In order to keep the electrical MOSFET characteristic, that is the switching behavior, unchanged with shrinking device length, the capacitance of the MOS structure has to be kept constant. The MOS capacitor is part of the MOSFET structure consisting of

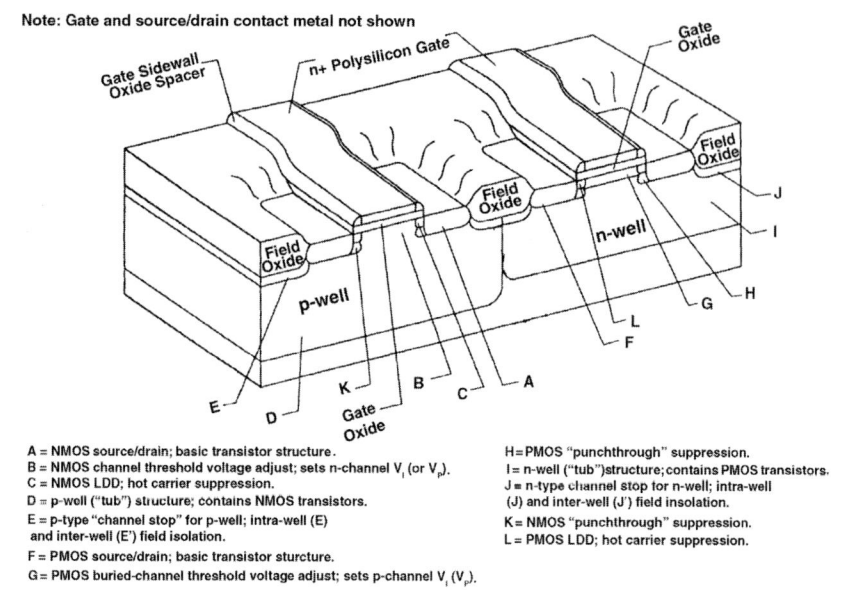

Figure 1.59 Schematic diagram of a C-MOSFET device.

A = NMOS source/drain; basic transistor structure.
B = NMOS channel threshold voltage adjust; sets n-channel V_t (or V_n).
C = NMOS LDD; hot carrier suppression.
D = p-well ("tub") structure; contains NMOS transistors.
E = p-type "channel stop" for p-well; intra-well (E) and inter-well (E') field isolation.
F = PMOS source/drain; basic transistor sturcture.
G = PMOS buried-channel threshold voltage adjust; sets p-channel V_t (V_p).
H = PMOS "punchthrough" suppression.
I = n-well ("tub")structure; contains PMOS transistors.
J = n-type channel stop for n-well; intra-well (J) and inter-well (J') field insolation.
K = NMOS "punchthrough" suppression.
L = PMOS LDD; hot carrier suppression.

the Si/SiO$_2$/gate metal structure. The switching behavior of the MOSFET is described by the threshold voltage V_T. This is the minimum voltage that has to be applied to the gate in order to generate the conducting channel (ON-position of the MOSFET). The voltage V_G that is applied to the gate is distributed into two parts, the potential drop in the oxide ϕ_{ox} and that in the Si (surface potential ϕ_s due to the charge in the scl of the Si):

$$V_G = \phi_{ox} + \phi_S \tag{1.38}$$

The potential drop ϕ_{ox} can be expressed by the capacitance of the MOS capacitor C_{ox} as:

$$\phi_{ox} = Q_C/C_{ox} = eN_A d_{scl}/C_{ox} \tag{1.39}$$

where Q_C is the charge on the gate. From the definition:

$$C_{ox} = \varepsilon_o \varepsilon_{ox}/t_{ox} \tag{1.40}$$

Note that here C_{ox} is given in units of F cm^{-2} (capacitance density), $t_{ox} = d_f$ (film thickness) of the oxide. Together with Equation 1.9 for the extension of the space charge layer, we get an expression for ϕ_{ox} in terms of the charge N_A in the scl:

$$\phi_{ox} = \frac{1}{C_{ox}} \sqrt{2eN_A \varepsilon \varepsilon_{ox}(V_{GB} + 2V_F)} \tag{1.41}$$

where V_F is the Fermi potential.

The threshold potential V_T is defined as the point where inversion starts. For this point it is $\phi_s = 2V_F$, because the inverted surface has to have the same concentration of minority carriers as the concentration of the majority carriers had before (when $V_{GB} = 0$). In other words, the Fermi level now has to have the same distance from the valence band as the distance to the conduction band was before (vice versa for a p-type semiconductor). Therefore, for the threshold voltage we end up with the expression:

$$V_T = 2V_F + \frac{1}{C_{ox}}\sqrt{2eN_A\varepsilon\varepsilon_{ox}(V_{GB} + 2V_F)} \qquad (1.42)$$

According to this equation the switching behavior of a MOSFET depends critically on the capacitance of the MOS-capacitor structure involved. For example, a decreasing C_{ox} would result in an increasing threshold voltage. Conversely, the clear tendency in microelectronics is to decrease V_T in order to reduce power consumption (less leakage at lower applied potentials). In order to keep C_{ox} constant with shrinking device dimensions (shrinking capacitor area) the SiO_2 film has to be made thinner. The scaling of the oxide film is one of the most critical issues facing the microelectronics industry. For the currently developed ULSI devices with gate lengths below 100 nm, an SiO_2 gate oxide thickness below 3.5 nm is necessary. At these dimensions the Si/SiO_2 interface region becomes a significant portion of the device dimensions. Interface properties such as interface roughness, charges at the interface (such as D_{it}), interfacial stress and tunneling will all affect device performance and reliability. In particular, tunneling is a principal problem, that is, independent of the interface quality, Fowler–Nordheim (FN) and direct tunneling start to dominate the leakage currents [269–271]. For an oxide voltage ϕ_{ox} below 3.2 V, the electron tunneling barrier changes from being triangular to trapezoidal (see Figure 1.60). Oxide tunneling then completely switches from FN to direct tunneling.

Logic CMOS technology may be able to tolerate the higher leakage currents due to the direct tunneling, but DRAM will not be able to tolerate them if $t_{ox} < 3.5$ nm. Consequently, the Si/SiO_2 system approaches its limits [273–275] and new solutions that allow for a further scaling of devices have to be found. One approach would be the use of a material with a higher dielectric constant such as the valve metal oxides mentioned previously. This would allow the gate-capacitance of the MOSFET to be increased without reaching the tunnel region for the gate oxide thickness. Another

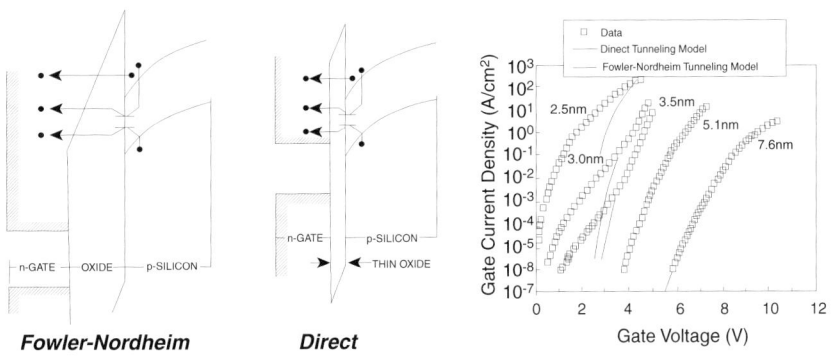

Figure 1.60 Comparison of FN and direct tunneling for shrinking gate oxide thicknesses.

approach is the use of a completely new design for the MOSFET structure. At present the MOSFET is printed laterally on top of the wafer surface. Alternatively, the MOSFET could be buried vertically into trenches that are dug in the wafer surface. This requires less space and therefore allows for higher densely packed structures without the need to shorten the gate length, and in turn allows the gate oxide thickness to be maintained. This option therefore allows keeping SiO_2 as the gate dielectric material. New results on both options are presented below. However, firstly, another important capacitor structure on integrated circuits, namely the storage capacitor on DRAMs (dynamic random access memories), is introduced.

1.4.1.2 Si/SiO$_2$ in DRAMs

DRAM is one of microelectronics industry's highest volume parts. Because of its emphasis on high density, low cost and longer retention times, the DRAM is ideal for setting a roadmap on minimum feature size, defect density and material quality. DRAM, therefore, can be considered the most important technology and research driver in electronics. It is thus ideally suited for deeper investigation into the details of current scientific issues regarding the use of dielectric materials, such as Si/SiO_2 and the valve metal oxides in electronics.

Random access memory, *RAM*, constitutes most semiconductor memory. Owing to its array type of structure, it allows any part of the memory to be read or written as fast as any other part. Conversely, serial access memories such as hard disks, floppy disks, magnetic tape, core memory or optical media (CD, DVD) can be very slow depending on the bit location that is being processed. The basic RAM storage element is the cell, which is duplicated once for every bit to achieve an array structure (two-dimensional lattices). Each cell can store one bit of information, '1' or '0'. Semiconductor RAMs can be categorized into roughly three groups: nonvolatile (NVRAM), static (SRAM) and dynamic (*DRAM*), each of which has several variants. All three types are typically used together. NVRAM is used for permanent storage of information that can be accessed in an ROM (read only memory) mode (e.g. BIOS chips for initialization of a computer). An SRAM cell stores data in a flipflop, consisting of several MOSFETs. It retains its information as long as power is applied to the cell, but loses it when power is removed. A DRAM cell is the smallest possible cell consisting of one MOSFET (access transistor) and one storage capacitor only (see Figure 1.61). The capacitor stores charge, which represents the bit, and the MOSFET transfers the

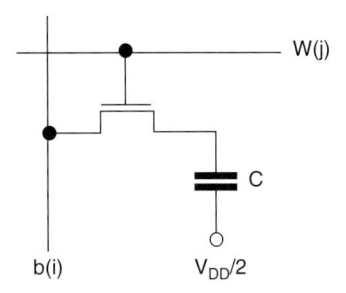

Figure 1.61 DRAM cell.

bit (charge) from and to the capacitor. Because the readout is destructive, the cell must be refreshed by immediately following every read operation with a write-back operation that re-writes the data which were just read. 'Dynamic' means that the cell must also be refreshed periodically, otherwise the cell would lose its information due to cell leakage. DRAMs are therefore optimized for low leakage currents. They are not optimized for speed as, for example, are the SRAMs. Owing to the more complicated cell structure the SRAMs are more expensive (one SRAM cell occupies 8–16 times more space than a DRAM cell) and are therefore used where speed is important, and DRAMs are used where cost efficient storage of large amounts of data is required. A faster DRAM or cheaper SRAM could eliminate the need to use both parts in one system [276] (mutual substitution).

1.4.1.2.1 **DRAM Operation** A DRAM cell consists of a storage capacitor and a transfer transistor (n-MOSFET) that acts as a switch. The presence of charge in the storage capacitor indicates a logical '1' and its absence a logical '0'. Cells are arranged in arrays of rows (word line w_j or W/L) and columns (bit line b_j or B/L) that are orthogonal to each other. Multiple sub-arrays replace a single large array to shorten the word and bit lines and thereby reduce the time to access a cell. For example, a 256 Mbit array typically consists of 16 × 16 Mbit sub-arrays. Word lines control the gate of the transfer transistor and the bit lines are connected to sense amplifiers. Adjacent bit line pairs can be considered as folded in the middle, broken and connected to a shared sense amplifier (folded bit-line design). Alternatively, amplifiers can be placed between two sub-arrays, thus connecting each sense amplifier to one bit-line in each array. This open bit-line design leads to more compact designs but offers less noise immunity than folded bit lines. When a word line is selected, all transfer transistors connected to that word-line are turned on and charge transfer occurs between the storage capacitors and the bit-lines crossing the word-line. In Figure 1.62

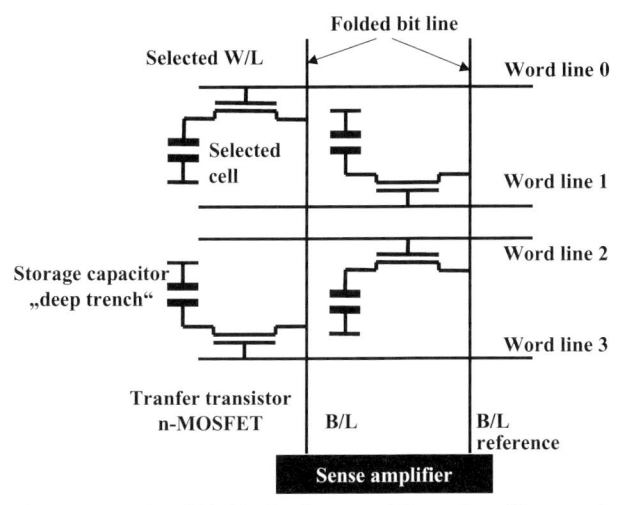

Figure 1.62 DRAM folded bit-line layout and illustration of its operation.

one example is shown. Before a read operation, B/L and B/L are shortened and all bit-line pairs are pre-charged to a voltage V_b, halfway between the internal power supply voltage V_{DD} and ground. To read the cell the selected word line (W/L 0 in this example) is raised to V_{DD} turning on all transfer transistors on that word-line. This does not affect any transistors on the other word lines. Each sense amplifier detects the polarity of charge transfer by measuring the voltage difference, ΔV_b, between B/L and the reference B/L, and thus determines whether the cell stored a logic '1' or '0'. The signal is very small (100–200 mV) because when the transistor turns on, charge redistributes between the small storage capacitor and the capacitance of the bit-line. The latter can be more than ten-times higher due to the large number of devices connected to it and other stray capacitances. The magnitude of the bit-line signal ΔV_b therefore depends on the ratio of the storage capacitance to the bit line capacitance, that is:

$$\Delta V_b = \frac{V_{DD}}{2} \frac{1}{1 + \frac{C_b}{C_s}} \qquad (1.43)$$

where ΔV_{DD} = internal power supply voltage, C_b = bit-line capacitance and C_s = storage node capacitance. As ΔV_{DD} and C_b are reduced by about the same factor, C_s must be kept above a critical level, typically 30–40 fF. The stored charge decays away quickly because of the inherent leakage of the cell. The cell retention time typically ranges from ms to some 100 ms. A periodic refresh is therefore necessary to restore the charge before its level drops below a critical value where a '1' is indistinguishable from a '0'. An SEM micrograph of the word- and bit-lines of a DRAM cell (0.25 μm pitch) is shown in Figure 1.63.

1.4.1.3 DRAM Storage Capacitor (Deep Trench)

The construction of a DRAM cell in 'deep trench' (DT) technology is shown in Figure 1.64. The top shows schemes and the bottom of the figure shows the corresponding SEM micrographs of the DRAM cell. The storage capacitor is connected to an access n-MOSFET by the BuriEd STrap, therefore this layout is known as the 'BEST'-cell. When the MOSFET is turned on by applying a voltage to the gate, charge can flow from the source of the MOSFET to the drain that is connected to

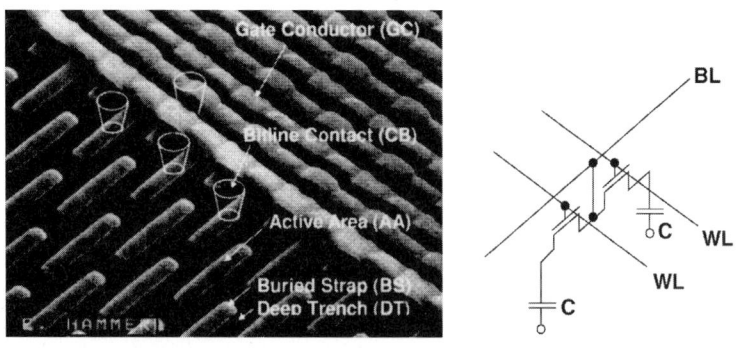

Figure 1.63 SEM and schematic representation of the DRAM word- and bit-line configuration.

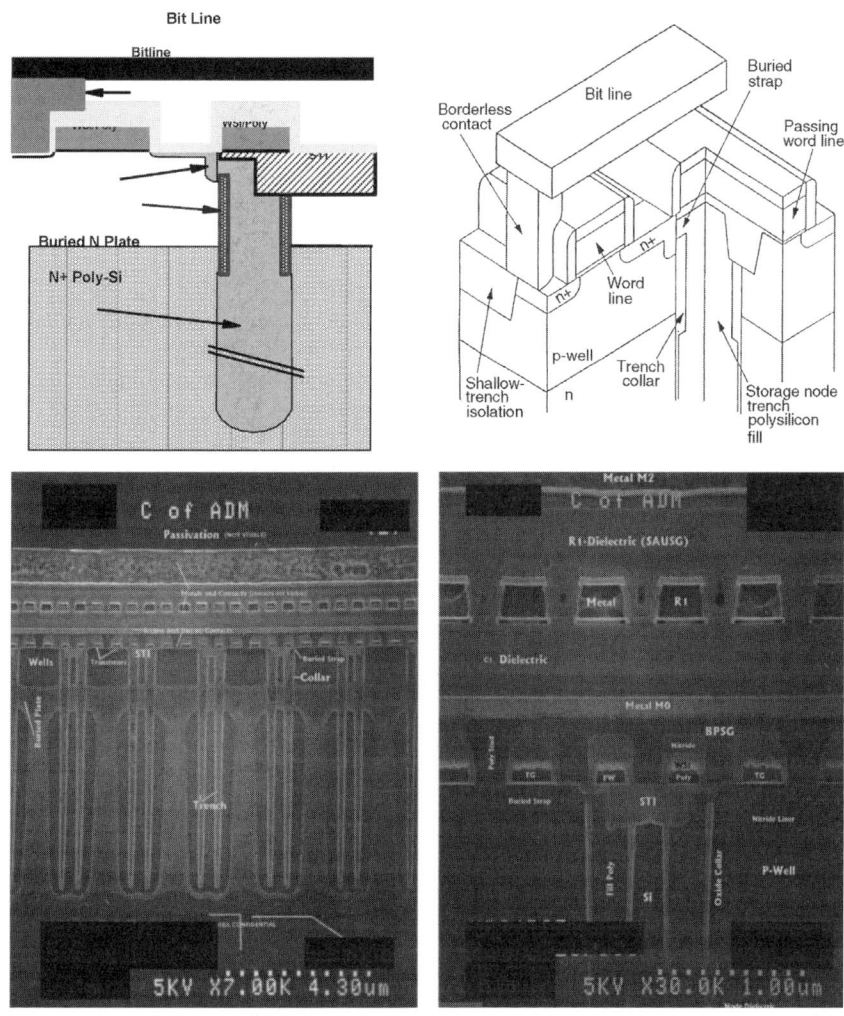

Figure 1.64 Cross-section of a 'Best'-cell DRAM (BuriEd STrap) showing the MOSFET access transistor and the storage trenches (deep trench-capacitor); 0.25 µm feature size (256 Mbit DRAM).

the bit-line. The source region of the MOSFET in turn is connected to the DT by the buried strap, which is made up of highly n-type doped, conducting polycrystalline Si. The DT itself is a simple capacitor structure consisting of a hole (trench), which is dug into the Si surface with a laterally oval shape. The DT is formed by reactive ion etching into the wafer. Subsequently, the bottom part of the trench is highly n-type doped to form a conducting capacitor plate. All trenches are connected by this buried n-plate. After formation of this plate, the dielectric layer is formed. Conventionally, this is done by a thermal oxidation of the trench sidewalls, that is, an SiO_2 dielectric film is formed. In order to maximize the capacitance, the node dielectric is made as thin as possible. After implementation of the node dielectric, an oxide collar has to be formed at the top portion of the DT. This thick oxide collar prevents the vertical transistor structure (see Figure 1.64) from turning on, which otherwise would result in a high vertical leakage (this vertical device is an undesired but unavoidable npn-transistor structure). Subsequently, the trench is filled with a conducting material such as highly doped Si and the buried strap is formed.

Note that there are always two DTs very close to each other. These DTs have to be isolated from each other by an STI (shallow trench isolation) process. There is also a 'passing word-line' right on top of the DT, which connects to a gate of a DT pair that is one plane deeper in the structure. A portion of the STI therefore also covers the top of the DT to generate insulation from this word-line conductor. The two DTs that are farther apart are connected by one active area containing two MOSFET structures, which share one bit-line contact (folded bit-line layout). In Figure 1.64 a cell of 0.25 μm feature size F is shown, that is, the channel length of the MOSFET access transistor is about that size as is the lateral top dimension of the DT. The depth of the DT is about 6 μm yielding an aspect ratio of 24 in this instance.

As already mentioned above, the minimum lateral feature size of future gigabit DRAM generations has to be shrunk well below 0.1 μm to keep pace with productivity demands. However, the bit storage capacitance does not scale but has to be kept constant or even increased to compensate for higher charge leakage if reduced dielectric layer thickness or new node materials are applied. A storage capacitance of up to 40 fF per cell will be necessary for future DRAM generations. Future DRAMs can only meet this target when several new measures are taken. With conventional materials and new processes a minimum feature size F of 0.10 μm in the best case can be achieved.

For even smaller feature sizes high ε_r materials for the storage capacitor dielectric become essential. Figure 1.65 shows the calculated capacitance per cell as a function of F for a DT with $1F$ by $2F$ lateral top-down dimensions. These dimensions are typical for current $8F^2$ and $6F^2$ array layouts (one DRAM cell laterally covers a space of xF^2). For the calculations, an oxy-nitride (ONO) dielectric layer with an oxide (SiO$_2$) equivalent thickness t_{eq} of 3.5 nm ($\varepsilon = 3.8$) was assumed. The process for formation of such an ONO film is described below. The depth of the vertical trench top isolation oxide (DT-collar) that reduces the active capacitor area was assumed to be 1.3 μm. Six different possibilities (curves) to achieve a capacitance increase are specified in Figure 1.65. These possibilities are discussed in some detail below. TiO$_2$ and Ta$_2$O$_5$ are considered as the high ε_r node materials.

1.4.1.3.1 Increase of Aspect Ratio for DT Si Etch up to 60
The maximum achievable DT depth is not limited by an Si Reactive Ion Etch (RIE) stop, but by hard mask erosion [277–279]. A boron doped silicon oxide (BSG) is used as the hard mask material (CVD deposition). The BSG is patterned by conventional photoresist technology. A photoresist to BSG RIE selectivity of above 4:1 is possible. In turn, the BSG hard mask to Si RIE selectivity was found to depend on the DT top dimension (lateral DT perimeter). The corresponding experimental relationship is shown in Figure 1.66 for a conventional Si RIE tool set. For each data point the required BSG

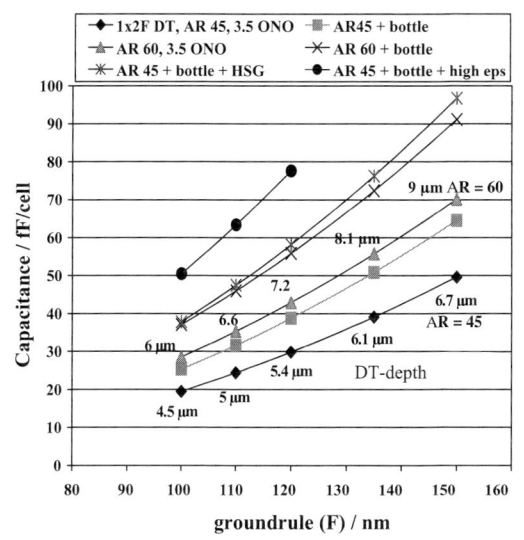

Figure 1.65 DT capacitance roadmap. The six curves represent different options: 1, Si-etch with an aspect ratio (AR) of 45, straight DT profile; 2, AR = 45 + bottle shape DT profile; 3, AR = 60; 4, AR = 60 + bottle; 5, AR = 45 + bottle + HSG (hemispherical silicon grains); 6, AR = 45 + bottle + high ε dielectric materials such as TiO_2, Ta_2O_5 and Al_2O_3.

layer thickness is given, assuming a DT Si RIE aspect ratio of 45 and a required remaining BSG thickness of 150 nm after Si RIE (process window).

Figure 1.67 shows, as an example, cross-sections of DT structures with 6 μm depth and a trench width of 0.135 μm (AR = 45). For this a 1.2 μm thick BSG hard mask stack was used, which is well below the capabilities of current tool sets. In order to ensure a clean DT profile a sidewall passivation layer has to be formed during Si RIE, requiring a well controlled ratio between etchant (e.g. HBr) and passivating (O_2) gases. We found the passive layer formation to be vertically inhomogeneous with thicker films formed in the upper DT portion. This results in a decreasing trench open area with increasing etch time (DT clogging) reducing the BSG to Si etch selectivity and the differential etch rate. Therefore, a multiple step RIE process (see patent Ref. [280]) was applied with an integrated removal of the passivation layer at certain times. With this scheme the differential etch rate and RIE selectivity were significantly enhanced. The corresponding data are also shown in Figure 1.68. An AR of up to 60 could be achieved. Figure 1.67 (right) shows a corresponding cross-section with 8.6 μm DT depth and 0.15 μm width. As the DT profile is less controlled for this process, post-engineering of the DT-sidewall profile is needed. For this, an inexpensive wet-clean is applied, as is explained next.

1.4.1.3.2 Bottle-shaped DT

For well controlled engineering of the lateral DT profile a new wet etch process employing diluted NH_4OH chemistry was developed

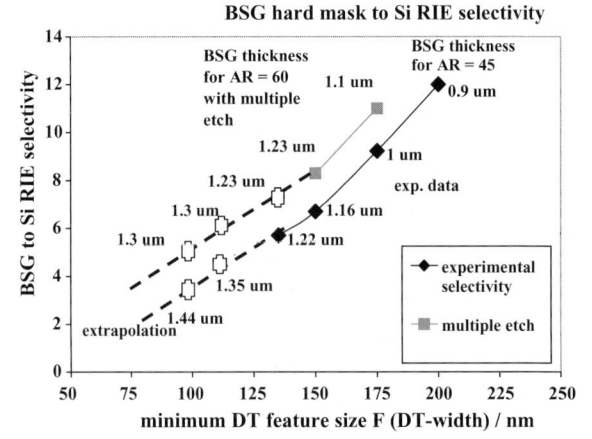

Figure 1.66 Measured hard mask (BSG) to DT Si RIE selectivity as a function of feature size for different etch processes (single and multiple etch). The hard mask thickness required to achieve a DT aspect ratio (AR) of 45 (single etch) or 60 (multiple etch) is given for each data point.

(see patent in Ref. [281]). In a first step the thin native oxide is removed from the surface and an H-terminated Si surface is generated applying diluted HF (200:1) for 60 s at RT (25 °C). Subsequently, the anisotropic etch is performed using 40:1 NH$_4$OH ($T = 25$–$45\,°C$; time $= 60$–$120\,s$). The following reactions take place:

$$Si + 3\,OH^- \rightarrow Si(OH)^{++} + 4e^-$$

$$4H_2O + 4e^- \rightarrow 4OH^- + 2H_2$$

Besides the formation of a smooth single crystalline DT sidewall, this process increases the DT area (bottle-shaped DT) in the bottom portion, therefore significantly

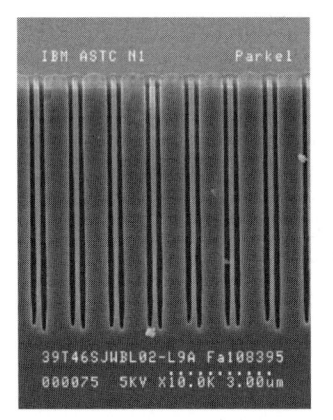

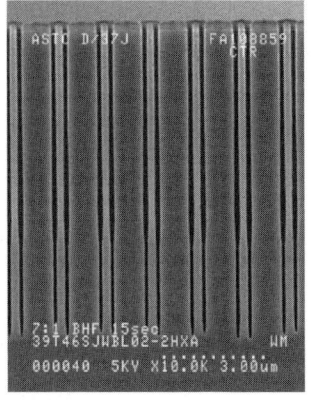

Figure 1.67 Left: DT Si etch with an aspect ratio AR of 45 (single step DT etch); DT depth 6 µm, DT width 0.135 µm. Right: two-step DT Si etch with AR = 60; DT depth 8.6 µm, width 0.15 µm.

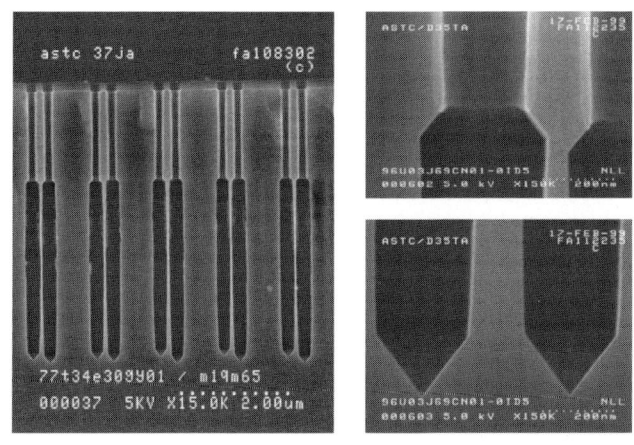

Figure 1.68 SEM of a bottle-shaped DT. The corresponding etch leaves a single crystalline high-quality surface.

increasing the capacitance. The process is self-aligned to the DT-collar oxide, that is protected by a thin nitride layer which is formed in advance, that is, the reactions in both steps are highly selective with respect to Si_3N_4. Depending on the array layout (DT to DT distance) a capacitance increase of up to 80% can be achieved (e.g. in an 8 F^2 layout). The corresponding curve in the roadmap of Figure 1.65 is based on a 6 F^2 layout. For this specific example, a capacitance increase of 30% was achieved. Figure 1.68 shows cross-sections of DT bottle profiles. Owing to its anisotropy this etch simultaneously takes care of another important problem that is associated with the texture dependence of Si/SiO_2 growth.

In Figure 1.69 an example of a conventional DT top-view is shown (oval shape) after collar oxide formation (black region) and DT fill with a highly doped poly-Si. The thermal oxide formed shows a strong dependence on the crystal orientation of the Si surface. This holds true for all oxidation modes, for example, dry and wet oxidation (in the 'wet'-oxidation mode there is a carefully adjusted addition of H_2O as steam to the oxidation chamber resulting in an increased oxidation rate). This also holds true for

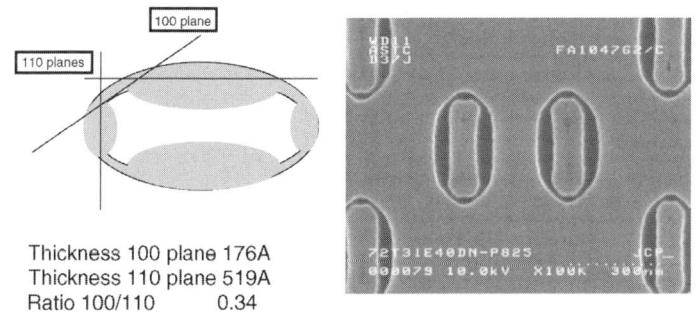

Thickness 100 plane 176A
Thickness 110 plane 519A
Ratio 100/110   0.34

Figure 1.69 Top-down view of a DT structure without applying the anisotropic etch. The oxide collar (dark region) and the DT fill can be seen.

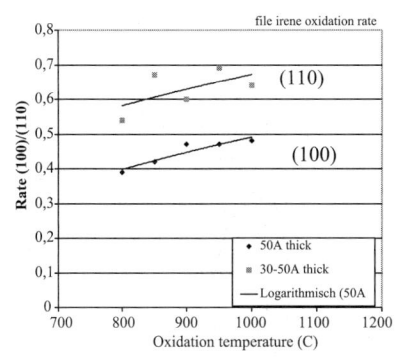

Figure 1.70 SiO$_2$ oxidation rates on different Si surfaces.

an electrochemical oxide formation. The film thickness ratio between the oxides grown on the (100) and (110) surfaces can not be suppressed completely.

As an example, Figure 1.70 shows the oxide thickness on different Si surfaces for various formation modes, different forming temperatures, respectively. It is understandable that this non-uniformity constitutes a significant reliability problem for the SiO$_2$ films. For the aforementioned gate oxides, this problem was not addressed as the corresponding oxidation is performed on planar single crystalline wafer surfaces. In fact, the strong texture dependence of Si/SiO$_2$ oxide formation is one of the main reasons why single crystalline wafer surfaces are used in microelectronics. As soon as vertical structures such as the DT are used, this issue returns. For instance, for the dielectric node formation within the DT mentioned previously, this phenomenon must be considered. The minimum node thickness therefore is determined by the (110) planes where the thinnest oxide grows. A varying node oxide thickness inside the DT therefore has to be tolerated and, fortunately, for capacitor application it can be. However, for a potential vertical MOSFET it can not be accepted as the V_T depends on the oxide thickness, as shown above. A varying oxide thickness therefore means an uncontrolled threshold voltage, which in principle opposes MOSFET operation.

Nevertheless, such exact types of structures, that is vertical MOSFET within a DT, are currently being developed [308]. The key solution to overcome this problem lies in the anisotropic etch, as presented here. The top-down view of a DT after anisotropic etch and subsequent oxide formation and fill is shown in Figure 1.71. The DT shows a perfect rectangular shape, that is, the (100) plane is etched faster than the (110) plane and therefore is completely eliminated eventually, thus leaving a single crystalline (110) surface everywhere in the DT. The void inside of the poly-fill does not cause any problems (e.g. no resistance increase of the poly-fill), but Weibull slope reliability measurements actually showed a significantly improved node reliability for the anisotropically etched DTs. Accordingly, the center picture in Figure 1.71, showing a TEM high resolution zoom of the SiO$_2$ dielectric film, indicates an excellent uniformity of the film, even at critical sites such as the tip of the DT. The anisotropic etch therefore not only establishes an important means of increasing the DT capacitance, but more importantly, a high quality, single crystal surface is provided, which increases the reliability of the oxides formed upon it.

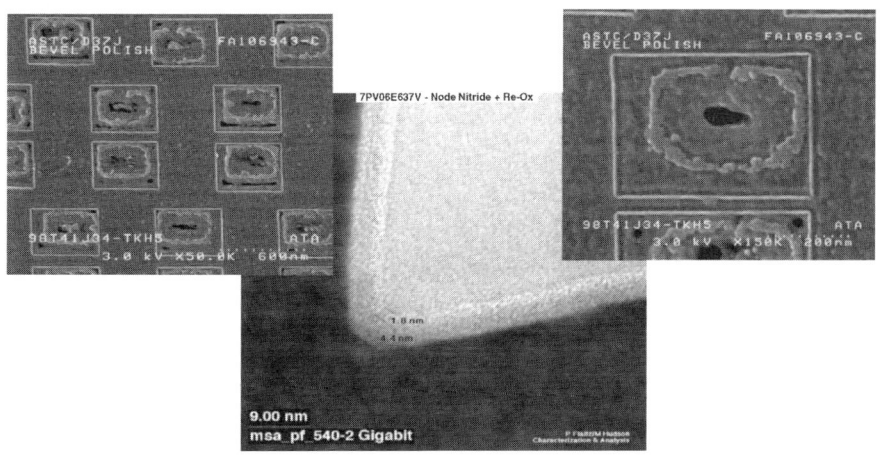

Figure 1.71 Top-down view of anisotropically etched DT showing a perfect rectangular shape with (110) surfaces only. In the center, a high resolution TEM of the DT tip is shown with a perfectly uniform SiO$_2$ node dielectric inside.

1.4.1.3.3 Hemispherical Silicon Grain HSG Formation

A further DT capacitance increase can be achieved by introducing a surface roughness on the trench sidewalls. For this, the well known hemispherical silicon grain (HSG) formation process can be used. This process requires deposition of a thin doped amorphous Si layer on top of the vertical trench walls. After lining, the structure is heated to initiate a nucleation (or seeding) process. In a subsequent annealing step the nuclei are grown to yield grains of the desired size (nuclei grow by mass transport of high mobility Si atoms), thus creating a rugged surface. For the DT, an exact and uniform growth of very small grains (below 25 nm) is essential to prevent any DT clogging. This requires a strict control of the process parameters therefore reducing the process windows.

Another principal problem with the use of rough surfaces such as HSG lies in the polycrystalline texture of the surface. Thus the texture dependence of oxide growth has to be considered again, causing the reliability problems mentioned. The only means to circumvent this problem is the use of deposition processes instead, or in addition to thermal or electrochemical dielectric film formation. Normally deposition rates do not depend on the texture. At present, a combination of thermal SiO$_2$ ($\varepsilon_r = 3.9$) and CVD deposited Si$_3$N$_4$ ($\varepsilon_r = 7.2$) is one of the most commonly used dielectrics when combined with a rugged electrode such as HSG and/or three-dimensional capacitor structures. This dielectric multi-layer stack is termed *ONO* (SiO$_2$/Si$_3$N$_4$/SiO$_2$) (oxide/nitride/oxide) or *ON* (SiO$_2$/Si$_3$N$_4$) (oxide/nitride). The ONO stack is prepared by successive thermal oxidation of the bottom polysilicon electrode, low-pressure chemical vapor deposition (CVD) of Si$_3$N$_4$, and thermal oxidation of the Si$_3$N$_4$ film. In order to reduce the SiO$_2$ equivalent thickness, the bottom oxide layer can be removed from the film structure to form the ON structure. According to studies, the leakage current properties of these ONO and ON films are dominated by the Poole–Frenkel current [282, 283]. However, when the oxide

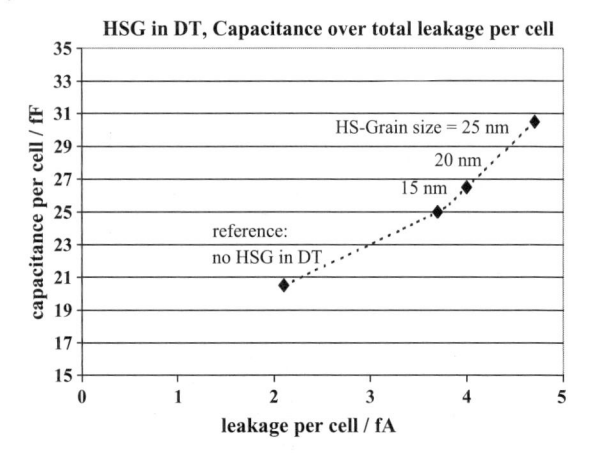

Figure 1.72 DT capacitance as a function of node leakage for a reference process without HSG, and three different HSG processes with three different HSG grain sizes, respectively. For the measurements DT test structures with $1 F \times 1.5 F$ top-down dimension ($F = 0.15\,\mu m$, DT depth $5\,\mu m$) were used.

equivalent thickness becomes less than 5 nm, the leakage current properties become dominated by direct tunneling current. This is undesirable as the direct tunneling current can only be suppressed by increasing the physical thickness of the capacitor dielectrics [284]. This limits the equivalent thickness of ON films to 5 nm.

Furthermore, a transition layer exists between the polysilicon electrode and the bulk Si_3N_4 film. This transition layer is formed when an initial layer of Si_3N_4 is deposited on the native oxide and the mixture of Si_3N_4, native oxide, and Si nuclei form an Si-rich and defective layer. As the Si_3N_4 film becomes ultra-thin, the oxide equivalent thickness of Si_3N_4 rapidly decreases due to the density reduction in the defective layer. In addition, the tunneling leakage current rapidly increases [285]. To overcome this problem, the native oxide can be changed to a nitrided layer by performing rapid thermal nitridation (RTN) on the polysilicon electrode surface before CVD of Si_3N_4 [284]. Figure 1.72 shows the effect of the HSG grain size on the capacitance and leakage currents of a DT using this type of ON film. With such a process, a capacitance gain of up to 50%, with node to buried plate leakage currents still meeting the specifications, can be achieved.

1.4.2
Alternative Dielectric Materials

As the downscaling limits of conventional SiO_2 and ONO dielectrics are reached, alternative dielectric films with higher dielectric constants ε_r are needed. The valve metal oxides TiO_2 (ε_r up to 170) and Ta_2O_5 ($\varepsilon_r = 25$) are among the candidates for the next generation of DRAMs. However, even high ε_r ferroelectric materials such as (Ba, Sr)TiO_3 have been investigated for this purpose. The integration of Ta_2O_5 has already been successfully realized within the 'stack capacitor' DRAM technology. In contrast

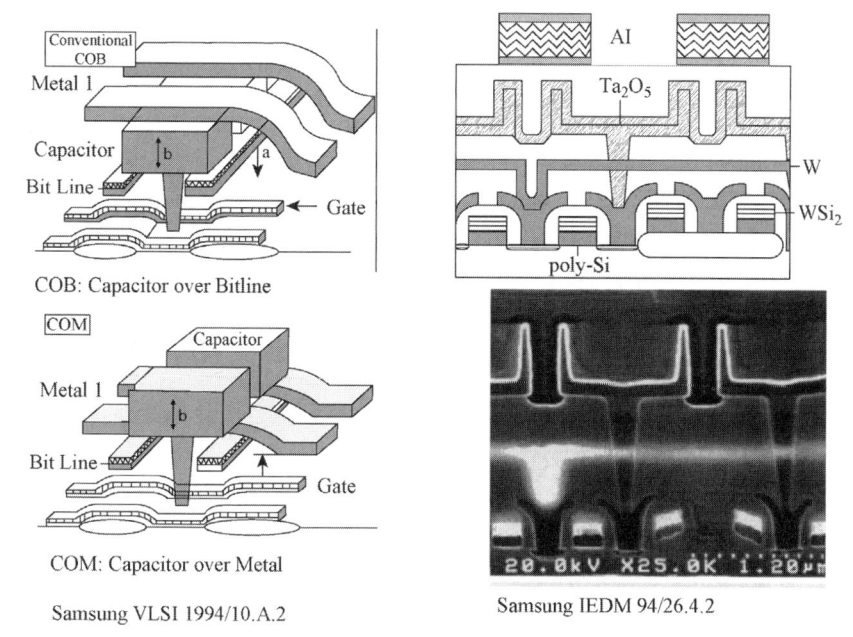

Figure 1.73 Examples for 'stack-capacitor' DRAMS. In contrast to the 'deep trench' capacitor that is etched into the Si below the access transistor, the stack capacitor is deposited on top of the access transistor. Ta_2O_5 is used as the dielectric film.

to the DT-DRAM technology, the stack storage capacitor is formed on top of the MOSFET access transistor and not below it. Consequently, the stack capacitor structure is not formed by an etch process but by deposition techniques. This limits the aspect ratio to about ten, whereas for the DT structures aspect ratios up to 100 were discussed above. Correspondingly, the active capacitor surface area of a stack capacitor is in principle smaller than for a DT capacitor and downscaling limits are reached earlier in stack capacitor technology.

Thus, it is no surprise that Ta_2O_5 was first introduced into the stack technology and that this technology has mainly driven alternative dielectric development in the past. In Figure 1.73 a typical stack storage capacitor structure using Ta_2O_5 as the dielectric material is shown. For formation of alternative dielectric layers, principally it is deposition techniques such as PVD/CVD that have to be applied. Usually the resulting Si/dielectric film interfaces are more disturbed than the thermally or electrochemically formed ones (more interface defects). Moreover, the adhesion of the deposited films or/and an interface diffusion of atoms from the dielectric into the bulk can be critical. In order to improve the quality of the Si/dielectric interface, often barrier layers have to be used between the Si and the dielectric film. One example for such a stack has already been presented above, namely the ONO or ON stack. In the following sections the most important options for alternative dielectric materials, namely Ta_2O_5, TiO_2 and the ferroelectrics, are introduced.

1.4.2.1 Ta$_2$O$_5$

Ta$_2$O$_5$ can replace SiO$_2$ or ONO dielectric films without major process integration changes. Many methods for depositing Ta$_2$O$_5$ films have been proposed. Among them, CVD methods such as LPCVD [286, 287], plasma-enhanced CVD (PECVD) [288, 289] and electron cyclotron resonance plasma-enhanced CVD (ECR PECVD) [292] are the most effective in depositing Ta$_2$O$_5$ films. CVD methods are desirable because Ta$_2$O$_5$ exhibits good step coverage [293]. High-temperature annealing is usually necessary to improve the quality of as-deposited tantalum oxide films prepared by all types of deposition methods, except for advanced methods such as PECVD and ECR PECVD [290, 291].

PECVD processes are based on the activation of a precursor by a high-energy plasma. The advantage of PECVD is that it provides both a high deposition rate and good step coverage [291]. The ECR PECVD process uses both microwave and a high magnetic field to create the plasma at low operating pressures. An ECR plasma decomposes the precursor so powerfully that high-quality films can be deposited at a low temperature. Furthermore, the low process pressure (down to 10^{-4} Torr) inhibits the homogeneous gas-phase reactions, and the low ion energy offers the benefit of greatly reduced damage to the device [290]. Unlike other CVD processes, the electrical properties of the ECR PECVD film do not need to be improved by annealing in an oxygen ambience at high temperatures, due to the crystallization of the tantalum oxide film and the growth of the interfacial silicon oxide layer [290]. RTN, which forms an Si$_3$N$_4$ barrier layer, is effective in preventing the formation of an interfacial SiO$_2$ layer for the LPCVD deposition of Ta$_2$O$_5$ but is ineffective for the ECR PECVD system. This is because the Si$_3$N$_4$ buffer layers are transformed into silicon dioxide during tantalum oxide deposition, due to the strong oxidation power of the ECR oxygen plasma [294].

Currently, titanium nitride (TiN) and tungsten (W) are being studied as buffer layers as they are expected to yield an interfacial oxide with higher dielectric constant in the ECR oxygen plasma [294]. As typical precursor materials for Ta and Ta$_2$O$_5$ CVD deposition, liquids such as Ta(OCH$_3$)$_5$, Ta(OC$_2$H$_5$)$_5$ and related alkoxides, and solid materials, such as TaCl$_5$ and related halides, are usually used. The advantage of liquid sources is the good controllability of flow rate over a wide range. However, the Ta$_2$O$_5$ deposited using liquid sources by thermal CVD or photo-CVD are subject to carbon contamination, which leads to poor electrical properties. On the other hand, the use of plasma-enhanced CVD and ECR CVD oxidizes the Ta(OC$_2$H$_5$)$_5$ so strongly that it dissociates under the applied electric field in the plasma and reacts with ionized oxygen molecules in the following chain reaction that forms Ta$_2$O$_5$:

$$Ta(OC_2H_5)_5 \rightarrow Ta + OC_2H_5$$

$$2O_2 \rightarrow 2O + O_2$$

$$Ta + O \rightarrow Ta_2O_5$$

$$OC_2H_5 + O_2 \rightarrow CO_2, CO, H_2O, C_2H_4, CH4, \ldots$$

Thus Ta_2O_5 is deposited and the carbon compounds are evacuated [288]. The only disadvantage is that a thick interfacial SiO_2 layer with a low dielectric constant will be formed during PECVD.

Solids such as $TaCl_5$ can also be used as source material for Ta_2O_5 deposition but have the disadvantage of poor controllability of the flow rate. One concern about as-deposited CVD Ta_2O_5 capacitors is the excessive leakage current that is exhibited, which is believed to be caused by a deficiency of oxygen in the as-deposited film. To fill the vacancies with oxygen to reduce leakage current and to improve the electrical properties of CVD Ta_2O_5 capacitors, various post deposition annealing techniques have been proposed, for example: annealing in O_2 [296], UV-O_3 [297] and dry O_2 [297, 298]; rapid thermal O_2 annealing (RTO) [293]; and rapid thermal annealing in N_2O (RTN_2O) [299, 300]. Among these annealing techniques, rapid thermal annealing (RTA) in N_2O is the most effective in suppressing leakage current [300]. However, the oxygen atoms from N_2O produce a thicker interfacial SiO_2 layer than conventional techniques with oxygen, because N_2O is a better oxidizing agent [301]. The formation of an interfacial oxide reduces the effective dielectric constant, which is a serious disadvantage for high dielectric capacitor applications. Thus, RTN treatment is required to suppress the oxidation during annealing [285].

During the dry-O_2 annealing, the densification of the Ta_2O_5 film due to hydrocarbon desorption decreases the equivalent thickness. At higher annealing temperatures, the growth of an SiO_2 layer at the interface between Ta_2O_5 and Si leads to an increase in the SiO_2 equivalent thickness. When no-RTN treatment is performed prior to Ta_2O_5 deposition, the oxide equivalent thickness is monotonously increased because of the oxidation of the Si surface.

An additional serious problem of Ta_2O_5 capacitors is caused by their exposure to high temperature, which unavoidably occurs in subsequent process steps. One example is the boro-phospho-silicate glass (BPSG) reflow process after capacitor formation in the stack technology. Here temperatures up to 600 °C occur. In DT technology this problem is even more severe as here several process steps with temperatures up to 1050 °C are applied. The Ta_2O_5 degradation in BPSG reflow with a top TiN electrode was shown to be due to TiN oxidation at the TiN/BPSG interface [302, 303]. The TiN oxidation leads to a volume expansion and build-up of compressive stress, which eventually leads to cracks in the Ta_2O_5 film. As a result, the leakage current is increased. This problem could be fixed by placing a thin layer of poly-Si between TiN and BPSG thus suppressing the TiN oxidation and therefore reducing the leakage current [303]. The Ta_2O_5 dielectric with the bi-layered TiN/poly-Si top electrode was applied successfully to the full process for 256-Mbit stack-DRAM devices. Degradation free Ta_2O_5 capacitors can be produced. As mentioned earlier, an interfacial oxide layer that has a low dielectric constant is inevitably grown during Ta_2O_5 deposition and the post-deposition annealing process. The bottom electrode also affects the interfacial oxide layers and the capacitance of the dielectric film.

It was found that for ECR PECVDA, different electrode materials have different incubation periods, which is the time required to cover up the substrate surfaces completely with Ta_2O_5. Furthermore, the incubation period for the Ta_2O_5 film deposition was related to the oxidation of the electrodes [304]. The effect of the interfacial oxide layer on the capacitance of the dielectric film is:

$$C = \varepsilon_0 \varepsilon_r \frac{A}{d^*_{Ta2O5} + d^*_{io}} \tag{1.44}$$

$$d^*_{Ta_2O_5} = d_{Ta_2O_5} \cdot \varepsilon_{SiO_2}/\varepsilon_{Ta_2O_5} \tag{1.45}$$

$$d^*_{io} = d_{io} \cdot \varepsilon_{SiO_2}/\varepsilon_{io} \tag{1.46}$$

where A is the capacitor area, ε_o is the permittivity of empty space and ε_{SiO_2}, $\varepsilon_{Ta_2O_5}$ and ε_{io} are the relative dielectric constants of SiO_2, Ta_2O_5 and the interfacial oxide (io), respectively; $d_{Ta_2O_5}$ and d_{io} are the thickness of the Ta_2O_5 film and the interfacial oxide layer, respectively. Therefore, when Ta_2O_5 is used as a capacitor material, the capacitance is actually determined by $(d^*_{Ta_2O_5} + d^*_{io})$. Thus, d_{io} should be reduced to a value as small as possible.

According to the literature, a platinum (Pt) bottom electrode does not form any interfacial oxide ($d^*_{io} = 0$) and thus offers the highest capacitance. However, at present it can only be deposited by physical vapor deposition (PVD) methods, which give poor step coverage, and therefore is excluded for an eventual DT process (high aspect ratio). On TiN and W bottom electrodes, interfacial oxide is formed with thicknesses of 0.9 and 2.1 nm, respectively, which are less than the ones formed on a pure Si electrode ($d^*_{SiO_2} = 4$ nm). Both TiN and W can be deposited by CVD, which gives good step coverage, and are currently already used in other DRAM process steps. Between the two, TiN has a lower d^*_{io} value and is very stable with Ta_2O_5 at high temperatures (700–900 °C). Thus, TiN is the most promising material as a bottom electrode when used for Ta_2O_5 dielectric films [304].

The electrode material also affects the leakage current of the Ta_2O_5 capacitors. It was found that among some common metal and nitride electrodes (W, Mo, Ti, WN, MoN, TiN and TaN), the voltage that induces a leakage current of $1 \mu A \, cm^{-2}$ (V_{crit}) increases with the work function of the electrode immediately after electrode formation and after annealing at 400 °C [305]. Thus, for a low-temperature annealing process, the leakage current in the Ta_2O_5 film at a given voltage decreases with increasing work function of the top electrode. This result indicates that the conduction mechanism is an electrode-limited type, and that the barrier height for the electrons at the top electrode/Ta_2O_5 interface limits the leakage current. For high-temperature annealing (800 °C), there is almost no correlation between the work function and the leakage current. It is assumed that after high-temperature annealing, the current is affected by the reaction between Ta_2O_5 and the electrode. It was found that TiN is the optimum electrode for low temperature processes (about 400 °C) while Mo or MoN are the optimum electrodes for high-temperature processes (about 800 °C) because they exhibit the smallest leakage current [305].

The downscaling limit of Ta_2O_5 capacitors is estimated to be around an equivalent oxide thickness of 1.5 nm [284]. Combined with polysilicon electrode techniques, such as HSG, Ta_2O_5 capacitors will be available for 1 Gb DRAMs and beyond.

Currently, Ta_2O_5 has already reached the stage of practical use and is the most likely choice for replacement of ONO/ON dielectrics after the 256 Mbit generation.

1.4.2.2 Ti/TiO$_2$

Another possible candidate as an alternative dielectric for DT technology is TiO_2 (see patent Ref. [306]). The dielectric constant of crystalline rutile is between 86 for a crystal orientation with the optical axis (c-axis) perpendicular to the surface and 170 for the c-axis parallel to the surface. This exceptionally high ε_r allows relatively thick dielectric layers (>30 nm) with an SiO_2 equivalent thickness t_{eq} still below that of pure SiO_2 or ONO to be employed. Therefore, in contrast to the previously described dielectrics, for TiO_2 a crystalline layer might be acceptable. (As explained in Section 1.3, ultra-thin dielectric, amorphous films must be used in order to keep the leakage current low.) The use of a crystalline film is beneficial in DT technology due to its better stability with respect to high temperatures. In DT technology several high temperature steps (up to 1050 °C) are required after node dielectric deposition.

The temperature stability is therefore a decisive advantage of rutile TiO_2 in comparison with the other mentioned dielectrics. In order to generate the required rutile phase high temperatures up to 1050 °C are necessary. In contrast to the stack capacitor technology, these temperatures are no problem in DT technology (which explains why TiO_2 is not used for stacked capacitors). With TiO_2 dielectrics, leakage currents below $10^{-10}\,A\,cm^{-2}$ have been reported in the literature [307]. For TiO_2 deposition, the following process scheme, illustrated in Figure 1.74, could be possible:

1. DT etch and buried plate formation as usual.

2. Si_3N_4 barrier layer formation by nitridation (thickness below 1 nm).

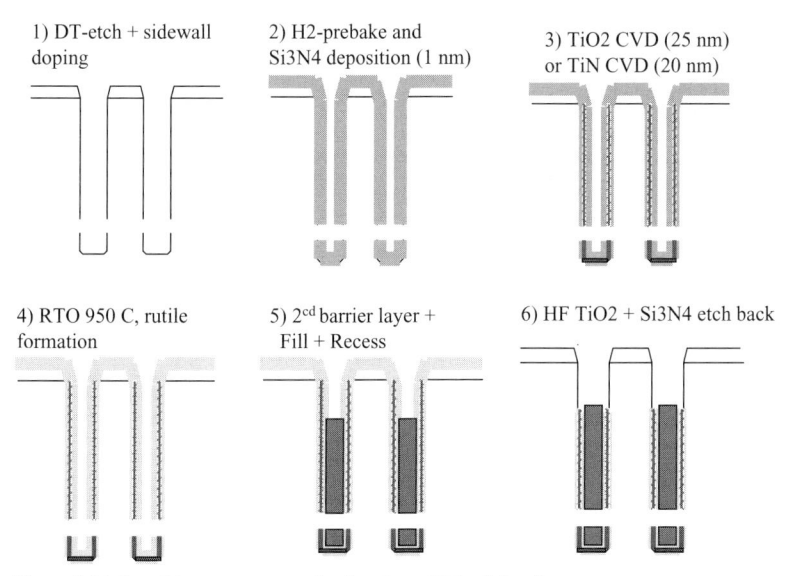

Figure 1.74 Possible process steps for forming a TiO_2 dielectric layer in a DT storage capacitor. Details are described in the text.

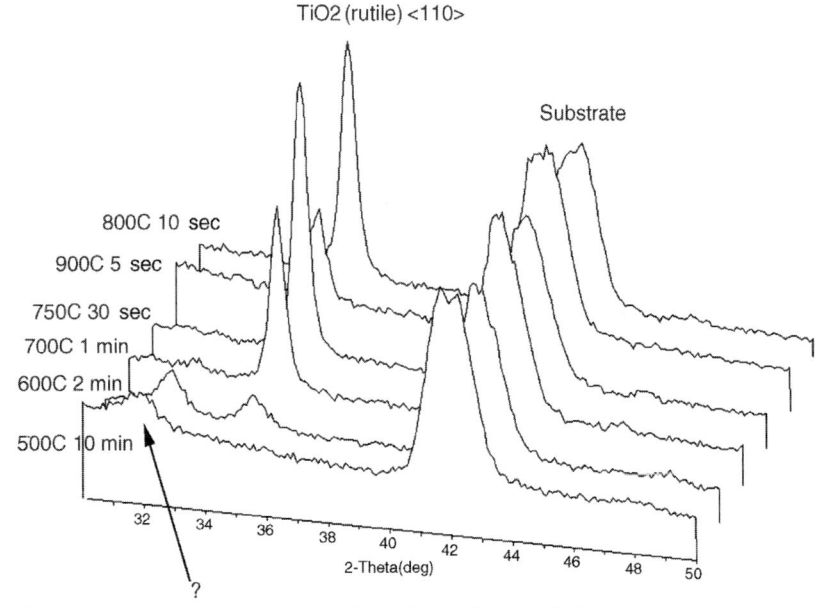

Figure 1.75 XRD-spectra. Temperature dependence of TiO$_2$ rutile formation.

3. CVD-TiN-deposition. Alternatively, a direct PECVD TiO$_2$ deposition is possible. However, the conformity of the TiN deposition is superior.

4. Oxidation above 950 °C to form the rutile phase. For this, a complete oxidation of the TiN has to be achieved. This can, for example, be checked by optical *in situ* methods such as SAME, which allows for the determination of the crystallographic orientation, as was explained in Section 1.3.

5. Second barrier layer (Si$_3$N$_4$) + DT fill with highly doped polysilicon + planarization by CMP (chemical mechanical polishing) + oxide collar recess as usual.

6. Etchback of TiO$_2$ in the collar portion using HF/glycerol.

The temperature dependence of rutile formation from such a TiN was measured by X-ray diffraction. The results are shown in Figure 1.75.

1.5
Summary and Conclusions

A whole variety of technologies exist for passive and integrated circuit (IC) capacitor fabrication. There is however one common denominator for all these different technologies: they all make use of deliberately chosen *valve metal, Si and ceramic oxides* as the dielectric thin films, which therefore *play a key role in electronics*. The reasons for this and the underlying research issues were discussed in this chapter.

In order to maximize the capacitance C, generally an increase of the capacitor plate surface area A, the use of higher dielectric constant ε_r materials and decreasing dielectric film thickness d_f is pursued (simple capacitor equation $C = \varepsilon_r A/d_f$). However, there is no single, cost effective technology available yet that simultaneously combines an optimum of all these parameters. The highest capacitance volume efficiency is achieved with electrolytic capacitors such as Ta capacitors. It was shown that the major reason for this is due to the application of a sinter process, which allows the formation of a three-dimensional, sponge-like capacitor plate structure with extremely large surface area (several $m^2 g^{-1}$). A thin dielectric film (15 to 200 nm) is formed directly on top of this porous substrate by electrochemical anodization.

For integration of capacitors into ICs, 'deep trench' structures can be used. The dielectric oxides have to be compatible with the integration process. The integration scheme of IC fabrication was described using DRAM technology as an example. The extremely sophisticated IC integration schemes usually exclude the application of fancy high ε_r materials, such as the ceramic perovskites. In order to keep the capacitance high, ultra-thin (nano) dielectric films must be employed. For SiO_2 or Si_3N_4, the film thickness rapidly approaches the minimum tunnel limit of <3 nm. Owing to the thin dielectric film thickness for IC and electrolytic capacitors, the oxides have to withstand extremely high electric fields even at moderate applied voltages (e.g. $E = 10^8 V m^{-1}$ for $d_f = 100$ nm and $U_{applied} = 10$ V). Moreover, they reliably have to endure thermal (typically -40 up to $125\,°C$) and mechanical stresses over a minimum device lifetime of between 5 and 20 years. The demands are thus for an optimized combination of material properties, which have been discussed and studied.

Some of the key properties of the various oxide systems are summarized and compared in the band models of Figure 1.76. In order to tolerate high electric field strengths for a long time, the dielectric film has to have an amorphous structure. This is due to the continuum of localized states within the mobility gap, which act as traps for the electrons and therefore prevent critical electron acceleration up to a level where dielectric breakdown (avalanche ionization) occurs. Therefore, all established ultra-thin film dielectrics, namely SiO_2, Al_2O_3 and Ta_2O_5 show uniform amorphous structures. This 'knock out' criterion for ultra-thin film application is also fulfilled by Nb_2O_5 and TiO_2. However, the latter systems tend to form sub-oxides (e.g. NbO, Nb_2O and the Magnéli phases in the case of TiO_2) resulting in non-uniform oxide properties. The oxygen vacancies act as donors in the mobility gap. Consequently, these oxides show a pronounced n-type semiconducting behavior.

By Schottky–Mott analysis of the bias dependent capacitance measurements a quantitative determination of the donor defect state concentration N_D is possible. For the anodically formed Nb_2O_5 and TiO_2 films, N_D is typically in the order of 10^{20} cm^{-3}. These systems therefore show higher leakage currents than the insulating Ta_2O_5, SiO_2 and Al_2O_3 films. Moreover, Nb_2O_5 and TiO_2 are prone to dislocation formation (local crystallization), which further increases the leakage current. In the case of TiO_2 even crystallization to the rutile or anastase phase can occur at high formation potentials above 50 V.

The valve metal oxide systems ZrO_2 and HfO_2 and most of the typical electro-ceramic perovskite compounds such as $BaTiO_3$ and $(Ba,Sr)TiO_3$ (BST) show an

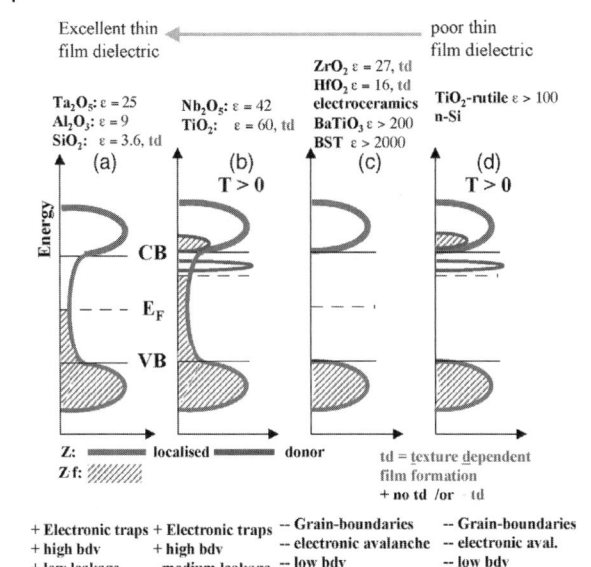

Figure 1.76 Band models of the various oxide systems. The oxides can be placed into four groups regarding their suitability as ultra-thin dielectric film. Performance degrades from left to right. (a) Ta_2O_5, Al_2O_3 and SiO_2 show an amorphous structure. This is essential for ultra-thin dielectric films, as the continuum of localized states in the mobility gap act as electron traps. This prevents electron avalanche breakdown resulting in highest bdv (breakdown voltage). (b) Nb_2O_5 and TiO_2 are also amorphous, but show a high donor concentration due to oxygen vacancies. These oxides therefore show n-type semiconducting behavior with an increased leakage. (c) ZrO_2, HfO_2 and perovskite electroceramics usually show crystalline structures with a high density of grain boundaries that act as high leakage paths. Owing to the reduced bdv, these materials are preferentially used for thicker film applications (>1 μm), e.g. MLCC technology. (d) Rutile TiO_2 with oxygen vacancies shows the worst combination of electronic properties for dielectric applications. It behaves as n-type Si. Z, density of states; Zf, occupied states; CB and VB, conduction and valence band edge; E_F, Fermi energy.

excellent insulating behavior with no bias dependence of capacitance at all. However, as illustrated in the band model of Figure 1.76, they usually form crystalline oxides, which are prone to dielectric breakdown at high field strengths. Therefore, it is difficult to integrate these systems into technologies such as integrated circuits where single digit nm thick films are needed. However, they are widely used for multilayer ceramic capacitors (MLCC) where thicker dielectric films (>1 μm) are acceptable.

The presented electrochemical measurements prove that formation conditions sensitively affect the electronic film properties. Consequently, the formation conditions can be used for an adjustment and control of electronic properties to a certain extent. For instance, in the case of Ti/TiO_2 changing from potentiodynamic to potentiostatic formation, conditions allows for significant reduction of the donor defect state concentration. Moreover, under potentiodynamic conditions, a pronounced texture dependence of oxide growth is observed, which can be significantly suppressed by potentiostatic formation. The texture dependence itself was investigated by means of the new nl-photoresist droplet method, which allows all types of electrochemical measurements to be performed at high lateral resolution, for example, on single substrate grains. Simultaneously, optical methods were applied to the same single substrate grains. With the new 'Spectroscopic Anisotropy Micro-Ellipsometry' method (SAME) crystallographic properties such as the surface orientation of substrate grains and crystallization of anodically formed films and their epitaxial relationship to the substrate grains could be determined. It was shown that SAME, along with the new nl-photoresist method, is a powerful tool for studying of texture dependent anodic oxide growth.

References

1 Kaesche, H. (1990) *Die Korrosion der Metalle*, 3rd edn, Springer, Berlin.
2 Michaelis, A., Kudelka, S. and Schultze, J.W. (2000) *Corros. Rev.*, **18**, 395.
3 Hurlen, T. and Hjornkjol, S. (1991) *Electrochim Acta*, **36**, 189.
4 Grätzel, M. and Rotzinger, F.P. (1985) *Chem. Phys. Lett.*, **118**, 474.
5 Burleigh, T.D. (1989) *Corrosion*, **45**, 465.
6 Kudelka, S.Ph.D. Thesis (1996) Shaker Publishing, University of Düsseldorf.
7 Michaelis, A.Ph.D. Thesis (1994) Shaker Publishing, University of Düsseldorf.
8 Leach, J.S.L. and Pearson, B.R. (1988) *Corros. Sci.*, **28**, 43.
9 Charlesby, A. and Polling, J.J. (1954) *Proc. Phys. Soc. London, Ser. B*, **67**, 201.
10 Davies, J.-A., Domeij, B., Pringle, J.P.S. and Brown, F. (1965) *J. Electrochem. Soc.*, **112**, 675.
11 Schultze, J.W. and Macanio, V.A. (1986) *Electrochim. Acta*, **33**, 355.
12 Young, L. (1954) *Trans. Faraday Soc.*, **50**, 159.
13 Macagno, V. and Schultze, J.W. (1984) *J. Electroanal. Chem.*, **180**, 157.
14 Randall, J.J., Bernard, W.J. and Wilkinson, R.R. (1965) *Electrochim. Acta*, **10**, 183.
15 Whitton, J.L. (1968) *Electrochem. Soc.*, **115**, 58.
16 Aladjem, A., Brandon, D.G. and Yahalom, J. (1970) *Electrochim. Acta*, **15**, 663.
17 Schweinsberg, M. (1998) Ph.D. Thesis, Shaker Publishing, University of Düsseldorf.
18 Patzelt, T. (1993) Ph.D Thesis, Shaker Publishing, University of Düsseldorf.
19 Young, L. (1956) *Trans. Faraday Soc.*, **52**, 502.
20 Lohrengel, M.M. (1993) *Mater. Sci. Eng. R*, **11**, 243.
21 Ebling, D. (1991) Ph.D. Thesis, University of Düsseldorf.
22 Jung, D.Y. and Metzger, N. (1985) *J. Electrochem. Soc.*, **132**, 386.
23 Virchheim, R. (1989) *Corr. Sci.*, **29**, 183.
24 Dyer, C.K. and Alwitt, R.S. (1978) *Electrochim. Acta*, **23**, 347.
25 Plumb, R.C. (1958) *J. Electrochem. Soc.*, **105**, 498.
26 Schnyder, B. and Kötz, R. (1992) *J. Electroanal. Chem.*, **339**, 167.
27 Diggle, J.W., Downie, T.C. and Goulding, C.W. (1969) *Chem. Rev.*, **69**, 365.
28 Bartels, C. (1982) Ph.D. Thesis University of Düsseldorf.
29 Homkjol, S. (1988) *Electrochem. Acta*, **33**, 337.
30 Brown, F. and Macintosh, W.D. (1973) *J. Electrochem. Soc.*, **Z20**, 1096.
31 Bartels, C., Schultze, J.W., Stimming, U. and Habib, M.A. (1982) *Electrochim. Acta*, **27**, 129.
32 Kirchhoff, M. Ph.D. Thesis, Shaker Publishing, University of Düsseldorf.
33 Irene, E.A. and Ghez, R. (1987) *Appl. Surf. Sci.*, **30**, 1.
34 Liu, Q., Wall, J.F. and Irene, E.A. (1994) *J. Vac. Sci. Technol., A*, **12**, 2625.

35 Poler, J.C., McKay, K.K. and Irene, E.A. (1994) *J. Vac. Sci. Technol., B*, **12**, 88.
36 Chongsawangvirod, S., Irene, E.A., Kalnitsky, A., Tay, S.P. and Ellul, J.P. (1990) *J. Electrochem. Soc.*, **137**, 3536.
37 Güntherschulze, A. and Betz, H. (1931) *Z. Elektrochem.*, **37**, 726.
38 Di Quarto, D., Di Paola, A. and Sunseri, C. (1981) *Electrochim. Acta*, **26**, 1177.
39 Möhring, A. University of Düsseldorf, invention of contact cell.
40 West, M. (ed.) (1977) *Lasers in Chemistry*, Elsevier, Amsterdam.
41 Friedrich, K.A. and Richmond, G.L. (1993) *Ber. Bunsenges.*, **97**, 389.
42 Dhamelincourt, P. and Bisson, P. (1977) *Microsc. Acta*, **79**, 267.
43 Acher, O., Bigan, E. and Drevillon, B. (1989) *Rev. Sci. Instrum.*, **60**, 65.
44 Kötz, R. (1993) *Ber. Bunsenges.*, **97**, 427.
45 White, C.D. (1990) *Anal. Chem.*, **62** (11), 1133.
46 Butler, A.M. (1983) *J. Electrochem. Soc.*, **130**, 2358.
47 Butler, A.M. (1984) *J. Electrochem. Soc.*, **131**, 2185.
48 Schultze, J.W., Bade, K. and Michaelis, A. (1991) *Ber. Bunsenges. Phys. Chem.*, **95**, 1349.
49 Leitner, K. and Schultze, J.W. (1988) *Ber. Bunsenges. Phys. Chem.*, **92**, 181.
50 Schultze, J.W. and Thietke, J. (1989) *Electrochim. Acta*, **34**, 1769.
51 Bade, K., Karstens, O., Michaelis, A. and Schultze, J.W. (1992) *Faraday Discuss. R. Soc. Chem.*, **94/6**.
52 Dignam, M.J. (1981) *Comprehensive Treatise of Electrochemistry* (ed. J.O.M. Bockris), Plenum Press, New York, pp. 247–306.
53 Güntherschulze, A. and Betz, H. (1934) *Z. -Phys.*, **92**, 367.
54 Verwey, E.J.W. (1935) *Physica*, **2**, 1059.
55 Cabrera, N. and Mott, N.F. (1948–1949) *Rep. Progr. Phys.*, **12**, 163.
56 Sato, N. and Cohen, M. (1964) *J. Electrochem. Soc. Ill*, 512.
57 Chao, C.Y., Lin, L.F. and Macdonald, D.D. (1981) *J. Electrochem. Soc.*, **128**, 1187.
58 Vetter, K.J. (1961) *Elektrochemische Kinetik*, Springer, Berlin.
59 Fehlner, F.P. and Mott, N.F. (1970) *Oxide. Met.*, **2**, 59.
60 Burstein, G.T. and Davenport, A.J. (1989) *J. Electrochem. Soc.*, **136**, 936.
61 Davenport, A.J. and Burstein, G.T. (1990) *J. Electrochem. Soc.*, **137**, 1496.
62 Mott, N.F. and Davis, E.A. (1979) *Electronic Processes in Nanocrystalline Solids*, Clarendon Press, Oxford.
63 Carlslav, H.S. and Jaeger, J.C. (1959) *Conduction of Heat in Solids*, Oxford United Press, New York.
64 Hillenkamp, F. (1989) *Microbeam Anal.*, 277.
65 Yoneyama, H., Kawai, K. and Kuwabata, S. (1988) *J. Electrochem. Soc.*, **128**, 1699.
66 Friedrich, F. and Raub, C.J. (1984) *Metalloberfläche*, **38**, 237.
67 Khan, H.R., Kittel, M. and Raub, C.J. (1988) *Surf. Coat. Technol.*, **35**, 215.
68 Benderski, V.A. and Velichko, G.I. (1982) *J. Electroanal. Chem.*, **140**, 1.
69 Schultze, J.W. (1982) *Stimming, Weise, Ber. Bunsenges. Phys. Chem.*, 233.
70 Di Quarto, F., Piazza, S., D'Agostino, R. and Sunseri, C. (1989) *Electrochim. Acta*, **34**, 321; Di Quarto, F., Piazza, S. and Sunseri, C. (1987) *Ber. Bunsenges. Phys. Chem.*, **91**, 437.
71 Di Quarto, F., Piazza, S. and Sunseri, C. (1989) *J. Chem. Soc. Faraday Trans. I*, **85**, 3309.
72 Sukamto, J.P.H., Smyrl, W.H., McMillan, C.S. and Kozlowski, M.R. (1992) *J. Electrochem. Soc.*, **130**, 265.
73 Sukamto, J.P.H., McMillan, C.S. and Smyrl, W.H. (1993) *Electrochim. Acta*, **38**, 15.
74 Michaelis, A. and Schultze, J.W. (1996) *Appl. Surf. Sci.*, **106**, 483.
75 Gärtner, W.W. (1959) *Phys. Rev.*, **116**, 84.
76 Di Quarto, F., Piazza, S. and Sunseri, C. (1993) *Electrochim. Acta*, **38**, 29.
77 Pai, D.M. and Enck, R.C. *Phys. Rev. B*, **11**, 5136.
78 Kudelka, S., Michaelis, A. and Schultze, J.W. (1995) *Electrochim. Acta*, **40**, 983.

79 Schultze, J.W., Bade, K. and Michaelis, A. (1991) *Ber. Bunsenges. Phys. Chem.*, **95**, 1349.

80 Kudelka, S., Michaelis, A. and Schultze, J.W. (1995) *Ber. Bunsenges. Phys. Chem.*, **99**, 1020.

81 Michaelis, A. and Schultze, J.W. (1996) *Thin Solid Films*, **274**, 82.

82 Azzam, R.M.A. and Bashara, N.M. (1977) *Ellipsometry and Polarized Light*, North-Holland, Amsterdam.

83 Erman, M. and Theeten, J.B. (1986) *J. Appl. Phys.*, **60** (3), 859.

84 Michaelis, A. and Schweinsberg, M. (1998) *Thin Solid Films*, **313**, 756.

85 Michaelis, A., Delplancke, J.L. and Schultze, J.W. (1995) *Mater. Sci. Forum*, **185–188**, 471.

86 Kudelka, S., Michaelis, A. and Schultze, J.W. (1996) *Electrochim. Acta*, **41**, 863.

87 Bard, A.J., Fan, F.R. and Mirkin, M.V. (1994) *Electroanal. Chem.*, **18**, 244.

88 Kudelka, S. (1996) Dissertation, Verlag Shaker, University of Düsseldorf.

89 Brunn, P.O., Fabrikant, V.I. and Sankar, T.S. (1984) *Quart. J. Mech. Appl. Math.*, **37**, 311.

90 Schultze, J.W., Kudelka, S., Davepon, B. and Krumm, R. (1997) In Passivity and its Breakdown, Electrochemical Society Proceedings Volume 97–26, pp. 725–739.

91 Kudelka, S. and Schultze, J.W. (1997) *Electrochim. Acta*, **42**, 2817.

92 Kudelka, S., Michaelis, A., Smailos, E. and Schultze, J.W. (1996) *Metalloberflaeche*, **50** (5), 369.

93 Meisterjahn, P. (1988) Ph.D. Thesis, University of Düsseldorf.

94 Whitton, J.L. (1968) *J. Electrochem. Soc.*, **115** (1), 58.

95 Khalis, N. and Leach, J.S.L. (1986) *Electrochim. Acta*, **31** (10), 1279.

96 Leach, J.S. and Pearson, B.R. (1984) *Electrochim. Acta*, **29** (9), 1271.

97 Weaver, J.H., Krafka, C., Lynch, D.W. and Koch, E.E. Optical Properties of Solids in Physics Data, Surf. Sci. Index, ed. E. Umbach, D. Menzel.

98 Schweinsberg, M., Michaelis, A. and Schultze, J.W. (1997) *Electrochim. Acta*, **42**, 3303.

99 Kukli, K., Ritala, M. and Leskela, M. (1999) *J. Appl. Phys.*, **86** (10), 5656.

100 Kattelus, H., Ronkainen, H., Kanniannen, T. and Skarp, J. (1988) Proceedings of 28th European Solid State Device Conference ESSDERC 98, Bordeaux, p. 444.

101 Kukli, K., Ihanus, J., Ritala, M. and Leskela, M. (1996) *Appl. Phys. Lett.*, **68**, 3737.

102 Kukli, K., Ihanus, J., Ritala, M. and Leskela, M. (1997) *J. Electrochem. Soc.*, **144**, 300.

103 Bartels, C. (1982) Ph.D. Thesis, University of Düsseldorf.

104 Smyth, D.M. (2000) *Solid State Ionics*, **129**, 5.

105 Modestov, A.D. and Davydov, A.D. (1999) *J. Electroanal. Chem.*, **460**, 214.

106 Dyer, C.K. and Leach, J.S.L. (1975) *Electrochim. Acta*, **20**, 151.

107 Heusler, K.E. and Schlüter, P. (1969) *Werkst. Korr.*, **20** (3), 195.

108 Di Quarto, F., Piazza, S. and Sunseri, C. (1990) *Electrochim. Acta*, **35** (1), 99.

109 Gray, K.E. (1975) *Appl. Phys. Lett.*, **27** (8), 462.

110 Di Quarto, F., Piazza, S., Dàgostino, R. and Sunseri, C. (1987) *J. Electroanal. Chem.*, **228**, 119.

111 Heusler, K.E. and Schulze, M. (1975) *Electrochim. Acta*, **20**, 37.

112 Smyth, D.M. and Tripp, T.B. (1966) *J. Electrochem. Soc.*, **113** (10), 1048.

113 Al-Kharafi, F.M. and Badaway, W.A. (1995) *Electrochim. Acta*, **40** (16), 2623.

114 El-Mahdy, G.A. (1997) *Thin Solid Films*, **307**, 141.

115 Rama Devi, D. and Sastry, K.S. (1991) *Bull. Electrochem.*, **7** (2), 73.

116 Badawy, W.A., Gad-Allah, A.G. and Rehan, H.H. (1987) *J. Appl. Electrochem.*, **17**, 559.

117 Störmer, H. and Gerthsen, D. University of Karlsruhe, LEM.

118 Vermilyea, D.A. (1965) *J. Appl. Phys.* **36** (11), 3663; Vermilyea, D.A. (1965) *J. Phys. Chem. Solids*, **26**, 133.
119 Gad-Allah, A.G., Abd El-Rahman, H.A. and Abou-Romia, M.M. (1988) *J. Appl. Electrochem.*, **18**, 532.
120 Treichel, H., Mitwalsky, A., Tempel, G., Zorn, G., Bohling, D.A., Coyle, K.R., Felker, B.S., George, M., Kern, W., Plane, A.P. and Sandler, N.P. (1995) *Adv. Mater. Opt. Electron.*, **5**, 163.
121 Ord, J.L., Hopper, M.A. and Wang, W.P. (1972) *J. Electrochem. Soc.*, **119** (4), 439.
122 Odynets, L.L. (1984) *Elektrokhimiya.*, **20** (4), 463.
123 Dyakonov, M.N., Kostelova, L.A., Novotelnova, A.V. and Khanin, S.D. (1984) *Sov. Tech. Phys. Lett.*, **10** (3), 115.
124 Malygin, V.V., Shatalov, A.Y. and Kuchin, S.V. (1968) *Elektrokhimiya*, **4** (10), 1194.
125 Huang, B.X., Wang, K., Church, J.S. and Li, Y. (1999) *Electrochim. Acta*, **44**, 2571.
126 Randall, J.J., Bernard, W.J. and Wilkinson, R.R. (1965) *Electrochim. Acta*, **10**, 183.
127 Ovcharenko, V.I. and Viet-Ba, L. (1973) *Elektrokhimiya.*, **9** (11), 1618.
128 Ord, J.L. and Lushiku, E.M. (1979) *J. Electrochem. Soc.*, **126** (8), 1374.
129 Young, L. (1960) *Can. J. Chem.*, **38**, 1141.
130 Hand, R.B., Ling, H.W. and Kolski, T.L. (1961) *J. Electrochem. Soc.*, **108** (11), 1023.
131 Ammar, I.A., Darwish, S. and Ammar, E.A. (1973) *Metalloberfläche Angew. Elektrochem.*, **27** (5), 163.
132 Ammar, I.A. and Ismail, I.K. (1972) *Metalloberfläche Angew. Elektrochem.*, **26** (10), 378.
133 Kalra, K.C., Singh, K.C. and Singh, M. (1997) *Indian J. Pure Appl. Phys.*, **35**, 39.
134 Cavigliasso, G.E., Esplandiu, M.J. and Macagno, V.A. (1998) *J. Appl. Electrochem.*, **28**, 1213.
135 Malyuk, Y.I. and Chernyakova, L.E. (1973) *Ukr. Khim. Zh.*, **39** (1), 45.
136 Bairachnyi, B.I., Andryushchenko, F.K. and Yaroshok, T.P. (1983) *Elektrokhimiya*, **19** (8), 1114.
137 Bokii, L.P., Danilyuk, Y.L., Dyanokov, M.N., Kotousova, I.S., Mirzoev, R.A., Muzhdaba, V.M., Rozenberg, L.A. and Khanin, S.D. (1979) *Elektrokhimiya*, **15** (9), 1307.
138 Cherki, C. (1971) *Electrochim. Acta*, **16**, 1727.
139 Melody, B., Kinard, T. and Lessner, P. (1998) *Electrochem. Solid State Lett.*, **1** (3), 126.
140 Heusler, K.E. (1970) *Z. Metallkde.*, **61** (11), 828.
141 Palatnik, L.S., Pozdeev, Y.L., Naboka, M.N. and Starikov, V.V. (1994) *Tech. Phys. Lett.*, **20** (5), 382.
142 Johansen, H.A., Adams, G.B. and Rysselberghe, P.V. (1957) *J. Electrochem. Soc.*, **104** (6), 339.
143 Matthews, C.G., Ord, J.L. and Wang, W.P. (1983) *J. Electrochem. Soc.*, **130**, 285.
144 Karla, K.C., Singh, K.C. and Singh, M. (1998) *J. Indian. Chem. Soc.*, **75**, 67.
145 Dyer, C.K. and Leach, J.S.L. (1978) *J. Electrochem. Soc.*, **125** (1), 23.
146 Vaskevich, A., Rosenblum, M. and Gileadi, E. (1995) *J. Electrochem. Soc.*, **142** (5), 1501.
147 Pergament, A.L. and Stefanovich, G.B. (1998) *Thin Solid Films*, **322**, 33.
148 Freitas, M.B.J.G. and Bulhoes, L.O.S. (1997) *J. Appl. Electrochem.*, **27**, 612.
149 Hornkjol, S. (1990) *Acta Chem. Scand.*, **44**, 404.
150 Boodts, J.F.C., Sluyters-Rehbach, M. and Sluyters, J.H. (1986) *J. Electroanal. Chem.*, **206**, 153.
151 Rama Devi, D., Kalyani, G. and Sastry, K.S. (1989) *J. Electrochem. Soc. India*, **38** (4), 291.
152 Nigam, R.K., Singh, K.C. and Maken, S. (1988) *Indian J. Chem.*, **27**, 390.
153 Levinskii, Y.V., Zaitsev, A.B., Bocharova, V.I., Novikova, S.M., Polyakov, Y.M. and Patrikeev, Y.B. (1991) *Porosh. Metallurg.*, **339** (3), 216.
154 Jackson, N.F. and Hendy, J.C. (1974) *Electromp. Sci. Technol.*, **1**, 27.
155 Kathirgamanathan, P. and Ravichandran, S. (1995) *Synth. Metals*, **74**, 165.

156 Shtasel, A. and Knight, H.T. (1961) *J. Electrochem. Soc.*, **108** (4), 343.
157 Schwartz, N., Gresh, M. and Karlik, S. (1961) *J. Electrochem. Soc.*, **108** (8), 750.
158 Ling, H.W. and Kolski, T. (1962) *J. Electrochem. Soc.*, **109** (1), 69.
159 Ling, H.W. and Kolski, T.L. (1963) *Electrochem. Techn.*, **1** (4), 92.
160 Pozdeev, Y., Carts-Europe 97 Proceedings.
161 Levinskii, Y.V., Zaitseev, A.B., Polyakov, Y.M. and Patrikeev, Y.B. (1990) *Porosh. Metallurg.*, **327** (3), 197.
162 Levinskii, Y.V., Zaitseev, A.B., Polyakov, Y.M. and Patrikeev, Y.B. (1990) *Porosh. Metallurg.*, **329** (5), 29.
163 Levinskii, Y.V., Zaitseev, A.B., Polyakov, Y.M. and Patrikeev, Y.B. (1990) *Porosh. Metallurg.*, **331** (7), 23.
164 Levinskii, Y.V., Zaitseev, A.B., Bocharova, V.I., Novikova, S.M., Polyakov, Y.M. and Patrikeev, Y.B. (1991) *Porosh. Metallurg.*, **341** (5), 61.
165 Lohrengel, M.M. (1991) *Electrochemical and Optical Techniques for the Study and Monitoring of Metallic Corrosion* (eds M.G.S. Ferreira and C.A. Melendres), Kluwer Press, Dordrecht, p. 69.
166 Young, L. (1961) *Anodic Oxide Films*, Academic Press, New York.
167 Udupa, H.V.K. and Venkatesan, V.K. (1974) *Encyclopedia of Electrochemistry of the Elements*, **vol.2** (ed. A.J. Bard), Marcel Dekker, New York, p. 53.
168 Bispinck, H., Ganschow, O., Wiedmann, L. and Benninghoven, A. (1979) *J. Appl. Phys.*, **18**, 113JF.
169 Vermilyea, D.A. (1953) *Acta Metall.*, **1**, 282.
170 Vetter, K.J. (1961) *Elektrochemische Kinetik*, Springer Verlag, Berlin, Göttingen, Heidelberg.
171 Schultze, J.W., Lohrengel, M.M. and Ross, D. (1983) *Electrochim. Acta*, **28**, 984.
172 Smith, D.J. and Young, L. (1983) *Thin Solid Films*, **101**, 11.
173 Heusler, K.E. and Schulze, M. (1975) *Electrochim. Acta*, **20**, 237.
174 Dewald, J.F. (1955) *J. Electrochem. Soc.*, **102**, 1.
175 Galizzioli, D. and Trasatti, S. (1973) *J. Electroanal. Chem.*, **44**, 367.
176 Lohrengel, M.M., Schubert, K. and Schultze, J.W. (1981) *Werkst. Korr.*, **32**, 13.
177 Tegart, W.J. (1956) *The Electrolytic and Chemical Polishing of Metals*, Pergamon Press, London.
178 Habib, M.A., Bartels, C., Schultze, J.W. and Stimming, U. (1982) *Electrochim. Acta*, **27**, l29.
179 Pettinger, B., Schöppel, H.-R., Yokoyama, T. and Gerischer, H. (1974) *Ber. Bunsenges. Phys. Chem.*, **78**, 1024.
180 DeGryse, R., Gomes, W.P., Cardon, F. and Vennik, J. (1975) *J. Electrochem. Soc.*, **122**, 711.
181 Schultze, J.W. and Mohr, S. (1981) *Dechema-Monograph.*, **90**, 213.
182 Schultze, J.W., Lohrengel, M.M., Ross, D. and Stimming, U. (1980) *Chem.-Ing.-Tech.*, **52**, 447.
183 Butler, M.A. and Ginley, D.S. (1978) *J. Electrochem. Soc.*, **125**, 228.
184 Pozdeev-Freeman, Y. (1999) *Carts-Europe*, 24.
185 Lin, J., Masaaki, N., Tsukune, A. and Yamada, M. (1999) *Appl. Phys. Lett.*, **74** (16), 2370.
186 Sikula, J., Pavelka, J., Vasina, P., Zednicek, T. and Hashiguchi, S. (1999) *Carts-Europe*, 147.
187 Shimizu, K. and Kobayashi, K. (1996) *Philos. Mag.*, **73B** (3), 461.
188 Desu, C.S., Joshi, P.C. and Desu, S.B. (1999) *Mater. Res. Innovat.*, **2**, 299.
189 Clem, P.G., Jeon, N.-L., Nuzzo, R.G. and Payne, D.A. (1997) *J. Am. Ceram. Soc.*, **80** (11), 2821.
190 Li, Y.-M. and Young, L. (1998) *Electrochim. Acta*, **44**, 605.
191 Lai, B.C. and Lee, J.Y. (1999) *J. Electrochem. Soc.*, **146** (1), 266.
192 Naamoune, F., Hammouche, A. and Kahoul, A. (1998) *J. Chim. Phys.*, **95**, 1640.
193 Levinskii, Y.V., Zaitsev, A.B. and Blagoveshchenskii, Y.B. (1992) *Porosh. Metallurg.*, **360** (12), 11.

194 Shimizu, K., Kobayashi, K., Thompson, G.E., Skeldon, P. and Wood, G.C. (1997) *J. Electrochem. Soc.*, **144** (2), 418.

195 Aoyama, T., Saida, S., Okayama, Y., Fujisaki, M., Imai, K. and Arikado, T. (1996) *J. Electrochem. Soc.*, **143** (3), 977.

196 Gusakov, A.G., Raspopov, S.A., Vecher, A.A. and Voropayew, A.G. (1993) *J. Alloys Comp.*, **201**, 67.

197 Haneczok, G., Weller, M. and Diehl, J. (1994) *J. Alloys Comp.*, **211/212**, 71.

198 Ivanovskii, A.L. and Mryasov, O.V. (1993) *Inorg. Mater.*, **29** (9), 1257.

199 Surov, Y.I., Troitskii, A.V., Povolotskii, E.G., Sherstobitova, O.M. and Onanova, N.Sh. (1991) *Izv. Akad. SSSR*, **27** (8), 1771.

200 Khanin, S.D. (1995) *Mater. Sci. Forum*, **185–188**, 573.

201 Gebhardt, E., Seghezzi, H.-D. and Keil, H. (1963) *Z. Metallkde.*, **54** (1), 31.

202 Nickerson, W. and Altstetter, C. (1973) *Scripta Metallurg.*, **7**, 377.

203 Pozdeev-Freeman, Y., Rodenberg, Y., Gladkikh, A., Karpovski, M. and Palevski, A. (1998) *J. Mater. Sci. Electron.*, **9**, 1.

204 Pozdeev-Freeman, Y., Gladkikh, A., Karpovski, M. and Palevski, A. (1998) *J. Electron. Mater.*, **27** (9), 1034.

205 Trifonov, I. (1980) *Acta Phys. Acad. Sci. Hung.*, **49** (1–3), 269.

206 Pringle, J.P.S. (1973) *J. Electrochem. Soc.*, **120** (10), 1391.

207 Wyatt, P.W. (1975) *J. Electrochem. Soc.*, **122** (12), 1660.

208 Ord, J.L., Clayton, J.C. and Wang, W.P. (1977) *J. Electrochem. Soc.*, **124** (11), 1671.

209 Young, L. (1977) *J. Electrochem. Soc.*, **124** (4), 528.

210 Kadary, V. and Klein, N. (1980) *J. Electrochem. Soc.*, **127** (1), 139.

211 Young, L. and Smith, D.J. (1979) *J. Electrochem. Soc.*, **126** (5), 765.

212 Propp, M. and Young, L. (1979) *J. Electrochem. Soc.*, **126** (4), 624.

213 Kimura, S.-I., Nishioka, Y., Shintani, A. and Mukai, K. (1983) *J. Electrochem. Soc.*, **130** (12), 2414.

214 Smith, D.J. and Young, L. (1982) *J. Electrochem. Soc.*, **129** (11), 2513.

215 Klein, G.P. (1973) *J. Electrochem. Soc.*, **120** (12), 1789.

216 Mohler, D. and Hirst, R.G. (1961) *J. Electrochem. Soc.*, **108** (4), 347.

217 Smyth, D.M. and Shirn, G.A. (1968) *J. Electrochem. Soc.*, **115** (2), 186.

218 Aoyama, T., Saida, S., Okayama, Y., Fujisaki, M., Imai, K. and Arikado, T. (1996) *J. Electrochem. Soc.*, **143** (3), 977.

219 Smyth, D.M., Shirn, G.A. and Tripp, T.B. (1963) *J. Electrochem. Soc.*, **110** (12), 1264.

220 Smyth, D.M. and Tripp, T.B. (1963) *J. Electrochem. Soc.*, **110** (12), 1271.

221 Smyth, D.M., Shirn, G.A. and Tripp, T.B. (1964) *J. Electrochem. Soc.*, **111** (12), 1331.

222 Smyth, D.M., Tripp, T.B. and Shirn, G.A. (1966) *J. Electrochem. Soc.*, **113** (2), 100.

223 Young, L., Yang, T.-M. and Backhouse, C. (1995) *J. Electrochem. Soc.*, **142** (10), 3479.

224 Young, L., Yang, T.-M. and Backhouse, C. (1995) *J. Electrochem. Soc.*, **142** (10), 3483.

225 Melody, B. and Kinard, T. (1993) *J. Electrochem. Soc.*, **140** (11), L162.

226 Tripp, T.B. and Foley, K.B. (1990) *J. Electrochem. Soc.*, **137** (8), 2528.

227 Climent, F., Capellades, R. and Gil, J. (1986) *J. Electrochem. Soc.*, **133**, 959.

228 Albella, J.M., Montero, I. and Martinez-Duart, J.M. (1984) *J. Electrochem. Soc.*, **131** (5), 1101.

229 Palatnik, L.S., Pozdeev, Y.-L. and Stavikov, V.V. (1994) *Tech. Phys. Lett.*, **20** (4), 276.

230 Duenas, S., Castan, H., Barbolla, J., Kola, R.R. and Sullivan, P.A. (1999) *Mater. Res. Soc. Symp. Proc.*, **567**, 371.

231 Kerrec, O., Devilliers, D., Groult, H. and Chemla, M. (1995) *Electrochim. Acta*, **40** (6), 719.

232 Oechsner, H. and Schoof, H. (1982) *Thin Solid Films*, **90**, 337.

233 Friedrich, C., Kritzer, P., Boukis, N., Franz, G. and Dinjus, E. (1999) *J. Mater. Sci.*, **34**, 3137.

234 Chang, J.P., Steigerwald, M.L., Fleming, R.M., Opila, R.L. and Alers, G.B. (1999) *Mater. Res. Soc. Symp. Proc.*, **574**, 329.

235 Di Quarto, F., La Mantia, F. and Santamaria, M. (2005) *Electrochim. Acta*, **50**, 5090.

236 Reisch, M. (1998) *Elektronische Bauelemente*, Springer, Berlin.

237 Hähn, R. and Retelsdorf, H.-J. (1984) *Erzmetall.*, **37** (9), 444.

238 Gupta, C.K. and Jena, P.K. (1964) *Trans. AIME*, **230**, 1433.

239 Argent, B.B. and Milne, J.C. (1960) *J. Less-Common Metals*, **2**, 154.

240 Schulze, K. (1981) *J. Metals*, 33.

241 Jere, G.V., Patel, C.C. and Krishnan, V. (1961) *Trans. AIME*, **221**, 866.

242 Krishnamurthy, N., Venkataramani, R. and Garg, S.P. (1984) *Refract. Hard Metals*, 41.

243 Wilhelm, H.A., Schmidt, F.A. and Ellis, T.G. (1966) *J. Metals*, 1303.

244 Laverick, C. (1988) *J. Less-Common Metals*, **139**, 107.

245 Okabe, T.H., Park, I., Jacob, K.T. and Waseda, Y. (1999) *J. Alloys Comp.*, **288**, 200.

246 Suzuki, R.O., Aizawa, M. and Ono, K. (1999) *J. Alloys Comp.*, **288**, 173.

247 Knott, W., Frommeyer, G., Klapdor, A. and Windbiel, D. (1998) *Z. Naturforsch.*, **53b**, 459.

248 Pozdeev-Freemann, Y. (2000) Proceedings of the TIC-meeting, San Francisco.

249 Vermilyea, D.A. (1957) *J. Electrochem. Soc.*, **104** (9), 542.

250 Pawel, R.E. and Campbell, J.J. (1964) *J. Electrochem. Soc.*, **111** (11), 1230.

251 Jackson, N. (1973) *J. Appl. Electrochem.*, **3**, 91.

252 Pozdeev-Freeman, Y. (1998) *Qual. Reliab. Eng. Int.*, **14**, 79.

253 Tierman, M. and Millard, R.J. (1983) Proceedings of 33rd Electronic Components Conference, Orlando, FL.

254 Pozdeev, Y. (1997) Proceedings of 17th Capacitor and Resistor Technology Symposium (CARTS '97), Jupiter, FL, 161.

255 Pozdeev-Freeman, Y., Gladkikh, A., Karpovski, M. and Palevski, A. (1998) *J. Mater. Sci.-Mater. Electron.*, **9**, 307.

256 Pozdeev-Freeman, Y. and Gladkikh, A. (1998) *J. Electron. Mater.*, **27**, 1034.

257 *CRC Handbook of Chemistry and Physics*, CRC Press, Boca Raton, FL, (1984).

258 Poznyak, S.K., Talapin, D.V. and Kulak, A.I. (2005) *J. Electroanal. Chem.*, **579**, 299.

259 Pozdeev, Y. (2000) US Patent 6,010,660, June 4.

260 Pozdeev-Freeman, Y. and Gladkikh, A. (1999) Proceeds of 20th Capacitor and Resistor Technology Symposium (CARTS 2000), Huntington Beach, CA, p. 140.

261 Samsonov, G.V. (ed.) (1978) *Handbook of Physico-Chemical Properties of Oxides (in Russian)*, Metallurgiya, Moscow.

262 Merker, U., unpublished results.

263 Munshi, M.Z.A. (1995) *Solid State Batteries and Capacitors*, World Scientific Press.

264 Bardeen, J. and Brattain, W.H. (1948) *Phys. Rev.*, **74**, 230.

265 Shockley, W. (1949) *Bell Syst. Tech. J.*, **43** (5), 28.

266 Deal, B.E. (1980) *J. Electrochem. Soc.*, **127**, 979.

267 Wolf, S. (1980–1999) *Silicon Processing*, **3 vols.**, Lattice Press, Sunset Beach, CA.

268 Widmann, D., Mader, H. and Friedrich, H. (1996) *Technologie hochintegrierter Schaltungen*, Springer, Berlin.

269 Ruge, I. and Mader, H. (1991) *Halbleiter Technologie*, Springer, Berlin.

270 Fowler, R.H. and Nordheim, L. (1928) *Proc. R. Soc. London Al*, **19**, 173.

271 Gundlach, K.H. (1966) *Solid State Electron.*, **9**, 949.

272 Sze, S.M. (1981) *Physics of Semiconductor Devices*, 2nd edn., John Wiley & Sons, New York.

273 Irene, E.A. (1988) *CRC Crit. Rev. Solid State Mater. Sci.*, **14**, 175.

274 Masejian, J. (1988) *The Physics and Chemistry of SiO_2, and the Si-SiO_2 Interface*, (eds C.R. Helms and B.E. Deal), Plenum Press, New York.

275 Eagleshain, D.J. (Dec 1994) *NMS Bull*, 57.

276 Kareh, B.E., Bronner, G.B. and Schuster, S.E. (1997) *Solid State Technol.*, 89.

277 Muller, K.P., Roithner, K. and Timme, H.J. (1995) *Microelectron. Eng.*, **27**, 457.
278 Muller, K.P. and Roithner, K. (1995) *Proc. Electrochem. Soc.*, **27**, 266.
279 Tabara, S. (1997) *Jpn. J. Appl. Phys.*, **36**, 2508.
280 Tews, H., Michaelis, A., Lee, B. and Schroeder, U. (2001) Selectively Etched Trench with Reduced Trench Widening, US Patent 99-0141.
281 Kudelka, S., Michaelis, A. and Többen, D. (2002) Bottle Shaped Trench Formation by Anisotropic Wet Etch, US Patent 98-0324.
282 Dwyer, J.J.O. (1973) *The Theory of Electrical Conduction and Breakdown in Solid Dielectrics*, Clarendon Press, Oxford, p. 137.
283 Tang, K.S., Lau, W.S. and Samufra, G.S. (1997) *Circuits Devices*, 27.
284 Ishitani, A., Lesaicherre, P., Kamiyama, S., Ando, K. and Watanabe, H. (1983) Trends in Capacitor Dielectrics for DRAMs. *IEICE Trans. Electron.*, **E76-C** (11), 1564.
285 Ando, K., Ishitani, A. and Hamano, K. (1992) *IEEE Electron. Device Lett.*, **13** (7), 372.
286 Shinriki, H. and Nakata, M. (1991) *IEEE Trans. Electron. Devices*, **38** (3), 455.
287 Isobe, C. and Saitoh, M. (1990) *Appl. Phys. Lett.*, **56**, 907.
288 Murawala, P.A., Sawai, M., Tatsuta, T., Tsuji, O., Fujita, S. and Fujita, S. (1993) *Jpn. J. Appl. Phys.*, **32** (1), 368.
289 Kim, S.O., Byun, J.S. and Kim, H.J. (1991) *Thin Solid Films*, **206**, 102.
290 Kim, I., Ahn, S.D., Cho, B.W., Ahn, S.T. and Lee, J.Y. (1994) *Jpn. J. Appl. Phys.*, **33**, 6691.
291 Nagahori, A. and Raj, R. (1995) *J. Am. Ceramic Soc.*, **78**, 1585.
292 Laviale, D., Oberlin, J.C. and Devine, R.A.B. (1994) *Appl. Phys. Lett.*, **65**, 2021.
293 Kamiyama, S. and Saeki, T. (1991) *IEEE Electron. Devices Meeting Tech. Digest*, 827.
294 Kim, I., Kim, J.S., Kwon, O.S., Ahn, S.T., Chun, J.S. and Lee, W.J. (1995) *J. Electron. Mater.*, **24**, 1435.
295 Olsson, C.-O.A., Verge, M.-G. and Landolt, D. (2004) *J. Electrochem. Soc.*, **151** (12), B652.
296 Kamiyama, S., Suzuki, H. and Watanabe, H. (1993) *IEEE Electron. Devices Meeting Tech. Digest*, 49.
297 Shinriki, H. and Nakata, M. (1991) *IEEE Electron. Devices Meeting Tech. Digest*, 455.
298 Matsui, M. and Oka, S. (1988) *Jpn. J. Appl. Phys.*, **27**, 506.
299 Sun, S.C. and Chen, T.F. (1994) *IEEE Electron. Devices Meeting Tech. Digest*, 333.
300 Lau, W.S. and Khaw, K.K. (1996) *Jpn. J. Appl. Phys.*, **35**, 2599.
301 Lau, W.S. and Chu, P.K. (1997) *Jpn. J. Appl. Phys.*, **36**, 661.
302 Shih, B.W., Chen, I.C., Banerjee, S., Brown, G.A. and Bohlmann, J. (1987) *IEEE Electron. Devices Meeting Tech. Digest*, 582.
303 Kwon, K., Kang, C., Park, S.O., Kang, H. and Ahn, S. (1996) *IEEE Electron. Devices Meeting Tech. Digest*, 919.
304 Kim, I., Chun, J.S. and Lee, W. (1996) *Mater. Chem. Phys.*, **44**, 288.
305 Matsuhashi, H. and Nishikawa, S. (1994) *Jpn. J. Appl. Phys.*, **33**, 1293.
306 Michaelis, A. (1998) Rutile Dielectric Material for Semiconductor Devices, US Patent 97-0192.
307 Lee, Y.I., Chan, K.K. and Brady, M.J. (1995) *J. Vac. Sci. Technol., A*, **13** (3), 596.
308 Michaelis, A. (2006) *Advanced Ceramic Oxides for Electronics*, Fraunhofer IRB Verlag, Stuttgart.
309 Kovacs, K., Kiss, G., Stenzel, M. and Zillgen, H. (2003) *J. Electrochem. Soc.*, **150** (8), B361.
310 Bondarenko, A. and Ragoisha, B. (2005) *J. Solid State Electrochem.*, DOI 10.1007.

2
Superconformal Film Growth

Thomas P. Moffat, Daniel Wheeler, and Daniel Josell

2.1
Introduction

Superconformal film growth is a technologically important process that results in the progressive reduction of surface roughness during film growth without the occlusion of voids. Historically, superconformal growth has provided electroplating in particular with a tremendous cost advantage over other film growth methods that required post-deposition processing in order to provide smooth surfaces for functional and/or decorative purposes [1–9]. Depending on the length scale of the surface features in question, several different mechanisms may be involved. The smoothing processes are usually associated with electrolyte additives that cause local inhibition or acceleration of the deposition rate and thereby provide stabilization of smooth planar growth fronts [1–4]. In the electroplating community this behavior has often been described as either leveling and/or brightening, with the responsible additives being referred to as either levelers or brighteners. Additive-derived leveling deals with smoothing surface features that have dimensions of 1–500 µm, while brightening refers to the production of specular surfaces by mechanisms that operate at optical length scales, 1–1000 nm [1–4]. More recently, a variant of the above plating technology, known as 'superfilling' (Figure 2.1), has become a key step in the Damascene process used in the production of on-chip interconnects for integrated circuits [10–12]. The same technology will likely also be important to 3D-integration of circuits by interchip vias, microelectromechanical (MEMS) devices, and other evolving technologies [13, 14]. 'Superfilling' has also been shown to be applicable to CVD processing ushering in a range of new possible applications [15].

In this chapter we will discuss some recent developments in this field as of 2004. Particular emphasis is given to quantitative studies of the connection between electrochemical kinetics and morphological evolution. No attempt has been made to review all of the relevant literature comprehensively, but rather our intent is to highlight a selection of exciting results that deal primarily with the filling of micro- and nanometer scale features.

Advances in Electrochemical Science and Engineering, Vol. 10.
Edited by Richard C. Alkire, Dieter M. Kolb, Jacek Lipkowski and Philip N. Ross
Copyright © 2008 WILEY-VCH Verlag GmbH & Co. KGaA, Weinheim
ISBN: 978-3-527-31317-6

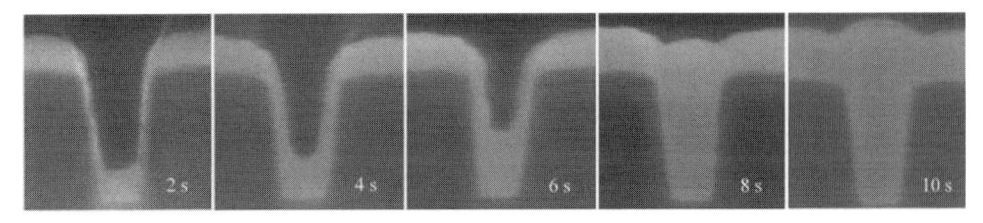

100 nm

Bottom-up filling as a function of time

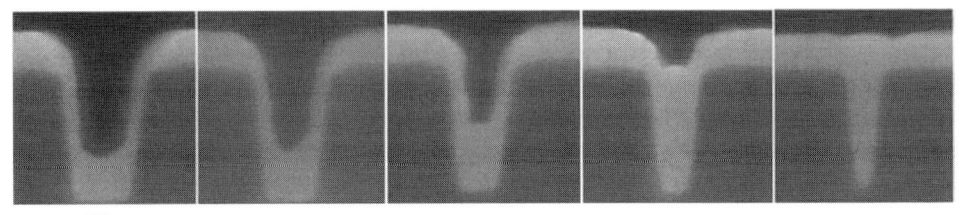

100 nm

Bottom-up filling as a function of aspect ratio

Figure 2.1 Bottom-up superfilling of sub-micrometer trenches by copper electrodeposition from an electrolyte containing PEG–SPS–Cl.

2.2
Destabilizing Influences

As with most phase transformations, establishing and maintaining a planar growth front is subject to a variety of destabilizing influences. Principal among these are the potential and compositional gradients that develop in the electrolyte during electrodeposition [1–4, 16–20]. For two perfectly flat coplanar electrodes of infinite extent, simple steady-state solutions for interface motion by deposition or dissolution are available. However, the slightest disturbance from this electrode geometry due to an interfacial asperity or field perturbation leads to local increases in the field gradients that destabilize the growth front, as indicated in Figure 2.2a. Density-driven convective [21] or electroconvective effects further complicate the picture [22]. All the above can combine to result in non-uniform growth accompanied by rapid amplification of the surface roughness. Several discussions of the current distribution associated with the coupling of electrical and compositional gradients are available in earlier volumes in this series [18, 20] as well as its antecedent [4, 16] and a related series [17]. A series of perturbation studies of interface stability for parallel plate electrochemical cells that consider both supported and unsupported electrolytes are also available [22–36]. At the same time, the dynamic scaling of interface roughness during growth under partial transport control has also received some attention [37–42]. The only stabilizing influence for chemically simple systems is provided by the potential dependent surface tension and surface diffusion. Nonetheless, several

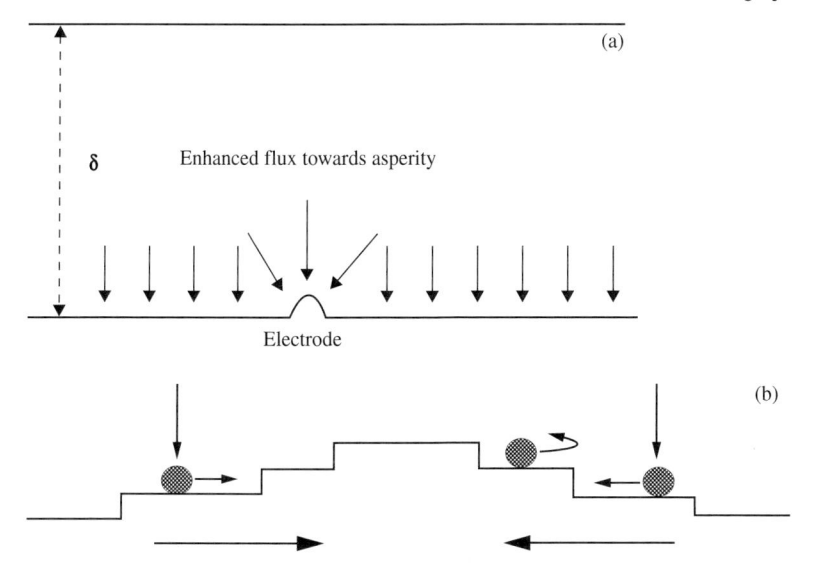

Figure 2.2 (a) The principal source of destabilization of planar growth results from the coupling of a rough surface with either potential or compositional field gradients. (b) Initial perturbations may arise from dynamic electrocrystallization effects. As indicated, a step-edge barrier to intra-layer transport leads to preferential attachment of adatoms to ascending step edges and reflection at descending step edges, which leads to the formation of three-dimensional islands versus layer-by-layer growth (adapted from Ref. [47]).

practical strategies are available to minimize the impact of the destabilizing forces. Supporting electrolytes can be used to minimize the electrical field gradients. Deposition processes can be performed at relatively slow rates, that is under interfacial charge transfer control, so that the concentration field gradients and related inhomogeneties are minimized; alternatively, pulsed potential or pulsed current deposition may be used to periodically relax the corresponding depletion gradients [36]. A first attempt at considering the influence of cyclic growth on the scaling behavior of surface roughness was published recently [44].

In addition to destabilization associated with field gradients in the electrolyte, certain microstructural elements can also give rise to rough surfaces. For example, step bunching during electrocrystallization, in addition to hillock formation, has been widely observed [45], and is most likely associated with the intrinsic step-edge barrier(s) on a given surface [46, 47], as indicated in Figure 2.2b. Such processes have received considerable study in the physical [46, 47] and chemical vapor deposition [48, 49] literature, while more limited reports on step energetics and dynamics in electrochemical systems have recently become available [51, 52]. The accommodation of strain energy associated with misfit stresses, grain boundary formation, etc., provides an additional perturbation source for triggering large-scale roughening during film growth [53, 54].

In this chapter we will describe certain schemes by which some of these destabilizing forces might be checked. Of particular interest is the use of additive adsorption phenomena to control the kinetics of charge transfer reactions in order to give rise to superconformal film growth.

2.3
Stabilization and Smoothing Mechanisms

At least four distinct smoothing and stabilization mechanisms have been identified, namely: (i) geometric leveling, (ii) an inhibitor-based, diffusion–adsorption–consumption mechanism of leveling, (iii) brightening by additive-induced grain refinement and (iv) brightening by shape change driven alterations in catalyst surface coverage [1–4]. Schematic drawings of the respective mechanisms are presented in Figure 2.3.

2.3.1
Geometric Leveling

If electrodeposition occurs under interfacial charge transfer control with negligible reactant depletion and no anisotropic electrocrystallization effects, a uniform current distribution, or growth velocity, can be anticipated on all planar surfaces. This inherently conformal growth mode will give rise to geometric leveling or filling of any surface profile free of reentrant sidewalls [4].

2.3.2
Inhibitor-based Leveling

In contrast to geometric leveling, additive-based 'true' leveling arises from the non-uniform accumulation of an inhibitor on the electrode surface. The differential coverage arises from variable access and consumption of the inhibiting species across a rough surface. For a dilute concentration of leveling agent, the diffusive flux to the surface is maximized at asperities and minimized at recesses. This often results in preferential blocking of the metal deposition at high points and selective filling of surface depressions. To avoid saturation of the surface coverage and thus sustain the gradients that are key to the leveling process, the adsorbed inhibitor must be consumed or deactivated at a finite rate. The leveling effect is maximized for features whose dimensions, depth and lateral spacing approach the hydrodynamic boundary layer thickness (typically, 100 µm).

Interest in leveling dates from at least 1935 and the diffusion–adsorption–consumption/reduction mechanism was reasonably well articulated by the early 1960s, and is closely associated with the work of Leidhesier, Edwards, Watson, Kardos, Foulkes, Rogers, Beacom, Riley, Krichmar and Kruglikov among others [1–9]. The mechanism was developed based on observations ranging from additive-induced electrode polarization to feature-filling experiments and spatially resolved microstructural and compositional analysis. By the mid-1980s, advanced computational resources enabled quantitative assessment of feature-filling experiments by shape change simulations [19, 55–58].

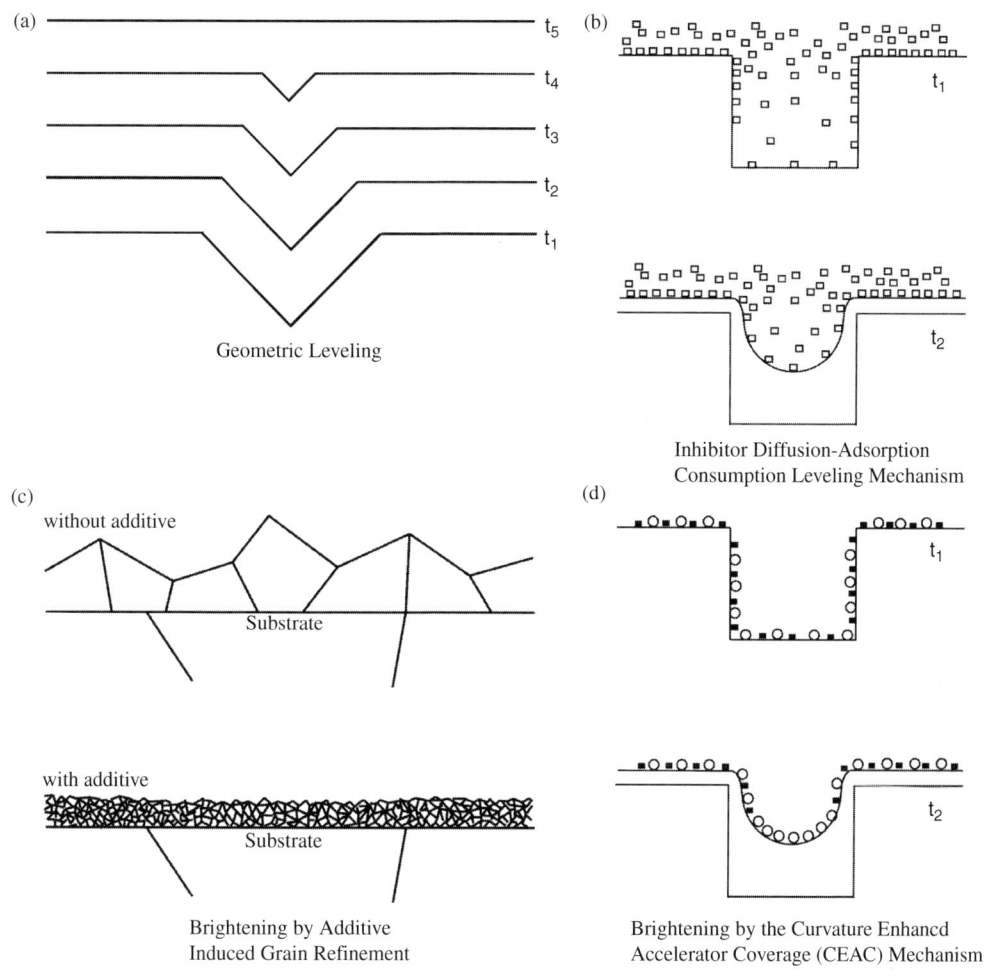

Figure 2.3 Schematic of the four stabilization methods: (a) geometric leveling during conformal deposition (at times t_1 through t_5) in a feature with a non-reentrant sidewall angle; (b) leveling derived from a spatially varying inhibitor flux due to gradients in the inhibitor concentration; (c) additive-induced grain refinement limiting the lateral dimensions associated with faceting; and (d) additive-induced brightening associated with competitive adsorption between an adsorbed inhibitor (black square) and a more strongly bound catalytic surfactant (open circles) that leads to enrichment of the catalyst on concave sections and accelerated bottom-up growth.

2.3.3
Brightening by Grain Refinement

Electrodeposits from simple salt electrolytes often exhibit macroscopically rough, faceted surfaces that are associated with epitaxial and/or columnar growth of large-grained polycrystalline materials. The addition of certain additives leads to substantial grain refinement, often into the nanometer range, and randomization of crystallite orientation. The combination leads to a reduction in facet size and in many instances the production of smooth, bright surfaces [1–9, 59–61]. However, it has been widely noted that grain refinement alone is not necessarily sufficient to obtain brightening [1–7].

2.3.4
Catalyst-derived Brightening

More recently, it has been recognized that a distinct smoothing or brightening mechanism can be derived from a mass balance of the competition between an inhibitor and catalyst for available surface sites. Of particular interest is the example of non-planar substrates. The essential idea behind the curvature enhanced accelerator coverage (CEAC) mechanism is that: (i) the growth velocity is proportional to the local accelerator, or catalyst, surface coverage and (ii) the catalyst remains segregated at the metal/electrolyte interface during metal deposition. For growth on non-planar geometries this leads to enrichment of the catalyst on advancing concave surfaces and dilation on convex sections, which, in combination, can give rise to distinct bottom-up filling of sub-micrometer features [12, 62–65]. The process becomes increasingly important at smaller (optical) length scales as the change in catalyst coverage of a growing surface for a given deposition rate is proportional to the local interface curvature.

The development of the CEAC model has largely been driven by the needs of the electronics industry. Interestingly, in the course of preparing this chapter, we discovered that the basis of this mechanism was actually first uncovered in the thesis work of Schulz-Harder [66] performed at the Technische Universität in Berlin in 1971. The first written report of that work beyond the university lies in a review paper by Kardos [3] followed subsequently by two poorly cited archival publications [67, 68]. As will be shown, the central importance of the CEAC growth mechanism to state-of-the-art copper metallization and printed circuit board manufacturing is undeniable. Thus, in light of the substantial industrial investment in copper plating technology, Schulz-Harder's work must be viewed as one of the most undervalued electroplating works of the 20th century. A similar observation pertains to the related 1990s rediscovery of spontaneous recrystallization of electrodeposited copper at room temperature, a phenomenon that was first noted in the 1940s [69, 70].

2.3.5
Stabilization Across Length Scales

In certain systems, a given additive can transition from exhibiting brightening to leveling behavior as a function of its concentration [1–4]. However, more often than not, practical plating baths contain a combination of levelers and brighteners in order to generate smooth surfaces over a wide range of length and time scales. For example, in copper metallization of integrated circuits, the competitive suppressor–brightener adsorption dynamics of the CEAC mechanism are central to filling of deep sub-micrometer features, while the addition of a leveler provides attenuation of the superfilling dynamic at longer length scales, thus reducing undesirable bump formation ('momentum plating') that otherwise occurs above features during the final stage of plating [71]. Coupling of inhibitor-based leveling with the CEAC brightening mechanism is thus a timely example of an industrially relevant multiscale modeling challenge.

In subsequent sections the models outlined above will be developed in a more quantitative manner within the context of specific processes and chemical systems. However, a key aspect of all the physical models outlined above is that they are not dependent on some idiosyncrasy of a given chemical system or detailed mechanism. For example, the inhibitor-based leveling models apply both to coumarin-induced leveling of nickel deposition and thiourea-derived smoothing during copper deposition [1–4]. Similarly, the CEAC brightening mechanism is effective for describing systems ranging from copper electrodeposition in an acid sulfate based systems [12, 62–67], to silver deposition from cyanide-based electrolytes [72–76] and halide catalyzed copper CVD [15, 77].

2.4
Additive Processes

The interplay between electrolyte additives and the metal deposition process underlies all the smoothing and stabilization processes outlined above. The additives can inhibit or accelerate the metal deposition reaction by a variety of different mechanisms that can be easily quantified by conventional electrochemical measurements. Understanding the physical and chemical basis behind the additive-perturbed metal deposition kinetics is a more complicated business that demands knowledge of interface structure and its correlation with function. Such information can be derived from complementary electrochemical surface science experiments. Indeed, between about 1985 and 2004 tremendous advances in the study of immersed interfaces were witnessed. Scanning tunneling microscopy (STM), X-ray scattering, vibrational spectroscopy and related *in situ* and *ex situ* methods have all yielded new insights into the structure and dynamics of metal surfaces and of the influence of adsorbates on surface structure and reactivity [78–83]. Simultaneously, creative engineering of surfaces at the nanometer scale has resulted in novel interfacial architectures or motifs with properties relevant to both the science and technology of electrodeposition [78–86].

In conventional additive-based electroplating, competitive adsorption for surface sites occurs coincident with substrate immersion and the onset of metal deposition. Additive adsorbate coverage evolves with metal deposition and is often accompanied, to some extent, by incorporation into the growing solid with associated influence on the microstructure and resulting physical properties. The adsorbates can have an accelerating or inhibitory effect on the metal deposition process. Inhibition manifests as an increase in electrode polarization with additive concentration while accelerating or depolarizing additives exhibit the opposite trend. Recent use of

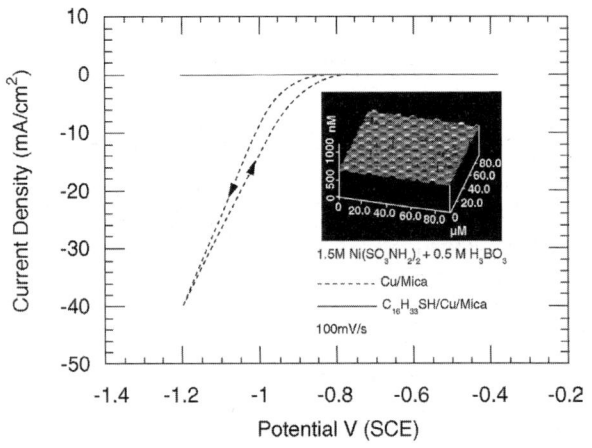

Figure 2.4 Voltammetry demonstrating that a self-assembled C_{16}-alkane thiol film effectively blocks nickel electrodeposition on a copper substrate. The inset shows that patterning methods, such as soft lithography, may be used to produce three-dimensional topographies by through-mask plating using an alkanethiol layer as a resist [96].

surfactant-engineered interfaces has been helpful in revealing the mechanisms by which molecules and other adsorbates affect the electrodeposition process.

In the simplest model, the extent of charge transfer inhibition is attributable to blockage of a certain number of active surface sites, θ_j, by adsorbed molecules, while metal deposition proceeds by the usual mechanism on the remaining portions of the surface [87–95].

$$i_M = i_{M\theta_j=0}(1 - \theta_j) \tag{2.1}$$

In one limit, a blocked electrode can be established by formation of a well-ordered self-assembled monolayer film. As shown in Figure 2.4, a self-assembled C_{16} alkanethiolate film on copper almost completely blocks the electrodeposition of nickel [96]. The thin dielectric layer simultaneously acts as a tunnel barrier to electron transfer and a transport barrier to ion transfer reactions as the solvated metal cations are unlikely to penetrate a well-packed, vertically-oriented alkane layer. Controlled patterning of such a monolayer film, for example by contact printing or perhaps self-assembly, followed by metal deposition can yield results analogous to simple 'through-mask' or template plating as shown in the inset AFM image in Figure 2.4. In principle, this can lead to pattern transfer at length scales approaching that of the molecular constituents of the layer. The blocking character of such films is a consequence of molecular functionality, adsorbate–adsorbate and adsorbate–substrate interactions and depends on the conditions under which the adlayer is formed. For alkanethiol films, changes in alkane chain length and end group lead to substantial changes in the ability to block metal deposition or dissolution reactions [97–111]. The schematic shown in Figure 2.5 hints at a number of possible outcomes during metal deposition [105–112].

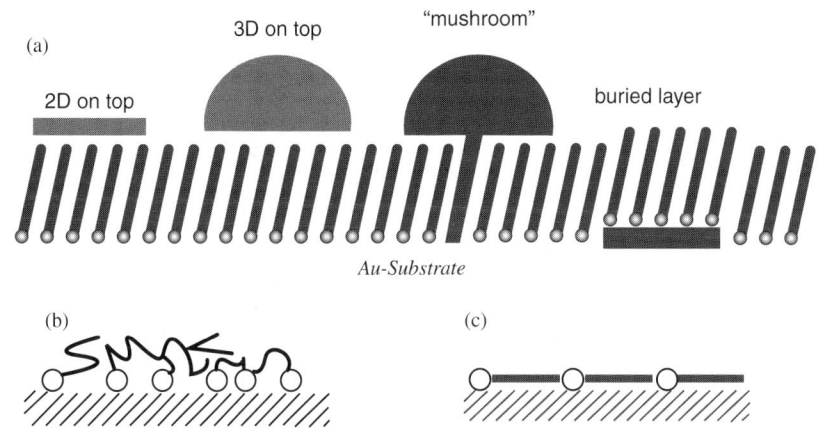

Figure 2.5 (a) Schematic of some possible outcomes during metal deposition on well organized monolayer films (adapted from Ref. [105]). (b) Many plating additives form less well ordered films. (c) In some instances the relevant adsorbates are most active at dilute coverage (adapted from Ref. [112]).

Deposition on top of the blocking layer or, more likely, overgrowth of the film, results in occlusion of the molecule and represents a limiting example of additive incorporation whereby the trapped molecule remains largely intact. This is analogous to some template-guided biomineralization reactions [113] and/or strain relieving lateral overgrowth processes [114]. In this instance, a continuous film is created by overgrowth from pinhole-like defects, resulting in a compact metal film covering the top of a non-wetting methyl-terminated alkane film. Indeed, this scheme has been proposed as a high-resolution method for surface replication [115, 116] where the role of the alkane film is to permit easy separation of the metal overplate from the underlying master.

At the other extreme of adsorbate behavior is the possibility of 'floating' or segregating to the free surface during metal deposition. Weakly bound hydrophobic surfactants, [e.g. sodium dodecyl sulfate (SDS)], that are often added to plating solutions as wetting or cleaning agents, are an example. More recently, it has been shown that even such simple surfactants can form interesting potential-dependent structures that can act as templates during metal deposition [117–120]. Significant segregation is also evident in additive systems that exhibit stronger adsorbate–substrate interactions. Indeed, the use of atomic surfactants to manipulate film growth has been described for both electrochemical [121–124] and vacuum [46, 47, 125] environments. The effects of anions such as halides, chalcogenides and/or low surface energy metals, Pb, Bi, Tl, have received considerable attention. Such adlayers can exert a strong influence on roughness evolution and surface anisotropy. This knowledge and approach also has much to offer towards the understanding of shape evolution of anisotropic nanoparticles by chemical reduction methods [126]. Segregating surfactant may also exert a strong influence on the kinetics of inner-sphere electron transfer or ion transfer reactions; a prototypical example being iodine-catalyzed dissolution of Pd [127] or chloride catalyzed Cu^{2+}/Cu^+ reduction during copper plating [128].

In practical plating environments, adsorbates display a full range of characteristics lying between and beyond the two limits described above. In some systems, the most active additive species may only be present at dilute, sub-monolayer coverages [129]. In contrast, compact molecular adlayers may be analogous to leaky dielectrics where θ_j becomes a measure of defect populations and activity, rather than site occupancy *per se*. Another important consideration is the lifetime of the active sites and their spatial distribution, that is, the statistical mechanics of defects in adsorbate layers might provide spatial averaging of the inhibition effect. While significant inroads towards addressing certain fundamental questions have been made, many longstanding questions persist, especially as they relate to practical systems; for example: does adsorption occur randomly or is there a strong influence of site bias, how rapidly do the adsorbates diffuse across the surface and how does this affect the metal deposition process? Other important aspects are the possibility of chemical interaction between the adsorbate and reaction intermediate associated with the metal deposition reaction, for example as is known to occur between Cu^+ and additives such as benzotriazole [59], etc. Additional complications might be dissociative adsorption of certain additives with some portion of the molecule being incorporated into the metal deposit, such as is the case for coumarin [130] and thiourea [131] additions in nickel-plating baths. In the case of multiple additives, both competitive and co-adsorption effects are known and the transient nature of such interactions needs to be considered. The potential dependence of adlayer structure, composition and dynamics has been examined in recent years and this knowledge is just beginning to impact thinking about technical electroplating problems [12, 79–83, 98–111, 124]. Likewise, the interaction between polymers and surfactants in aqueous solutions is relevant towards understanding many additive systems [132]. Less well developed, in the context of metal electrodeposition, is the possibility of multiple surface phases or 2D-phase separation effects and their effect on deposition kinetics and morphological evolution. Indeed, the formative demonstration of metal deposition in 3D-templates formed by bimodal phase separation of surfactant-laden electrolytes is a topic of substantial current interest [133].

For most commercially relevant plating systems there is limited information available on the phase and coverage of adsorbed additive species. In the absence of such molecular knowledge, most of our discussion of superconformal film growth will presume surface reactivity is a linear function of the fractional surface coverage, θ_j, of the relevant rate influencing adsorbate species, as suggested in Equation 2.1. This postulate has been widely used to interrelate electrochemical measurements with shape change simulations. Continued application of modern surface analytical tools for evaluating additive surface chemistry and structure as a function of deposition conditions will help elucidate the underlying physical basis of the various process models described below.

2.4.1
Adsorption Kinetics

The adsorption of ions and molecules on metallic surfaces is usually structure sensitive [78–83, 123, 134]. Furthermore, the dynamics of organization and the resulting microstructures or domain patterns are usually path sensitive. Tractable modeling or simulation of polycrystalline film growth in commercially important processes presently requires approximation (i.e. neglect) of such effects. Additive adsorption is typically modeled to be either an equilibrium process, in accord with the Langmuir isotherm, or, alternatively, to be constrained by finite Langmuir kinetics given by:

$$\frac{d\theta_j}{dt} = k_{ads} C_j (1 - \theta_j) - k_{des} \theta_j \qquad (2.2)$$

where the equilibrium condition is given by $d\theta_j/dt = 0$. The Gibbs energy of adsorption is taken to be independent of coverage, although a variety of modifications, such as the Frumkin isotherm, are available as the need arises [92–95]. Adsorption at charged interfaces, which is usually potential dependent, is reflected in the activation dependent rate constants for adsorption, k_{ads}, and desorption, k_{des}. In many plating applications additive adsorption involves convolution with solvent displacement, or even homogeneous chemical reactions, so that the associated adsorption rate constant is not elementary in nature [135]. The actual surface coverage may be quantified by radiotracer methods or surface analytical techniques. The latter may also reveal the spatial distribution and chemical nature of the respective adsorbates. If the adsorbates are electroactive the coverage may also be examined by electroanalytical methods.

If the adsorption rate is high, or the additive concentration low, substantial depletion of the additive concentration occurs next to the electrode. In a supported electrolyte, the interfacial concentration of the additive, C_i, is determined by mass balance between adsorption, desorption and the available diffusional flux:

$$k_{ads} C_j (1 - \theta_j) - k_{des} \theta_j = \frac{D}{\Gamma} \left(\frac{C_j - C_o}{\delta} \right) \qquad (2.3)$$

where D is the diffusion coefficient, δ is the hydrodynamic boundary layer thickness and Γ the saturation adsorbate coverage.

2.4.2
Surface Segregation versus Consumption Processes

For metal deposition, the adsorbates must either: (1) segregate onto the freshly created surface, (2) become buried or occluded in the solid by overgrowth of the neighboring surface, (3) undergo a deactivation process that results in either accumulation of the inactive species on the surface, desorption into the electrolyte or incorporation into the growing solid or (4) some combination of the above. Exactly how the adsorbates behave can have a profound effect on morphological and microstructural evolution and the resulting physical properties of the solid.

2.4.2.1 Adsorbates Segregated onto Growing Surface

Sustained accumulation and segregation or 'floating' of an irreversibly adsorbed species on a growing surface eventually leads to coverage saturation. In contrast, at sub-saturation values, surface segregation has noteworthy consequences for non-planar electrode geometries. This phenomenon lies at the heart of the CEAC model of brightening [62–67]. Namely, in the case of fixed initial coverage of a single adsorbate species, incremental motion of concave surface segments results in a decrease in the local area and thereby an increase in the local coverage of the adsorbate, while dilution occurs on convex sections as indicated in the schematic figure shown in Figure 2.6 [66]. Under these conditions the evolution of local adsorbate coverage, θ_j, on a moving surface is determined from the normalized rate of area change according to:

$$\frac{d\theta_j}{dt} = -\frac{1}{area}\frac{d(area)}{dt}\theta_j = \kappa v \theta_j \tag{2.4}$$

where the second equality comes from the definition of the local curvature κ and growth velocity v normal to the surface.

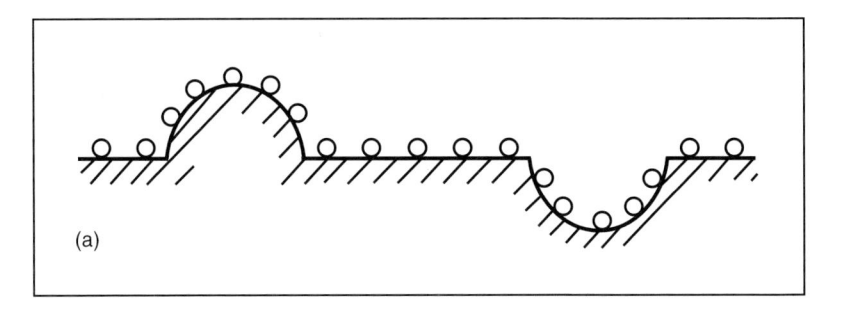

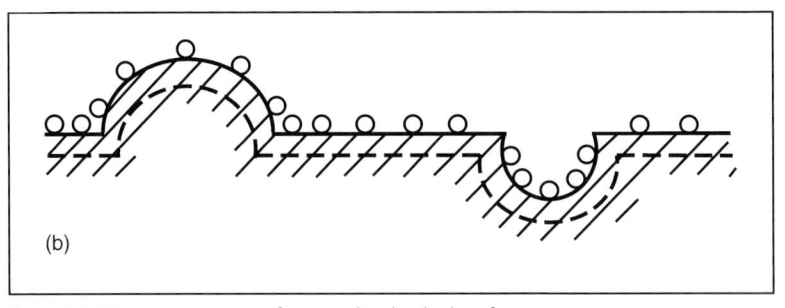

Figure 2.6 Mass conservation of a strongly adsorbed surfactant results in enrichment on an advancing concave surface due to area reduction, while the opposite is true for a convex profile, $t_b > t_a$ (source Ref. [67]).

For a surface saturated with two different species, the picture is complicated by the nature of the adsorbate interactions. Growth on a concave surface must result in either compression of adlayer species or their removal from the interface either by desorption into the electrolyte, multilayer formation or incorporation into the solid. In the limiting case of one species being more strongly bound to the surface, area reduction accompanying motion of a concave surface would result in expulsion of the more weakly bound species or, alternatively, the formation of a multilayered structure. In contrast, movement of a convex surface simply opens up new surface sites for additive adsorption from solution. In either case the general situation is analogous to phase evolution that occurs during compression or expansion of multicomponent films on a Langmuir trough or classical dropping mercury electrode. In addition to compositional changes, structural rearrangements are possible that can affect the permeability and charge transfer resistance of the adlayer; none of these effects have been considered quantitatively in electroplating models published to date.

For atomic adsorbates, the criteria for effective segregation at the solid/liquid interface focuses on species: (a) with minimal solid solubility in the deposit matrix, that is Tl and Pb in Ag or Cu; (b) that do not form 3D-intermetallic phases; (c) that lower the overall surface energy of the system; or (d) that exhibit some degree of surface mobility. In the case of molecular additives, the presence of a charged functional group appears to be a characteristic that further encourages sustained segregation at the growth front [136].

Importantly, any process that results in lateral coverage gradients of either the additive or metal adatom will be countered by surface diffusion. Such surface transport is known to be strongly influenced by both potential and electrolyte composition through associated impact on the structure and composition of the surface. For example, anions are known to lead to substantial enhancement of metal adatom transport with diffusion coefficients ranging from $2 \cdot 10^{-16}$ up to $8 \cdot 10^{-13}$ cm^2 s^{-1} [137, 138].

2.4.2.2 Adsorbates Incorporated into Growing Deposit

Adsorbates that do not segregate to the surface during deposition may be incorporated into the growing deposit, thereby affecting the microstructure and physical properties of the solid. Mass balance on the moving planar interface requires that:

$$\frac{d\theta_j}{dt} = -\frac{vC_{j,\text{solid}}}{\Gamma_j} \tag{2.5}$$

where v is the interface velocity of the growing solid and C_j is the concentration of the incorporated additive.

A variety of consumption rate laws have been proposed. One of the simplest, and perhaps most intuitive, is that adsorbate consumption is proportional to coverage, θ_j, and metal deposition rate, i, according to [139]:

$$\frac{d\theta_j}{dt} = -k_{\text{inc}}\theta_j i \tag{2.6}$$

where, of course, the deposition rate itself can be a function of θ_j [140].

In one limiting case the inhibiting molecule is immobilized upon adsorption and subsequently buried as an inclusion during metal deposition. In this instance the rate constant, k_{inc}, is simply related to the neighboring metal deposition rate to obtain [140]:

$$k_{inc} = \frac{M_W}{h\rho zF} \tag{2.7}$$

by geometry, where h corresponds to the vertical height of the immobilized additive, M_w is the mean atomic weight of the deposit, ρ its density, z the number of equivalents and F is Faraday's constant.

For dilute additive solutions, consumption can easily become limited by diffusion of the adsorbate to the interface. A particularly tractable situation occurs at the dilute limit of the Langmuir isotherm where C_j is proportional to the surface coverage, and in the limit of diffusion controlled adsorption C_{solid} is directly proportional to the additive flux and thus the bulk electrolyte concentration. Such transport limited incorporation was reported in some radiotracer studies of thiourea incorporation in nickel and copper plating in the 1950–1960s [1–4, 17, 130, 131, 141–146]. Consistently, a common observation was that the additive concentration in the solid was proportional to the additive concentration at the interface and inversely proportional to metal deposition rate, i, [1–4, 130, 131, 141, 142] such that:

$$C_{j,solid} = \frac{zFk_{inc}C_j}{\Omega i} \tag{2.8}$$

where Ω is the specific volume of the solid and the current efficiency is 100%.

Other consumption rate laws have since been reported where adsorbate consumption is a non-linear function of coverage with a complicated potential dependent rate constant, k_{inc}, where the rate of consumption decreases with the metal deposition rate over a certain range of potential [12]. As in the case of adsorption processes, such complicated relationships indicate that the experimentally derived rate laws do not reflect elementary rate processes, but rather convolve the effects of chemical reaction and/or surface phase transitions. The interaction between benzotriazole and transiently generated Cu^+ during copper deposition provides an excellent example of a coupled chemical reaction that is directly associated with additive incorporation and bright deposits [59]. Indeed, it is important to recognize that the equilibrium concentration of the reactive Cu^+ in acid Cu^{2+}/Cu technical plating electrolytes more often than not exceeds the total concentration of all the additives in the electrolyte, and is thus expected to play a significant role in additive dynamics, particularly at low overpotentials.

In the 1950–1970s, wet chemical and radiotracer methods were widely used to quantify additive incorporation in electrodeposits. More recently, modern surface analytical tools such as SIMS, AES, XPS and RBS analysis have been implemented. However, there have been few attempts to correlate electrochemical measurements with additive incorporation into the solid. Simultaneous monitoring of the quenching of reactivity and microstructural evolution of electrodes with pre-engineered adsorbate structures (e.g. underpotential deposited layers, monomolecular films) offers particular promise in this regard [12, 97–111, 136].

2.4.2.3 Deactivation of Adsorbate

In addition to incorporation into the deposit, adsorbed molecular species may also undergo fragmentation that may or may not correspond to deactivation of the additive. A well studied case is the conversion of the leveling agent coumarin into meliotic acid, that is ascribed to electrochemical reduction process occurring at ten times the rate at which coumarin is incorporated in the deposit [130]. Potential driven, η, deactivation processes may be described by

$$\frac{d\theta_j}{dt} = -k_{red}\theta_j \qquad (2.9)$$

where $k_{red} = k^o_{red}\exp(b\eta)$ [57, 147, 148].

Radiotracer analysis is particularly effective for tracking the fate of additives, because molecular fragmentation can typically be expected to result in an impurity element distribution different from that of the original molecules. For example, in nickel plating in the presence of thiourea, complete incorporation of the molecule should give a carbon/sulfur ratio of 1.0, while experiments indicate a ratio ranging from 0.01 to 0.7 depending on the deposition conditions [130, 131, 141, 142]. As with additive incorporation, the decomposition process can be a sensitive function of the co-adsorbates [145, 149, 150].

In any study of additive plating care must be taken to account for parasitic reactions, such as homogeneous decomposition of the additive in the electrolyte, oxidation at the anode, etc. For example, the thiol/disulfide chemistry associated with certain S-bearing molecules has given rise to significant difficulties with process control over the years [151–160]. The judicious choice of additives combined with appropriate cell design and operating conditions can minimize or entirely avoid some of these difficulties [151–158]. Nevertheless, in practice it has become common to use electroanalytical means to monitor the stability of electroplating baths [159–161].

2.4.3 Adsorbate Evolution

In order to fully describe the progression of adsorbate coverage during electrodeposition, mass balance is imposed by summing the individual processes, j: adsorption and desorption (Equation 2.2), geometry (Equation 2.4), consumption (Equations 2.5 and 2.6) and reduction (Equation 2.9):

$$\frac{d\theta_j}{dt}_{total} = \sum_l \left(\frac{d\theta_j}{dt}\right)_l \qquad (2.10)$$

An assessment of a variety of expressions for the steady-state coverage of single additive systems under both potentiostatic or galvanostatic metal deposition is available [140]. Perhaps the most interesting among these is the possibility of

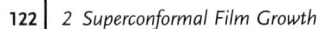

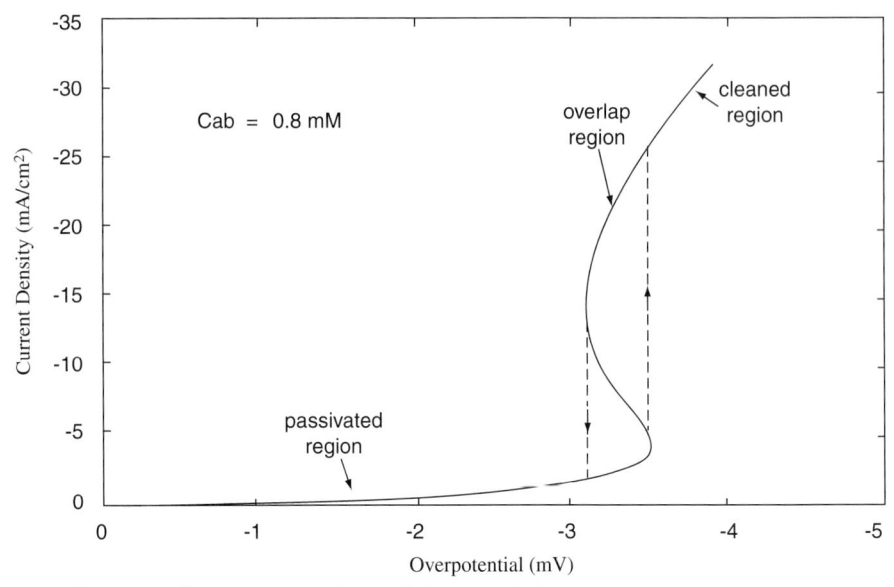

Figure 2.7 Voltammetric curve for metal deposition in the presence of a dilute inhibitor. The hysteresis reflects the possibility of a potential regime where multiple steady-state values for inhibitor surface coverage are possible. This arises from the competition between transport of the blocking species to the interface and its coverage- and rate-dependent consumption (source Ref. [140]).

multiple steady-state solutions that arise when the coverage evolution equation involves higher order ($q > 1$) dependences in surface coverage, θ^q. An example of this is given in Figure 2.7 for potentiostatic deposition where the additive flux is under mass transport control. The solid line is the analytical solution and the dotted lines correspond to the anticipated experimental η–i result. In this example, depassivation is associated with adsorbate consumption (or desorption) that leads to rapid acceleration of the metal deposition rate that outruns the ability of the blocking additive to accumulate at the interface. Analogous hysteretic behavior is widely observed in systems where reactions occur under mixed, interface reaction–reactant transport, control [162]. Multicomponent systems that involve synergistic and competitive interactions among adsorbates can lead to even more complicated behaviors, which are only beginning to be examined.

2.4.4
Impact on Microstructure

For epitaxial growth, strongly segregating adsorbates are known to influence the surface anisotropy and roughness. A variety of effects have been observed by optical microscopy, while more recently scanning probe microscopy has begun to provide insight into the atomistic details of the process [45, 78–83, 124]. The possible role of

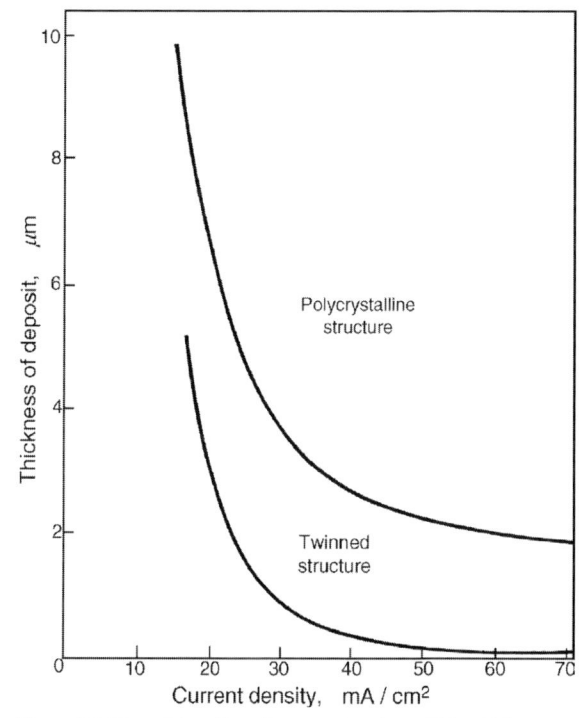

Figure 2.8 Map of the effect of current density and film thickness on microstructure development that is representative of copper deposition from a simple acid cupric sulfate solution (source Ref. [165]).

adsorbates in blocking or pinning step flow have been discussed [163, 164], although more recently detailed experimental and theoretical studies of the role of adsorbates on step geometry and motion have been pursued [51, 52]. Based on thin film growth studies (PVD, MBE) performed in vacuum, the influence of adsorbates on the magnitude of the step edge barrier to interlayer mass transport and thereby step bunching and roughness evolution appears to be particularly important [47].

Deposition at faster rates often leads to polycrystalline films as a result of stacking faults and related defects, as indicated in Figure 2.8 [165]. The onset of polycrystalline growth occurs more rapidly at higher current densities. The role of non-coherent nucleation and the contribution of additives to this process have also been explored [166–168]. Certain molecular species yield dramatic alterations of the microstructure associated with incorporation of the molecule and/or its products [166–216]. Resulting grain refinement, a common example, leads to increases in hardness, decreases in conductivity and changes in magnetic coercivity. In certain instances brightening is observed. The mechanism by which such grain refinement occurs has received some attention [17, 168, 169, 173, 180, 182]. Information on the effect of pH, current density and anion adsorption on texture development is available [184, 185]. Likewise, quantitative models of texture development and the impact of additives are beginning to appear [186, 187]. The prototypical case of sulfide formation, associated with thiourea adsorption and decomposition [189, 190] has been shown to result in significant grain refinement [60] and high values of stored energy [191]. A wide range of studies on the effects of thiourea in copper plating are also available in the Russian literature [9]. Recent STM studies [102, 103, 105, 129, 192] have revealed how such adsorbates lead to dramatic changes in the nucleation and growth of copper. In other systems, the additives may be incorporated in their molecular form with the 'grain size' of the resulting composite being dependent on the concentration of incorporated molecules [59, 193]. In such systems, the original molecule can be recovered from the deposit by wet chemical analysis [59, 193]. For the deposition of reactive metals, the effect of additives on parasitic reactions, such as hydrogen evolution, along with interfacial pH shifts must also be considered [194–196].

Microstructural studies (X-ray, optical, electron and force microscopy) complemented by chemical analysis (SIMS, radiotracer, spectroscopy, etc.) of the distribution of additives offer perhaps the most direct path to understanding how film growth dynamics are impacted by adsorbates. Monitoring the effect of processing conditions on physical properties provides an additional, albeit indirect, avenue for studying the effect of additive chemistry on incorporation. For example, the grain-refinement action of additives during electrodeposition is often followed by recovery and recrystallization that can proceed at significant rates even at room temperature [69, 70, 197–212]. This is particularly so for materials with homologous temperatures of 1100 °C or less, such as electroplated copper used for on-chip interconnects. The recrystallization process may be monitored in real time either directly by ion-beam or electron channeling resolved imaging of the evolving grain structure (EBSD, etc.) or indirectly by monitoring changes in physical properties, such as hardness, electrical resistance or stress state with time. The recrystallization transients are a sensitive function of the processing conditions such as deposition rate, film thickness, electrolyte additive composition, etc. [197–212, 243]. Several studies have indicated that, for systems with multiple additives, interesting cross interactions occur that give rise to significant changes in the additive incorporation behavior [213, 214]. Unfortunately, proprietary concerns have limited the amount of information that is available for providing a one-to-one correlation between the deposition parameters and the resulting microstructure and relaxation processes.

Another particularly interesting effect associated with additive electroplating is the spontaneous formation of periodic laminated deposits due to spatial–temporal variations in adsorbate coverage and incorporation during growth [143, 215]. The sensitivity of such lamination to additive flux was utilized to reveal the non-uniform incorporation of leveling additives that occurs during filling of grooved electrodes, as is evident in the classic micrographs shown in Figure 2.9 [4, 5, 216]. Regions of high additive flux, that is the free-surfaces are laminated while the recessed regions where the additive flux is reduced exhibit a non-laminated 'fibrous' structure. Similar observations of lamination in pulse plated samples speaks to the potential or rate dependence of adsorbate incorporation [5].

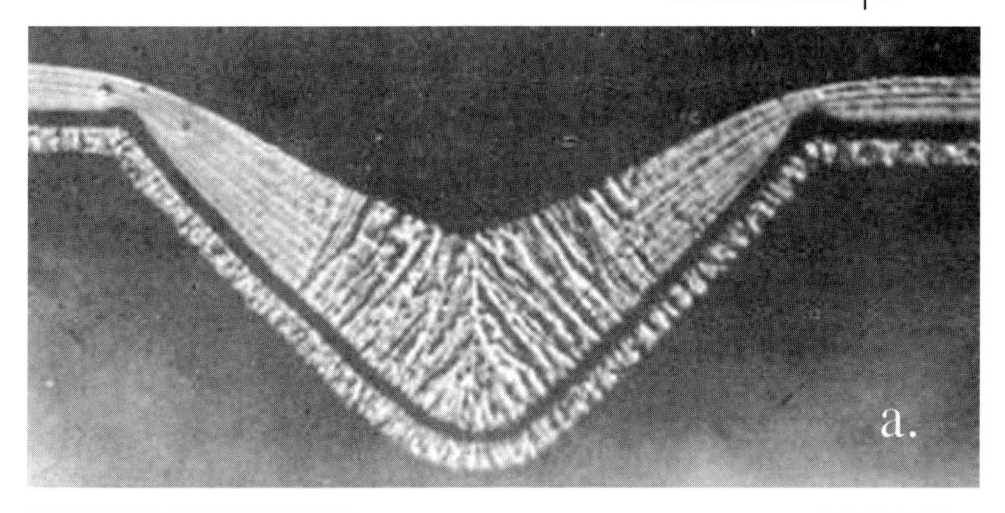

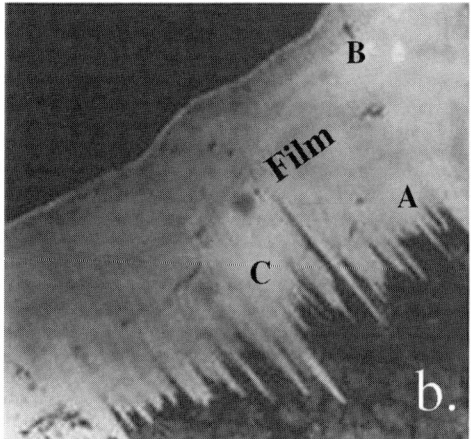

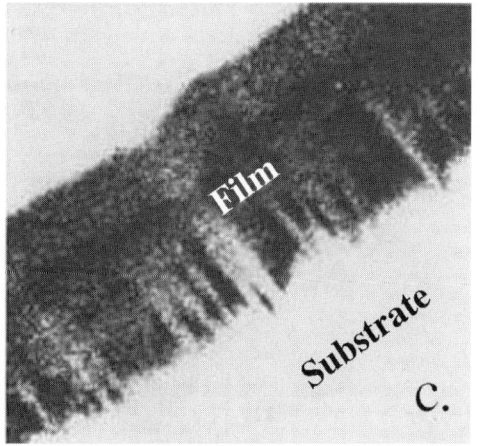

Figure 2.9 (a) Leveling of a 37.5 μm deep groove during nickel deposition from an electrolyte containing 2-butyne-1,4-diol. Preferential etching of the metallographic cross-section reveals that the leveling agent gives rise to a laminated deposit on planar sections while the decreased flux of the additive into the recessed groove yields a fibrous deposit (source Ref. [4]). (b, c) Related evidence of preferred additive co-deposition on high points is available from radiotracer studies. The optical micrograph of an oblique cross-section shows thiourea induced leveling during nickel deposition. (c) An autoradiograph of (b) showing that minimal incorporation occurs in the recessed sections of the film. Note the banding contrast and lateral uniformity once planarity is attained (source Ref. [143]).

2.4.5
Quantifying Adsorbate Inhibition of Metal Deposition

The addition of inhibitors to Ni, Cu, Sn, Pb, ... plating electrolytes results in a decrease in the metal deposition rate at fixed potential or, equivalently, an increase in electrode polarization for a fixed applied current density as indicated in Figure 2.10

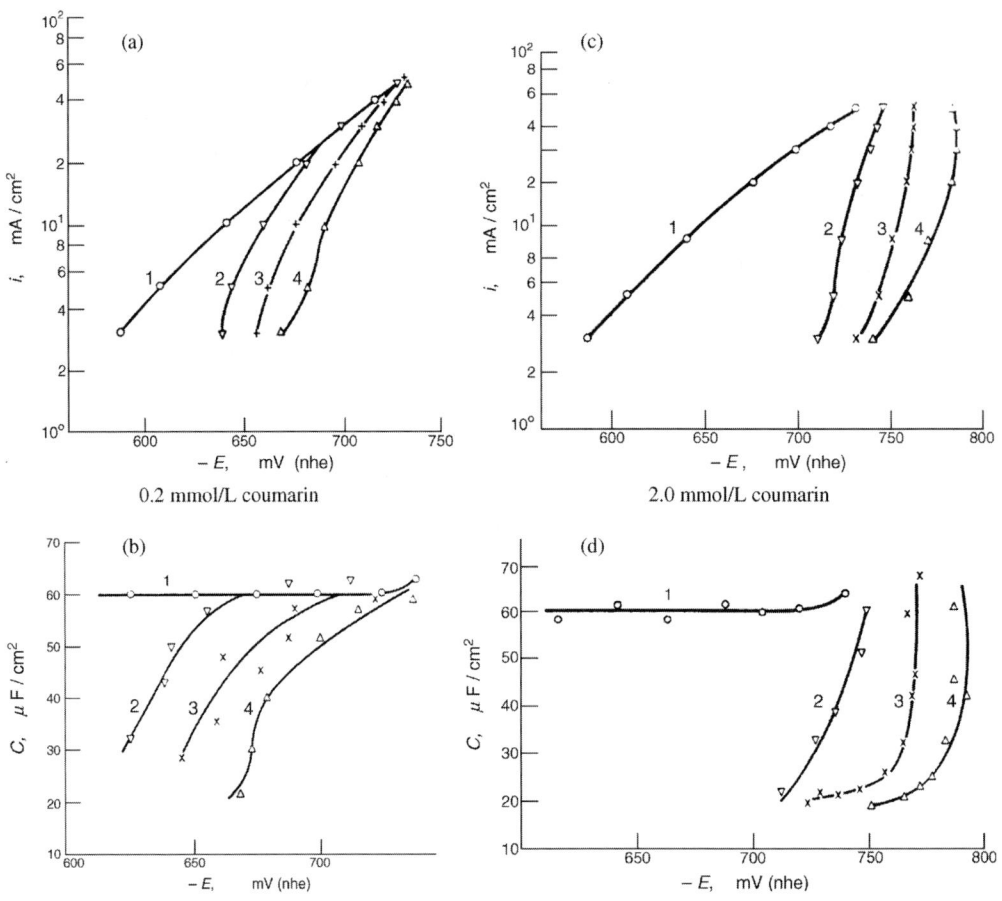

Figure 2.10 The influence of inhibitor flux on electrode polarization is revealed by η–i curves collected for coumarin concentrations of (a) 0.2 mmol L^{-1} and (c) 2.0 mmol L^{-1} and enumerated hydrodynamic conditions: (1) additive-free independent of rotation rate; (2) 400 rpm; (3) 1540 rpm; and (4) 4000 rpm. Simultaneous capacitance measurements (b and d, respectively) provide a measure of the impact on interface composition and structure that can be parameterized in terms of an effective surface coverage (source Ref. [218]).

1–4, 58, 217–226. Both qualitative and quantitative insight can be garnered from transient η–i, i–t and η–t measurements in quiescent or stirred solutions, while measurements of steady-state behavior are best performed under well-defined hydrodynamic conditions. Typically, a rotating disc electrode (RDE), or a related method, is used to specify and/or modulate the hydrodynamic boundary layer thickness, δ. With an RDE the boundary layer is specified by

$$\delta = 1.61 D^{2/3} \omega^{-1/2} v^{1/6} \qquad (2.11)$$

where ω is the angular rotational speed of the disk and v is the kinematic viscosity of the electrolyte, a typical value being in the region of 0.01 cm^2 s^{-1}. The resulting control of transport enables the extent of inhibition to be probed as a function of inhibitor flux to the surface [217, 218]. Simultaneously, the inhibitor surface coverage is monitored by studying its effect on the double layer structure; the inhibiting molecules displace adsorbed water molecules leading to a change in the capacitance such that the associated surface coverage can be obtained from

$$\theta = \frac{C_0 - C_\theta}{C_0 - C_{sat}} \quad (2.12)$$

where C_0, C_θ and C_{sat} are the capacitances at zero coverage, θ_j and saturation coverage, respectively [218]. A useful example of the utility of these combined methods is provided by studies of the coumarin–nickel system. The variation of inhibitor coverage and its effect on the metal depositon rate are revealed by η–i curves collected under controlled hydrodynamic conditions in parallel with capacitance measurements summarized in Figure 2.10a–d [218]. Likewise, an even more complete analysis is available by examining the full impedance [227–229] and electrohydrodynamic [230] response of the electrode. The latter exhibits a low frequency phase shift of 180° consistent with the observation that the metal deposition rate decreases as the coumarin flux increases, as indicated in Figure 2.11.

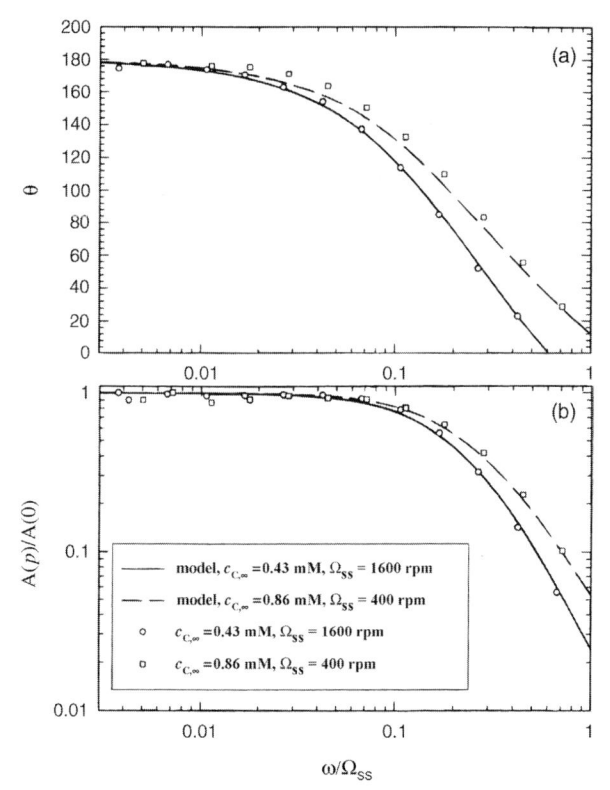

Figure 2.11 Electrohydrodynamic impedance spectroscopy of coumarin inhibited nickel deposition. Note at low perturbation frequency the 180° phase shift (θ) between the nickel deposition rate and the inhibitor flux to the surface (source Ref. [230]).

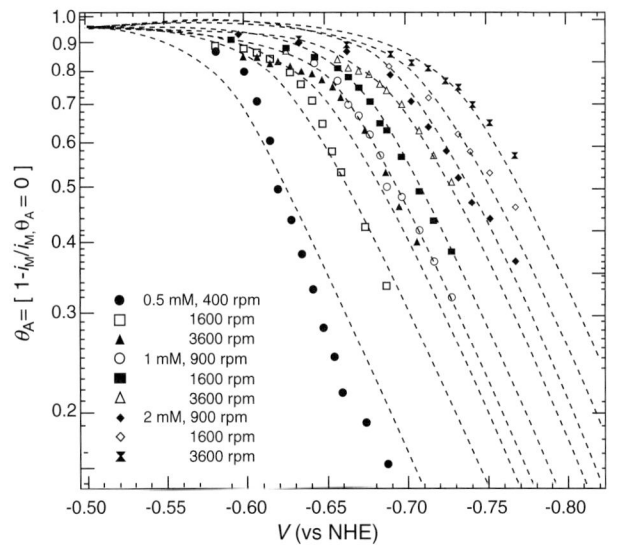

Figure 2.12 Surface coverage of coumarin inhibitor as a function of the potential for various coumarin concentrations and electrode rotation rates. The dashed lines correspond to a fit to the leveling model outlined in the text (source Ref. [58]).

The effect of the inhibitor coverage on the deposition kinetics can also be mapped by referencing the additive-perturbed deposition kinetics to the additive-free case, in accord with Equation 2.1 [58]. Recent results analyzed in this manner are shown in Figure 2.12, where the coverage is a distinct function of the additive concentration in the bulk electrolyte and the hydrodynamic boundary layer thickness [58]. The dotted lines are simulations based on a steady-state model [57] that considers metal deposition to proceed through an adsorbed intermediate M^+_{ads}, which competes with coumarin for available surface sites:

$$M^{2+} + e^- \leftrightarrow M^+_{ads} \qquad k_{1,M} \qquad (2.13)$$

$$M^+_{ads} + e^- \leftrightarrow M \qquad k_{2,M} \qquad (2.14)$$

The coverage of the reaction intermediate θ_M, the inhibitor coumarin, θ_A, and the number of free sites, θ_0 sum to unity:

$$\theta_M + \theta_A + \theta_0 = 1 \qquad (2.15)$$

In the model [57], the coumarin coverage is defined by a Langmuir isotherm (Equation 2.2) with electrochemical reduction of the adsorbed coumarin (Equation 2.9) being balanced by diffusion of coumarin to the interface (Equation 2.3). Incorporation of the coumarin in the solid was considered to be negligible. The results of the model have been summarized by three characteristic dimensionless groups, λ, γ and μ:

$$\lambda = \frac{k_{2,M}}{k_{1M} + k_{2M}} \qquad \gamma = \frac{C_b}{K} \qquad \mu = \frac{DC_b}{k_A \delta} \qquad (2.16)$$

The dimensionless group λ reflects the relative rate of the two serial elementary reactions that describe metal deposition. Where $\lambda \to 1$ the second reaction is fast, resulting in minimal surface concentration of the intermediate metal species, θ_M, whereas for $\lambda \to 0$ the surface coverage of the intermediate, θ_M, can become substantial if there are sufficient sites that are not blocked by the inhibitor.

The dimensionless group γ is given by the ratio of bulk inhibitor concentration to the Langmuir adsorption constant. In the absence of consumption or competition from the adsorbed metal intermediate, $\gamma = 1$ corresponds to an inhibitor coverage of 0.5. Thus significant inhibition is expected when $C_b \geq K$ ($\gamma \geq 1$).

The dimensionless group μ is the ratio of the intrinsic rate of inhibitor mass transport to the rate of reduction, or more generally deactivation, of the adsorbed inhibitor. For large values of μ, mass transport effects are minimal as the surface concentration approaches the bulk value. For smaller values, mass transport limitations become significant, resulting in depletion of the inhibitor concentration near the electrode surface and a consequent decrease in coverage. The results for the model [57] are summarized in Figure 2.13, where the steady-state inhibitor coverage is plotted versus μ with the product $\lambda\gamma$ as a parameter. An important simplification follows from the fact that while the metal deposition rate depends on all three dimensionless groups, the ratio of the $i_M/i_{M\theta=0}$ depends only on μ and the product $\gamma\lambda$. Strong inhibition ($\theta > 0.5$) is evident when $\gamma\lambda > 1$. When $\mu > 1$ the inhibitor concentration at the interface approaches its bulk value while for $\mu < 1$ mass transport effects become important. Optimal leveling occurs for $\mu = 1$, which corresponds to the onset of diffusional control for inhibitor accumulation. A similar dimensionless

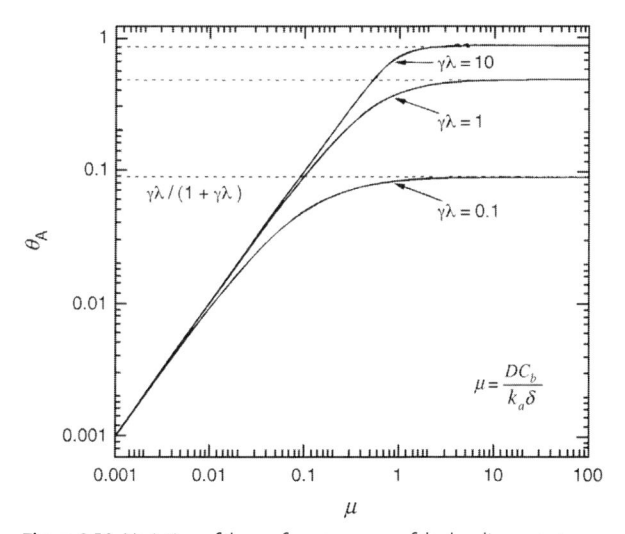

Figure 2.13 Variation of the surface coverage of the leveling agent θ_A with the dimensionless parameter μ. Maximum leveling power occurs when μ approaches unity (source Ref. [57]).

formulation of leveling was published in the Russian literature in 1980 for describing leveling of a sinusoidal surface profile [231]. Kinetic models as those outlined above provide a sound basis for shape change simulation of leveling phenomenon, as will be detailed in Section 2.5.3 [57, 58].

2.4.6
Co-adsorption Effects

In addition to the single additive site blocking mechanism outlined above, there are several systems that require cooperative interactions for inhibition to occur. Such cooperative interactions may also lead to enhancement of passivation in certain instances. For example, the combination of *p*-toluenesulfonamide and coumarin in nickel plating [218] yields enhanced inhibition over that provided by the individual constituents. With copper electroplating, benzotriazole [232] and poly(ethylene glycol) (PEG) [12, 136, 233–244] are widely used inhibitors that are strongly influenced by anion co-adsorption. In fact, the inhibition of copper deposition by PEG depends entirely upon co-adsorption of halide. As shown in Figure 2.14, the addition of PEG alone does not significantly perturb the metal deposition kinetics, while the addition of halide alone actually results in an increase or catalysis of the deposition process. However, when PEG and Cl^- are added in combination, the metal deposition kinetics can be reduced by almost two orders of magnitude, that is, the exchange current decreases from 1.4 to 0.039 mA cm^{-2}. Even greater inhibition is available by substituting bromide for chloride. In contrast to the reduction of the exchange current density, only a small variation in the Butler–Volmer charge transfer coefficient is evident following PEG–Cl addition. This suggests that the mechanism, or rate-limiting step, of the copper reduction reaction remains largely unaltered by

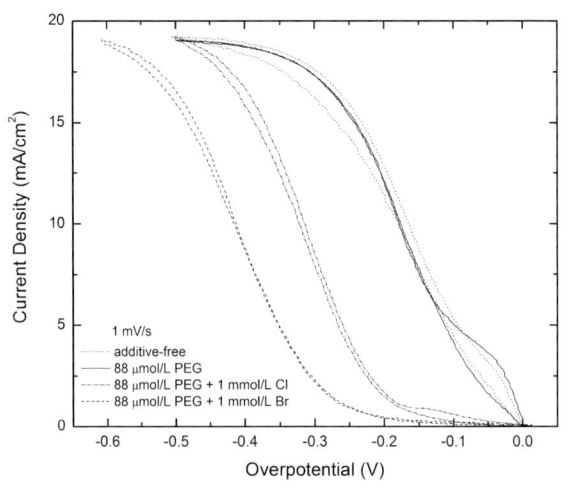

Figure 2.14 Slow sweep rate η–i curves showing halide-dependent PEG inhibition of copper electrodeposition.

the PEG–Cl$^-$ barrier layer; the blocking layer simply limits the access of aquo-cupric ions to the metal surface.

Halide adsorption on copper has received significant attention, which is well summarized in a review [134]. At potentials relevant to technical plating, chloride displaces sulfate and/or bisulfate from the interface forming a variety of ordered adlayers [80, 134, 245–247]. At least for slow deposition rates, halide exhibits a significant propensity to float or segregate on the surface during metal deposition from dilute CuSO$_4$ solutions containing 1 mM Cl [134, 246, 247]. In more concentrated CuSO$_4$ solutions, the equilibrium Cu$^+$ can exceed the solubility product for CuCl, leading to precipitation and a decrease in the ability of PEG to block the metal deposition reaction by co-adsorption [248]. In contrast, copper plating in PEG–Cl electrolytes containing less than 0.2 mmol L^{-1} chloride exhibits a distinct potential dependent breakdown of the passivating layer, and sweep rate dependent hysteretic η–i behavior is observed, as shown in Figure 2.15. The hysteresis reflects the consequences of chloride displacement by consumption (or desorption) being significant relative to the kinetics of its accumulation at the interface. This is an experimental manifestation of the simple model [140] outlined in Figure 2.7, whereby control of the metal deposition rate resides with chloride coverage at the interface. A summary of the dependence of inhibition on the chloride concentration is given in Figure 2.16a. Similar effects are seen if the PEG concentration is made very dilute while the chloride activity is maintained in the 0.2–1 mmol L^{-1}, range as shown in Figure 2.16b.

The tenacity of the as-formed PEG–Cl blocking layer has been demonstrated by the sustained inhibition of blocked electrodes following transfer into additive-free

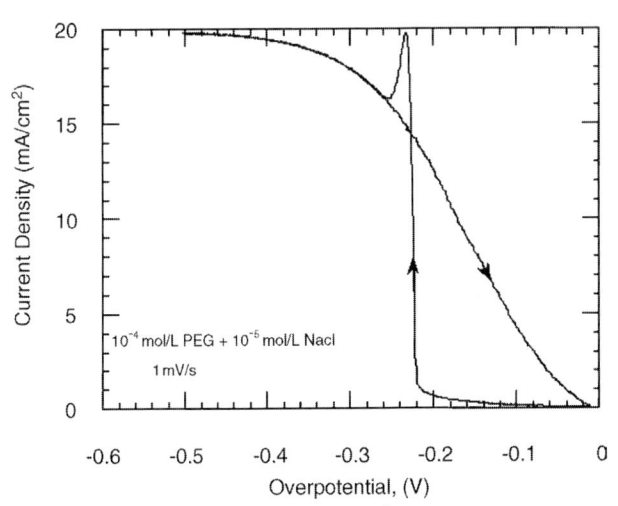

Figure 2.15 In the presence of $\approx$10 µmol L^{-1} Cl$^-$, the blocking PEG–Cl film is metastable as revealed by sharp potential-dependent depassivation. At the critical potential the inhibiting layer is disrupted and metal deposition accelerates to the transport-controlled limit.

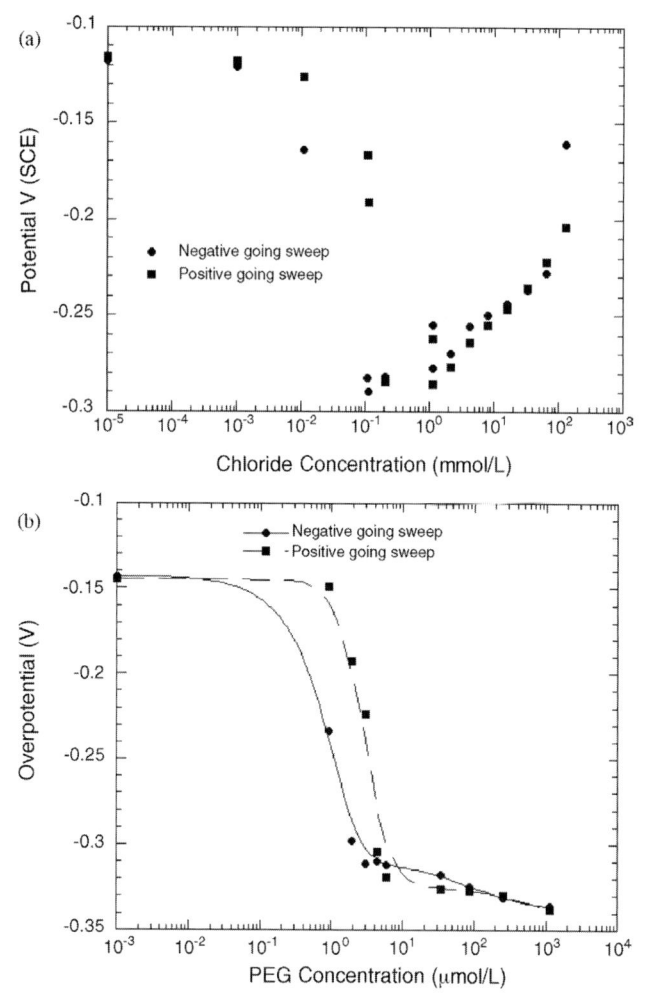

Figure 2.16 Inhibition of the copper deposition reaction provided by co-adsorption of PEG–Cl is a strong function of the Cl and PEG concentration. This dependence is characterized by sampling the potential for a particular current density (10 mA cm^{-2}) during slow sweep rate voltammetry (1 mV s^{-1}). (a) For an electrolyte containing 88 μmol L^{-1} PEG, maximum sustained inhibition occurs in the range of 0.2–1 mmol L^{-1} Cl$^-$. Higher Cl concentrations lead to precipitation of CuCl and a breakdown in blocking behavior. (b) For an electrolyte containing 1 mmol L^{-1} Cl$^-$, stable inhibition is evident for PEG concentrations above 8 μmol L^{-1}.

electrolytes [234, 241]. The longevity of electrode passivation under these circumstances is greatly increased by chloride addition to the additive-free electrolyte, reflecting a degree of reversibility associated with the chloride adsorption.

Variation of the molecular weight of PEG reveals that the onset of significant inhibition occurs for molecules with more than ≈10 monomer units (n) [234]. A quartz crystal micro-gravimetric (EQCM) study of the addition of PEG

(MW = 3350 g mol^{-1}) to a copper plating electrolyte containing ≈1.4 mmol L^{-1} Cl$^-$ and 1 mmol L^{-1} Cu^{2+} reveals a frequency shift Δf proportional to mass gain, of ≈5–8 Hz [241]. The shift is only evident in the presence of chloride and saturates for a chloride activity of 1 mmol L^{-1} Cl$^-$. The theoretical sensitivity of a 5 MHz crystal is 56 Hz cm^2 µg^{-1}, thus a 5–8 Hz shift corresponds to a film mass of 0.09–0.14 µg cm^{-2}. Considering the bulk density of 3400 g mol^{-1} PEG (1.204 g cm^{-3}), yields an average film thickness of between 0.7 and 1.2 nm, indicating that of the order of a monolayer of PEG is required for inhibition [241].

Further support is provided by a recent ellipsometry study, performed in the absence of Cu^{2+}, indicating that a 0.6 nm PEG layer is co-adsorbed with Cl at potentials relevant to copper plating [249]. EQCM experiments [241] indicate that PEG–Cl co-adsorption has a weak potential dependence; varying between 5 and 8 Hz in the −0.45 to −0.75 V MSE potential window. A weak power law dependence on the molecular weight of PEG, $\Delta f \approx n^{0.34}$, was reported. The exponent was examined as a possible diagnostic for evaluating various possible conformations of the adsorbed PEG molecules [241].

The cooperative nature of inhibition in this system is ascribed to halide mediated PEG adsorption that may also involve interactions with the Cu$^+$ reaction intermediate. Indeed, the earliest studies of PEG–Cl inhibition discuss the possible formation of various polyether–cuprous chloride compounds ranging from PEG helically wound around a CuCl core to crown-ether like moieties bound to the copper surface via chloride [233, 234]. Analogous arguments of complex formation were made based on experimental measurements of the physical properties of solutions containing a high PEG/Cu$^{+/2+}$ ratio [239, 240].

In order to delve deeper into the structure and composition of the PEG–Cl layer, several researchers have performed *in situ* Raman spectroscopy. Early work inferred that at least two different forms of PEG exist on the electrode surface [236]. For an open circuit condition a featureless Raman spectra was observed, while at more negative potentials several PEG related vibrational modes were evident. This particular study [236] was complicated by the use of 14 mmol L^{-1} Cl$^-$, which usually results in CuCl precipitation under open circuit conditions. The latter is known to weaken the inhibition provided by PEG compared with slightly more dilute chloride solutions [248]. Subsequent work in 2 mol L^{-1} H$_2$SO$_4$ containing 0.28 mmol L^{-1} Cl$^-$ unambiguously revealed the simultaneous presence of chloride and PEG on the surface at potentials relevant to copper plating [238]. The chloride signal decreased with decreasing potential while the polyether modes became stronger. Most recently, a study of the CuSO$_4$–H$_2$SO$_4$–KCl system provided a more detailed view of the potential dependent interactions between copper, its ions, chloride and PEG [250]. The intensity of the Cu–Cl mode decreases with decreasing potential, while the strongest polyether mode was only weakly dependent on potential. A novel strong mode at 260 cm^{-1} was observed during copper electrodeposition that was attributed to a three coordinate polyether–chloride cuprous complex. The signal intensity passed through a maxima at ≈−0.543 V MSE and was only observed if all the constituents, Cu ions, Cl and PEG, were in solution [250]. Importantly this mode was absent in an earlier study [238] that did not include copper ions. In contrast, a recent

EQCM study demonstrated that PEG and Cl^- stabilize the formation of Cu^+ at potentials near open circuit [248]. Accordingly, it was concluded that neither Cu^+ nor Cu^{2+} from the electrolyte are necessarily required to form the inhibiting surface layer at more negative potentials [248]. Thus, some debate remains as to the importance of Cu^+ to the formation and longevity of the PEG–Cl blocking layer [241, 248–250]. However, if Cu^+ is not required, the physical basis by which 'non-ionic' PEG interacts with the halide covered electrode needs to be rationalized.

2.4.7
Catalysis of Metal Deposition

In contrast to the inhibiting adsorbates discussed at length in the preceeding sections, some adsorbates exert a depassivating, or depolarizing catalytic effect on the deposition rate. A simple example is the addition of Cl^- to $CuSO_4$ solutions where adsorbed Cl^- catalyzes the Cu^{2+}/Cu^+ reaction as shown in Figure 2.17 [128, 251]. Likewise, sulfide and selenide ions have also been reported to catalyze copper deposition [9, 252]. A more complicated example is thiourea, which, at low sub-micromolar

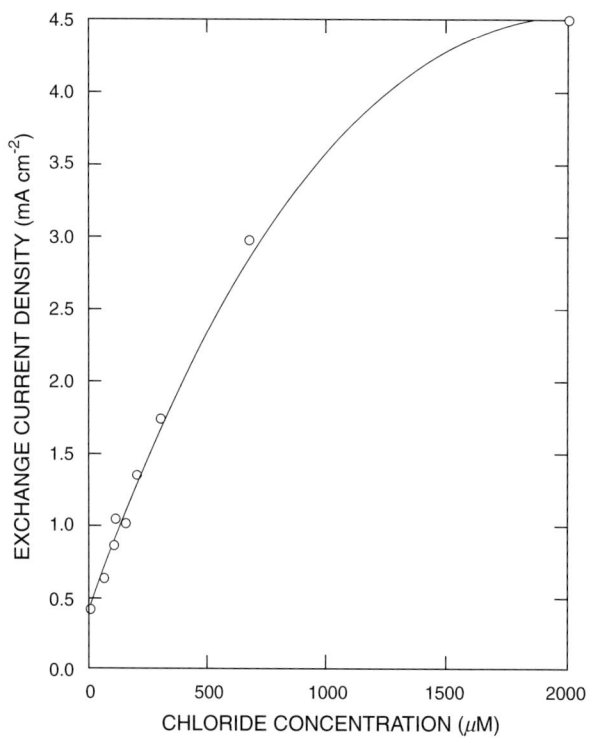

Figure 2.17 Chloride catalysis of the copper deposition reaction is quantified by the increase in the exchange current density of the rate determining Cu^{2+}/Cu^+ reaction with increasing chloride concentration (source Ref. [128]).

concentrations, catalyzes nickel deposition while it inhibits deposition at higher concentrations [3, 216]. In many instances catalysis is provided by an additive that disrupts the inhibition that is otherwise inherent in a given plating chemistry. As an example, in coinage metal deposition from cyanide-based electrolytes, disruption of the inhibiting cyanide based overlayer by the 'catalytic' species, that is, certain chalcogenides and upd (underpotential deposition) sp-metals, leads to an acceleration of the deposition rate [72–76, 121, 253–261]. In other systems a catalytic agent is added to an electrolyte that also contains an inhibitor which is distinct from the metal salt addition; this is the case for bright, acid copper baths as detailed in the following section [12, 62, 66, 67, 136, 243, 262, 264]. At least two different views of the catalytic action have been discussed, one deals with simple displacement of the inhibiting species by the accelerating catalyst as outlined above, while the other invokes the catalyst being involved in a catalytic redox cycle [223, 254, 263] (Figure 2.17).

2.4.8
Activation of Blocked Electrodes by Competitive Adsorption of a Catalyst

A particularly interesting situation exists when inhibitor(s) and catalyst(s) are present in the same electrolyte. A variety of outcomes are possible depending on the relative concentration and adsorption strength of the respective adsorbates and the nature of the adsorbate–adsorbate interactions. By judicious choice of additives, the dynamic effects of competitive adsorption on metal deposition can yield bright deposits by superconformal film growth over a wide range of feature lengths. Representative electrolytes contain a dilute concentration ($\approx 10^{-7} < x < 10^{-4}$ mol L^{-1}) of a strongly adsorbing surfactant catalyst that slowly displaces a less strongly adsorbed blocking layer which formed rapidly upon immersion of the electrode. The key metrics of the activation process are rising i–t [12, 136] or decreasing η–t [265, 266] transients, and hysteretic η–i curves [12, 62, 136, 161, 237, 243, 264, 265], coincident with the formation of a bright compact deposit. The magnitude and time constant associated with transients reflect the dynamic range, that is, passive to active, available for a given additive system.

Two representative examples of this behavior are reflected in two distinct different chemical systems, namely: (a) copper deposition from an acid sulfate electrolyte containing the co-inhibitors PEG–Cl and a bi-functional catalytic species SPS–Cl [12, 136, 243, 264]; and (b) silver deposition from a cyanide electrolyte where inhibition is provided by adsorption of silver cyanide species and catalysis is achieved through adsorption of selenocyanate, SeCN$^-$ [72–75]. Similar behavior is evident in some electrolytes used for the deposition of bright soft gold films [121, 180, 255–261, 267].

The correlation between the electrochemical transients observed on planar electrodes and superconformal film growth in sub-micrometer features indicates that the electrochemical metrics can be used for rapid screening of potential 'superfilling' electrolytes. A more detailed and focused discussion of measurements and modeling of the SPS–PEG–Cl additive system for copper plating follows with the notion that the approach used is directly applicable to many other systems.

Immersion of a copper electrode into a quiescent plating solution, containing 0.2–1.5 mmol L^{-1} Cl, 88 μmol L^{-1} PEG (3400 g mol^{-1} MW) and 1–50 μmol L^{-1} sulfonate-terminated disulfide (SPS), exposes the surface to a high concentration of surfactant polyether species, resulting in the rapid formation of the PEG–Cl blocking layer as outlined above [12, 136, 243]. Subsequently, the more dilute catalytic SPS species disrupts and displaces the PEG-based passivating layer, leading to an acceleration of the metal deposition reaction. This sequence is evident in the hysteretic voltammetric curves shown in Figure 2.18a [12]. The

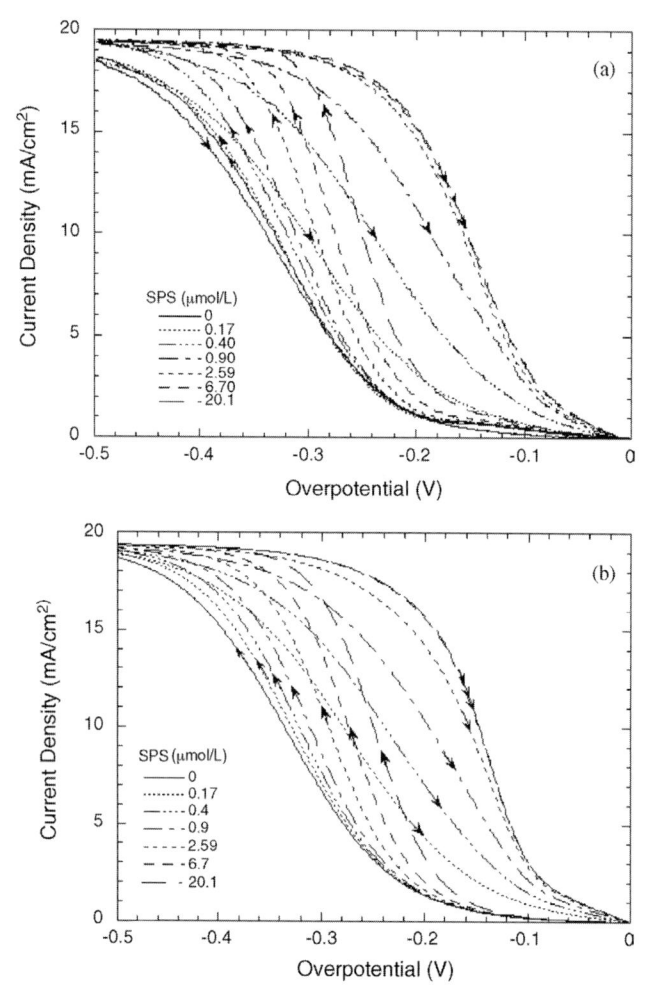

Figure 2.18 (a) Hysteretic voltammograms are obtained with copper electrodes when dilute SPS is added to an acid 0.24 mol L^{-1} CuSO$_4$ electrolyte containing the inhibiting PEG–Cl species. For the 1 mV s^{-1} sweep rate, the response on the return sweep is effectively saturated beyond the addition of 2.59 μmol L^{-1} SPS. (b) Simulations based on the competitive adsorption model outlined in Section 2.4.10, where the blocking PEG–Cl layer is disrupted and displaced by the accumulating SPS [12].

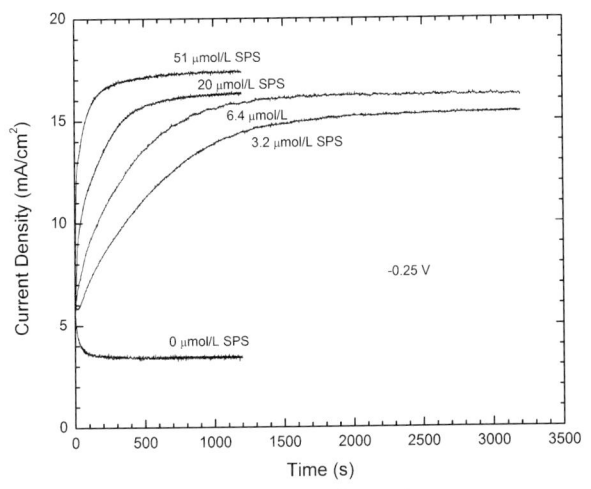

Figure 2.19 Rising chronoamperometric transients characterizing the activation of PEG–Cl inhibited copper electrodes induced by gradual accumulation of SPS adsorbate at −0.25 V [12].

metal deposition rate at a given potential on the negative-going sweep increases with increasing SPS concentration. Further additions of SPS result in progressively higher deposition rates on the negative-going potential sweep; the responses on the return scans are nearly identical, indicating surface saturation with the catalytic adsorbate. Significant brightening of the mechanically abraded Cu electrode, evident after one voltammetric cycle, clearly shows that the hysteretic behavior derives from a change in interfacial chemistry, as opposed to an increase in surface area.

Similar behavior is observed in chronoamperometric experiments as shown in Figure 2.19 [12, 136]. In the absence of SPS, a steady-state current is observed once the hydrodynamic boundary layer has been established ($\approx$30 s). In the presence of SPS the current increases in a manner that qualitatively corresponds to following a vertical trajectory across the hysteretic i–η curves shown in Figure 2.18a. For a given overpotential, the rise time decreases monotonically with SPS concentration as seen in Figure 2.19. Films grown in the absence of SPS were dull and rough while SPS additions resulted in bright films even though deposition took place at a faster rate, that is, in the presence of steeper cupric ion depletion gradients. For chronopotentiometry, decreasing transients are observed [265, 266].

The electrochemical transients, hysteretic η–i, rising i–t and decreasing η–t may be simply and quantitatively explained in terms of disruption and displacement of the rapidly formed PEG–Cl blocking layer by the more gradual adsorption and formation of the catalytic SPS–Cl species [12, 62, 66, 67, 136, 243, 264, 268, 269]. Additional complications arise when considering nucleation of copper on a foreign substrate [270, 271].

2.4.9
Catalyst Function and Consumption

The effect of dialkyl- and diaryl-disulfide additions on copper plating from a simple acidified cupric sulfate electrolyte has been examined [272, 273]. Anionic sulfonate terminated molecules were generally found to slightly perturb the copper deposition rate, while diaryl additions with cationic amino groups resulted in significant inhibition of the metal deposition reaction. A catalytic role of sulfided electrodes in copper plating from thiourea solutions has also been noted [274, 275]. The extent of catalysis was shown to be a sensitive function of the pre-treatment conditions [275]. Pre-sulfiding copper electrodes was also shown to eliminate the inhibition otherwise provided by an OS-20 wetting agent [274].

In the SPS–PEG–Cl system the activation provided by SPS may manifest itself in at least two ways: the disulfide or thiolate group tethers the molecule to the surface, while the charged sulfonate terminal group disrupts and prevents reformation of the passivating PEG-based layer [12, 136]. Related analytical studies of the displacement of non-ionic polymer films by adsorption of anionic surfactants on hydrophobic substrates are available [276]. Much more can be learned about the competitive interaction by using electrodes with predefined adsorbate architectures. For example, the important role of the sulfonate end-group in SPS was demonstrated by examining the rate of copper deposition on various catalyst-derivatized electrodes in a catalyst-free PEG–Cl electrolyte [12, 136, 277]. As shown in Figure 2.20, thiols or disulfides

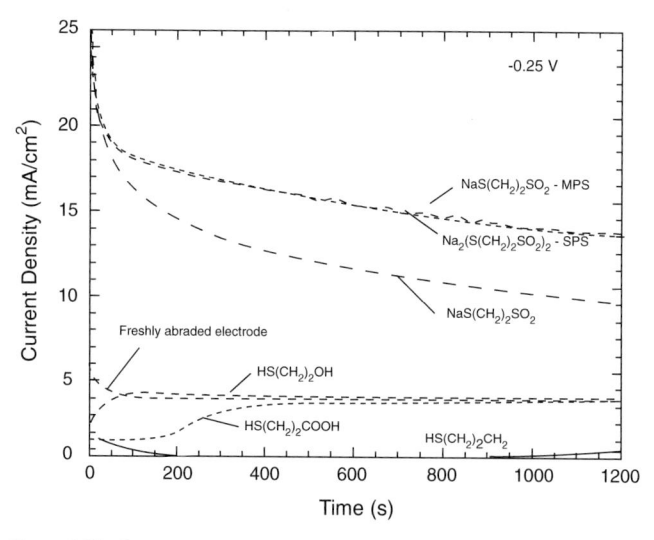

Figure 2.20 Chronoamperometric transients of copper deposition on various thiol and disulfide derivatized copper electrodes. The transients were measured in an acidified $CuSO_4$ electrolyte containing 88 µmol L^{-1} PEG and 1 mmol L^{-1} Cl. The sustained catalytic activity provided by thiol/disulfides molecules with sulfonate terminal groups is evident, in contrast to the transient enhancement of inhibition provided by molecules with OH, COOH or CH_3 terminal groups [12].

with a charged sulfonate terminal group yield significant and sustained catalysis of the metal deposition rate, indicating that they remain segregated on the surface of the deposit and prevent formation of the passivating PEG–Cl film [136]. The extent of disruption is a monotonic function of the initial catalyst coverage, θ_{SPS}.

In contrast to SPS, electrodes derivatized with molecules containing alternative end groups resulted in increased inhibition. In the case of $-CH_3$-terminated molecules, the increased inhibition is sustained for hundreds of seconds, while electrodes modified with $-OH$ or $-COOH$ terminal groups are quickly deactivated, presumably by 'mushroom-like' overgrowth of the metal at defects in the monolayer film that leave the molecule buried at the substrate–electrodeposit interface (e.g. Figure 2.5).

The same experiments can be used to characterize the catalyst consumption kinetics, independent of the ambiguities associated with the adsorption process [12, 136]. Chronoamperometric experiments reveal that electrodes derivatized for 60 s in 0.5 mmol L^{-1} SPS exhibit notable resilience against passivation by PEG–Cl, particularly at higher overpotentials. For comparison, transients for freshly abraded electrodes show the baseline towards which the behavior of the modified electrodes evolves as the catalyst is consumed. Deactivation at -0.05 V (Figure 2.21a) occurs with a time constant of ≈ 200 s, while at -0.25 V (Figure 2.21b) significant catalysis is sustained for a period that is at least an order of magnitude longer [12, 136].

The ability of the catalyst to remain on the surface during rapid metal deposition indicates that it is able to segregate, or float, on the growing metal surface to a significant extent. In contrast, the more rapid consumption at slower growth rates suggests a competing chemical reaction as the origin for the deactivation process. The decaying current transients associated with catalyst consumption can be modeled as a higher order reaction, that is, with exponent $q > 1$, in catalyst coverage [12]:

$$\frac{d\theta_{SPS}}{dt} = -k_{inc}\theta_{SPS}^q \tag{2.17}$$

This may imply that significant lateral interactions occur between catalyst molecules, possibly representing the first stage of copper sulfide formation. The potential dependence of the reaction order, q, and the rate constant, k_{inc}, are obtained by fitting the data as shown in Figure 2.21a,b with the values obtained depending only on the kinetic parameters, α_{SPS} and i_{SPS}^o, obtained from fitting metal deposition on the catalyst saturated surface [12]. The reaction order, q, is fairly well described as a linear function of potential while the rate constant k_{inc} exhibits a more complicated potential dependence, passing through a maximum near -0.1 V (Figure 2.22) as described by a generalized asymmetric peak function.

As is evident in Figure 2.22, significant dispersion exists between the different series of experiments, which is most likely related to some uncontrolled aspect of the derivatization experiments, possibly including the effects of uncontrolled chloride co-adsorption in the different experiments. Nevertheless, the experiments provide an order of magnitude estimate of the consumption parameters and, equally important, reveal the potential dependence of the catalyst deactivation process. The potential

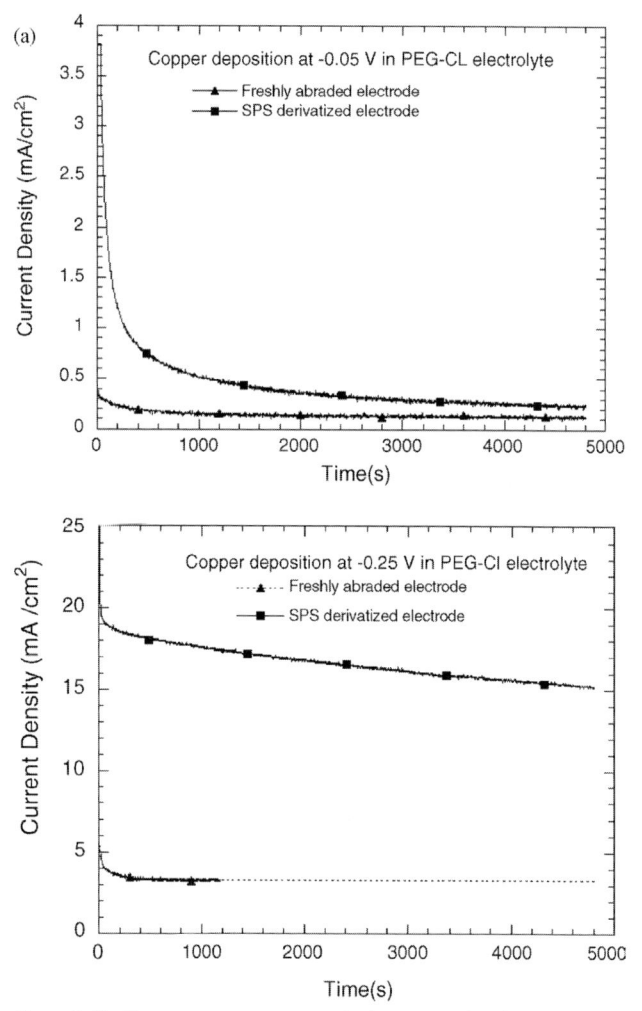

Figure 2.21 Chronoamperometry reveals the potential and time dependence of the catalytic effect of SPS derivatization. (a) At smaller overpotentials, pre-adsorbed catalyst on the electrode is rapidly deactivated. (b) In contrast, the catalytic effect is sustained for thousands of seconds at higher overpotentials. Practical plating is performed under the latter conditions [136].

dependence is arguably counterintuitive as the maximum deactivation rate occurs when the growth rate is low. Interestingly, rotating ring-disc studies of copper deposition reveal that the interfacial concentration of the solvated Cu^+ intermediate also goes through a maximum at small, non-zero overpotentials, ≈ -0.02 V, falling below the experimental background by ≈ -0.15 V [278, 279]. The similarities of the potential dependencies suggest that the enhanced consumption process at small overpotentials is probably coupled with Cu^+ activity at the interface [12], perhaps in a manner analogous to the known interactions between O_2 and Cu^+ [280].

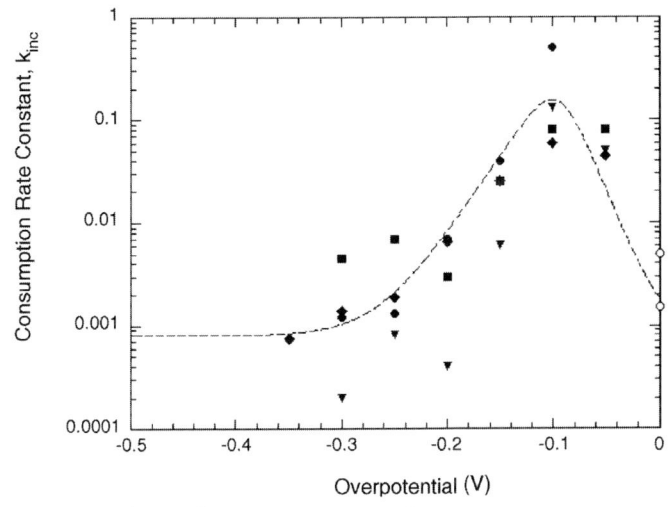

Figure 2.22 The SPS deactivation or incorporation rate constant, $k_{inc}[s^{-1}]$, is a strong function of overpotential, exhibiting a peak value near −0.1 V. The data shown were derived from four separate sets of experiments as denoted by the different symbols [12].

Integration of the chronoamperometric transients for SPS-derivatized electrodes allows the level of SPS incorporation to be estimated. Assuming that the full monolayer of catalyst is incorporated (i.e. no desorption) and using the still incomplete deactivation transients shown in Figure 2.21a,b to obtain the deposit thickness, yields a conservative (upperbound) estimate of the impurity levels [12]. The sulfur content in the deposits is estimated to be 4 and 200 ppm S for growth at −0.25 and −0.05 V, respectively. The potential dependence and magnitude are in reasonable agreement with published SIMS data on impurity content, an example of which is given in Figure 2.23; care must be taken in such comparisons to make sure that recrystallization has not affected the results of the post-deposition analysis [210–212].

2.4.10
Quantifying the Effects of Competitive Adsorption on Metal Deposition

A detailed understanding of competitive adsorption and its effect on metal deposition reactions clearly remains to be established. However, in order to proceed to shape change simulations, simplification by way of chemical process models can be implemented as in the example that follows.

In the SPS–PEG–Cl system the central role of the SPS-based catalyst is to open channels in the PEG–Cl blocking layer, thereby allowing the Cu^{2+}/Cu^+ reaction to proceed unhindered. This effectively accelerates the metal deposition reaction without requiring any change in the Cu^{2+}/Cu reduction mechanism, that is, activation of a blocked electrode. Accelerated copper deposition occurs in the proximity of SPS adsorption sites. Chloride adsorbed in neighboring sites may also

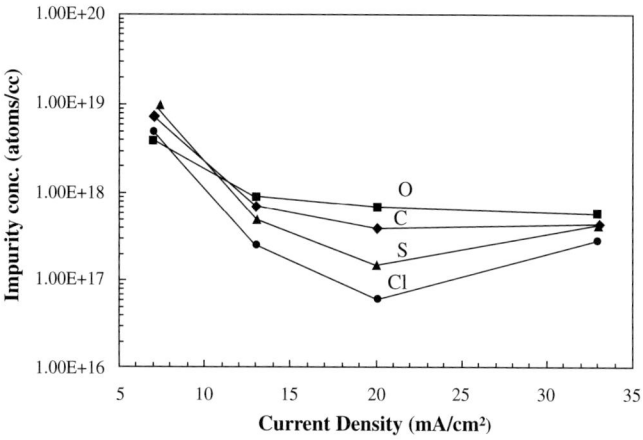

Figure 2.23 Impurity concentration in the bulk of a copper deposit is a function of the applied current density. For reference, the atom density of bulk copper is $8.49 \cdot 10^{22}$ cm^{-3} (2 1 1).

exert its well known catalytic properties for copper deposition, thereby expanding the effective area catalyzed. Nonetheless, for simplicity, the total number of available surface sites is taken here as the sum of the coverages of the inhibiting PEG–Cl species, θ_{PEG-Cl}, and the catalytic SPS–Cl species, θ_{SPS-Cl} [12]:

$$\theta_{SPS-Cl} + \theta_{PEG-Cl} = 1 \tag{2.18}$$

Chloride is known to form compact ordered monolayers on copper surfaces under open circuit conditions in simple halide solutions; for example, $c(2 \times 2)$ Cl on Cu (100) [80, 134, 246, 247]. However, little is known about the adsorption geometry of the inhibiting PEG–Cl or the catalytic SPS–Cl complex, both highly topical questions for further research.

2.4.10.1 Site Dependence of Charge Transfer Kinetics

The effect of the competitive adsorption on the kinetics of the copper deposition reaction is described as the product of the fractional interface cupric ion concentration $C_{Cu}^{2+}/C_{Cu}^{2+\infty}$ and the coverage dependent charge transfer rate constants, k_i using:

$$i(\theta_{SPS}, \eta) = \frac{C_{Cu^{2+}}}{C_{Cu^{2+}}^{\infty}} [k_{PEG}(1 - \theta_{SPS}) + k_{SPS}\theta_{SPS}] \tag{2.19}$$

and Equation 2.18 is used to eliminate explicit reference to the PEG coverage:

$$k_j = 2i_j^o \left[\exp\left(\frac{-\alpha_j F\eta}{RT}\right) - \exp\left(\frac{(1-\alpha_j) F\eta}{RT}\right) \right] \tag{2.20}$$

The exchange current, i_j^o corresponds to the rate-determining step for copper deposition on surface sites j and a factor of two is used to account for the overall two-electron Cu^{2+}/Cu process. The values of i_{PEG}^o and the transfer coefficient α_{PEG} are readily determined from η–i and i–t experiments performed in a catalyst-free PEG–Cl electrolyte. Evaluation of i_{SPS}^o and α_{SPS} is more difficult. It involves examining the response of an SPS-derivatized electrode in the same PEG–Cl catalyst-free electrolyte with appropriate consideration given to catalyst consumption, as described in the previous section [12]. The interface concentration of Cu^{2+} can be related, assuming steady-state hydrodynamic conditions, to the diffusion-limited current density, i_L through:

$$\frac{C_{Cu^{2+}}}{C_{Cu^{2+}}^{\infty}} = 1 - \frac{i}{i_L} \quad \text{and} \quad i_L = \frac{2FD_{Cu^{2+}} C_{Cu^{2+}}^{\infty}}{\delta} \tag{2.21}$$

2.4.10.2 Catalyst Evolution

Evolution of the SPS-based catalyst on a planar surface is dictated by both adsorption and consumption. These are quantified using the effective potential dependent rate constants for SPS adsorption k_{ads} and incorporation k_{inc}, respectively, with a power law exponent q for the consumption. For catalyst precursor concentration C_{SPS} one obtains:

$$\frac{d\theta_{SPS}}{dt} = k_{ads}(1 - \theta_{SPS})C_{SPS} - k_{inc}\theta_{SPS}^q \tag{2.22}$$

at the planar metal/electrolyte interface [12].

2.4.10.3 SPS Adsorption from the Electrolyte

In an electrolyte containing a dilute SPS concentration, disulfide adsorption occurs coincident with displacement of the less strongly bound PEG-based inhibitor. The adsorption or exchange process can be described by irreversible Langmuir kinetics [112, 135, 281]. The effective rate constant, k_{ads}, is potential dependent, the rate increasing exponentially with overpotential η according to

$$k_{ads} = k_0 \exp\left(-\frac{\alpha_{ads} F \eta}{RT}\right) \tag{2.23}$$

The symmetry of the activation barrier, α_{ads}, and the pre-exponential term, k_0, are determined by fitting the experimental η–i curves on the planar substrates over the range of SPS concentration (e.g. Figure 2.18b). The potential dependence requires further comment [12].

Potential dependent halide adsorption, including order–disorder phenomena, is well known [134, 246, 247], and it is reasonable to expect that the breakdown of the PEG–Cl blocking layer might also be potential dependent. In a similar fashion, the composition and structure of thiols and disulfides adsorbed on gold from simple electrolytes have been shown to be potential dependent [282]. In the present example, it is also possible that the approach of SPS to the electrode surfaces is screened by complexation with the potential dependent concentration of Cu^+ that is generated at the electrode. Importantly, the equilibrium Cu^+ concentration in the additive-free system (i.e. Equations 2.1 and 2.2) is of the order of $\approx 400\,\mu mol\,L^{-1}$ and Cu^+ is known to form complexes with all the additives under consideration [239, 279, 280, 283–285]. Furthermore, the equilibrium Cu^+ concentration decreases exponentially with potential, that is, 60 mV per decade of concentration [283–285]. Thus, the increasing rate of SPS adsorption with overpotential may partly reflect the decreasing interfacial Cu^+ concentration scavenging SPS (in addition to chloride, oxygen [280, 283], etc.) as it approaches the interface. Further work will be required to elucidate the relative contributions of these processes to the potential dependence of k_{ads}.

As the rate constant k_{ads} increases rapidly with overpotential, the adsorption process must eventually become constrained at higher overpotentials by diffusion of SPS in the electrolyte. Mass balance between the catalyst accumulating on the interface and diffusing through the electrolyte is given by:

$$k_{ads}(1-\theta_{SPS})C_{SPS} = \frac{D_{SPS}}{\Gamma_{SPS}}\frac{(C_{SPS}^\infty - C_{SPS})}{\delta} \tag{2.24}$$

Rearranging, one obtains the SPS concentration adjacent to the interface:

$$C_{SPS} = \frac{C_{SPS}^\infty}{1+\dfrac{k_{ads}\Gamma_{SPS}\delta(1-\theta_{SPS})}{D_{SPS}}} \tag{2.25}$$

to be used in Equation 2.22.

Iteratively fitting all the voltammetry and chronoamperometry data for SPS-derivatized electrodes yields a set of values for i_{SPS}^0, α_{SPS}, k_{inc} and q that simultaneously provide a good description of all the experimental results, for example, Figures 2.18, 2.19 and 2.21.

With the parameters, i_{PEG}^0, α_{PEG}, i_{SPS}^0, α_{SPS}, $k_{inc}(\eta)$ and $q(\eta)$ derived from experiments using either freshly abraded or SPS-derivatized electrodes in SPS-free PEG–Cl electrolyte, the hysteretic η–i data for the full SPS–PEG–Cl electrolyte were used to fit the catalyst adsorption parameters k_0 and α_{ads} [12]. The negative going voltammetric sweep is particularly sensitive to the adsorption parameters, while the saturated return scan is dominated by the potential dependence of catalyst consumption, k_{inc}.

It should be recognized that the derived parameter set is not unique; because of the scatter in experimental consumption transients (i.e. Figure 2.22), equally good fits to the η–i and i–t SPS–PEG–Cl data yield a locus of $k_{inc}(\eta)$ and $q(\eta)$ parameters [12]. Further study and refinement of the derivatization procedure should help clarify the source of the experimental dispersion. In addition, different combinations of parameters yield distinct predictions for the catalyst evolution so that surface analytical experiments should help narrow the selection. Despite these uncertainties, a correct description of the SPS–PEG–Cl system clearly includes both potential dependent adsorption and deactivation processes. This conclusion is reinforced by multicycle voltammetry where the potential dependence of adsorption that is most significant at large overpotentials is convolved with consumption that dominates catalyst evolution at low overpotentials.

As shown in Figure 2.24, repetitive potential cycling reveals that the second and subsequent negative going η–i sweeps do not follow the path of the preceding return sweep beyond −0.05 V. This reflects the onset of significant deactivation

Figure 2.24 Variation of the voltammetric switching potential reveals the strong effect of potential-dependent adsorption and consumption of the SPS-derived catalyst (adapted from Ref. [12]).

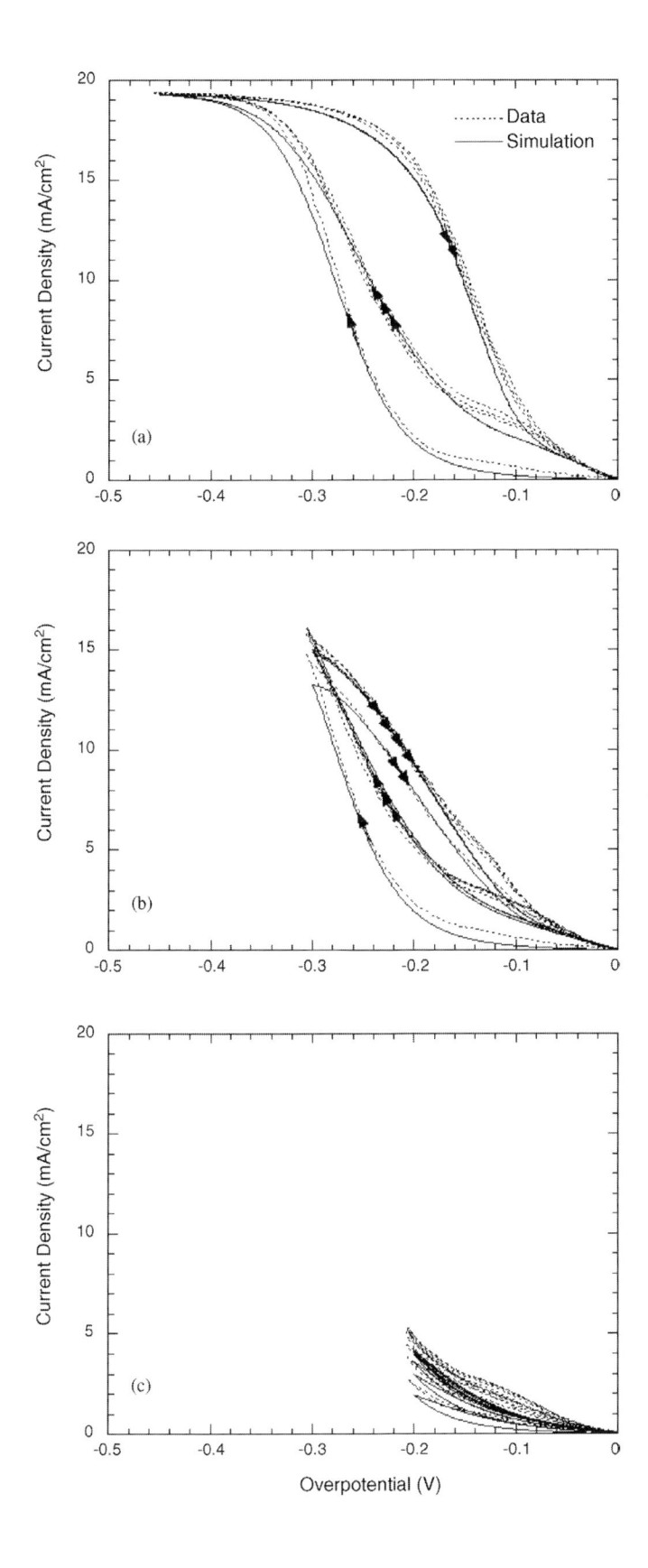

starting by this overpotential; an observation that is consistent with the chronoamperometric deactivation transients summarized in Figure 2.22. A steady-state η–i response is attained generally by completion of the second cycle, although the precise number of cycles depends on the SPS concentration and other parameters, including the switching potential and scan rate (not shown). The potential dependence of SPS adsorption is most clearly revealed by changing the switching overpotential while the sweep rate (1 mV s^{-1}) and total scan time (2400 s) are held constant. The high metal deposition rates that characterize the return sweep for a switching overpotential of -0.45 V are not accessible if the sweep is reversed at less than -0.25 V. Voltammograms at intermediate values are consistent with this trend. These results demonstrate that the rate of displacement of the inhibiting PEG–Cl layer by SPS adsorption is an increasing function of overpotential; voltammetric cycling to larger overpotentials permits a larger increase in the catalyst surface coverage that manifests as higher currents on the return sweep [12, 136].

Interest in using low acid electrolytes has developed and care must be taken to assay what, if any, effect pH will have on the optimization of the above chemistry [286, 287]. Likewise, attention must be given to the correct evaluation of kinetic parameters in the presence of significant ohmic losses associated with the lower conductivity electrolytes.

2.5
Interface Motion and Morphological Evolution

Progress in understanding morphological evolution during electroplating has advanced substantially in recent years. The activity involves both experimental characterization of electrode shape change and simulations thereof. The quality of shape change simulations has increased in step with the steady advances in computational power [18, 64, 288] and predictions can be performed over a wide length scale [289, 290]. At the same time, advances in lithographic patterning and various forms of microscopy have facilitated higher throughput experimental studies of the filling and smoothing of sub-micrometer features, thereby providing insight into surface brightening phenomena. It is worth noting that in the 1960s and 1970s, the same experimental approach was used to study leveling phenomena by metallographic examination of cross-sectioned electrodeposits formed on grooved substrates [1–5]. The patterning in that instance was derived from phonograph or gramophone master templates, micromachining, embossing, etc., and then and today reflect the science–technology push–pull that characterizes most technological endeavors. Forging a connection between kinetics measurements on planar electrodes and morphological evolution was, and still is, an important aspect of research in this field.

2.5.1
Shape Change Simulations

A quantitative description of feature filling requires a shape change algorithm to convert the quantitative description of the effect of adsorbates on metal deposition

2.5 Interface Motion and Morphological Evolution

on planar electrodes into simulations of morphological evolution. For simple geometries, this may yield an analytical form although more commonly numerical evaluation of the construct is required. Regardless of the number and type of additives, the formalism should: (a) provide a description of the functional relationship between the metal deposition rate and additive adsorbate coverage; (b) ensure complete adsorbate mass balance, including accumulation from the electrolyte, the effects of segregation on non-planar surface segments and adsorbate deactivation or incorporation in the solid, etc.; and (c) account for the temporal and/or spatial variation of concentration (and potential) fields within the electrolyte and how this relates to the metal deposition rate and evolution of additive surface coverage.

For the micrometer and sub-micrometer feature filling applications detailed in this chapter, negligible overpotential variation occurs on the feature length scale [1–4, 291]. In contrast, such effects can become fairly important at the work piece or wafer length scale; careful engineering of the electrochemical cell, electrode contacts, seed layer resistance and supporting electrolyte is required in order to obtain a uniform current distribution across larger electrodes. A selection of reports detailing various strategies for effective plating tool design are available in the literature [1–4, 19, 288, 292–301].

In all simulations, interface motion is related to the metal deposition rate through the requirement that the time-derivative of the local surface position, $\vec{S}$, namely the local growth velocity, v,

$$\frac{d\vec{S}}{dt} = \vec{v} \qquad (2.26)$$

is related to local current density by Faraday's law (assuming 100% current efficiency):

$$\vec{v} = \frac{i(\theta_j, \eta)\Omega\,\vec{n}}{nF} \qquad (2.27)$$

where $\vec{n}$ is the local unit normal to the surface pointing into the electrolyte, while n is the number of equivalents. Reiterating three key equations, the current density (i.e. deposition rate) is proportional to the metal ion concentration in the adjacent electrolyte and adsorbate coverages:

$$i(\theta_j, \eta) = \frac{C_{M^+}}{C_{M^+}^\infty} \sum_j k_j \theta_j \qquad (2.28)$$

As discussed in Section 2.4, a linear dependence of growth rate on adsorbate coverage (s) is presumed where the rate constant(s) k_j are described by potential-activated Butler–Volmer kinetics

$$k_j = n i_j^0 \left[\exp\left(\frac{-\alpha_j F\eta}{RT}\right) - \exp\left(\frac{(1-\alpha_j)F\eta}{RT}\right) \right] \qquad (2.29)$$

and the individual i_j^0 and α_j are determined from metal deposition rates on electrodes saturated with the respective adsorbate, $\theta_j \to 1$.

The most general form of adsorbate evolution is given by the sum of the individual processes, j, detailing additive adsorption, desorption, incorporation, deactivation, reduction, area change, etc.:

$$\frac{d\theta_j}{dt} = \sum_l \left(\frac{d\theta_j}{dt}\right)_l \tag{2.30}$$

Co-adsorption and competitive interactions within a given mechanism may be accounted for by a linear combination of terms, although higher order and cross-term dependences may also be applicable.

As deposition proceeds, depletion of the metal ion(s) and additive(s) in the electrolyte must be accounted for in the context of the given electrode geometry. A complete description involves solution of the time-dependent diffusion equation:

$$\frac{\partial C_j}{\partial t} = D_j \nabla^2 C_j \tag{2.31}$$

However, in many models the time independent Laplacian form:

$$\nabla^2 C_j = 0 \tag{2.32}$$

is used, consistent with the assumption of a pseudo steady-state deposition process.

The distinction between the two forms has been examined in some detail as sub-micrometer feature filling and relaxation of the diffusive boundary layer occur on the same time scale [64, 302]. Boundary conditions are required at the top, bottom and sides of the domain as shown in Figure 2.25; a statement of the initial conditions is also required for the time-dependent form. Where temporal evolution is considered, the relaxation of the boundary layer is usually arrested at the same value, δ, corresponding to the experimentally relevant hydrodynamic conditions used in

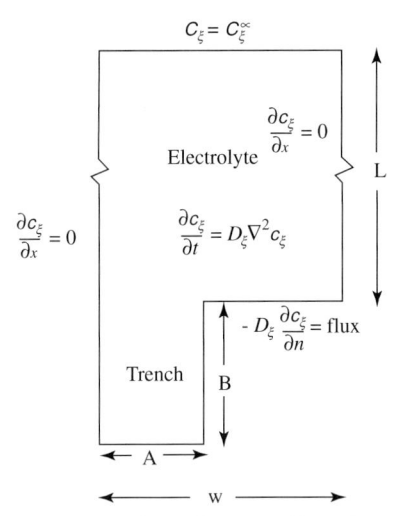

Figure 2.25 Solution domain and boundary conditions for trench filling simulations (source Ref. [64]).

time-independent solutions. In many instances the trench or via spacing is assumed to be periodic in the plane of the electrode permitting a zero lateral flux boundary condition, for all constituents, to be imposed midway between features, as required by symmetry. While the modeled trench spacing is frequently large compared with other dimensions, so as to avoid interactions between the associated depletion gradients, some studies have focused explicitly on modeling the effects of such overlap in order to assess the effect of pattern size and arrangement on feature filling.

Mass balance considerations provide the boundary condition between the electrolyte and growing solid. For the metal ion/metal deposition reaction the diffusive flux of the metal ion from the electrolyte equals the rate of deposition on the surface:

$$-D_M \frac{\partial C_M}{\partial n} = v \left(\frac{1}{\Omega} - \frac{1}{\Omega_L} \right) \qquad (2.33)$$

where the second term in the parentheses accounts for the molar volume Ω_L of metal ions in the electrolyte, a term that is small relative to that of the metal (Ω) and thus usually ignored. The mass conservation statement for an adsorbate depends on the details of the adsorption model, but may be summarized as the balance between the diffusive additive flux to the surface and the adsorption processes that perturb the catalyst concentration in the adjacent electrolyte:

$$-D_j \frac{\partial C_j}{\partial n} = \Gamma_j \left[\left(\frac{d\theta}{dt} \right)_{ads} + \left(\frac{d\theta}{dt} \right)_{des} \right] \qquad (2.34)$$

Shape change algorithms vary from simple analytical descriptions to a variety of numerical models of increasing sophistication. For example, numerical front tracking or string models [12, 62] and surface triangularization (3D-simulations [303]) methods that evaluate morphological evolution by tracking the interface and locally varying adsorbate coverage in the presence of uniform concentration fields, or some close approximation thereof, allow for the relatively rapid evaluation of morphological evolution. Boundary element [55–57, 304, 305] and related methods [302, 306, 307], and more recently the level-set [64, 308] method, provide for a more robust description that tracks interface position and chemistry, while consistently evolving the concentration fields in the electrolyte using finite difference schemes.

During deposition in trenches and other recessed patterned features, metal ion depletion gradients within such features generally act to the detriment of filling. If the feature aspect ratio exceeds a certain value, a seam or void can form along the center of the structure depending on the extent of metal ion depletion. However, in certain instances, such depletion effects are of secondary importance to superconformal filling of sub-micrometer features. In such situations, certain simplifying approximations may be used to efficiently ascertain the important trends in the mechanism and to map a wide range of parameter space. More often than not, this involves circumventing the computationally expensive evaluation of the Laplacian associated with the metal ion and additive(s) gradients in the electrolyte. As a first cut, simplification of the work piece geometry towards a more symmetrical shape can enable an analytical description [1–4]. This is effective as long as the general shape is sustained during deposition.

Another approach is to maintain the original electrode shape and map the diffusional flux from a simplified geometry onto the actual work piece. For example, depletion across a 150 μm wide hydrodynamic boundary layer is approximated as spatially uniform while the local gradients within a 0.5 μm trench of interest are ignored or further approximated. Such techniques have been usefully employed in both analytical and interface tracking algorithms for feature filling using the curvature enhanced accelerator coverage mechanism [12, 62, 63]. Related simplifications were also used in earlier attempts to predict the magnitude of the so-called 'leveling power' of certain electrolytes [231]. As with any approximation, care must be taken to evaluate the range and limits of applicability imposed by such choices.

In spite of the success of some simplified feature filling models, a complete and robust quantitative assessment of general feature filling requires that the relevant metal ion and additive gradients be properly evaluated. Computer codes must calculate the concentration fields within the electrolyte, including time evolution where appropriate, subject to boundary conditions on the moving metal/electrolyte interface. Studies to date have focused on comparing predicted shape evolution with experiment. The generated codes explicitly track interface position and adsorbate coverage. As such, it should be straightforward to follow additive incorporation in the growing solid although this aspect of feature fill modeling presently remains largely unexplored.

The stability of a planar interface towards roughening is closely related to feature filling. Indeed, as outlined in Section 2.2, the destabilizing influence of metal ion depletion gradients on general morphological evolution is widely recognized. It is also well known that electrolyte additives can exert a strong influence on roughness evolution, and this aspect of smoothing mechanisms has been studied by linear perturbation methods [309, 310], dynamic scaling [38–43] and numerical evolution of arbitrary surface profiles [311]. In contrast to feature filling simulations, the linear stability analysis involves examining the resilience of a planar interface to a perturbation of variable wavelength and infinitesimal amplitude. For analytical simplicity, this usually involves a sinusoidal perturbation. A limitation of such analysis is the possibility that the impact of higher order terms may be lost due to linearization. Numerical models do not suffer this shortcoming, but the results are constrained to the specific geometry examined. The utility of scaling analysis to characterize and model roughness evolution [43, 46, 47] is a well known subject in the MBE and PVD film growth community. However, modeling electrochemical and chemical deposition faces additional technical challenges, largely absent from those formulations, including the impact of concentration gradients and adsorbates on the metal deposition process. Some formative steps in addressing these challenges have been taken [39, 42, 129].

2.5.2
Geometric Leveling

In the absence of reactant depletion gradients, that is film growth under charge transfer control, and absent significant electrocrystallization anisotropy, a uniform

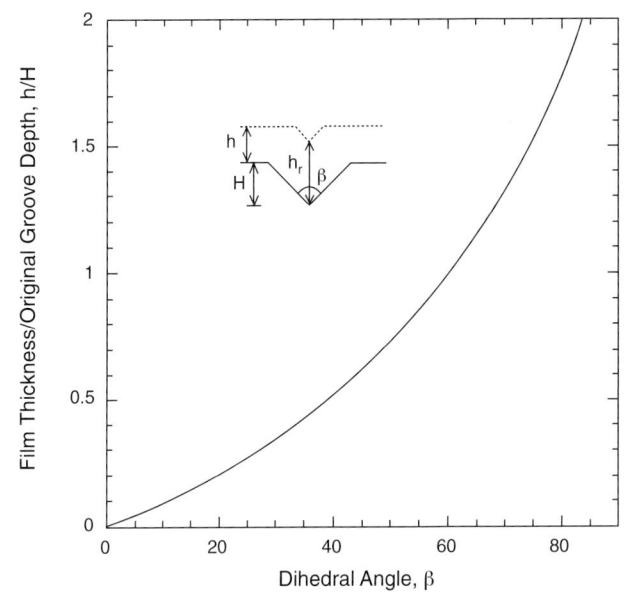

Figure 2.26 Conformal deposition in a V-notch groove gives rise to geometric leveling. As shown, the dihedral angle β of the feature substantially impacts the deposit thickness in the field, h, that is required to completely fill a groove of depth H.

current distribution (i.e. growth velocity) can be anticipated on all planar surfaces [1–4]. For a grooved surface such as shown in Figure 2.26, conformal deposits of thickness h on the planar surface results in greater thickness h_r at the bottom midplane of the groove as given by:

$$h_r = \frac{h}{\sin(\beta/2)} \quad (2.35)$$

Thus, a groove of initial depth H would be completely filled once the deposit thickness on the surrounding planar field h satisfies:

$$h = \frac{H}{[1/\sin(\beta/2)] - 1} \quad (2.36)$$

The sharp dependence of the ratio h/H on the dihedral angle β of the sidewalls is shown graphically in Figure 2.26. An example of geometric leveling is provided by nickel deposition in a 30 μm deep semicircular groove from an additive-free $NiSO_4$–$NiCl_2$–H_3BO_3 electrolyte [58]. As shown in Figure 2.27, for the semicircular geometry no leveling occurs until a thickness equivalent to the initial depth of the hemicylinder is deposited and a V-notch is formed. At this point, geometric leveling begins with formation of the groove corresponding to a discontinuity in the time derivative of the mid-plane height of the surface. The contour lines correspond to a

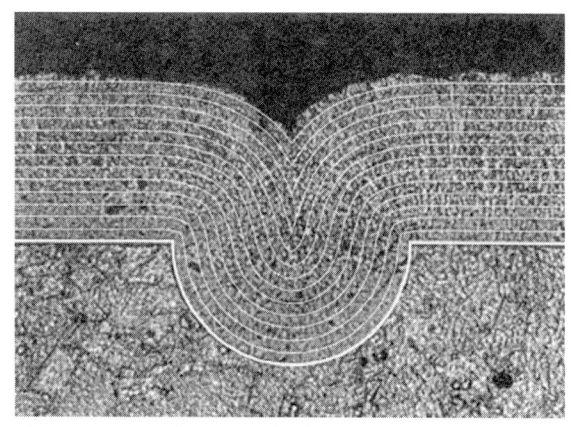

Figure 2.27 Optical micrograph of 45 μm thick nickel film deposited in a 30 μm deep semicircular groove. The film was grown at 10 mA cm^{-2} and 900 rpm from an additive-free NiSO$_4$ NiCl$_2$ H$_3$BO$_3$ electrolyte. The white lines represent simulations for geometric leveling corresponding to successive deposition of nickel layers 3.5 μm thick. Note significant leveling only occurs after a V-notch has formed (source Ref. [58]).

simulation based on kinetic parameters (rate constants and transfer coefficients) derived from measurements on planar electrodes in the same solution [58]. The excellent agreement between experiment and simulation is congruent with the assumption of uniform current distribution across the surface profile.

A similar filling sequence has been observed during conformal deposition in sub-micrometer trenches with sloping sidewalls [243, 312, 313]. Initially, deposition is conformal but once the bottom corners meet to form a V-notched geometry, geometrical leveling leads to rapid filling of the feature. For a triangular notch, or flat-bottom trapezoidal trench or via, the growth surface is planar upon completion of the geometrical leveling and remains so during subsequent deposition. Of course, other things being equal, as β is reduced, reactant depletion effects within the groove eventually become significant, and sub-conformal feature filling results. Quantitative analysis in this instance requires coupling the feature geometry with the developing concentration field gradients.

A similar analysis has been applied to filling faceted notches by a CVD process [314]. The influence of kinetic roughening on feature-filling efficiency must also be evaluated. A predictive formalism for assaying the effectiveness of conformal growth in filling features based on simple scaling parameters measured on a planar substrate is available although, as noted above, close attention needs to be given to depletion effects [315].

Finally, when examining other smoothing mechanisms it is important to note that geometric leveling can contribute substantially to feature filling whenever sloping sidewalls are present [243, 312, 313]. Indeed, engineering features with slightly tilted sidewalls is known to greatly enhance the void-free filling capability and reliability of certain chemistries.

2.5.3
Inhibitor-based Leveling

2.5.3.1 Feature Filling

True leveling is associated with adsorption of an inhibitor that blocks the local metal deposition reaction. The balance between inhibitor accumulation and deactivation determines the local steady-state surface coverage. For significant leveling to occur, inhibitor deactivation or consumption at the surface must happen at a high enough rate that the local coverage is determined by diffusion of the leveler to the given surface profile. In this instance, the greater flux to asperities and other high points, relative to that going to recesses leads to a spatially distributed inhibitor coverage. The enhanced attenuation of the metal deposition rate at high points gives rise to superconformal filling of surface recesses. A variety of leveling agents have been identified for different systems, although most attention has been focused on nickel and copper [1–9, 55–58, 316–319].

Quantitative studies of leveling define a term known as 'leveling power' (LP) for a given surface profile, such as a groove. Slight variations in the definition used by various researchers have hindered comparison between data sets. Perhaps the most widely applied definition is given by the difference between the currents on peaks and valleys normalized by the global or average current density:

$$LP = \frac{i_{peak} - i_{recesses}}{i_{avg}} \quad (2.37)$$

Understanding of the relationship between leveling performance, mass transport and polarization conditions was largely in place by the 1960s. The studies of Kardos and Foulke [1–4] focused on the influence of variations in the diffusion layer thickness on the deposition rate and its effect on feature filling. Watson and Edwards [4, 17, 216] examined the influence of varying leveling agent concentration on the same. By doing so, they were both able to relate the LP to electrolyte composition. Measurements of the effect of additives on the polarization behavior of a planar surface enabled the following expression to be evaluated:

$$d\eta = \left(\frac{\partial \eta}{\partial C_a}\right)_i dC_a + \left(\frac{\partial \eta}{\partial i}\right)_{C_a} di \quad (2.38)$$

at the feature length scale $d\eta = 0$, and if the additive is under transport controlled accumulation then $dC_a = kC_a$. Then combining Equations 2.37 and 2.38, yields:

$$LP = \frac{kC_a \left(\frac{\partial \eta}{\partial C_a}\right)_i}{i \left(\frac{\partial \eta}{\partial i}\right)_{C_a}} = -\frac{kC_a}{i}\left(\frac{\partial i}{\partial C_a}\right)_\eta \quad (2.39)$$

where C_a is the leveler concentration, η the overpotential, i the associated average current density and k is determined from the Laplacian solution for the metal deposition flux to the given electrode geometry [4, 17, 216]. The formalism was able to predict the optimum leveling agent concentration for nickel plating in the presence of coumarin or thiourea. One limitation of this description is that neither the surface coverage nor transport conditions at the evolving interface were explicitly defined or evaluated [57, 58].

Subsequent work by Kruglikov and coworkers [217, 316] explored the action of levelers under well defined hydrodynamic conditions. The analogy between the diffusive flux distribution and the primary current distribution (i.e. Laplace equation) was used for calculating the leveling power. The expression for the attenuation of the amplitude of a sinusoidal profile was shown to be:

$$LP = \frac{4\pi H}{\lambda} \left(\frac{\partial \ln i}{\partial \ln \delta} \right)_\eta \quad (2.40)$$

where i is the current density, δ is the boundary layer thickness (fixed value with respect to the mean surface height), and H and λ are the amplitude and wavelength, respectively, of the sinusoidal profile. The results were shown to be applicable to surface profiles with wavelengths less than or equal to the boundary layer thickness and for amplitudes that were a small fraction of the wavelength. Interferometric measurements may be used to monitor the filling of such features [320]. If the inhibiting leveling agent is consumed at the diffusion limited rate, then Equation 2.40 is equivalent to Equation 2.39 [17, 316].

Krichmar and coworkers [17, 321–323] performed similar studies examining the attenuation of sinusoidal surface profiles when adsorption of the leveling agent was under transport control. Subsequently, a more detailed dimensionless analysis was performed with full consideration given to mass balance of the leveling additive [147, 324]. Using parameters constrained by experiment, a maximum in the effective 'leveling power' or smoothing rate was predicted as a function of the ratio of the flux of the inhibitor to that of the metal cation, as shown in Figure 2.28 [324]. Likewise, the dependence of leveling power on the normalized current density was examined as a function of three dimensionless parameters: (a) the fractional coverage of the inhibitor, (b) the ratio of the leveler and metal cation fluxes and (c) a term reflecting the magnitude of the Langmuir adsorption constant.

The above analytical solutions are of limited utility for describing the filling of more complicated features. The need for a more precise description for leveling in complex geometries, combined with the rapid advances in computational power, fostered the first numerical shape change simulations in the 1990s. A sequence of three publications detailed progressively more sophisticated leveling models. In the first model the flux of leveling agent was transport limited and the surface overpotential was assumed to depend on both the current density and transport-limited leveler flux [55]. An overpotential relationship was interpolated from experimental results drawn from the literature [231] in order to simulate coumarin leveling of a V-shaped groove profile. In the second model, the effect of the leveling agent was described in terms of fractional blocking of the surface by the adsorbed species [56]. In this situation the additive was incorporated at a rate proportional to the local current density and the leveling agent flux was again assumed to be transport limited. The simulations predicted an optimum in leveling power as a function of inhibitor flux. Qualitative agreement with the available experimental results [217] was reported.

$$\alpha = \frac{D_{leveler} V_{leveler} C_{leveler}}{D_{M+} V_{M+} C_{M+}}$$

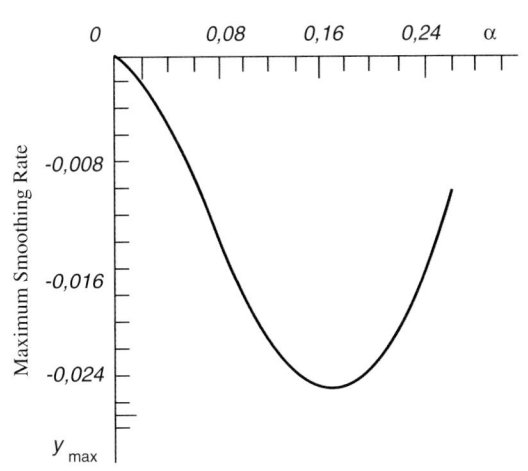

Figure 2.28 The leveling or smoothing power is a function of the ratio of the leveler flux to that of the metal as characterized by the respective diffusion coefficients D_j, molar volume v_j and concentrations C_j (source Ref. [324]).

A shortcoming of these early models was the absence of a causal relationship between additive concentration and inhibitor coverage. This was explicitly included in a subsequent pair of publications that provided a direct one-to-one comparison between theory [57] and feature filling experiments [58] of coumarin leveling during nickel deposition in hemi-cylindrical and V-notch geometries. The kinetic parameters used for the shape change simulations were determined by experiments on planar electrodes. According to the inhibition model (previously outlined in Section 2.4.5), coumarin adsorption is described by a Langmuir isotherm with the ratio of the transport limited diffusional flux and deactivation rate constant of the coumarin being given by the dimensionless parameter μ ($= DC_b/k_A\delta$) (Equation 2.16 and Figure 2.13). A comparison between experiment and simulation (overlying contours) of the filling of a hemi-cylindrical groove is shown in Figure 2.29. Good agreement is evident with a slight asymmetry in the experimental results being attributable to perturbation of the boundary layer associated with fluid flow over the geometry [58, 325–327].

Similar effects are observed in through mask-plating and etching [328–330]. Leveling performance over a range of processing conditions, with a current density of 5 or 50 mA cm^{-2} and rotation rate varying from 100 to 400 rpm, is summarized in Figure 2.30 as a function of the dimensionless parameter μ [58]. The leveling power (LP) was evaluated from the measured groove depth i_{field}/i_{valley} at a specified time and found to exhibit a maximum near $\mu = 1$. Values significantly higher result in an insignificant gradient in leveler concentration due to saturation of the surface coverage, while much lower μ values lead to insignificant leveler coverage across the feature.

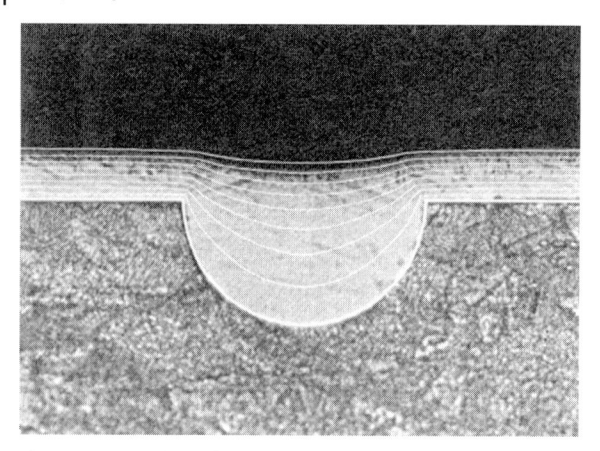

Figure 2.29 Leveling of a 30 μm deep semicircular groove by nickel deposition, at 10 mA cm^{-2} and 400 rpm, from an NiSO$_4$–NiCl$_2$–H$_3$BO$_3$ electrolyte containing 1 mmol L^{-1} coumarin. Rapid leveling is apparent with feature filling completed after only 15 μm of nickel have been deposited on the surrounding field. The contour lines, spaced 1.5 μm apart on the field, represent simulations of leveling for $\mu = 1$, $\gamma\lambda = 36$ and a boundary layer thickness δ that is 0.8 of the original groove depth (source Ref. [58]).

A significant difference between leveling models is the manner by which the inhibiting adsorbates are deactivated. In general, leveler coverage evolution is obtained from mass balance allowing for adsorption, desorption and deactivation:

$$\Gamma_{lev}\frac{d\theta_{lev}}{dt} = Ck_{ads}(1 - \theta_{lev}) - k_{des}\theta_{lev} - \sum_j W\theta_{lev} \tag{2.41}$$

The deactivation term, W, may correspond to potential dependent reduction, k_{red}, fragmentation or consumption (as in Equation 2.9). Alternatively, consumption may scale with the local current density $W = k_{deac}i$ as in Equation 2.6, or some other functional relationship. One must consult the literature or perform experiments in order to determine the nature of the gradient-inducing W consumption term for a particular electrolyte-additive system. The form of the deactivation term will influence feature filling behavior, particularly when the current density varies significantly along the surface profile, as in situations where leveling is likely to be most significant. The various consumption forms will also yield different impurity profiles in the deposit and thereby different physical characteristics.

Simulations of the role of leveling agents in 'through-mask' or 'template' filling of features in the range of 10 μm have been explored using steady-state kinetic models with the consumption function $W = k_{kinc}i$. Consistent with the diffusion–adsorption–consumption model of leveling, the addition of levelers results in strongly hindered growth in regions such as the perimeter between patterned conductors and insulators that are most affected by geometrically enhanced diffusion [18, 288, 331].

Levelers have traditionally been used to fill features with characteristic dimensions, typically tens of micrometers, which are a substantial fraction of the thickness of the $\approx$100 μm hydrodynamic boundary layer [1–5]. However, the role of leveling in

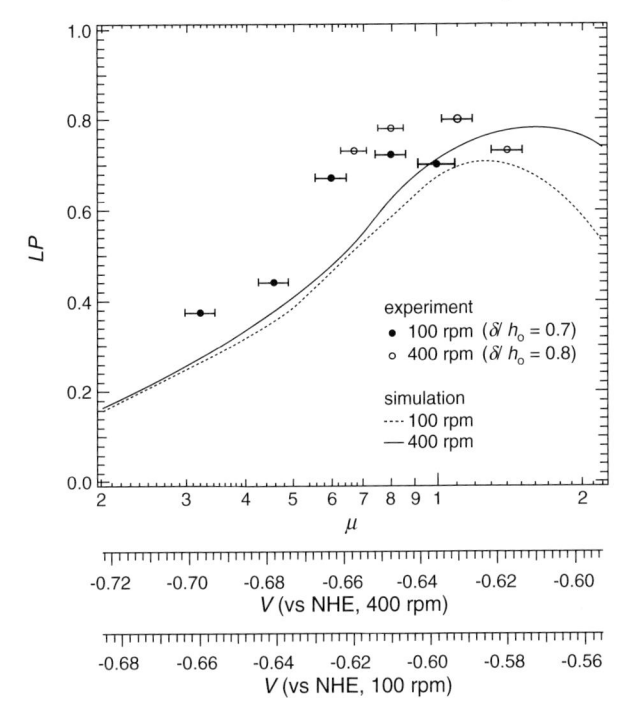

Figure 2.30 Summary of experimental and simulated data for coumarin-induced leveling during nickel deposition in a semicircular groove. The data are shown as a function of the dimensionless group μ along with the relevant experimental parameters. In this instance (unlike Equation 2.37) the leveling power (LP) is defined as the ratio of $(i_{valley} - i_{peak})/i_{valley}$. The results correspond to deposition of 3 μm of nickel on the planar field, or 0.1 of the original groove depth (source Ref. [58]).

the filling of sub-micrometer features on a conductive substrate has also been explored. Work on this problem was motivated by the 1998 IBM report [10] of void-free filling of sub-micrometer features in the 'Damascene process'. Understanding of the feature filling process was significantly impeded by the proprietary nature and number of additives used in the first electrolytes. The known presence of at least one rate-suppressing additive and the nominal consistency with a leveling-type mechanism [55–58] made it almost inevitable that the first models for superfilling would utilize some form of leveling construct. However, modeling with the previously developed formalism ($W = k_{inc}i$) was unable to generate the rate differentiation within the cavity required to describe the observed shape evolution [10, 332]. As the most relevant length scale for leveling is the thickness of the hydrodynamic boundary layer, ≈100 μm, it was perhaps not surprising that conventional single additive leveling models were ineffective at the sub-micrometer Damascene feature scale. Nevertheless, in an attempt to obtain simulation results that approached the experimentally observed interface profiles, an empirical expression was adopted that included a power law dependence on the ratio of the leveler/metal ion flux that was independent of previous experimental results.

However, even with this empirical form, the simulations [10, 306, 332] were unable to capture important aspects of the experimentally observed shape change [243, 333–335] in second generation electrolytes that included: (a) an incubation period, (b) a square as opposed to V-notch interface contour during bottom–up filling and (c) bump formation above filled features. In hindsight, the lack of agreement was not surprising as the second-generation electrolyte chemistries did not contain levelers as defined above. Rather, the 'superfilling' process evolved from a competition between suppressors and accelerators that is aptly described by the CEAC mechanism, as will be discussed in Section 2.5.4 [12, 62–64].

Nevertheless, levelers are now being used in the third-generation electrolytes for attenuating the undesired bump formation that often accompanies superfill and thereby facilitating the post-plating chemical-mechanical planarization process [71, 336]. Thus, assessment of leveling, specifically the interaction with the 'superfilling' mechanism at sub-micrometer dimensions, remains an important topic. Subsequently, other more transparent leveling models were described including one [305] based on the leveling model outlined in Section 2.4.5. Simulations based on diffusion-limited deposition of a leveling precursor were also presented [305]. Likewise, leveling simulations for sub-micrometer feature filling using a simple dimensionless ratio, the rate of inhibitor consumption to its diffusive flux to the surface all scaled to the depth of the feature, are available [337]. The simulations provide qualitative insight into the functional relationship between leveler concentration and shape evolution. A weakness of this particular study is that the concentrations of all electrolyte species at the mouth of the feature are set equal to the bulk electrolyte value. This boundary condition was extended into the experimentally relevant regime in a subsequent study [302].

In most practical plating applications it can take 10–30 s to establish the hydrodynamic boundary layer, whereas filling times for <100 nm wide features are typically in the range of 10 s. Many levelers and/or suppressors are macromolecules that have diffusion coefficients that can be as much as two orders of magnitude lower than those of relevant small molecules, such as certain catalyst and metal cations [338, 339]. As a consequence, competition among such additives may substantially impact feature filling during the first 10 s. A study of the impact of transient diffusion of a single leveler on the filling of sub-micrometer trenches has been presented for both potentiostatic and galvanostatic deposition [302].

Leveling 100 nm wide trenches is modeled in Figure 2.31 using both a quasi-steady-state simulation that neglects this transient period and a domain decoupled approach that simulates the impact of the transient. The development of sloping sidewalls, predicted in either case, is an important signature of such a leveling process. This effect is evident in some of the latest generation of commercial Damascene plating baths that contain levelers in addition to accelerators and suppressors. However, it should be noted that similar effects may also occur through spatially varying adsorption of the suppressor during the transient period, particularly for an electrolyte with a dilute suppressor concentration. Comparable effects have also been anticipated if a viscous film, that is thick relative the features of interest, develops on the electrode [340]. In general, if the characteristic time for

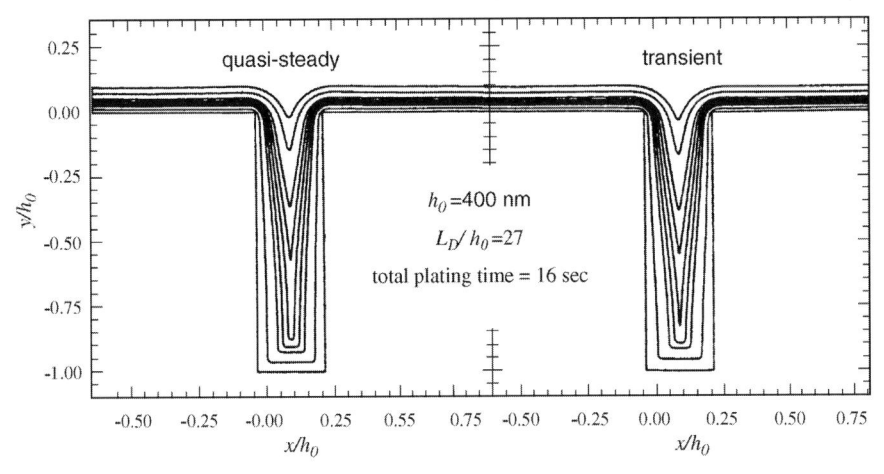

Figure 2.31 Simulations of leveling during metal deposition in 100 nm wide trenches. The leveling agent is assumed to be a large molecule with a diffusion coefficient of 10^{-7} cm^2 s^{-1}. Filling occurs in 16 s, on the order of the boundary layer relaxation time, so care must be taken to properly account for transient effects. Two different solutions are compared. On the left is the quasi-steady-state approach where the steady-state value of the current density was used at every intermediate step. On the right, a domain decoupled approach was used in order to assess the impact of the transient effects on feature filling (source Ref. [302]).

suppressor accumulation on the surface approaches the trench fill-time, then significant variation of suppressor coverage, and the deposition rate across the surface profile, is possible [302].

As feature sizes continue to shrink, at least two additional factors will need to be considered in leveling models. The first is the effect of area change as outlined in Section 2.4.2: if the inhibiting adsorbate has a finite residence time on a curved surface, interface propagation will necessarily lead to changes in the local coverage in accord with Equation 2.4. In this instance, mass balance of the adsorbed leveling agent is given by:

$$\Gamma_{lev}\frac{d\theta_{lev}}{dt} = Ck_{ads}(1 - \theta_{lev}) - k_{des}\theta_{lev} - W\theta_{lev} + \kappa v \theta_{lev} \qquad (2.42)$$

If the curvature effect is significant relative to other terms for leveler evolution, this might lead to a weakening of the leveling power. Similar weakening may also become evident as the lateral mobility or surface diffusion of the leveling adsorbate becomes significant for the feature length scale and fill time being considered, the effect expected to be more important with decreasing metal deposition rate. As feature dimensions reach deeply into the nanometer regime, the discrete nature of the molecules will become evident. In particular, large molecules will have limited access to high aspect ratio regions due to steric hindrance [341]. At such length scales, the thermodynamic effect of curvature on the metal deposition and stability will also become important.

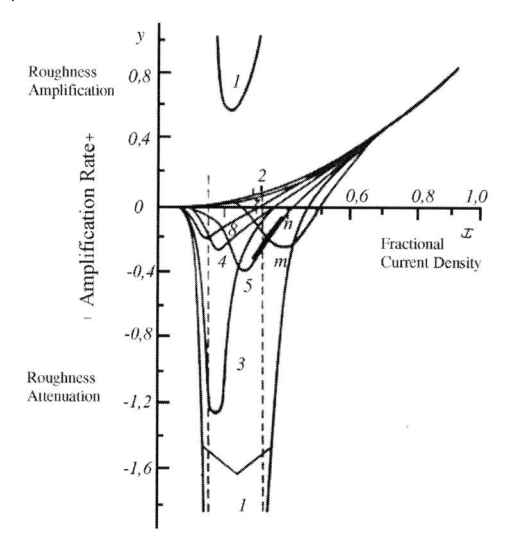

Figure 2.32 Damping/amplification of a sine wave electrode profile as a function of the applied current density normalized to the diffusion limited value. The numbers correspond to different dimensionless parameter inputs that define the relative strength of additive adsorption, consumption and transport. Importantly, for some conditions, the attenuation rate can actually exceed the amplification rate at $i/i_{lim} = 1$ (e.g. curve 3 around $i/i_{lim} \approx 0.18$ and curve 1 for currents just above and below the low-current unstable region) (source Ref. [324]).

2.5.3.2 Stability Analysis

In addition to feature filling, leveling agents also provide stabilization to planar growth fronts. Indeed, the perturbation approach commonly used to evaluate interface stability [23–37] is completely analogous to the analytical leveling descriptions that model the attenuation or growth of small amplitude sinusoidal perturbations. Figure 2.32 shows damping of surface roughness for a range of metal depletion gradients (given by the normalized current density) that are less than the diffusion-limited metal deposition rate [324]. Under favorable circumstances, the smoothing velocity may even exceed the maximum rate of roughening associated with diffusion-limited deposition. Once the LP has been determined using a sinusoidal surface profile, the result may be extended to the prediction of smoothing of an arbitrary surface after deposition of an average thickness h_{avg} according to:

$$\ln \frac{R_0}{R} = \frac{2\pi}{\lambda_{mean}} (LP) h_{avg} \qquad (2.43)$$

where R_0 and R are the rms roughness before and after deposition, λ_{mean} is the mean wavelength of the surface irregularities and LP is determined by experiments using the sinusoidal surface profile [316].

The prediction of a threshold current above which a leveler is unable to stabilize growth has also been identified in a recent linear stability study that includes two additional terms not previously considered, namely: (a) complex formation between the additive and the metal cation and (b) interaction between the polar additives and the inhomogeneous electric field associated with the surface perturbations [310]. Compared with additive-free systems, a significant domain of stability can be obtained with increasing additive concentration, additive incorporation or equilibrium adsorbate coverage.

As noted in Sections 2.4.5 and 2.4.6, the interplay between additive consumption, (incorporation, deactivation, etc.) and mass transport limited accumulation can give rise to multiple steady-state coverage values over finite ranges of potential. The opportunities, and, equally important, the difficulties of operating in this metastable regime have been discussed [140]. Under such conditions, it is possible for an infinitesimal decrease in the boundary layer thickness to cause a finite change in the surface coverage and a shift from the active to the passive state. Under idealized conditions, the tip of a nascent roughness element can be passivated while the rest of the surface remains active. For an inverted geometry the opposite could be true; a recessed region would be active while neighboring planar regions remain passive. The possible utility of this strategy to superfill sub-micrometer features with copper has been explored in the PEG–Cl system using processing parameters that correspond to metastable behavior as exemplified in Figure 2.15. For patterned substrates, selective deposition occurs in the bottoms of sub-micrometer trenches, most probably related to a decrease in the local chloride activity required to form the blocking PEG layer [342].

2.5.4
Catalyst-derived Brightening

2.5.4.1 Feature Filling

Movement of a non-planar interface necessarily results in changes in the local area that can impact adsorbate coverage and/or chemistry as described by the CEAC mechanism outlined in Section 2.4.2 [12, 62–68]. As with leveling, use of a simple electrode geometry permits an analytical description of the process to be obtained. For example, a substrate with micromachined hemi-cylindrical features allows the geometry-driven changes in surfactant coverage to be simply described by changes in radius of the feature, or more generally the curvature [66, 67]. For an unsaturated surface, surfactant coverage increases on advancing concave sections and decreases on convex sections. If the surfactant is catalytic towards metal deposition, this geometrical effect will naturally give rise to accelerated deposition on concave sections followed by convex bump formation above the final feature, as indicated in the schematic drawing shown in Figure 2.33a.

The first experimental demonstration of what was subsequently labeled the CEAC mechanism was the bottom-up filling of a 30 μm deep V-notch groove in a three additive copper plating electrolyte containing chloride, a proprietary depolarizing catalytic agent and a proprietary polarizing or inhibiting species [66, 67]. The bottom-up filling dynamic was observed for specimens that were aged in the electrolyte prior to electroplating. Aging resulted in subsequent depolarization of the electrode potential during electroplating due to preferential adsorption of the depolarizing agent or catalytic species on the surface. The catalytic or memory effect was sustained

2 Superconformal Film Growth

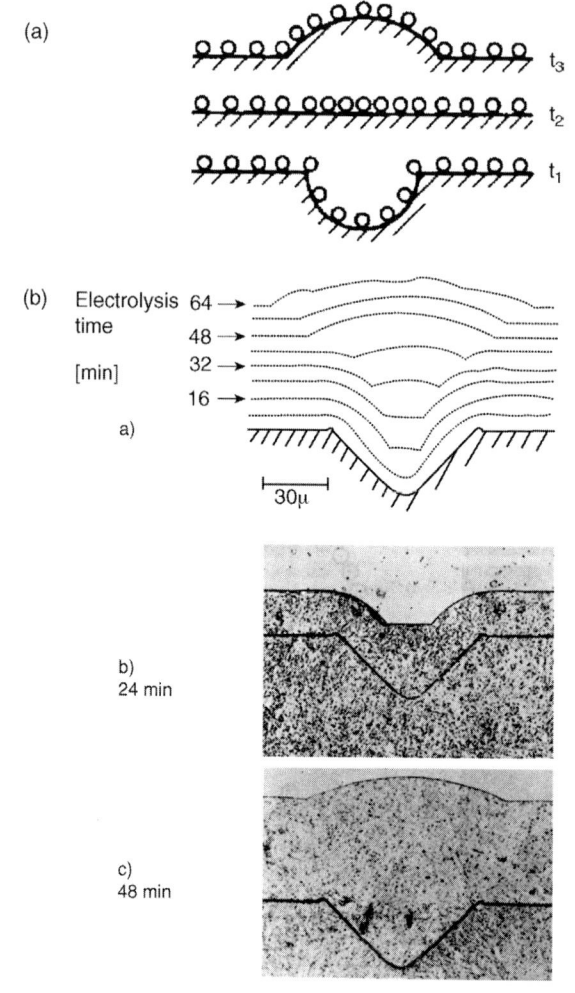

Figure 2.33 (a) If a catalytic surfactant species is evenly distributed across an electrode with a hemi-cylindrical groove at time t_1, subsequent metal deposition will lead to an enrichment of the catalyst on the concave section and accelerated filling of the groove. Once planarity is obtained (t_2) the floating catalyst that is concentrated on the surface above the original feature leads to subsequent bump formation (t_3) (source Ref. [67]). (b) One of the first observations of the catalyst enhanced accelerator coverage mechanism (source Ref. [66]).

for some tens of minutes, indicating the effective segregation of the depolarizing catalyst on the growing surface. The magnitude of the effect was shown to be a function of the catalyst coverage in terms of aging time in addition to relative catalyst concentration in the electrolyte. Too low a catalyst concentration resulted in subconformal deposition due to insufficient catalyst accumulation on the surface. Likewise, too high a concentration resulted in saturation of the surface whereby further geometrically derived differentiation was not possible. For an optimum concentration, the bottom-up filling dynamic was followed by an inversion of the interface curvature and bump formation, as shown in Figure 2.33b. An analytical

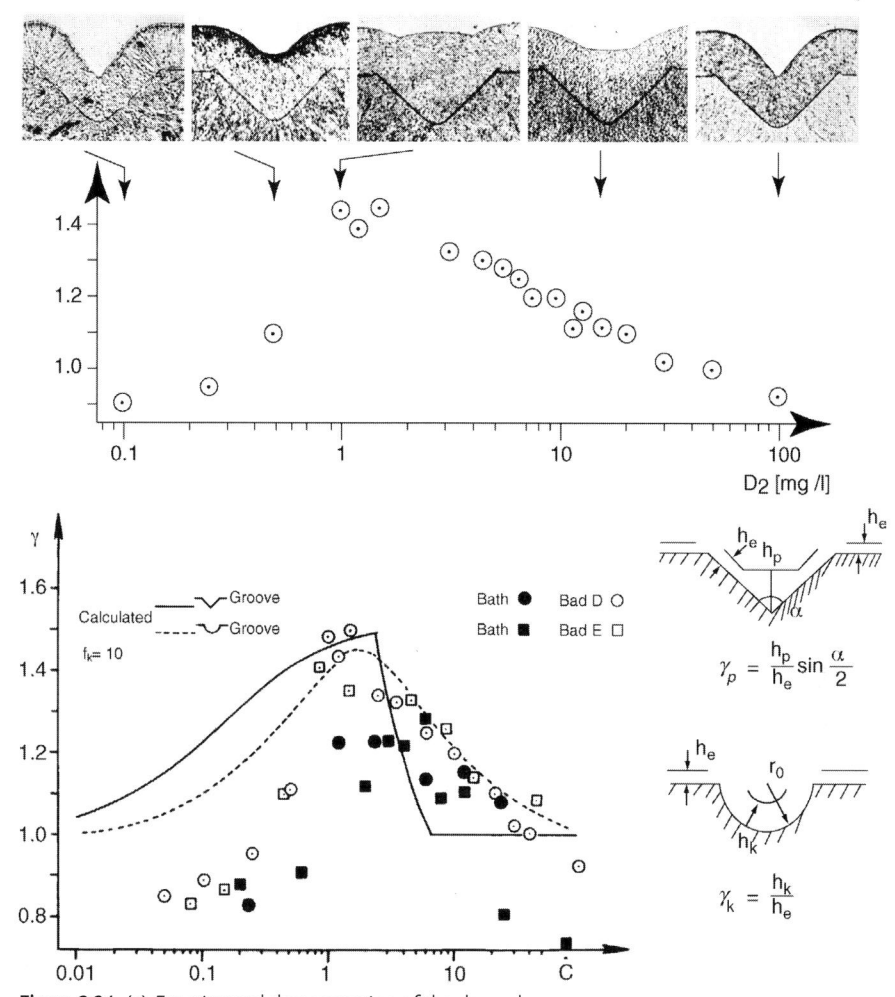

Figure 2.34 (a) Experimental demonstration of the dependence of feature filling on catalyst concentration in the electrolyte. (b) Experimental results for filling V-notch and hemi-cylindrical grooves are compared with a theoretical model (source Ref. [66]).

$$\gamma_p = \frac{h_p}{h_e} \sin \frac{\alpha}{2}$$

$$\gamma_k = \frac{h_k}{h_e}$$

model was developed for shape evolution on a hemi-cylindrical electrode as was an approximate description for a V-notch microgroove. Comparison with experiments using commercial proprietary additives is shown in Figure 2.34.

From an historical perspective, it is interesting to note that there are subsequent references to bump formation or 'superleveling' in the former-Soviet literature, although no mechanistic explanation of the phenomena was offered [317–319]. The observation of bumps above filled features was also the first indication that leveling models could not explain superfilling of sub-micrometer features in the Damascene copper process [333–335]. Driven by this observation the area change mechanism [66–68] was rediscovered in 2000–2001 by researchers as the CEAC mechanism [12, 62, 65] trying to understand the superfilling of sub-micrometer features first reported in 1998 by the IBM group [10]. The PEG–SPS–Cl acid copper system was identified as a prototypical electrolyte for superfilling. As outlined in Sections 2.4.8–2.4.10) for appropriate additive concentrations, [SPS]/[PEG] < 1, the electrochemical response of a planar electrode is effectively described by a competitive adsorption model, whereby the more strongly adsorbing SPS derived catalyst displaces the rapidly formed inhibiting PEG layer from the electrode. The growth velocity is described simply as a function of the SPS coverage (Equation 2.19).

For shape change simulations a general statement of catalyst evolution on the moving interface includes SPS adsorption, consumption (Equation 2.22) and area changes associated with the interface motion (Equation 2.4).

$$\Gamma_{SPS}\frac{d\theta_{SPS}}{dt} = k_{ads}(1-\theta_{SPS})C_{SPS} - k_{inc}\theta_{SPS}^q + v\kappa\theta_{SPS} \qquad (2.44)$$

where κ describes the geometry of the surface profile [12]. For superfilling applications, the curvature term should control system behavior. As described earlier, electroanalytical and surface analytical measurements on planar electrodes ($\kappa \approx 0$) may be used to define the adsorption and consumption parameters for a given additive package.

A range of algorithms has been used for CEAC shape change simulations of sub-micrometer feature filling [62–65]. The first generation invoked a string model to track interface position and catalyst coverage while depletion effects within the trench were ignored [62]; later versions of this model accounted for average depletion across the boundary layer between the free surface and bulk electrolyte [12]. Two different initial conditions were explored: (a) fixed catalyst coverage at $t = 0$, corresponding to catalyst derivatization prior to plating in an electrolyte containing PEG–Cl and (b) catalyst adsorption via Langmuir kinetics corresponding to electrodeposition in an electrolyte containing SPS–PEG–Cl [12, 62, 343].

Superfill in both trench and via geometries was studied, including the impact of sidewall tilt [12]. The simulations recovered the unique shape transitions associated with bottom-up superfilling, namely: (1) the incubation period of near conformal growth associated with accumulation of significant catalyst, followed shortly thereafter by (2), the formation and rapid advance of 45° inclined segments from the bottom corners, which then (3) intersect, leading to further catalyst enrichment and (4), a transition to a nearly flat bottom contour [12, 62]. This sequence was followed by rapid bottom-up filling as the advancing, nearly flat bottom surface collected catalyst from the slower moving sidewalls, leading eventually to bump formation as the fast moving bottom surface reached the top of the trench before the sidewalls could impinge. The simulations were also able to accurately predict fill versus fail conditions as a function of feature size, overpotential and trench aspect ratio.

A second generation model coupled the level set method for interface tracking with a finite difference scheme, allowing concurrent tracking of adsorbate coverage and complete evaluation of depletion effects within and beyond the trench [64, 308]. Similar models have also been used to evaluate feature filling as a function of separation and related patterning effects [307]. For optimal superfilling, the shape transitions predicted using the full solution were qualitatively indistinguishable from that of the string model [62, 64]. Simulations of failure to fill differ only in that the level set construct yields voiding due to depletion of metal ion concentration within the trench, while the string model indicates seam formation due to sidewall impingement. Several of these simulation codes are now available in the public domain (http://www.ctcms.nist.gov/fipy/).

In parallel with the development of numerical CEAC solutions, several analytical treatments for trench and via filling were established [63, 313, 344, 345]. One simple model for trench filling assumes that material is transferred from eliminated sidewall area and redistributed across the bottom surface as it advances upward, leading to the observed bottom-up acceleration through positive feedback [63]. If the bottom surface reaches the top of the original feature before sidewall impingement, it is said to be successfully filled. This analytical model provides a reasonably accurate description of the shape profiles observed prior to bump formation during optimized bottom-up trench filling.

The same approach has also been applied to describe bottom-up filling of vias [344]. Because of the assumptions employed, the constructs inherently fail to describe profiles for sub-optimal feature filling or for bump formation, which often accompanies successful bottom-up filling. However, the model permits rapid exploration of the parameter space associated with optimal trench superfilling. For example, the concentration and potential dependence of trench filling as a function of feature size or aspect ratio can be examined in less time than it takes to generate a single solution using the original string ('tens of minutes') or level set ('hours') models running on a machine with a clock speed of ≈ 1 GHz [63].

In addition, an exact algebraic solution for the incubation period of bottom-up filling has been described for the situation when catalyst is pre-adsorbed and negligible consumption occurs during subsequent metal deposition [344]. As with the string model, this analytical solution captures the essential aspects of the shape transitions that accompany the trench superfilling CEAC dynamic. The analytical solution also provides a metric for evaluating the accuracy of numerical models and associated computer codes.

Experimentally, the most unambiguous demonstration of the role of the CEAC mechanism in trench superfilling is provided by a two-step process that involves derivatization of a patterned copper seed layer with a sub-monolayer quantity of catalyst, followed by copper plating in a catalyst-free electrolyte that contains PEG–Cl [343]. Five different catalyst pre-treatments were examined involving 30 s immersion of the patterned electrode in a stagnant 1.8 M H_2SO_4 solution, containing 0.5, 5, 50, 500 or 1000 μmol L^{-1} of the catalyst precursor, either SPS or MPS, followed by transferring to an SPS-free PEG–Cl electrolyte for copper deposition at an overpotential of -0.25 V.

In the following discussion, specimens are identified by the catalyst precursor concentrations used in the derivatization step. The feature filling images shown in Figure 2.35 [343] reveal several distinct shape transitions and related phenomena. The acceleration of the deposition rate provided by the disulfide (or thiolate) catalysts is evident from the decrease in the feature filling time from ≈ 200 s for a specimen derivatized in 0.5 μmol L^{-1} SPS to ≈ 40 s for derivatization in 500 μmol L^{-1} SPS. For the specimens derivatized in 0.5 μmol L^{-1} SPS, deposition proceeds conformally and

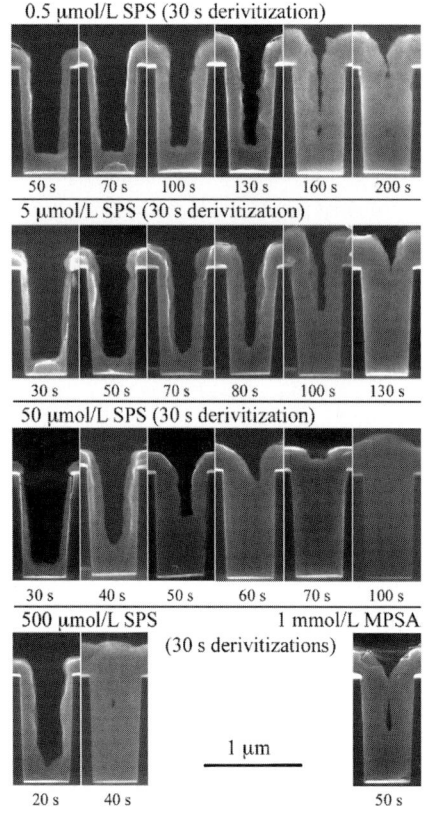

Figure 2.35 Sequential feature filling images for trenches that were pre-treated with SPS/MPS catalyst prior to copper plating for indicated times in a PEG–Cl electrolyte at −0.25 V. The conditions used for electrode derivatization are indicated (source Ref. [343]).

the surface roughens with increasing film thickness. Eventually, this results in void formation, when the sidewalls impinge, and a deep cusp forms above the trench. Identical behavior was observed for substrates which had not been derivatized, that is, $\theta_{SPS} = 0$ (not shown). For the 5, 50 and 500 µmol L^{-1} SPS derivatizations, the initial increment of growth is still nominally conformal. However, it is followed by accelerated growth at the bottom corners, a characteristic of the inception of superconformal growth. This yields the V-shaped bottom profile visible in the specimens plated for 70, 40 and 20 s, following derivatizations in 5, 50 and 500 µmol L^{-1} SPS, respectively.

Subsequently, catalyst enrichment at the apex of the V-shaped bottom leads to further acceleration and conversion to a flat bottom profile. Between 80 and 100 s the 5 µmol L^{-1} derivatized specimens exhibit rapid bottom-up filling, the hallmark of the 'superfilling process'. However, by 130 s the sidewalls impinge just before the rapidly advancing trench bottom reaches the top of the feature. In contrast, the 50 µmol L^{-1} specimens exhibit near optimum superfilling behavior with rapid bottom-up filling occurring between 50 and 70 s with negligible sidewall motion. An inversion of the growth front

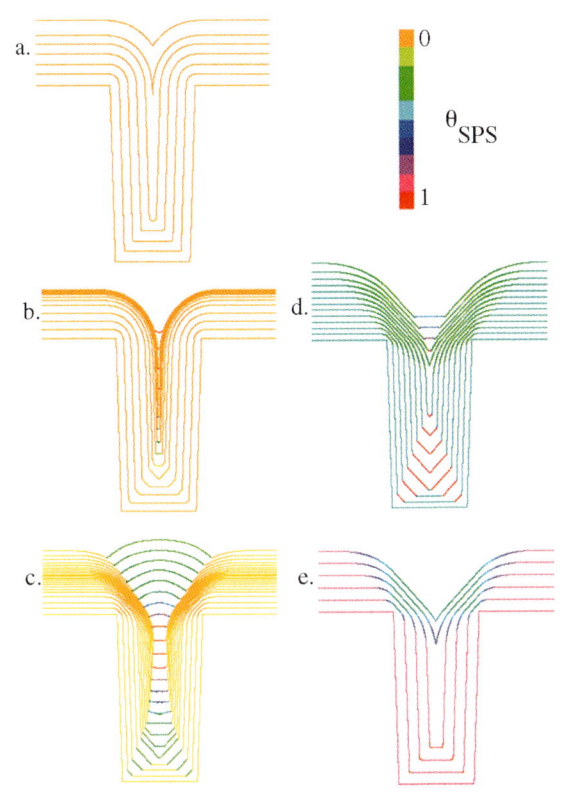

Figure 2.36 Simulations of feature filling of catalyst pre-treated electrodes. Interface motion is displayed using colorized contour lines to reflect the local catalyst coverage. Each simulation corresponds to a different initial catalyst coverage, $\theta_{SPS}^{t=0}$: (a) 0.000 54, (b) 0.0054, (c) 0.054, (d) 0.44 and (e) 0.88. The values correspond to those anticipated for the derivatization treatments specified in Figure 2.35. The feature filling times corresponding to the last growth contour are: (a) 177, (b) 113, (c) 85, (d) 39 and (e) 24 s (source Ref. [12]).

curvature is evident at 70 s, and by 100 s a large bump is seen above the trench. For the 500 µmol L^{-1} SPS and 1 mmol L^{-1} MPS specimens, void formation is clearly evident, this being more severe in the latter case. These experiments indicate that superconformal filling of sub-micrometer features can be derived from the evolution of a sub-monolayer quantity of a surface-confined catalyst, in the absence of the possibility of transport and/or homogeneous chemistry effects within the electrolyte.

Shape change simulations of these experiments are shown in Figure 2.36 [12]. They were implemented using the interface tracking code. The two-step process of catalyst derivatization followed by electroplating in a catalyst-free electrolyte allows simplification of the catalyst evolution equation (Equation 2.44) by eliminating the first term representing SPS adsorption from the copper plating electrolyte. The electrolyte/metal interface at various stages of filling is delineated using contour lines that are colorized to reflect the local catalyst coverage. With increasing coverage, the colors range from orange ($\theta_{SPS} = 0$) through yellow, green, blue and red ($\theta_{SPS} = 1$). The simulations use a uniform cupric ion concentration within the feature, the value being reduced from the bulk concentration, consistent with the depletion of cupric ion across the hydrodynamic boundary layer, δ. The consequences of ignoring depletion within the trench are discussed below. The initial thiolate coverage of the SPS derivatized electrodes, required for the simulations, were estimated from surface analytical measurements [346]. Derivatization for 30 s in 0.5, 5, 50 or 500 µmol L^{-1} SPS or 1000 µmol L^{-1} MPS yields initial fractional catalyst coverages of approximately 0.000 54, 0.0054, 0.054, 0.42 and 0.88, respectively. The corresponding simulations are shown in Figure 2.36a–e, respectively.

Deposition in trenches with an initial catalyst coverage of 0.000 54 is predicted to be conformal, as shown in Figure 2.36a. Because of geometric leveling associated with the sloping sidewalls, the trench is still predicted to be filled in the absence of sidewall roughness and ignoring the cupric ion depletion associated by the increasing aspect ratio of the unfilled region. Inclusion of these latter effects might be expected to result in prediction of an occluded void, such as that observed in the corresponding experiment (0.5 µmol L^{-1}, 30 s derivatization in Figure 2.35).

Increasing the initial catalyst coverage an order of magnitude is predicted to result in superfilling behavior as shown in Figure 2.36b; enrichment of the catalyst begins on the bottom corners leading to significant acceleration of the copper deposition rate and the formation of a V-notch shape. Further enrichment accompanies the shape transition from a V-notch to a flat bottom when the opposing corners meet at the bottom. The higher catalyst coverage leads to a marked increase in the upward velocity of the bottom surface. Further increase in catalyst collection on that surface continues through accumulation of the catalyst previously adsorbed on the sidewall area that is being eliminated by the rapid upward motion of the bottom surface. The narrowed bottom surface is predicted to escape the trench just before impingement of the sidewalls. In practice, the very large aspect ratio of the unfilled region during the final stages of filling leads to significant cupric ion depletion that might prevent filling, which would be consistent with the cusped surface profile evident in the experimental results for the 5 µmol L^{-1} SPS, 30 s derivatized electrode shown in Figure 2.35.

Increasing the initial catalyst coverage to 0.053 is predicted to result in near optimal superfilling behavior, as shown in Figure 2.36c. Enrichment at the bottom corners leads to rapid formation of first the V-notch and then the nearly flat bottom. As in the other simulations, the advancing bottom surface accelerates further as it collects catalyst from the eliminated sidewall areas. It also exhibits subtle oscillations in shape associated with the continual translation of catalyst from the sidewalls to the bottom surface. The oscillations are also detailed in the analytical solution of feature filling. The catalyst coverage approaches unity as the bottom surface reaches the top of the trench.

Further deposition results in bump formation above the trench that in practice is often referred to as 'momentum' plating. From this point on, the expanding surface area associated with further advance of the now convex section results in progressive dilution of the local catalyst coverage. The simulation also reveals the importance of catalyst dilution on the convex corners at the top of the trench during an earlier stage of feature filling; the decreasing deposition rate on these sections permits greater access of Cu^{2+} to the highly catalyzed bottom surface. This is a substantial concern, for example, for the 50 μmol L^{-1} SPS, 30 s sample in Figure 2.35, where deposition in the field over the trench is already proceeding at >50% of the diffusion limited current density. Significantly, this indicates that the geometrically differentiated surface reactivity, as predicted by the CEAC model, is capable of driving bottom-up filling even in the presence of the substantial metal ion concentration gradient that accompanies rapid metal deposition, which is retarding deposition the further down the trench one goes.

The colorized simulations of near optimal superfilling are particularly helpful in explaining the results of post deposition surface analytical work on bumped surfaces. Specifically, while the catalyzed bottom surface can reach coverage values approaching unity during filling, the bottom surface at this point has yet to reach planarity with the free surface making the geometry beyond the scope of conventional quantitative surface analysis. When the feature is completely filled and better suited for post deposition analysis, dilution effects associated with the area increase (note the convex surface geometry) result in a lateral variation of catalyst coverage that is less than 0.2. Measuring such variations in surface coverage at the sub-micrometer level is non-trivial using most standard surface analytical tools, possibly explaining the lack of reports detailing differences between the sulfur concentration on overfill bumps versus the neighboring planar field.

Increasing the initial catalyst coverage to 0.42 or 0.88 is predicted to result in a reversion to failure to superfill the trench (Figure 2.36d and e). For a starting coverage of 0.42, catalyst enrichment on the advancing concave corners approaches unity and a V-notch geometry is rapidly established in agreement with the experiment on the 500 μmol L^{-1}, 30 s derivatized SPS electrodes (Figure 2.35). However, the simulation also gives a clear visualization of why the CEAC mechanism nonetheless fails to fill this feature. Specifically, even with near-saturation coverage attained on the advancing concave corners, the coverage on the bottom surface cannot exceed ≈2.5-times the starting coverage on the sidewalls. With further increase in the differential deposition rate being impossible, the sidewalls impinge before the bottom surface is able to escape the trench.

For an initial catalyst coverage of 0.88, the deposition rate is predicted to be effectively saturated near the start of metal deposition with geometrically driven changes in catalyst coverage on the concave surfaces being minimal (Figure 2.36e). Deposition within the trench proceeds conformally, although motion of the top convex corners of the trench is still inhibited due to the catalyst dilution effects described earlier.

Though not modeled here, significant depletion of the Cu^{2+} ion within the trench is expected at the high catalyst coverages attained with the 500 μmol L^{-1} SPS, 30 s and 1 mmol L^{-1} MPS, 30 s derivatized specimens. Such depletion yields a gradient in the sidewall deposition rate, with faster deposition occurring at the top of the feature [64]. This results in earlier impingement at this location and void formation such as observed in both experimental specimens (Figure 2.35) rather than the seams predicted in Figure 2.36d,e. Simulations using the full level set code recover the formation of voids under these non-optimum filling circumstances [64, 308]. Interestingly, the impingement events lead to the formation of a new concave surface above the impinged sidewalls, such as that visible in Figure 2.36d,e, at which catalyst enrichment (and dilution once the surface inverts) can again occur through the CEAC mechanism. Thus, even bump formation cannot be taken as a robust indication of successful bottom-up superfilling.

In conventional Damascene processing, electroplating is performed in an electrolyte containing both an inhibitor and catalyst. Galvanostatic control is often used in many commercial processes and programs implementing controlled variation of current (e.g. current stepping, with lower starting currents) have also been implemented. One objective of such processes is repair of seed-layers prior to superfilling. Nonetheless, such growth programs may also result in uptake of adsorbed catalyst prior to significant metal deposition in a manner analogous to the electrode derivatization experiments described above.

Trench and via filling experiments and simulations have also been performed for electrodeposition on wafer fragments in an SPS–PEG–Cl electrolyte. The experiments are compared with shape change simulations in Figure 2.37 by plotting the heights of the trench bottoms along the mid-lines of the features as a function of the deposition time. The experimental interface morphologies are qualitatively indistinguishable from those observed in the derivatization experiments. The predicted time-dependence of the geometrical transitions are analogous to those observed in the derivatization simulations. A vertical trajectory is indicative of seam formation. Good agreement with the experiments is evident, although the simulations indicate that

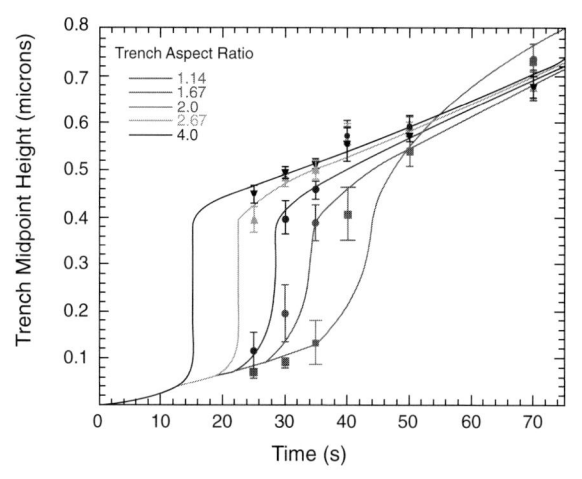

Figure 2.37 Trench-filling experiments (data points) are compared with shape change simulations (curves) by tracking the evolving height of the deposit along the trench centerline. A vertical transition such as is evident at ≈15 s for trenches with an aspect ratio of 4, indicates seam or void formation (source Ref. [12]).

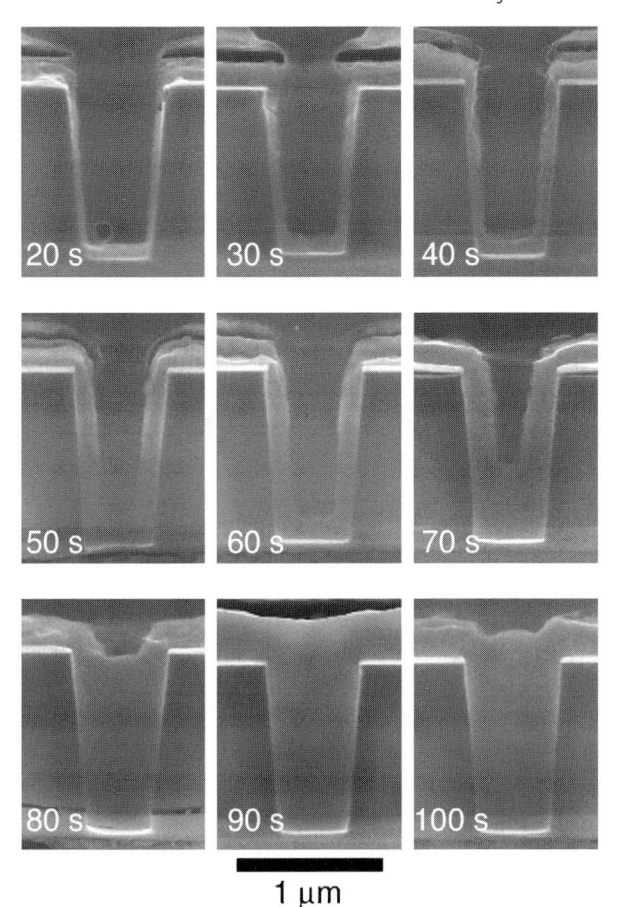

Figure 2.38 Time sequence for superfilling a via. Film growth is conformal for the first 60 s followed by accelerated deposition on the lower surface and rapid bottom-up filling (source Ref. [313]).

the threshold aspect ratio for superfilling is close to 1.6, while experiments suggest reliable filling up to an aspect ratio of 2 for these conditions. As features become smaller and feature filling times shorter, accurate understanding and evaluation of the initial conditions are expected to become increasingly important for obtaining good agreement between experiment and simulation.

Deposition in vias, unlike trenches, is a three-dimensional problem by virtue of the non-zero curvature of the cylindrical sidewalls [313, 344]. Filling experiments were conducted on cylindrical vias with a 4.5° slope of the sidewalls from the vertical. Field emission scanning electron microscopy (FE-SEM) images of the via filling are shown in ten second increments in Figure 2.38. Deposition is essentially conformal for the first 60 s with acceleration of the bottom surface by $\approx$70 s and complete super-conformal filling of the via by $\approx$90 s. The filling process is summarized in Figure 2.39 by plotting the height of the deposit along the centerline of the via as a function of

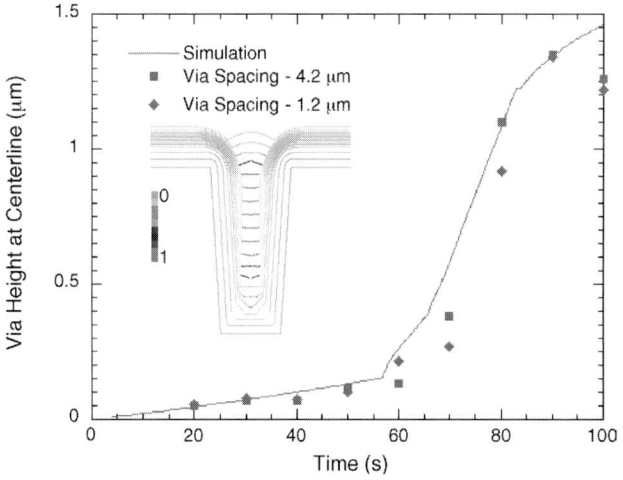

Figure 2.39 Tracking the interface mid-height position during via filling provides a comparison between simulation and experiment (Figure 2.38). The inset shows the simulated 'bottom-up' feature filling, with contours colorized to reflect the local θ_{SPS} coverage (source Ref. [12]).

deposition time. The solid line is the corresponding string model prediction. Agreement between experiment and simulation is excellent, especially considering the neglect of mass-transport derived depletion of cupric ion within the via.

In practice, sub-micrometer features often have outward sloping sidewalls, either due to limitation of the lithographic process or perhaps by design to enhance the probability of robust feature filling by introducing a component of geometric leveling. The beneficial impact of this sidewall tilt on filling is evident in Figure 2.39. Robust filling is enhanced both by the more rapid enrichment of catalyst at the small-radius via bottom and the slower acceleration and longer path of the sidewalls towards the via top. In contrast, simulations that are identical, with the exception of a vertical sidewall geometry, exhibit only marginal superfilling, with the bottom surface just barely escaping from imminent sidewall impingement. This is also evident through comparison of the deposit thickness on the sidewall at the top of the feature just prior to completion of feature filling and the diameter towards the bottom of the via.

The impact of surface diffusion on the curvature-derived enrichment and dilution processes that are central to the CEAC mechanism has not been evaluated. For the processing conditions explored thus far, the sharp angles evident in the copper feature-filling images suggest that such effects are not significant in this system. However, systems where the metal deposition rate is much lower and/or the catalyst has significant surface mobility will alter and eventually compromise the CEAC dynamic.

The generality of the CEAC mechanism has been demonstrated by extension to at least two other chemical systems, selenocyanate catalyzed silver deposition from a cyanide electrolyte [72–75] and iodine catalyzed CVD of copper from Cu(I)(hfac)(vtms) and related compounds [15, 77]. As shown in Figure 2.40, a one-to-one correlation between the SeCN-coverage and the silver deposition rate was established

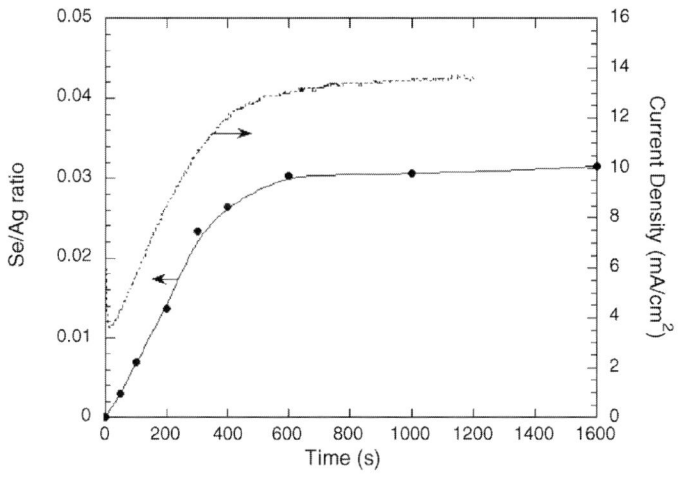

Figure 2.40 The correlation between acceleration of the silver deposition rate and the accumulation of a selenium catalyst on the surface is apparent for deposition at an overpotential of −0.35 V. The XPS data are delineated as solid circles with the line being only a guide to the eye (source Ref. [72]).

with SeCN-adsorption also yielding hysteretic η–i curves analogous to those previously described for the superfilling copper electrolytes. Furthermore, derivatization of silver-seeded trenches in a KSeCN electrolyte followed by plating in a simple silver cyanide electrolyte also resulted in bottom-up filling as shown in Figure 2.41. Reasonable agreement was also found between simulations and experiments of feature filling in a KSeCN–KCN–KAg(CN)$_2$ electrolyte.

The possibility of using electroless deposition to superfill sub-micrometer features has also been explored[347–349]; successful filling was recently reported for an alkaline EDTA-complexed electrolyte containing SPS and PEG as additives [349]. However, the tilted sidewalls in the lower half of the features, combined with the absence of kinetic data, make a mechanistic assessment of feature filling difficult.

Finally there is a report of superfilling using an electrolyte containing a mixture of two polyethers of differing molecular weights, of 2000 and 200 g mol^{-1}, respectively [350]. The short chain polyether does not substantially inhibit the metal deposition but may compete with the longer chain polymer for surface sites in terms of the CEAC mechanism. Similarly, successful superfilling has also been reported using 2-mercaptopyridine (2-MP) in the presence of PEG–Cl [351]. The molecule most probably binds to the copper surface through the sulfur group while the pK_a of the pyridine nitrogen is 1.07 so it is protonated at pH = 0 [352]. Further work will be required to understand feature filling in the Cl–PEG/PEG and Cl–PEG–2MP systems.

2.5.4.2 Stability Analysis

The phenomenon of brightening is ascribed to smoothing of surface features at length scales comparable to optical wavelengths, that is, sub-micrometer dimen-

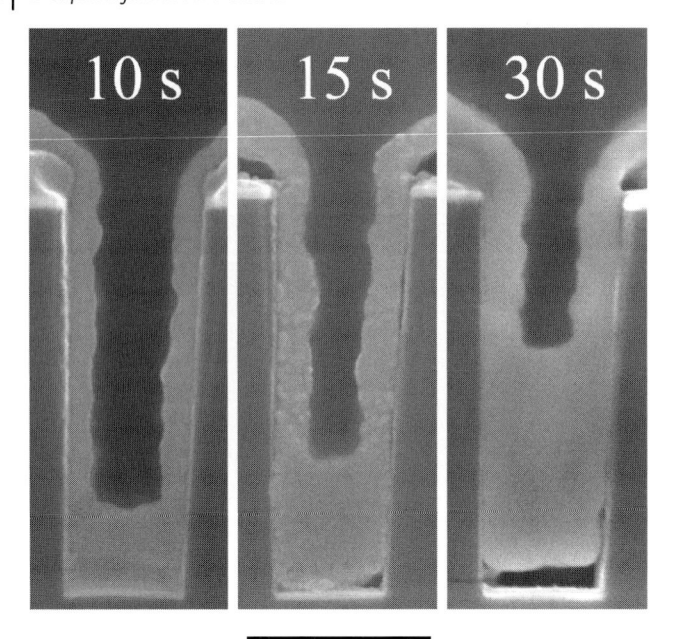

Figure 2.41 Silver deposition in trenches derivatized with an SeCN⁻ catalyst. The three specimens were derivatized for 10, 15 and 30 s, respectively, in a 20 μmol L^{-1} KSeCN electrolyte followed by electrodeposition for 20 s in 0.23 mol L^{-1} KAg(CN)$_2$ and 3.4 mol L^{-1} KCN at an overpotential of −0.2 V. The correlation of bottom-up filling with increased derivatization time (i.e. adsorbate coverage) is evident (source Ref. [75]).

sions. It is generally recognized that under steady-state conditions the diffusion–adsorption–consumption model will not strongly affect thin film deposition when the feature dimensions are orders of magnitude smaller than the boundary layer thickness associated with the concentration gradients. In contrast, for a given deposition rate, the curvature term in the CEAC formalism increases in magnitude as the length scales decrease.

A first-order perturbation study of the CEAC formalism in the presence of both accumulation and consumption of adsorbed catalyst provides an assessment of the stabilization provided by the mechanism. Stabilization towards a small amplitude sine wave perturbation of the electrode height and steady-state catalyst coverage was examined as a function of catalyst concentration in the electrolyte and the growth potential [309]. As shown in Figure 2.42, for deposition at an overpotential of −0.3 V the greatest stabilization is provided by a catalyst concentration of 0.1 μmol L^{-1} where planarity is maintained for perturbation wavelengths ($\lambda_{crit} = 2\pi/\omega_{crit}$) up to 210 μm, a value that exceeds the boundary layer thickness. The breakdown threshold λ_{crit}

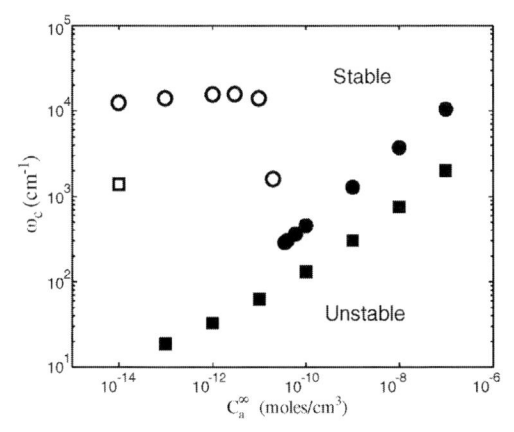

Figure 2.42 The dependence of the critical wavenumber on the concentration of the catalyst (accelerator) precursor SPS in the SPS–PEG–Cl copper plating system. Two different overpotentials were examined, $\eta = -0.3\,V$ (circles) and $\eta = -0.2\,V$ (squares). A perturbation is stable if its wavenumber satisfies $\omega > \omega_{crit}$; the stable range is increased by two orders of magnitude in the presence of sufficient catalyst. The closed symbols correspond to the boundary between real (monotonic) decay/growth modes while the open symbols correspond to the boundary between complex (oscillatory) decay/growth modes (source Ref. [309]).

decreases with further increase of catalyst concentration due to the growing metal depletion gradient associated with the higher corresponding steady-state adsorbate coverage; a diminished threshold of $\lambda_{crit} = 6\,\mu m$ is predicted for a catalyst concentration of $100\,\mu mol\,L^{-1}$.

The system is also sharply destabilized, $\lambda_{crit} = 4.5\,\mu m$, when the catalyst concentration falls below $0.1\,\mu mol\,L^{-1}$. The stabilization threshold shifts to longer wavelengths with decreasing overpotential because of the decreased cupric ion gradient at the lower deposition rate. The analysis predicts damped oscillatory behavior under certain conditions. This would appear to be analogous to the sequence of bottom-up feature filling, overfill bump formation and decay back to the planar growth front that has been observed in all superfilling processes, even though the experiments correspond to a perturbation that is well beyond that considered in the linear stability analysis. Intriguingly, the linear stability analysis predicts instability in the absence of the competition between catalyst adsorption and consumption. For example, stabilization is not predicted for non-zero initial catalyst coverage in the absence of significant accumulation or consumption during deposition. In contrast, experiments with derivatized electrodes exhibit substantial smoothing for a significant period of time.

The origin of this brightening was investigated by numerical analysis [311]. It was shown that surfaces that maintain fixed global coverage of a catalyst develop oscillating perturbations with finite, but extremely small, amplitudes. As predicted in the linear stability analysis, the system is unstable. However, because the oscillations are extremely small in amplitude, the instability does not lead to significant roughening, consistent with the experimental observation of surface brightening.

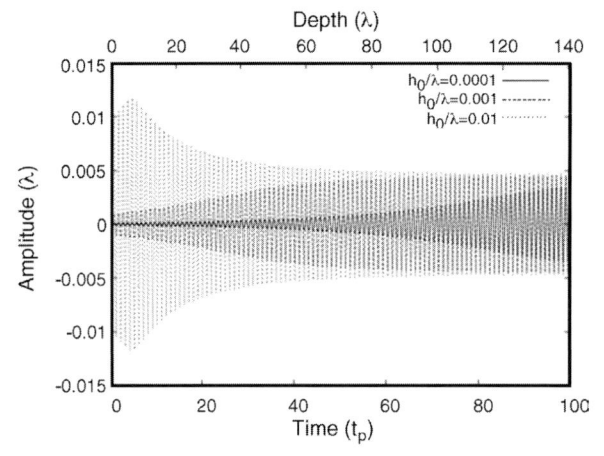

Figure 2.43 Finite amplitude, oscillatory perturbations can exist in the presence of pre-adsorbed, conserved catalyst. The existence of finite amplitude oscillatory instabilities is shown by plotting the oscillatory amplitude for initial perturbation amplitudes 0.01, 0.001 and 0.0001 (scaled by the perturbation wavelength λ) versus time (scaled by the oscillatory period t_p); the amplitudes of all three oscillatory perturbations converge onto the same value (source Ref. [311]).

The apparent 'stability' of the oscillating perturbations is demonstrated for a particular growth condition by the convergence of larger and smaller amplitude perturbations to a single intermediate value in Figure 2.43. Interestingly, a small contribution of catalyst surface diffusion may act to dampen these oscillations.

2.5.5
Bridging the Length Scales

Understanding the performance of materials systems over a wide range of length scales is a topic of active interest that is being spurred on by the ever-increasing computational power of our age. Superconformal film growth and the metallization of 300 mm wafers represents a technologically relevant test-bed for developing this engineering paradigm, with objectives ranging from obtaining uniform metal deposition over an entire wafer to ensuring that sub-100 nm features are void and defect-free and have the desired microstructure and related physical properties.

The overlap between the diffusion–adsorption–consumption model of leveling and the CEAC model of brightening is a topic of particular significance. Successful integration of these two processes for integrated circuit metallization requires optimization of the processing parameters to ensure initial domination by the CEAC mechanism, for bottom-up fill, followed by adsorption of a leveler to attenuate the catalytic action and prevent the undesirable bumping or momentum plating as shown in Figure 2.44. The time scale implicit in this description is set by the feature filling process itself. In semiconductor metallization it can be as short as several seconds. In contrast, thick film deposition or electroforming of macroscopic objects can take hours. The importance of properly accounting for all pertinent phenomena

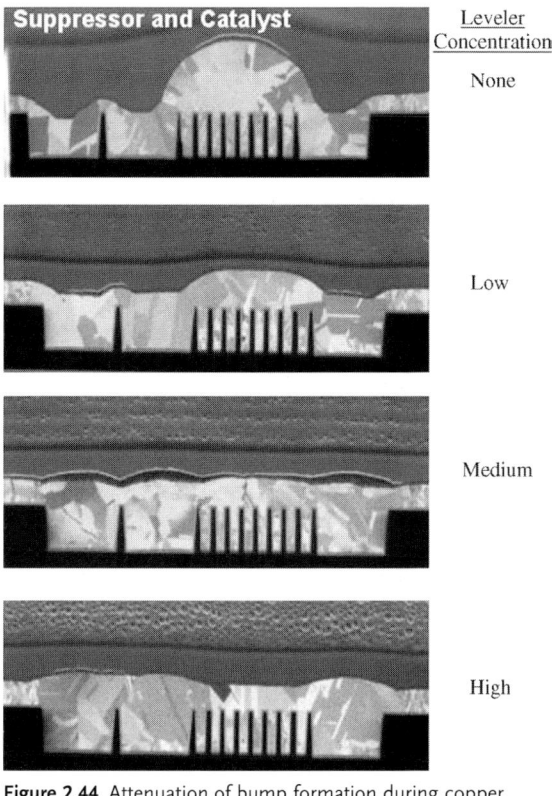

Figure 2.44 Attenuation of bump formation during copper deposition by the addition of a leveler to an electrolyte containing a suppressor and a brightener: top, suppressor and brightener; the next three images are the same but with increasing leveler concentration, as indicated (images provided courtesy of C. Witt, Cookson Electronics, Inc.).

can be shown by extended deposition using only the suppressor–brightener PEG–SPS–Cl chemistry. As visible in Figure 2.45, even though the local interface is very smooth as predicted by the CEAC mechanism, instabilities arising through density driven convection can result in significant pattern formation at the length scale of 100 μm. Such long wavelength, large amplitude instabilities can be sharply attenuated by the addition of levelers, a strategy that was adopted many years ago by the plating industry.

A brief survey of the literature suggest levelers for copper plating ranging from simple quaternary ammonium salts to polyethyleneimine (PEI) to dye molecules such as crystal violet [71, 161, 237, 262]. Interestingly, all of these leveling molecules are positively charged in the plating electrolyte. It is natural to speculate that the ion pairing between the positive charge of the leveler and the negatively charged sulfonate end group of the adsorbed catalyst may account for the quenching of the catalyst activity [353]. Future work will explore such issues in more detail. With a

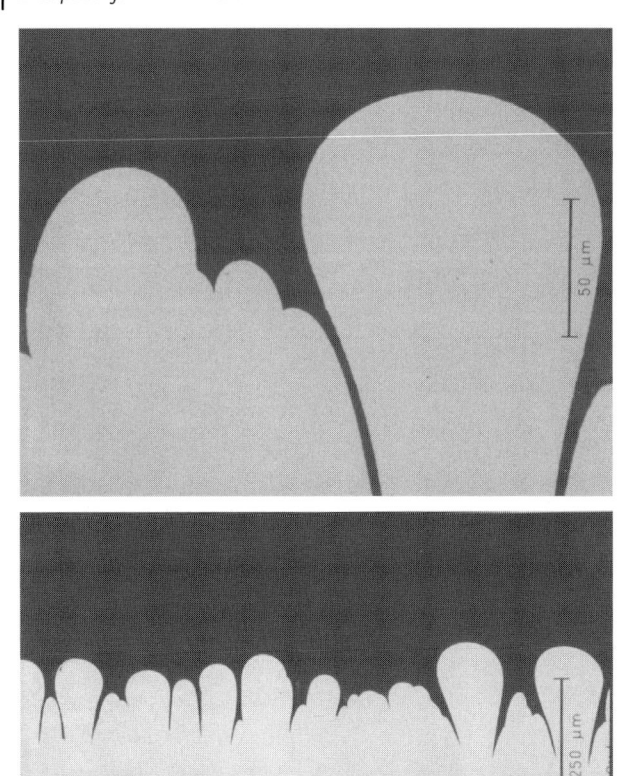

Figure 2.45 Extended plating in an SPS–Cl–PEG electrolyte yields a smooth, 'bright' surface on the sub-1 μm length scale through the CEAC mechanism. However, coupling with the density driven convective flow fields leads to pattern formation on the diffusive length scale of 100 μm. The addition of a leveler helps attenuate the latter effect (not shown).

quantitative description available for many of the relevant processes, an opportune moment is upon us for assessing constructs that couple the smoothing mechanisms associated with the CEAC and diffusion–adsorption–consumption leveling theory. Such an achievement will certainly advance us on our way towards the goal of understanding and predicting relationships between processing conditions and morphological evolution. More difficult will be correlating this with an understanding of microstructural development in these systems.

2.6
Conclusions and Outlook

The quantitative connection between the effect of additive adsorption and consumption on the metal deposition kinetics on planar electrodes and micrometer to submicrometer feature filling has come into full bloom since around 1997. Quantitative predictions of the impact of leveling and brightening agents over a wide range of length scales are now possible. Electrode derivatization represents a powerful tool for acquiring quantitative understanding of the influence and fate of individual electrolyte additives during metal deposition. Indeed, the influence of monolayer chemistry on extended metal deposition is central to the successful filling of recessed submicrometer surface features [12, 62–67]. Future studies using such well defined adsorbate structures promise to illuminate the role of additives in microstructural development.

References

1 Kardos, O. (1974) *Plating*, **61**, 129.
2 Kardos, O. (1974) *Plating*, **61**, 229.
3 Kardos, O. (1974) *Plating*, **61**, 316.
4 Kardos, O. and Foulke, D.G. (1962) Applications of Mass Transfer Theory: on Small-Scale Profiles, in *Advances in Electrochemistry and Electrochemical Engineering* vol. 2, (ed. C.W. Tobias), John Wiley & Sons, New York, p. 145.
5 Raub, E. and Muller, K. (1967) *Fundamentals of Metal Deposition*, Elsevier, New York.
6 Oniciu, L. and Muresan, L. (1991) *J. Appl. Electrochem.*, **21**, 565.
7 Franklin, T.C. (1987) *Surf. Coat. Technol.*, **30**, 415.
8 Franklin, T.C. (April 1994) *Plat. Surf. Finish.*, **62**.
9 Loshkarev, Y.M. and Govorova, E.M. (1998) *Prot. Met.*, **34**, 399.
10 Andricacos, P.C., Uzoh, C., Dukovic, J.O., Horkans, J. and Deligianni, H. (1998) *IBM J. Res. Dev.*, **42**, 567.
11 Andricacos, P.C. (1999) *Interface (Electrochem. Soc.)*, **8**, 32.
12 Moffat, T.P., Wheeler, D., Edelstein, M. and Josell, D. (2005) *IBM J. Res. Dev.*, **49**, 19, and references cited therein.
13 Sun, J.-J., Kondo, K., Okamura, T., Oh, S.J., Tomisaka, M., Yonemura, H., Hoshino, M. and Takahashi, K. (2003) *J. Electrochem. Soc.*, **150**, G355.
14 Melnick, B.M., Cale, T.S., Zaima, S. and Ohta, T. (eds) (2003) *Advanced Metallization Conference 2002*, Materials Research Society, Warrendale, PA.
15 Josell, D., Kim, S., Wheeler, D., Moffat, T.P. and Pyo, S.G. (2003) *J. Electrochem. Soc.*, **150**, C368.
16 Ibl, N. (1962) Applications of Mass Transfer Theory: The Formation of Powdered Metal Deposits, in *Advances in Electrochemistry and Electrochemical Engineering*, vol. 2 (ed C.W. Tobias) John Wiley & Sons, New York, p. 49.
17 Despic, A.R. and Popov, K.I. (1972) Transport Controlled Deposition and Dissolution of Metals, in *Modern Aspects of Electrochemistry*, vol. 7 (eds B.E. Conway and J.O'M. Bockris), Plenum Press, New York.
18 Dukovic, J. (1994) Current Distribution and Shape Change in Electrodepositon of Thin Films for Microelectronic Fabrication, in *Advances in Electrochemical Science and Engineering*, vol. 3 (eds

H. Gerischer and C.W. Tobias), Wiley-VCH, Weinheim, pp. 117.

19 Dukovic, J.O. (1990) *IBM J. Res. Dev.*, **34**, 693.

20 Barkey, D.P. (2002) Structure and Pattern Formation in Electrodeposition, in *Advances in Electrochemical Science and Engineering*, vol. 7 (eds R.C. Alkire and D.M. Kolb), Wiley-VCH, Weinheim, p. 151.

21 Wilke, C.R., Eisenberg, M. and Tobias, C.W. (1953) *J. Electrochem. Soc.*, **100**, 513.

22 Rosso, M., Chassaing, E., Fleury, V. and Chazalviel, J.-N. (2003) *J. Electroanal. Chem.*, **559**, 165.

23 Haataja, M., Srolovitz, D.J. and Bocarsly, A.B. (2003) *J. Electrochem. Soc.*, **150**, C699.

24 Krishnan, R., Johns, L.E. and Narayanan, R. (2002) *Electrochim. Acta*, **48**, 1.

25 Sundstrom, L.O. and Bark, F.H. (1995) *Electrochim. Acta*, **40**, 599.

26 Monroe, C. and Newman, J. (2004) *J. Electrochem. Soc.*, **151**, A880.

27 Chen, C.-P. and Jorne, J. (1991) *J. Electrochem. Soc.*, **138**, 3305.

28 Barkey, D.P., Mueller, R.H. and Tobias, C.W. (1989) *J. Electrochem. Soc.*, **136**, 2199.

29 Barkey, D.P., Mueller, R.H. and Tobias, C.W. (1989) *J. Electrochem. Soc.*, **136**, 2207.

30 Pritzker, M.D. and Fahidy, T.Z. (1992) *Electrochim. Acta*, **37**, 103.

31 Aogaki, R. and Makino, T. (1984) *J. Chem. Phys.*, **81**, 2154. ibid (1984) **81**, 2164.

32 Aogaki, R. and Makino, T. (1984) *J. Electrochem. Soc.*, **131**, 40.

33 Aogaki, R. (1982) *J. Electrochem. Soc.*, **129**, 2442.

34 Aogaki, R. and Makino, T. (1981) *Electrochim. Acta*, **26**, 1509.

35 Aogaki, R., Kitazawa, K., Kose, Y. and Fueki, K. (1980) *Electrochim. Acta*, **25**, 965.

36 Despic, A.R. and Popov, K.I. (1971) *J. Appl. Electrochem.*, **1**, 275.

37 de Bruyun, J.R. (1996) *Phys. Rev. E.*, **53**, R5561.

38 Huo, S. and Schwarzacher, W. (2001) *Phys. Rev. Lett.*, **86**, 256.

39 Cuerno, R. and Castro, M. (2001) *Phys. Rev. Lett.*, **87**, 236103.

40 Iwasaki, H. and Yoshinoba, T. (1993) *Phys. Rev. B*, **48**, 8282.

41 Schilardi, P., Medez, S., Salvarezza, R.C. and Arvia, A.J. (1998) *Langmuir*, **14**, 4308.

42 de Leon, P.F.J., Albano, E.V., Salvarezza, R.C. and Solari, H.G. (2002) *Phys. Rev E*, **66**, 042601.

43 Schwarzacher, W. (2004) *J. Phys.: Condens. Matter*, **16**, R859.

44 Raychaudhuri, S., Shapir, Y., Foster, D.G. and Jorne, J. (2001) *Phys. Rev. E*, **64**, 051604.

45 Ramasamy, P. (1993) *Handbook of Crystal Growth*, vol. 1 (ed. D.T.J. Hurle), Elsevier, New York, p. 1123, (and numerous references cited therein).

46 Pimpinelli, A. and Villain, J. (1998) *Physics of Crystal Growth*, Cambridge University Press, New York.

47 Michely, T. and Krug, J. (2004) Islands, Mounds and Atoms, Patterns and Processes, in *Crystal Growth Far from Equilibrium, Springer Series in Surface Sciences*, Springer, Berlin.

48 Vladimirova, M., Pimpinelli, A. and Videcoq, A. (2000) *J. Cryst. Growth*, **220**, 631.

49 Pimpinelli, A., Tonchev, V., Videcoq, A. and Vladimirova, M. (2002) *Phys. Rev. Lett.*, **88**, 206103.

50 Bales, G.S., Redfield, A.C. and Zangwill, A. (1989) *Phys. Rev. Lett.*, **62**, 776.

51 Ibach, H. and Schmickler, W. (2003) *Phys. Rev. Lett.*, **91**, 016106.

52 Giesen, M. (2001) *Prog. Surf. Sci.*, **68**, 1.

53 Gao, H. and Nix, W.D. (1999) *Annu. Rev. Mater. Sci.*, **29**, 173.

54 Floro, J.A., Chason, E., Cammarata, R.C. and Srolovitz, D.J. (2002) *MRS Bull.*, **27** (1), 19.

55 Dukovic, J.O. and Tobias, C.W. (1990) *J. Electrochem. Soc.*, **137**, 3748.

56 Jordan, K.G. and Tobias, C.W. (1991) *J. Electrochem. Soc.*, **138**, 1251.

57 Madore, C., Matlosz, M. and Landolt, D. (1996) *J. Electrochem. Soc.*, **143**, 3927.

58 Madore, C. and Landolt, D. (1996) *J. Electrochem. Soc.*, **143**, 3936.

59 Prall, J.K. and Shreir, L.L. (1964) *Trans. Inst. Met. Finish.*, **41**, 29.

60 Ke, B., Hoekstra, J.J., Sison, B.C. and Trevich, D. (1959) *J. Electrochem. Soc.*, **106**, 382.

61 Schmidt, W.U., Alkire, R.C. and Gewirth, A.A. (1996) *J. Electrochem. Soc.*, **143**, 3122.

62 Moffat, T.P., Wheeler, D., Huber, W.H. and Josell, D. (2001) *Electrochem. Solid State Lett.*, **4**, C26.

63 Josell, D., Wheeler, D., Huber, W.H., Bonevich, J.E. and Moffat, T.P. (2001) *J. Electrochem. Soc.*, **148**, C767.

64 Wheeler, D., Josell, D. and Moffat, T.P. (2003) *J. Electrochem. Soc.*, **150**, C203.

65 West, A.C., Mayer, S. and Reid, J. (2001) *Electrochem. Solid State Lett.*, **4**, C50.

66 Schulz-Harder, J. (1971) *Dissertation, Ein neuer Mechanismus des Einflusses von Badzusatzen auf die Stromdichteverteilung bei der Metallabscheidung auf rauhen Kathoden und seine experimentelle Begrundung*, Technischen Universitat Berlin.

67 Osterwald, J. and Schulz-Harder, J. (1975) *Galvanotechnik*, **66**, 360.

68 Osterwald, J. (1976) *Oberflache-Surface*, **17**, 89.

69 Lingk, C. and Gross, M.E. (1998) *J. Appl. Phys.*, **84**, 5547.

70 Cook, M. and Richards, T.L. (1944) *J. Inst. Met.*, **70**, 159.

71 Zheng, B., Step, E., Chen, J., Emami, R., Wang, Z.A., Nayak, R., Taylor, T. and Dixit, G. (2002) *Advanced Metallization Conference 2001* (eds A.J. McKerrow, Y. Shacham-Diamand, S. Zaima and T. Ohba), MRS, Warrendale, PA, p. 197.

72 Moffat, T.P., Baker, B., Wheeler, D., Bonevich, J.E., Edelstein, M., Kelly, D.R., Gan, L., Stafford, G.R., Chen, P.J., Egelhoff, W.F. and Josell, D. (2002) *J. Electrochem. Soc.*, **149**, C423.

73 Josell, D., Baker, B., Witt, C., Wheeler, D. and Moffat, T.P. (2002) *J. Electrochem. Soc.*, **149**, C637–C641.

74 Baker, B.C., Freeman, M., Melnick, B., Wheeler, D., Josell, D. and Moffat, T.P. (2003) *J. Electrochem. Soc.*, **150**, C61.

75 Baker, B.C., Witt, C., Wheeler, D., Josell, D. and Moffat, T.P. (2003) *Electrochem. Solid State Lett.*, **6**, C67–C69.

76 Ahn, E.J. and Kim, J.J. (2004) *Electrochem. Solid State Lett.*, **7**, C118.

77 Sham, K.-C., Lee, H.-B., Kwon, O.-K., Park, H.-S., Koh, W. and Kang, S.-W. (2002) *J. Electrochem. Soc.*, **149**, G109.

78 Bard, A.J., Abruna, H.D., Chidsey, C.E., Faulkner, L.R., Feldberg, S.W., Itaya, K., Majda, M., Melroy, O., Murray, R.W., Porter, M.D., Soriaga, M.P. and White, H.S. (1993) *J. Phys. Chem.*, **97**, 7147.

79 Kolb, D.M. (2001) *Angew. Chem., Int. Ed. Engl.*, **40** (7), 1162.

80 Wandelt, K. and Thurgate, S. (eds) (2003) *Solid-Liquid Interfaces; Macroscopic Phenomena-Microscopic Understanding, Springer Topics in Applied Physics*, vol. 85, Springer, Berlin.

81 *Electrolytic Metal Deposition, Fundamental Aspects and Applications* (eds A.S. Dakkouri and D.M. Kolb), Proceedings of the 5th Ulmer Elektrochemische Tage, Reprint from *Z. Phys. Chem.* **208**, Part 1–2 (1999).

82 Lorenz, W.J. and Plieth, W. (eds) (1998) *Electrochemical Nanotechnology, In situ Local Probe Techniques at Electrochemical Interfaces*, Wiley-VCH, New York.

83 Lipkowski, J. and Ross, P.N. (eds) (1999) *Imaging of Surfaces and Interfaces*, Wiley-VCH, New York.

84 Xia, Y.N., Rogers, J.A., Paul, K.E. and Whitesides, G.M. (1999) *Chem. Rev.*, **99**, 1823.

85 Michel, B., Bernard, A., Bietsch, A., Delamarche, E., Geissler, M., Juncker, D., Kind, H., Renault, J.-P., Rothuizen, H., Schmid, H., Schmidt-Winkel, P., Stutz, R. and Wolf, H. (2001) *IBM J. Res. Dev.*, **45**, 697.

86 Geissler, M., Schmid, H., Bietsch, A., Michel, B. and Delamarche, E. (2002) *Langmuir*, **18**, 2374.

87 Delahay, P. and Trachtenberg, I. (1957) *J. Am. Chem. Soc.*, **79**, 2355.

88 Schmid, R.W. and Reilley, C.N. (1958) *J. Am. Chem. Soc.*, **80**, 2087.

89 Delahay, P. and Trachtenberg, I. (1958) *J. Am. Chem. Soc.*, **80**, 2094.
90 Delahay, P. and Fike, C.T. (1958) *J. Am. Chem. Soc.*, **80**, 2639.
91 Fisher, H. (1960) *Electrochim. Acta*, **2**, 50.
92 Parsons, R. (1969) *J. Electroanal. Chem.*, **21**, 35.
93 Lipkowski, J. and Galus, Z. (1975) *J. Electroanal. Chem.*, **61**, 11.
94 Lipkowski, J. and Galus, Z. (1979) *J. Electroanal. Chem.*, **98**, 91.
95 Bard, A.J. and Faulkner, L.R. (2001) *Electrochemical Methods, Fundamentals and Applications*, John Wiley and Sons, New York.
96 Moffat, T.P. and Yang, H. (1995) *J. Electrochem. Soc.*, **142**, L220.
97 Eliadis, E.D., Nuzzo, R.G., Gewirth, A.A. and Alkire, R.C. (1997) *J. Electrochem. Soc.*, **144**, 96.
98 Petri, M., Kolb, D.M., Memmert, U. and Meyer, H. (2003) *Electrochim. Acta*, **49**, 175.
99 Petri, M., Kolb, D.M., Memmert, U. and Meyer, H. (2003) *Electrochim. Acta*, **49**, 183.
100 Epple, M., Bittner, A.M., Kuhnke, K., Kern, K., Zheng, W.-Q. and Tadjeddine, A. (2002) *Langmuir*, **18**, 773.
101 Esplandiu, M.J., Hagenstrom, H. and Kolb, D.M. (2001) *Langmuir*, **17**, 828.
102 Hangenstrom, H., Esplandiu, M.J. and Kolb, D.M. (2001) *Langmuir*, **17**, 839.
103 Esplandiu, M.J., Schneeweiss, M.A. and Kolb, D.M. (1999) *Langmuir*, **15**, 7802.
104 Hagenstrom, H., Schneeweiss, M.A. and Kolb, D.M. (1999) *Langmuir*, **15**, 2435.
105 Kolb, D.M. (2002) The Initial Stages of Metal Deposition as Viewed by Scanning Tunneling Microscopy, in *Advances in Electrochemical Science and Engineering*, vol. 7 (eds R.C. Alkire and D.M. Kolb), Wiley-VCH, Weinheim, p. 107.
106 Cavalleri, O., Bittner, A.M., Kind, H. and Kern, K. (1999) *Z. Phys. Chem.*, **208**, 107.
107 Cavalleri, O., Kind, H., Bittner, A.M. and Kern, K. (1998) *Langmuir*, **14**, 7292.
108 Cavalleri, O., Gilbert, S.E. and Kern, K. (1997) *Surf. Sci.*, **377**, 931.
109 Cavalleri, O., Gilbert, S.E. and Kern, K. (1997) *Chem. Phys. Lett.*, **269**, 479.
110 Nishizawa, M., Sunagawa, T. and Yoneyama, H. (1997) *Langmuir*, **13**, 5215.
111 Lakshmann Sarma, R. and Nageswar, S. (1982) *J. Appl. Electrochem.*, **12**, 329.
112 Schwartz, D.K. (2001) *Annu. Rev. Phys. Chem.*, **52**, 107.
113 (2004) New Developments in Bio-related Materials. *J. Mater. Chem.*, **14**, 2059.
114 Myers, T.H., Feenstra, R.M., Shur, M.S. and Amano, H. (eds) (1999) GaN and Related Alloys-1999, MRS Proceedings, vol. 595.
115 Azzaroni, O., Schilardi, P.L. and Salvarezza, R.C. (2003) *Electrochim. Acta*, **48**, 3107.
116 Azzaroni, O., Schilardi, P.L. and Salvarezza, R.C. (2002) *Appl. Phys. Lett.*, **80**, 1061.
117 Petri, M. and Kolb, D.M. (2002) *Phys. Chem. Chem. Phys.*, **4**, 1211.
118 Eliadis, E.D. and Alkire, R.C. (1998) *J. Electrochem. Soc.*, **145**, 1218.
119 Manne, S. and Gaub, H.E. (1995) *Science*, **270**, 1480.
120 Burgess, I., Jeffrey, C.A., Cai, X., Szymanski, G., Galus, Z. and Lipkowski, J. (1999) *Langmuir*, **15**, 2607.
121 McIntyre, J.D.E. and Peck, W.F., Jr. (1976) *J. Electrochem. Soc.*, **123**, 1800.
122 Adzic, R.R. (1984) Electrocatalytic Properties of the Surfaces Modified by Foreign Metal Adatoms, in *Advances in Electrochemistry and Electrochemical Engineering*, vol. 13 (eds H. Gerishcher and C.W. Tobias), John Wiley and Sons, New York, pp. 159.
123 Stickney, J.L., Rosaco, S.D. and Hubbard, A.T. (1984) *J. Electrochem. Soc.*, **131**, 260.
124 Adzic, R.R. (2001) Electrocatalysis on Surfaces Modified by Metal Monolayers Deposited at Underpotentials, in *Encyclopedia of Electrochemistry* (eds E. Gileadi and M. Urbakh), (Series Editors AJ Bard M Stratmann), Wiley-VCH, Weinheim, (chap. 4.3 and references cited therein).

125 Egelhoff, W.F. and Steigerwald, D.A. (1989) *J. Vac. Sci. Technol. A*, **7**, 2167.
126 Xia, Y., Yang, P., Sun, Y., Wu, Y., Mayers, B., Gates, B., Yin, Y., Kim, F. and Yan, H. (2003) *Adv. Mater.*, **15**, 353, and reference cited therein.
127 Sashikata, K., Matsui, Y., Itaya, K. and Soriaga, M.P. (1996) *J. Phys. Chem.*, **100**, 20027.
128 Nagy, Z., Blaudeau, J.P., Hung, N.C., Curtiss, L.A. and Zurawski, D.J. (1995) *J. Electrochem. Soc.*, **142**, L87.
129 Schilardi, P.L., Azzaroni, O. and Salvarezza, R.C. (2000) *Phys. Rev. B*, **62**, 13098.
130 Edwards, J. and Levett, M.J. (1964) *Trans. Inst. Met. Finish.*, **41**, 157.
131 Edwards, J. (1962) *Trans. Inst. Met. Finish.*, **39**, 33.
132 Jonsson, B., Lindman, B., Holmberg, K. and Kronberg, B. (1999) Surfactants and Polymers, in Aqueous Solutions, John Wiley and Sons, New York.
133 Attard, G.S., Bartlett, P.N., Coleman, N.R.B., Elliott, J.M., Owen, J.R. and Wang, J.H. (1997) *Science*, **278**, 838.
134 Magnussen, O.M. (2002) *Chem. Rev.*, **102**, 679.
135 Jung, L.S. and Campbell, C.T. (2000) *J. Phys. Chem. B*, **104**, 11168.
136 Moffat, T.P., Wheeler, D. and Josell, D. (2004) *J. Electrochem. Soc.*, **151**, C262.
137 Gonzalez-Velasco, J. (1998) *Surf. Sci.*, **410**, 283.
138 Andreasen, G., Nazzarro, N., Ramirez, J. and Salvarezza, R.C. (1996) *J. Electrochem. Soc.*, **143**, 466.
139 Krichmar, S.I. (1965) *Elektrokhimiya*, **1**, 858.
140 Roha, D. and Landau, U. (1990) *J. Electrochem. Soc.*, **137**, 824.
141 Edwards, J. (1964) *Trans. Inst. Met. Finish.*, **41**, 140.
142 Edwards, J. (1964) *Trans. Inst. Met. Finish.*, **41**, 169.
143 Rogers, G.T., Ware, M.J. and Fellows, R.V. (1960) *J. Electrochem. Soc.*, **107**, 677.
144 Beacom, S.E. and Riley, B.J. (1959) *J. Electrochem. Soc.*, **106**, 309.
145 Beacom, S.E. and Riley, B.J. (1960) *J. Electrochem. Soc.*, **107**, 785.
146 Beacom, S.E. and Riley, B.J. (1961) *J. Electrochem. Soc.*, **108**, 758.
147 Krichmar, S.I. and Pronskaya, A.Y. (1976) *Elektrokhimiya*, **12**, 22. (1976) *Sov. Electrochem.*, **12**, 20.
148 Kruglikov, S.S., Gamburg, Y.D. and Kudryavtsev, N.T. (1967) *Electrochim. Acta*, **12**, 1129.
149 Edwards, J. (1962) *Trans. Inst. Met. Finish.*, **39**, 52.
150 Edwards, J. and Levett, M.J. (1964) *Trans. Inst. Met. Finish.*, **41**, 147.
151 Healy, J.P., Pletcher, D. and Goodenough, M. (1992) *J. Electroanal. Chem.*, **338**, 167.
152 Horkans, J. and Dukovic, J.O. (2002) in Electrochemical Processing in ULSI Fabrication III (eds P.C. Andricacos, P.C. Searson, C. Reidsema-Simpson, P. Allongue, J.L. Stickney, G.M. Oleszek), vol. 2000-8, *The Electrochemical Society Proceedings Series*, Pennington, NJ, p.103.
153 Zhukauskaite, N.A., Lazauskene, A.Yu. and Malinauskas, A.A. (1989) *Prot. Met.*, **25**, 244.
154 Kim, J.J., Kim, S.K. and Kim, Y.S. (2003) *J. Electroanal. Chem.*, **542**, 61.
155 Kim, S.-K. and Kim, J.J. (2004) *Electrochem. Solid-State Lett.*, **7**, C98.
156 Koh, L.T., You, G.Z., Li, C.Y. and Foo, P.D. (2002) *Microelectron. J.*, **33**, 229.
157 Koh, L.T., You, G.Z., Lim, S.Y., Li, C.Y. and Foo, P.D. (2001) *Microelectron. J.*, **32**, 973.
158 Moffat, T.P., Baker, B., Wheeler, D. and Josell, D. (2003) *Electrochem. Solid State Lett.*, **6**, C59.
159 Freitag, W.O., Ogden, C., Tench, D. and White, J. (1983) *Plating*, **70**, 55.
160 Pavlov, M., Shalyt, E., Bratin, P. and Tench, D.M. (2003) in *Copper Interconnects, New Contact Metallurgies, Structure and Low-k Interlevel Dielectrics II*,

vol. 2003-10 (eds G.S. Mathad, V. Bakshi, H.S. Rathore, K. Kondo and C. Reidsema-Simpson), The Electrochemical Society, Pennington, NJ, p. 53.

161 Plieth, W. (1992) *Electrochim. Acta*, **37**, 2115.

162 Krischer, K. (2003) Nonlinear Dynamics in Electrochemical Systems, in *Advances in Electrochemical Science and Engineering* (eds R.C. Alkire and D.M. Kolb), Wiley-VCH, New York.

163 Bertocci, U. (1968) *Surf. Sci.*, **9**, 18.

164 Price, P.B., Vermilyea, D.A. and Webb, M.B. (1958) *Acta Metall.*, **6**, 524.

165 Bebczuk de Cusminsky, J. (1970) *Electrochim. Acta*, **15**, 73.

166 Kozlov, V. and Bicelli, L.P. (1996) *J. Cryst. Growth*, **165**, 421.

167 Kozlov, V.M. and Bicelli, L.P. (1997) *J. Cryst. Growth*, **177**, 289.

168 Kozlov, V.M., Bicelli, L.P. and Timoshenko, V.N. (1998) *J. Crystal. Growth*, **183**, 456.

169 Stoychev, D.S., Tomov, I., Vitanova, I. and Rashkov, St. (1978) *Surf. Technol.*, **7**, 433.

170 Vermilyea, D.A. (1959) *J. Electrochem. Soc.*, **106**, 66.

171 Aust, K.T., Hibbard, G., Palumbo, G. and Erb, U. (2003) *Z. Metallkd.*, **94**, 1066.

172 ElSherik, A.M., Erb, U. and Page, J. (1997) *Surf. Coat. Technol.*, **88**, 70.

173 Elsherik, A.M. and Erb, U. (1995) *J. Mater. Sci.*, **30**, 5743.

174 Velinov, V., Vitkova, S. and Pangarov, N. (1977) *Surf. Technol.*, **6**, 19.

175 Nikolic, N.D., Rakocevic, Z. and Popov, K.I. (2001) *J. Electroanal. Chem.*, **514**, 56.

176 Pearson, T. and Dennis, J.K. (1990) *Surf. Coat. Technol.*, **42**, 69.

177 Kelly, J.J., Tian, C. and West, A.C. (1999) *J. Electrochem. Soc.*, **146**, 2540.

178 Setty, T.H.V. and Wilman, H. (1966) *Electrochim. Acta*, **11**, 297.

179 Yoshihara, S., Ueno, M., Nagae, Y. and Fijishima, A. (1988) *J. Electroanal. Chem.*, **243**, 475.

180 Bindra, P., Light, P., Freudenthal, P. and Smith, D. (1989) *J. Electrochem. Soc.*, **136**, 3616.

181 Okinaka, Y. and Hoshino, M. (1998) *Gold Bull.*, **31**, 3.

182 Nakahara, S. and Okinaka, Y. (1981) *J. Electrochem. Soc.*, **128**, 284.

183 Denise, F. and Leidheiser, H. (1953) *J. Electrochem. Soc.*, **100**, 490.

184 Karayannis, H.S. and Patermarakis, G. (1995) *Electrochim. Acta*, **40**, 1079.

185 Ye, X., De Monte, M., Celis, J.P. and Roos, J.R. (1992) *J. Electrochem. Soc.*, **139**, 1592.

186 Li, D.Y. and Szpunar, J.A. (1997) *Electrochim. Acta*, **42**, 37.

187 Li, D.Y. and Szpunar, J.A. (1997) *Electrochim. Acta*, **42**, 47.

188 Rao, S.T. and Weil, R. (1979) *Trans. Inst. Met. Finish.*, **57**, 97.

189 Turner, D.R. and Johnson, G.R. (1962) *J. Electrochem. Soc.*, **109**, 798.

190 Johnson, G.R. and Turner, D.R. (1962) *J. Electrochem. Soc.*, **109**, 918.

191 Hofer, E.M. and Hintermann, H.E. (1965) *J. Electrochem. Soc.*, **112**, 167.

192 Holzle, M.H., Apsel, C.W., Will, T. and Kolb, D.M. (1995) *J. Electrochem. Soc.*, **142**, 3741.

193 Dye, J.L. and Klingenmaier, O.J. (1957) *J. Electrochem. Soc.*, **104**, 275.

194 Harris, L. (1973) *J. Electrochem. Soc.*, **120**, 1034.

195 Deligianni, H. and Romankiw, L.T. (1993) *IBM J. Res. Dev.*, **37**, 85.

196 Cooper, J.J.W.C., Dreisinger, D.B. and Peters, E. (1995) *J. Appl. Electrochem.*, **25**, 642.

197 Stoychev, D.S., Tomov, I.V. and Vitanova, I.B. (1985) *J. Appl. Electrochem.*, **15**, 879.

198 Tomov, I.V., Stoychev, D.S. and Vitanova, I.B. (1985) *J. Appl. Electrochem.*, **15**, 887.

199 Stoychev, D.S., Tomov, I., Vitanova, I. and Rashkov, St. (1978) *Surf. Technol.*, **7**, 433.

200 Rashkov, S. and Stoichev, D.S. (1978) *Surf. Technol.*, **6**, 155.

201 Harper, J.M.E., Cabral, C., Andricacos, P.C., Gignac, L., Noyan, I.C., Rodbell, K.P. and Hu, C.K. (1999) *J. Appl. Phys.*, **86**, 2516.

202 Detavernier, C., Rossnagel, S., Noyan, C., Guha, S., Cabral, C. and Lavoie, C. (2003) *J. Appl. Phys.*, **94**, 2874.

203 Brongersma, S.H., D'Haen, J., Ansteels, K., DeCeunick, W., Vervoot, I. and Maex, K. (2003) *Mater. Sci. Forum*, **426-4**, 2485.

204 Horkans, J., Cabral, C., Rodbell, K.P., Parks, C., Gribelyuk, M., Malhotra, S. and Andricacos, P.C. (2002)in Electrochemical Processing in ULSI Fabrication and Semiconductor/Metal Depositon III, C (eds J.L. Andricacos, P.C. Stickney, C. Searson, G.M. Reidsema-Simpson, Oleszek), vol. 2000-8P, The Electrochemical Society Proceedings Series, Pennington, NJ, p.110.

205 Pantleon, K., Jensen, J.A.D. and Somers, M.A.J. (2004) *J. Electrochem. Soc.*, **151**, C45.

206 Kerr, E., Vervoort, I., Sarens, A. and Maex, K. (2002) *J. Mater. Res.*, **17**, 582.

207 Lagrange, S., Brongersma, S.H., Judelewicz, M., Sarens, A., Vervoort, I., Richards, E., Palmans, R. and Maex, K. (2000) *Microelectron. Eng.*, **50**, 449.

208 Brongersma, S.H., Vervoort, R.E., Bender, H., Vandervorst, W., Lagrange, S., Beyer, G. and Maex, K. (1999) *J. Appl. Phys.*, **86**, 3642.

209 Yoon, M.-S., Park, Y.-J. and Joo, Y.-C. (2002) *Thin Solid Films*, **408**, 230.

210 Malhotra, S.G., Locke, P.S., Simon, A.H., Fluegel, J., De Haven, P., Hemmes, D.G., Jackson, R. and Patton, E. (2000) *Proceedings of the 1999 Advanced Metallization Conference* (eds M.E. Gross, T. Gessner, N. Kobayashiand Y. Yasuda), Materials Research Society, Warrendale, PA, p. 77.

211 Baker, B., Pena, D., Herrick, M., Chowdhury, R., Acosta, E., Simpson, C.R. and Hamilton, G. (1999) *in Electrochemical Processing in ULSI Fabrication and Semiconductor/Metal Deposition II* (eds P.C. Andricacos, P.C. Searson, C. Reidsema-Simpson, P. Allongue, J.L. Stickney and G.M. Oleszek), The Electrochemical Society Proceedings Series, Pennington, NJ.

212 Andricacos, P.C., Parks, C., Cabral, C., Wachnik, R., Tsai, R., Malhotra, S., Locke, P., Fluegel, J., Horkans, J., Kwietniak, K., Uzoh, C., Rodbell, K.P., Gignac, L., Walton, E., Chung, D. and Geffken, R. (1999) *in Electrochemical Processing in ULSI Fabrication and Semiconductor/ Metal Deposition II* (eds P.C. Andricacos, P.C. Searson, C. Reidsema-Simpson, P. Allongue, J.L. Stickney and M. Oleszek), The Electrochemical Society Proceedings Series, Pennington, NJ.

213 Kang, M. and Gewirth, A.A. (2003) *J. Electrochem. Soc.*, **150**, C426.

214 Mirkova, L. and Rashkov, St. (1982) *Chr. Nanev, Surf. Technol.*, **15**, 181.

215 Weil, R. and Jacobus, W.N. (1966) *Plating*, **53**, 102.

216 Watson, S.A. and Edwards, J. (1957) *Trans. Inst. Met. Finish.*, **34**, 167.

217 Kruglikov, S.S., Kudriavtsev, N.T., Vorobiova, G.F. and Antonov, A.Y. (1965) *Electrochim. Acta*, **10**, 253.

218 Kruglikov, S.S., Kudryavtsev, N.T. and Sobolev, R.P. (1967) *Electrochim. Acta*, **12**, 1263.

219 Schneider, H., Sukava, A.J. and Newby, W.J. (1965) *J. Electrochem. Soc.*, **112**, 568.

220 Sukava, A.J., Schneider, H., McKenney, D.J. and McGregor, A.T. (1965) *J. Electrochem. Soc.*, **112**, 571.

221 Chu, A.K.P. and Sukava, A.J. (1969) *J. Electrochem. Soc.*, **116**, 1188.

222 Roth, C.C. and Leidheiser, H. (1953) *J. Electrochem. Soc.*, **100**, 553.

223 Farndon, E.E., Walsh, F.C. and Campbell, S.A. (1995) *J. Appl. Electrochem.*, **25**, 574.

224 Martyak, N.M. and Seefeldt, R. (2004) *Electrochim. Acta*, **49**, 4303.

225 Zavarine, I.S., Khaselev, O. and Zhang, Y. (2003) *J. Electrochem. Soc.*, **150**, C202.

226 Muresan, L., Oniciu, L., Froment, M. and Maurin, G. (1992) *Electrochim. Acta*, **37**, 2249.

227 Cheng, C.C. and West, A.C. (1997) *J. Electrochem. Soc.*, **144**, 3050.

228 Epelboin, I. and Wiart, R. (1971) *J. Electrochem. Soc.*, **11**, 1577.

229 Chassaing, E., Joussellin, M. and Wiart, R. (1983) *J. Electrochem. Soc.*, **157**, 75.
230 Cheng, C.C. and West, A.C. (1998) *J. Electrochem. Soc.*, **145**, 560.
231 Kruglikov, S.S., Kudryavtsev, N.T., Medvedev, G.I. and Izmailova, T.M. (1972) *Prot. Met.*, **8**, 668.
232 Biggin, M.E. and Gewirth, A.A. (2001) *J. Electrochem. Soc.*, **148**, C339.
233 Hill, M.R.H. and Rogers, G.T. (1978) *J. Electroanal. Chem.*, **86**, 179.
234 Yokoi, M., Konishi, S. and Hayashi, T. (1984) *Denki Kagaku*, **52**, 218.
235 White, J.R. (1987) *J. Appl. Electrochem.*, **17**, 977.
236 Healy, J.P., Pletcher, D. and Goodenough, M. (1992) *J. Electroanal. Chem.*, **338**, 155.
237 Wunsche, M., Dahms, W., Meyer, H. and Schumacher, R. (1994) *Electrochim. Acta*, **39**, 1133.
238 Hope, G.A. and Brown, G.M. (1996) *Proceedings of the 6th International Symposium on Electrode Processes*, vol. 96-8 (eds A. Wieckowski and K. Itaya), The Electrochemical Society, Pennington, NJ, p. 215.
239 Stoychev, D. and Tsvetanov, C. (1996) *J. Appl. Electrochem.*, **26**, 741.
240 Stoychev, D. (1998) *Trans. Inst. Met. Finish.*, **76**, 73.
241 Kelly, J.J. and West, A.C. (1998) *J. Electrochem. Soc.*, **145**, 3472.
242 Kelly, J.J. and West, A.C. (1998) *J. Electrochem. Soc.*, **145**, 3477.
243 Moffat, T.P., Bonevich, J.E., Huber, W.H., Stanishevsky, A., Kelly, D.R., Stafford, G.R. and Josell, D. (2000) *J. Electrochem. Soc.*, **147**, 4524.
244 Wu, B.H., Wan, C.C. and Wang, Y.Y. (2003) *J. Appl. Electrochem.*, **33**, 823.
245 Brown, G.M. and Hope, G.A. (1996) *J. Electroanal. Chem.*, **405**, 211.
246 Moffat, T.P. (1995) in *Nanostructured Materials in Electrochemistry*, vol. 95-8 (eds P.C. Searson and G.J. Meyer), The Electrochemical Society, Pennington, NJ, p. 225.
247 Moffat, T.P. (1999) in *Electrochemical Processing in ULSI Fabrication and Semiconductor/Metal Deposition II*, vol. 99-9 (eds P.C. Andricacos, P.C. Searson, C. Reidsema-Simpson, P. Allongue, J.L. Stickney and G.M. Oleszek), The Electrochemical Society, Pennington, NJ, p. 41.
248 Doblhofer, K., Wasle, S., Soares, D.M., Weil, K. and Ertl, G. (2003) *J. Electrochem. Soc.*, **150**, C657.
249 Walker, M.L., Richter, L.J. and Moffat, T.P. (2005) *J. Electrochem. Soc.*, **152**, C403.
250 Feng, Z.V., Li, X. and Gewirth, A.A. (2003) *J. Phys. Chem. B.*, **107**, 9415.
251 Yokoi, M., Konishi, S. and Hayashi, T. (1983) *Denki Kagahu*, **51**, 460.
252 Hill, M.R.H. and Rogers, G.T. (1976) *J. Electroanal. Chem.*, **68**, 149.
253 Rogers, G.T. and Taylor, K.J. (1975) *Electrochim. Acta*, **20**, 695.
254 Rogers, G.T. and Taylor, K.J. (1975) *Electrochim. Acta*, **20**, 703.
255 Eisenmann, E.T. (1978) *J. Electrochem. Soc.*, **125**, 717.
256 Davidovic, D. and Adzic, R.R. (1988) *J. Serb. Chem. Soc.*, **53**, 499.
257 Poskus, D., Agafonovas, G. and Jurgaitiene, I. (1997) *J. Electroanal. Chem.*, **425**, 107.
258 Bek, R.Yu. (2002) *Russ. J. Electrochem.*, **38**, 1237.
259 Bek, R.Yu. and Shuraeva, L.I. (2003) *Russ. J. Electrochem.*, **39**, 229.
260 Bek, R.Yu. and Shuraeva, L.I. (2003) *Russ. J. Electrochem.*, **39**, 872.
261 Bozzini, B. and Fanigliulo, A. (2002) *J. Crystal. Growth*, **243**, 190.
262 Reid, J. (2001) *Jpn. J. Appl. Phys.*, **40** (4B), 2650.
263 Zakauskaite, N. and Malinauskas, A. (1988) *Elektrokhimiya*, **24**, 1694.
264 Reid, J. and Mayer, S. (1999) in *Advanced Metallization Conference* (eds M.E. Gross, T. Gessner, N. Kobayashi and Y. Tasuda), MRS, Warrendale, p. 53.
265 Reid, J., Gack, C. and Hearne, S.J. (2004) *Electrochem. Solid State Lett.*, **34**, 291.
266 Tan, M. and Harb, J.N. (2003) *J. Electrochem. Soc.*, **150**, C420.

267 Davidovic, D. and Adzic, R.R. (1988) *Electrochim. Acta*, **33**, 103.
268 Dow, W.-P., Huang, H.S. and Lin, Z. (2003) *Electrochem. Solid State Lett.*, **6**, C134.
269 Pearson, T. and Dennis, J.K. (1990) *J. Appl. Electrochem.*, **20**, 196.
270 Michailova, E., Vitanova, I., Stoychev, D. and Milchev, A. (1993) *Electrochim. Acta*, **38**, 2455.
271 Peykova, M., Michailova, E., Stoychev, D. and Milchev, A. (1995) *Electrochim. Acta*, **40**, 2595.
272 Stoychev, D., Vitanov, I., Bujukliev, R., Petkova, N. and Polarliev, I. (1992) *J. Appl. Electrochem.*, **22**, 978.
273 Stoychev, D., Vitanov, I., Bujukliev, R., Petkova, N. and Polarliev, I. (1992) *J. Appl. Electrochem.*, **22**, 978.
274 Gerenrot, Y.E., Gol'din, L.Z. and Landis, V.V. (1978) *Elektrokhimiya*, **14**, 1083. (1979) *Sov. Electrochem.*, **14**, 941.
275 Kuznetsova, L.A. and Kovarskii, N.Y. (1981) *Elektrokhimiya*, **17**, 1633; (1982) *Sov. Electrochem.*, **17**, 1359.
276 Casford, M.T.L., Davies, P.B. and Neivandt, D.J. (2003) *Langmuir*, **19**, 7386, and references cited therein.
277 Herzog, G. and Arrigan, D.W.M. (2003) *Anal. Chem.*, **75**, 319.
278 Yoon, S., Schwartz, M. and Nobe, K. (December 1994) *Plat. Surf. Finish.*, **81**, 65, and reference cited therein.
279 Yokoi, M., Konishi, S. and Hayashi, T. (1984) *Denki Kagaku*, **52**, 218.
280 Barkey, D., Oberholtzer, F. and Wu, Q. (1998) *J. Electrochem. Soc.*, **145**, 590.
281 Xu, S., Cruchon-Dupeyrat, S.J.N., Garno, J.C., Liu, G.-Y., Jennings, G.K., Yong, T.-H. and Laibinis, P.E. (1998) *J. Phys. Chem.*, **108**, 5002.
282 Schweizer, M., Hagenstrom, H. and Kolb, D.M. (2001) *Surf. Sci.*, **490**, L627.
283 Bertocci, U. and Turner, D. (1974) *Encyclopedia of Electrochemistry of the Elements*, vol. 2 (ed. A.J. Bard), Marcel Dekker, New York, p. 1.
284 Gileadi, E. and Tsionsky, V. (2000) *J. Electrochem. Soc.*, **147**, 567.
285 Bertocci, U. (1966) *Electrochim. Acta*, **11**, 1261.
286 Vicenzo, A. and Cavalotti, P.L. (2002) *J. Appl. Electrochem.*, **32**, 743.
287 Lee, D.W., Pyo, S.G., Ko, C.J., Lee, M.H., Kim, S. and Lee, J.G. (2004) *J. Electrochem. Soc.*, **151**, C204.
288 Dukovic, J.O. (1993) *IBM J. Res. Dev.*, **37**, 125.
289 Drews, T.O., Webb, E.G., Ma, D.L., Alameda, J., Braatz, R.D. and Alkire, R.C. (2004) *AIChE J.*, **50**, 226.
290 Park, T., Tugbawa, T., Boning, D., Chidambaram, C., Borst, G. and Shin, G. (2004) *J. Electrochem. Soc.*, **151**, C418.
291 Takahashi, M. and Gross, M.E. (1999) *J. Electrochem. Soc.*, **146**, 4499.
292 McHugh, P.R., Wilson, G.J. and Chen, L. (1999) in *Electrochemical Processing in ULSI Fabrication and Semiconductor/Metal Deposition II*, vol. 99-9 (eds P.C. Andricacos, P.C. Searson, C. Reidsema-Simpson, P. Allongue, J.L. Stickney and M. Oleszek), The Electrochemical Society Proceedings Series, Pennington, NJ, p. 71.
293 Landau, U. (1999) *Electrochemical Processing in ULSI Fabrication and Semiconductor/Metal Deposition II*, vol. 99-9 (eds P.C. Andricacos, P.C. Searson, C. Reidsema-Simpson, P. Allongue, J.L. Stickney and M. Oleszek), The Electrochemical Society Proceedings Series, Pennington, NJ, p. 25.
294 Dordi, Y., Landau, U., Lakshmikanthan, J., Stevens, J., Hey, P. and Lipin, A. (1999) in *Electrochemical Processing in ULSI Fabrication III* (eds P.C. Andricacos, J.L. Stickney, P.C. Searson, C. Reidsema-Simpson and M. Oleszek), The Electrochemical Society Proceedings Series, Pennington, NJ, p. 123.
295 Mehdizadeh, S., Dukovic, J., Andricacos, P.C., Romankiw, L.T. and Cheh, H.Y. (1990) *J. Electrochem. Soc.*, **137**, 110.
296 Deligianni, H., Dukovic, J.O., Walton, E.G., Contolini, R.J., Reid, J. and Patton, E. (1999) in *Electrochemical Processing in ULSI Fabrication and Semiconductor/*

Metal Deposition II (eds P.C. Andricacos, P.C. Searson, C. Reidsema-Simpson, P. Allongue, J.L. Stickney and M. Oleszek), The Electrochemical Society Proceedings Series, Pennington, NJ.

297 Klocke, J., McHugh, P., Wilson, G., Ritari, K., Roberts, M. and Ritzdorf, T. (2003) *Advanced Metallization Conference 2002*, MRS, Warrendale, PA, p. 373.

298 Cao, Y., Lee, J.M. and West, A.C. (2003) *Plat. Surf. Finish.*, **90**, Nov. 40.

299 DeBecker, B. and West, A.C. (1996) *J. Electrochem. Soc.*, **143**, 486.

300 Choi, Y.-S. and Kang, T. (1996) *J. Electrochem. Soc.*, **143**, 480.

301 Matlosz, M., Vallotton, P.H., West, A.C. and Landolt, D. (1992) *J. Electrochem. Soc.*, **139**, 752.

302 Chalupa, R., Cao, Y. and West, A.C. (2002) *J. Appl. Electrochem.*, **32**, 135.

303 Bär, E., Lorezn, J. and Ryssel, H. (2003) *Electrochem. Soc. Proc.*, **2003–10**, 21.

304 Cao, Y., Taephaisitphongse, P., Cahlupa, R. and West, A.C. (2001) *J. Electrochem. Soc.*, **148**, C466.

305 Soukane, S., Sen, S. and Cale, T.S. (2002) *J. Electrochem. Soc.*, **149**, C74.

306 Georgiadou, M., Veyret, D., Sani, R.L. and Alkire, R.C. (2001) *J. Electrochem. Soc.*, **148**, C54.

307 Im, Y.H., Bloomfield, M.O., Sen, S. and Cale, T.S. (2003) *Electrochem. Solid State Lett.*, **6**, C42.

308 Josell, D., Wheeler, D., Huber, W.H. and Moffat, T.P. (2001) *Phys. Rev. Lett.*, **87**, 016102.

309 McFadden, G.B., Coriell, S.R., Moffat, T.P., Josell, D., Wheeler, D., Schwarzacher, W. and Mallett, J. (2003) *J. Electrochem. Soc.*, **150** (9), C591–C599.

310 Haataja, M., Srolovitz, D.J. and Bocarsly, A.B. (2003) *J. Electrochem. Soc.*, **150**, C708.

311 Wheeler, D., Moffat, T.P., McFadden, G.B., Coriell, S. and Josell, D. (2004) *J. Electrochem. Soc.*, **151**, C538.

312 Gill, W.N., Duquette, D.J. and Varadarajan, D. (2001) *J. Electrochem. Soc.*, **148**, C289.

313 Josell, D., Baker, B., Witt, C., Wheeler, D. and Moffat, T.P. (2002) *J. Electrochem. Soc.*, **149**, C637.

314 Ratsch, C. and Zhangwill, A. (1991) *Appl. Phys. Lett.*, **58**, 405.

315 De Virgiliis, A., Azzaroni, O., Salvarezza, R.C. and Albano, E.V. (2003) *Appl. Phys. Lett.*, **82**, 1953.

316 Kruglikov, S.S. and Smirnova, T.A. (1973) Proceedings of the 8th Congress of the International Union for Electrodeposition and Surface Finishing, p.105.

317 Gerenrot, Y.E., Vaisburd, L.A. and Mokrinich, N.A. (1972) *Zashch. Metall.*, **8**, 81.

318 Gerenrot, Y.E., Vaisburg, L.A. and Sych, E.D. (1972) *Zashch. Metall.*, **8**, 338.

319 Gerenrot, Y.E. and Vaisburd, L.A. (1973) *Elektrokhimiya*, **9**, 1332.

320 Nanev, C., Mirkova, L. and Dicheva, K. (1988) *Surf. Coat. Technol.*, **34**, 483.

321 Krichmar, S.I. (1965) *Elektrokhimiya*, **1**, 858.

322 Krichmar, S.I. (1965) *Russ. J. Phys. Chem.*, **39**, 320.

323 Krichmar, S.I. and Pronskaya, A.Y. (1965) *Russ. J. Phys. Chem.*, **39**, 387.

324 Krichmar, S.I. (1981) *Elektrokhimiya*, **17**, 1444.

325 Mirkova, L., Nanev, C. and Rashkov, R. (1988) *Surf. Coat. Technol.*, **34**, 471.

326 Georgiadou, M. and Alkire, R. (1993) *J. Electrochem. Soc.*, **140**, 1340; **140**, 1343.

327 Chen, Q., Modi, V. and West, A.C. (1998) *J. Appl. Electrochem.*, **28**, 579.

328 Alkire, R.C., Deligianni, H. and Ju, J.-B. (1990) *J. Electrochem. Soc.*, **137**, 818.

329 Georgiadou, M., Mohr, R. and Alkire, R.C. (2000) *J. Electrochem. Soc.*, **147**, 3021.

330 Nilson, R.H. and Griffths, S.K. (2003) *J. Electrochem. Soc.*, **150**, C401.

331 Matsuda, I., Asa, F. and Osaka, T. (2003) *Electrochemistry*, **71**, 913.

332 Deligianni, H., Dukovic, J.O., Andricacos, P.C. and Walton, E.G. (1999) in *Electrochemical Processing in ULSI Fabrication and Semiconductor/Metal Deposition II*, vol. 99-9 (eds P.C.

Andricacos, J.L. Stickney and G.M. Olezak), The Electrochemical Society Proceedings, Pennington, NJ, p. 52.
333 Reid, J. and Mayer, S. (2000) *in Advanced Metallization Conference 1999* (eds M.E. Gross, T. Gessner, N. Kobayashi and Y. Yasuda), MRS, Warrendale, PA, p. 53.
334 Ritzdorf, T., Fulton, D. and Chen, L. (2000) *in Advanced Metallization Conference 1999* (eds M.E. Gross, T. Gessner, N. Kobayashi and Y. Yasuda), MRS, Warrendale, PA, p. 101.
335 Richard, E., Vervoort, I., Brongersma, S.H., Bender, H., Beyer, G., Palmans, R., Lagrange, S. and Maex, K. (2000) *in Advanced Metallization Conference 1999* (eds M.E. Gross, T. Gessner, N. Kobayashi and Y. Yasuda), MRS, Warrendale, PA, p. 149.
336 Taephaisitphongse, P., Cao, Y. and West, A.C. (2001) *J. Electrochem. Soc.*, **148**, C492.
337 West, A.C. (2000) *J. Electrochem. Soc.*, **147**, 227.
338 Albright, J.G., Paduano, L., Sartorio, R., Vergara, A. and Vitagliano, V. (2001) *J. Chem. Eng. Data*, **46**, 1283.
339 Quickenden, T.I. and Xu, Q. (1996) *J. Electrochem. Soc.*, **143**, 1248.
340 Gerenrot, Y.E. and Leichkis, D.L. (1977) *Elektrokhimiya*, **13**, 341.
341 Gomez, M.M., Vara, J.M., Hernandez, J.C., Salvarezza, R.C. and Arvia, A.J. (2001) *Electrochim. Acta*, **47**, 405.
342 Hayase, M., Taketani, M., Aizawa, K., Hatsuzawa, T. and Hayabusa, K. (2002) *Electrochem. Solid State Lett.*, **5**, C98.
343 Moffat, T.P., Wheeler, D., Witt, C. and Josell, D. (2002) *Electrochem. Solid State Lett.*, **5** (12), C110–C112.
344 Josell, D., Wheeler, D. and Moffat, T.P. (2002) *Electrochem. Solid State Lett.*, **5**, C49.
345 Josell, D., Moffat, T.P. and Wheeler, D. (2004) *J. Electrochem. Soc.*, **151** (1), C19–C24.
346 Moffat, T.P., Chen, P.J., Castillo, A., Baker, B., Egelhoff, W.F. and Richter, L. unpublished results.
347 Smy, T., Tan, L., Drews, S.K., Brett, M.J., Shacham-Diamand, Y. and Desilva, M. (1997) *J. Electrochem. Soc.*, **144**, 2115.
348 Kim, J.J., Cha, S.H. and Lee, Y.-S. (2003) *Jpn. J. Appl. Phys.*, **42**, L953.
349 Shingubara, S., Wang, Z., Yaegashi, O., Obata, R., Sakaue, H. and Takahagi, T. (2004) *Electrochem. Solid State Lett.*, **7**, C78.
350 Chang, S.-C., Shieh, J.-M., Lin, K.-C., Dai, B.-T., Wang, T.-C., Chen, C.-F., Feng, M.-S., Li, Y.-H. and Lu, C.-P. (2002) *J. Vac. Sci. Technol. B*, **20**, 1311.
351 Chiu, S.-Y., Shieh, J.-M., Chang, S.-C., Lin, K.-C., Dai, B.-T., Chen, C.-F. and Feng, M.-S. (2000) *J. Vac. Sci. Tecnnol B*, **18**, 2835.
352 Alonso, C., Pascual, M.J. and Abruna, H.D. (1997) *Electrochim. Acta*, **42**, 1739.
353 Tulpan, A. and Ducker, W.A. (2004) *J. Phys. Chem. B*, **108**, 1667.

3
Transition Metal Macrocycles as Electrocatalysts for Dioxygen Reduction

Daniel A. Scherson, Attila Palencsár, Yuriy Tolmachev, and Ionel Stefan

3.1
Introduction

The prospects of replacing Pt as an electrocatalyst for dioxygen reduction in fuel cell cathodes by materials based on more abundant and less expensive elements, continues to provide a strong impetus to the search for novel metal alloys, oxides and molecular species capable of promoting this reaction to yield water as a product, at potentials approaching thermodynamic values [1]. The first reports of dioxygen reduction mediated by a metal macrocycle electrocatalyst were published by Jasinski, who described enhancements in the activity of pressed Ni powder electrodes in alkaline solutions through incorporation of cobalt phthalocyanine within their structure [2, 3]. Since then, scientists have engaged in systematic studies of this phenomenon, prompted in part by the occurrence of an iron porphyrin, yet another metal macrocycle, in the active site of cytochrome-c oxidases [4, 5], a family of enzymes responsible for the reduction of dioxygen to water in living organisms (see Figure 3.1) without releasing toxic intermediates, such as peroxide or superoxide.

Along with phthalocyanines [6] and other less popular multidentate chelate structures [7], porphyrins are amenable to wide degrees of functionalization [8, 9], allowing both the electronic and structural (steric) properties of the metal-based electrocatalytic microenvironment to be carefully modulated. Pioneering efforts in molecular architecture primarily by the groups of Collman at Stanford, CA, USA [10] and Chang at Michigan State, MI, USA [11], led to the synthesis of face-to-face porphyrins in which the distance between the two rings could be adjusted fairly rigidly. This strategy culminated in the discovery of face-to-face Co–Co porphyrins, which were found to reduce dioxygen to water without producing solution phase peroxide, at potentials far more positive than those observed for single ring analogues [12]. Unfortunately, the stability of macrocycles is compromised by the formation of oxygen-based radical intermediates, which can attack the ring while in operation, generating species that exhibit much reduced activity for dioxygen reduction. Rather unexpectedly, however, the adsorption on high-area carbons of

Advances in Electrochemical Science and Engineering, Vol. 10.
Edited by Richard C. Alkire, Dieter M. Kolb, Jacek Lipkowski and Philip N. Ross
Copyright © 2008 WILEY-VCH Verlag GmbH & Co. KGaA, Weinheim
ISBN: 978-3-527-31419-5

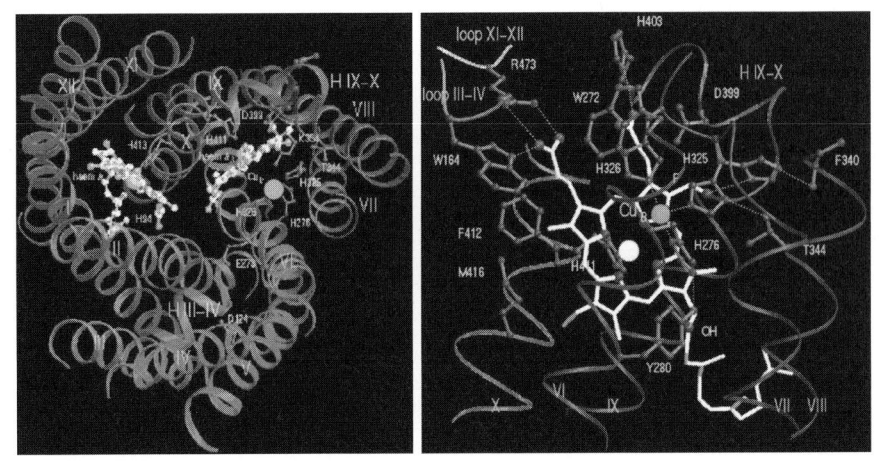

Figure 3.1 Three-dimensional diagram of the cytochrome-c oxidase of *Paracoccus denitrificans* (left panel). Only the edge of the two porphyrins (represented with white balls and sticks) can be seen in this panel. Shown in the right panel is a more detailed view of the porphyrin-based active site, which incorporates a Cu ion (grey circle) in close proximity to the Fe center (white circle) of one of the porphyrins of the heme cofactor (source ref. [5]).

certain transition metal macrocycles involving Fe, Co and a few other metal centers (and also a host of other metal containing precursors), followed by subsequent thermal treatment in an inert atmosphere at temperatures high enough to destroy much of the chelating ring, have yielded rather encouraging performances [13]. These findings may lead to the discovery of new, non Pt-based materials with potentially high technical relevance.

This chapter provides a critical review of transition metal macrocycles, both in intact and thermally activated forms, as electrocatalysts for dioxygen reduction in aqueous electrolytes. Fundamental aspects of electrocatalysis, oxygen reduction and transition metal macrocycles will be highlighted in this brief introduction, which should serve as background material for the subsequent more specialized sections.

3.1.1
Electrocatalysis

The kinetics of electrochemical reactions are often modified by the nature of the electrode material, and by the presence of atomic and molecular species either adsorbed on the surface or in the bulk solution [14]. Electrocatalysis is primarily concerned with the study of this phenomenon and, particularly, with the factors that govern enhancements in the rates of redox processes. Implicit in this general statement is the ability of the species responsible for these effects, or electrocatalyst, or the electrode itself, to carry out the reaction numerous times before undergoing possible deactivation. Electrocatalytic processes in which the electrode simply serves as a source or sink of electrons to generate solution phase species that display specific reactivity toward a given (solution phase) reactant define the scope of homogeneous electrocatalysis, whereas those mediated by the electrode surface, either bare or modified by adsorbed species, fall within the field of heterogeneous electrocatalysis. Situations could arise in which the electrocatalyst may be present in both phases. In fact, establishing the strict absence of adsorbed species on electrode surfaces for reactions presumed to occur via homogeneous electrocatalysis poses a challenging problem in itself. There are instances in which an adsorbed species blocks surface sites at which one or more unwanted reactions may proceed, while leaving unaffected other sites where a desired process would ensue. Such gains in specificity can, from a more general perspective, also be regarded as a form of electrocatalysis.

The potency of an electrocatalyst is usually defined in terms of the potential required to carry out a specific process at a prescribed rate, or current, per unit area of electrode. Platinum, for example, promotes hydrogen evolution and hydrogen oxidation in aqueous electrolytes, at very high rates at potentials very close to the thermodynamic redox potential for the reaction $H^+(aq) + e^- \leftrightarrow \frac{1}{2}H_2(g)$, that is, small overpotentials, η. Hence, it is a far more potent electrocatalyst than, for example, Hg or carbon, for which the onset for either reaction occurs at potentials far removed from that predicted value, that is, large η.

The efficacy of electrocatalyts can also be measured in terms of turnover numbers, as is customary in conventional chemical catalysis. In particular, for an electrocatalytic species present on the surface at a coverage of $2 \times 10^{-10}\,\mathrm{mol\,cm^{-2}}$, which is fairly typical for a single adsorbed monolayer of a macrocycle, such as a porphyrin on carbon, a current density of $5\,\mathrm{mA\,cm^{-2}}$ corresponds to a turnover number of $60\,\mathrm{s^{-1}}$. Confinement of molecular electrocatalysts to electrode surfaces brings about a number of advantages, including an increase in the local concentration of active species beyond their solubility limits. In addition, it may open new reactions pathways involving, for example, coordination of the reactant and/or intermediates, to the metal centers of two neighboring macrocycles, affording optimized conditions for the delivery of multiple electrons before any intermediates can escape into the bulk solution.

3.1.2
Dioxygen Reduction in Aqueous Electrolytes: General Aspects

The electrochemical reduction of dioxygen may be regarded as the archetypical and indeed most widely studied multielectron transfer process in electrochemistry [1]. From a quantum mechanical point of view, the ground state electronic configuration of dioxygen is a triplet, where each of the two-fold degenerate, antibonding π^* orbitals is occupied by a single electron. Injection of electrons into these orbitals would thus lead to a net destabilization of the O–O bond, which, depending on the detailed reaction pathway, would generate, in aqueous electrolytes, superoxide, peroxide, or, as highly desired for energy conversion applications, water, as the final product.

From a very general perspective, the overall reduction of dioxygen in aqueous electrolytes proceeds via the two major pathways shown in Scheme 3.1 for acid

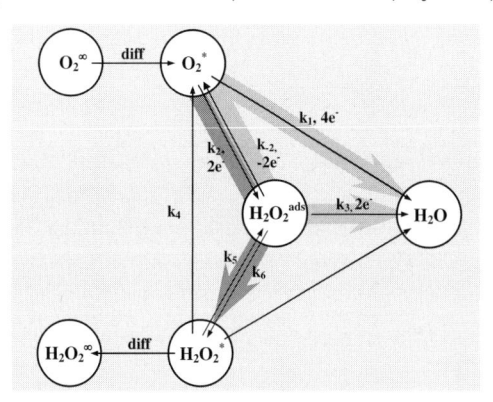

Scheme 3.1 Simplified mechanistic pathways for the heterogeneous reduction of dioxygen in aqueous acidic electrolytes. The definition of the rate constants follows that given in Ref. [16].

media, where the superscripts ∞, * and ads, refer to the bulk, near electrode region and adsorbed state, respectively:

1. a direct four-electron process, which yields water as the product (see light grey arrows in Scheme 3.1) without generation of detectable solution phase intermediates;

2. a two-electron pathway that produces solution phase peroxide (see dark grey arrow in the same scheme), a species that can be further reduced to water or undergo disproportionation regenerating O_2.

In this scheme the k_i values represent rate constants, which are generally potential dependent. It must be emphasized that the four-electron pathway does not imply the transfer of four electrons in a single step; rather, it underscores the fact that all intermediate species, such as, but not restricted to peroxide, remain bound to the electrode surface yielding, upon further reduction, water as the sole product. Also depicted in Scheme 3.1 are mass transport processes (diff) responsible for the replenishment of O_2 and removal of solution phase peroxide next to the interface, and the adsorption and desorption of the peroxide intermediate, for which the rate constants are labeled as k_5 and k_6, respectively. Not shown, for simplicity, is the one-electron reduction of dioxygen to superoxide, a radical species that exhibits moderate lifetime in strongly alkaline electrolytes [15].

The standard reduction potentials of the most relevant half-reactions involved in the four- and two-electron reduction of dioxygen in acid and alkaline aqueous media are listed in Table 3.1. It follows from these values, that, under full thermodynamic control, the equilibrium concentration of peroxide at the reversible potential for the four-electron reduction of oxygen in acid media, that is, 1.23 V, is of the order of 10^{-18} M. Hence, a stepwise reduction of dioxygen to yield currents of about $1 \, A \, cm^{-2}$ will require values for the standard heterogeneous rate constants for

Table 3.1 Standard reduction potentials for reactions of relevance to the two- and four-electron reduction pathways of dioxygen in acid and alkaline aqueous solutions.

Solution	Two-electron peroxide pathway		Four-electron pathway	
	Redox reaction	$E°$ (V vs. NHE)	Redox reaction	$E°$ (V vs. NHE)
Acid	$O_2 + 2H^+ + 2e^- \rightarrow H_2O_2$	0.68	$O_2 + 4H^+ + 4e^- \rightarrow 2H_2O$	1.229
	$H_2O_2 + 2H^+ + 2e^- \rightarrow 2H_2O$	1.77		
Alkaline	$HO_2^- + H_2O + 2e^- \rightarrow 3OH^-$	−0.065	$O_2 + 2H_2O + 4e^- \rightarrow 4OH^-$	0.401
	$O_2 + H_2O + 2e^- \rightarrow HO_2^- + OH^-$	0.867		

oxygen reduction to peroxide that greatly exceed the collision limit. Although this problem could be mitigated by introducing a catalyst capable of decomposing peroxide (see below), the rates at which such process should occur to meet the desired targets would be very high.

Contributions by numerous research groups over the past few decades have provided ample illustrations of the complexities associated with the electrochemical reduction of dioxygen, as evidenced by the sensitivity of both the energetics and mechanistic pathways to the nature of the electrode, including its surface microstructure, and the presence of adsorbed atomic and molecular species. Extensive use has been made of the rotating disk electrode (RDE), a forced convection system for which the fluid dynamics are well understood, and, particularly its variant, the rotating ring-disk electrode (RRDE, see Figure 3.2), to gain insight into the factors that control the mechanism of dioxygen reduction and determine the various rate constants involved.

As indicated in the right panel in this figure, rotation about the main axis of the cylinder forces the solution toward the flat end that contains the inlaid disk and ring electrodes. As the surface is closely approached, the solution spreads outward, forming a uniformly accessible diffusion layer along the surface of the disk. The RRDE has proven exceedingly useful in studies of O_2 reduction, as solution phase

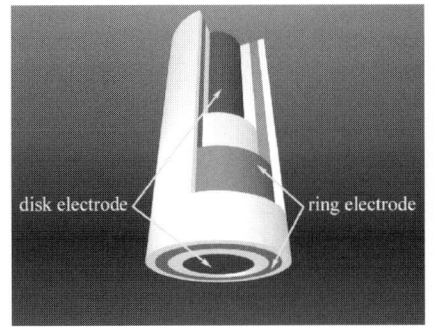

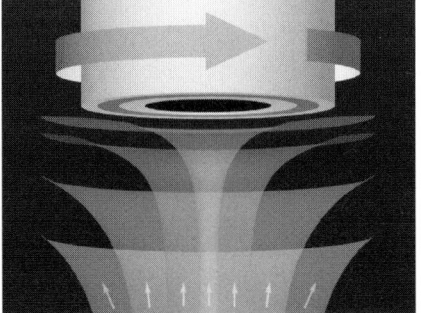

Figure 3.2 Schematic diagram of a rotating ring disk electrode (left panel) and a few associated streamline surfaces (right panel).

peroxide generated at the disk can be quantitatively detected at the concentric ring electrode polarized at a potential at which its oxidation is carried out under diffusion limited conditions [16]. It can, furthermore, be shown, that the expression

$$\varepsilon = 2I_{ring}/(I_{disk}N+I_{ring}) \qquad (3.1)$$

where I_{ring} and I_{disk} are the ring and disk currents, respectively, and N is the collection efficiency, a parameter that depends strictly on the specific geometry of the RRDE assembly, provides a quantitative measure of the extent to which the overall reaction proceeds via a two-electron pathway. In particular, $\varepsilon = 1$ for quantitative peroxide generation, that is, $I_{disk}N = I_{ring}$, and $\varepsilon = 0$ for strict four-electron reduction to yield water. Product specificity can also be defined by

$$n = 4 - 2I_{ring}/(I_{disk}N) \qquad (3.2)$$

the effective number of electrons transferred per dioxygen molecule, that is, $n = 4$ and 2, for water and peroxide, respectively. Values of N are often determined experimentally by dividing the measured I_{disk} for a simple reaction devoid of mechanistic complications, such as the reduction or oxidation of hexacyanoferrate, by I_{ring} for the reverse reaction recorded under diffusion limited conditions.

Platinum and some of its alloys, currently the electrocatalysts of choice for low-temperature, hydrogen-based, fuel cells, along with Ag and Pd, promote reduction of O_2 via both four- and two-electron pathways. This behavior is unlike that found for graphite and other forms of carbon and metals, such as Hg, on which O_2 reduction proceeds strictly via a two-electron pathway yielding solution phase peroxide. In fact, the onset potentials for the reduction of O_2 on carbonaceous materials, such as the basal plane of highly ordered pyrolytic graphite, HOPG(bp), ordinary pyrolytic graphite (OPG) and glassy carbon (GC) are very negative, and, therefore, are ideally suited for homogeneous and heterogeneous electrocatalytic studies. Evidence for a strict two-electron reduction of O_2 on OPG in alkaline media has been obtained by monitoring I_{disk} while scanning the disk potential, E_{disk}, at relatively small scan rates, v, known as dynamic polarization measurements, using an Au ring-OPG disk RRDE (Au|OPG). For these experiments the ring was polarized at $E_{ring} = 0.1$ V versus SCE, a value positive enough for the oxidation of solution phase peroxide generated at the disk to proceed under diffusion control (see Figure 3.3).

As shown in the figure, I_{disk} (upper panel) and I_{ring} (lower panel) are, within a constant, virtual mirror images of one another, yielding values of ε (see Equation 3.1) very close to unity. The hump or pre-wave centered at about -0.5 V versus SCE is related to the presence of redox active functional groups on the carbon surface [17], a further discussion of which is beyond the scope of this chapter. In addition, carbons display very small interfacial capacities, of the order of a few $\mu F\,cm^{-2}$, which are virtually (see below) independent of the applied potential, high overpotentials for both hydrogen and oxygen evolution, and also relatively high resistance to electrochemical oxidation. These characteristics translate into small and featureless voltammetric background currents extending beyond 1.1 V, making carbon an ideal electrode material for electrochemical studies of redox active adsorbed species in general.

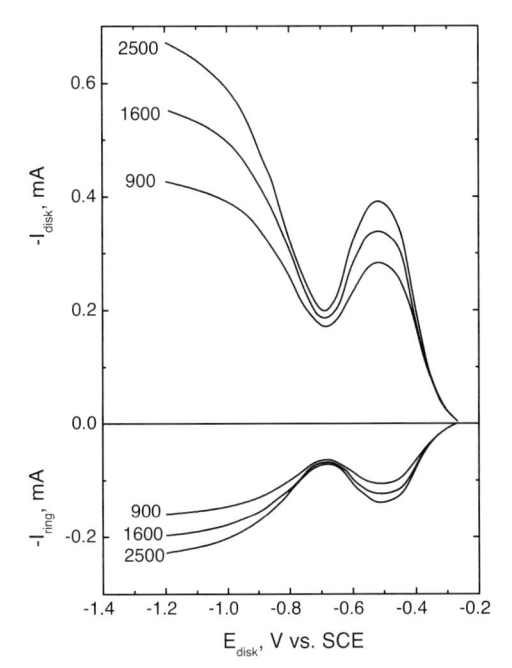

Figure 3.3 Dynamic polarization curves (scan rate $v = 10$ mV s^{-1}) for O_2 reduction on an OPG disk (area = 0.196 cm^2) of an Au|OPG RRDE ($N = 0.38$) in O_2 saturated 0.1 M NaOH recorded at rotation rates of $\omega = 900$, 1600 and 2500 rpm (upper panel). Shown in the lower panel are ring currents, I_{ring}, as a function of E_{disk} recorded with the Au ring polarized at $E_{ring} = 0.1$ V vs. SCE while the potential of the disk was being scanned [17].

The reduction of dioxygen in aqueous O_2 saturated alkaline solutions on Au single crystal electrodes exhibits a pronounced dependence on the surface microstructure. This behavior is clearly evidenced by the RDE dynamic polarization curves in 0.1 M NaOH shown in Figure 3.4, from which the electrocatalytic activity is found to follow the sequence Au(1 0 0) > Au(1 1 0) > Au(1 1 1). Rather startlingly, the activity of Au (1 1 1) for small η, even surpasses that of Pt(1 1 1) at the same pH (see dotted curve in this figure) under otherwise virtually the same experimental conditions [18, 19].

For molecular electrocatalysts otherwise, and especially transition metal macrocycles, the electrocatalytic activity is often modified by subtle structural and electronic factors spanning the entire mechanistic spectrum, that is, from strict four-electron reduction, as for the much publicized cofacial di-cobalt porphyrin, in which the distance between the Co centers was set at about 4 Å [12], to strict two-electron reduction, as in the monomeric (single ring) Co(II) 4,4′,4″,4‴ -tetrasulfophthalocyanine (CoTsPc) [20] and Co(II) 5,10,15,20-tetraphenyl porphyrin (CoTPP) [21]. Not surprisingly, nature has evolved highly specific enzymes for oxygen transport, oxygen reduction to water, superoxide dismutation and peroxide decomposition.

From a practical point of view, the performance of O_2 cathode materials that display poor or no activity for hydrogen peroxide reduction, such as carbon in alkaline

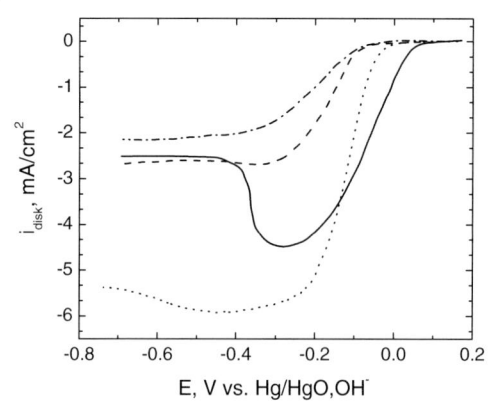

Figure 3.4 Dynamic polarization curves ($v = 50\,mV\,s^{-1}$) for oxygen reduction on Au(1 0 0) (—), Au(1 1 1) (– – –), Au(1 1 0) (– – – –) RDEs in O_2 saturated 0.1 M NaOH, recorded at $\omega = 1600$ rpm [18]. Shown in dotted lines are data collected for a Pt (1 1 1) RDE in O_2 saturated 0.1 M KOH, otherwise under the same conditions [19].

electrolytes, can be greatly enhanced by addition of a catalyst capable of promoting hydrogen peroxide decomposition (see reaction with rate constant k_4 in Scheme 3.1).

Techniques based on the RRDE afford very useful diagnostic tools for assessing such properties quantitatively. In one such tactic, the catalyst is applied to the disk electrode and the amount of O_2 generated via (heterogeneous) chemical decomposition is detected by a ring electrode capable of reducing O_2. Shown in Figure 3.5, for example, are results obtained at steady state in 1 mM H_2O_2 aqueous solutions at two different pH for Fe(III) protoporphyrin IX chloride (FePPIXCl), a model Fe porphyrin

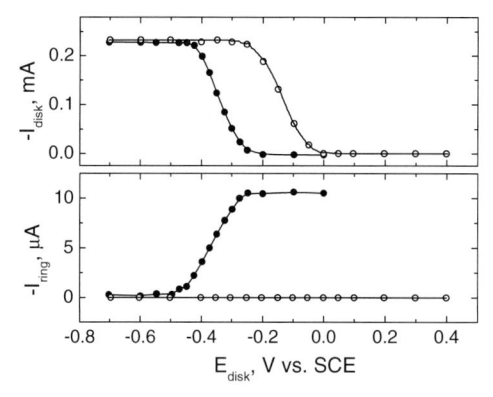

Figure 3.5 Steady state polarization curves recorded in 1 mM H_2O_2 aqueous solutions of pH = 1 (open circles) and pH = 8 (solid circles) using FePPIX adsorbed at monolayer coverages on a carbon disk and a CoPl coated carbon ring electrode of an RRDE arrangement at $\omega = 100$ rpm (see text and original reference for details). The ring in each instance was polarized at $E_{ring} = -0.3$ V vs. SCE, a potential negative enough for oxygen reduction to peroxide to proceed under diffusion limited conditions [22].

adsorbed on a carbon disk electrode of an RRDE arrangement, incorporating a ring electrode modified with a unique cobalt(II) *meso*-porphyrin diethyl ester, herein referred to as CoPI, (see Section 3.3) that can reduce O_2 to peroxide under pure diffusion control when polarized at $E_{ring} = -0.3$ V versus SCE [22]. At pH = 1 (open circles), $I_{ring} = 0$ before (and after) the onset of peroxide reduction on the disk. This behavior clearly reflects the inability of the FePPIX|OPG interface to promote spontaneous peroxide decomposition. In contrast, at pH = 8 (close to physiological conditions), the ring detects O_2 at potentials ahead of the onset of peroxide reduction on the disk, a clear indication that the interface at this pH (and also up to pH = 12, not shown) can decompose peroxide to yield O_2. At more negative disk potentials, however, peroxide undergoes reduction and, not surprisingly, I_{ring} drops down to negligible levels.

It must be stressed at the outset, that even for the best electrocatalysts so far identified, the potentials at which O_2 reduction to water is found to proceed at reasonably high rates are a few hundred mV negative to the thermodynamic value, consistent with the energies of activation for some of the mechanistic steps involved being very high. Theoretical insights into some of these aspects has been gained, due in part to the wider availability of user-friendly algorithms to perform high level quantum mechanical calculations, including, more recently, density functional theory [23], and also to the advent of new generations of computers with ever increasing speed and performance.

3.1.3
Transition Metal Macrocycles

3.1.3.1 General Characteristics
Macrocycles represent a wide class of large ring compounds incorporating several donor atoms, such as nitrogen, oxygen and sulfur, which can bind to a metal ion, often forming species that exhibit high thermodynamic stability, and chemical inertness. Of special relevance to this work, are macrocycles containing nitrogen coordinating sites, including cyclams (see A, Scheme 3.2), and the fairly planar, rigid and highly conjugated porphyrazines (C), phthalocyanines (D) and porphyrins (E–L) [7], which are amenable to various types of functionalization. In particular, addition of methyl pyridinium, or phenyl moieties incorporating sulfonic (E) or carboxylic groups, to the *meso*-positions of porphine (B), (which do not require the rather unstable and difficult to synthesize porphine as a reagent) impart the resulting porphyrins solubility in aqueous solutions over a wide pH range. Yet another type of modification involves direct axial binding of a substituent to the central metal, which may lift the metal out of the ring plane and, in some instances, change its oxidation and spin states. This is the situation for the μ-oxo type macrocyclic dimer in G, Scheme 3.2, in which a single oxygen bridge serves to covalently link the Fe centers of the two different porphyrins.

Whereas phthalocyanines have found wide use in technical and industrial applications, porphyrins have received comparatively greater attention in academic circles, motivated in part by the relative ease with which both

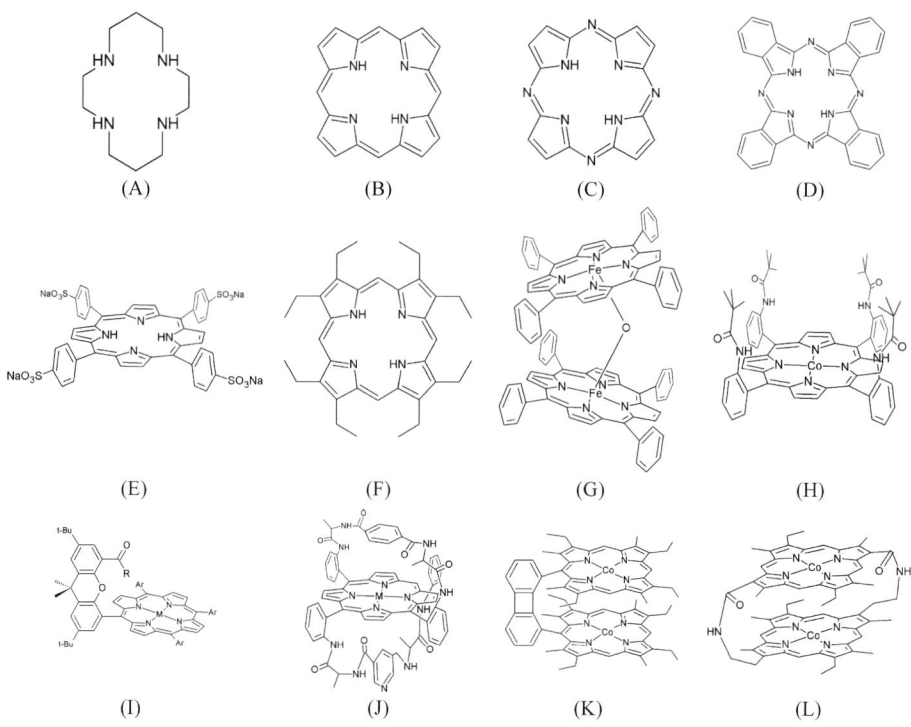

Scheme 3.2 Illustrative examples of metallated and unmetallated macrocycles.

meso- (e.g. E and H in Scheme 3.2) and β-pyrrolic substitutions (see e.g. F and L) can be effected.

For example, attachment of functionalized phenyl groups to the *meso*-positions of porphine (E), affords means of constructing rather exotic molecular structures (see Scheme 3.2) popularized by such colloquial terms as picket fence (H), hangman (I), basket handle (J), pacman (K) and face-to-face (L), to graphically describe their shape. Indeed, novel synthetic principles are being widely exploited to control, in a systematic fashion, electronic and structural properties of the MN_4 center microenvironment of porphyrins, not readily achievable with other macrocycles. Most prominent among these factors are the redox potentials of the metal center, which are believed to be paramount in conferring these materials electrocatalytic properties for O_2 reduction. Rather serendipitously, electrochemistry can be effectively utilized as a tool to generate redox related species. This approach not only avoids often laborious conventional syntheses, but allows characterization to be performed *in situ*, that is, in the same cell under potential control, employing electronic, vibrational and structural probes.

Insight into the assignment of redox transitions can be gained from quantum mechanical calculations, which can help correlate the energies of the empty and occupied orbitals with the onsets of reduction and oxidation of specific redox active species. Lever and coworkers have pointed out that for many metal phthalocyanines for which the metal lies in the plane of the ring, the difference in the magnitude of the peak potentials (or more precisely the energy) attributed to the first ring oxidation and first ring reduction seem to be very close to the energy associated with the $\pi \rightarrow \pi^*$ transition responsible for the so-called Q-band, that is, 1.5 V, or 650–700 nm [24]. In the case of transition metal phthalocyanines, the metal d-levels lie between the lowest unoccupied (LUMO) and highest occupied molecular orbitals (HOMO) of the ring; hence, the expected metal redox potentials may not only be close to those of the ring, but also affect the redox potentials of the ring itself. Under such conditions, peaks cannot be strictly ascribed as predominantly metal or ring in character.

There are instances, however, in which gross assignments can be made based on an analysis of voltammetric measurements. For example, addition of a nitrogenated Lewis base, such as, pyrazine, to a solution of a Co or Fe macrocycle will lead to axial coordination and thus to larger changes in the potentials of redox peaks involving higher metal-, compared with ring-character. Aspects of the electrochemistry and electrocatalysis of transition metal macrocycles in non-aqueous electrolytes, a topic not covered in this chapter, are reviewed elsewhere [24, 25].

3.1.3.2 Electrocatalytic Properties Toward Oxygen Reduction

From a general perspective, various factors may be expected to control the energetics and mechanistic pathways associated with the electrocatalytic activity of transition metal macrocycles for dioxygen reduction. These include, but are not limited to, their intrinsic (in the absence of O_2 in solution), electrochemical, for example redox potentials and heterogeneous electron transfer rates, and acid–base properties. The latter includes structural modifications induced by changes in pH, such as dimer formation, and also their affinity and chemical reactivity toward not only dioxygen, but also its reduction intermediates, for example superoxide and peroxide. Results reported by numerous laboratories indicate that the vast majority of single-ring macrocycles mediate dioxygen reduction via the generalized sequence of steps shown in Scheme 3.3, where M(X)P denotes the macrocycle and the number in parenthesis, X, the formal oxidation state of its metal center and the upper and lower case refer to electrochemical and chemical reactions, respectively.

The first step in this pathway (A) involves the electrochemical reduction of the inactive, oxidized form of the macrocycle, denoted without loss of generality as $M(III)P^+$, to yield the reduced, electrocatalytically active form of the species, M(II)P.

$M(III)P^+ + e^- \rightarrow M(II)P$ A

$M(II)P + O_2 \rightarrow M(II)PO_2$ a

$M(II)PO_2 + M(II)P \rightarrow M(II)PO_2^- + M(III)P^+$ b

$M(II)PO_2 + M(II)P \rightarrow [M(II)P]_2O_2$ c

$M(II)PO_2 + e^- \rightarrow M(II)PO_2^-$ B

$M(II)PO_2^- + H^+ \rightarrow M(III)P^+ + HO_2^-$ d

$M(II)PO_2^- + 3e^- + 4H^+ \rightarrow 2H_2O + M(II)P$ C

Scheme 3.3 Generalized mechanism for the macrocyclic-mediated reduction of dioxygen. Chemical and electrochemical processes are labeled in lower and upper case, respectively.

The latter then reacts with O_2 to form the adduct $M(II)PO_2$ (a) which, as indicated, can be further reduced chemically by a second $M(II)P$, present either in solution phase or adsorbed on the electrode surface, (b). Alternatively, $M(II)PO_2$ can be reduced by the electrode itself to form $[M(II)P]_2O_2^-$ (B), which may react with yet another $M(II)P$ species to yield a dimeric dioxygen adduct, $[M(II)P]_2O_2$ (c), dissociate to generate peroxide (d), or be subsequently reduced electrochemically to give water as the final product (C). Systematic studies of the factors that control the stability and reactivity of such adducts can provide much insight into the mechanisms by which macrocycles reduce dioxygen in electrochemical environments [26].

In the case of homogeneous electrocatalysis, all species in Scheme 3.3 are strictly in the solution phase, and, hence, the rates of the reactions involved can, in principle, be measured by conventional methods, including stopped-flow. In particular, Endicott and coworkers [27, 28] examined the formation of solution-phase dioxygen adducts of trans-$[Co([14]aneN_4)(OH_2)_2]^{2+}$, that is, the reduced form of Co(cyclam) (see A, Scheme 3.2, with M = Co), which are stable enough to be characterized by standard electrochemical and spectroscopic techniques. In contrast, most unprotected Fe(II) porphyrins, both as solids or in solution phase, react rapidly and irreversibly with dioxygen to form μ-oxo type dimeric species (see G, Scheme 3.2) [8]. In fact, the existence of certain highly reactive metal macrocycle–O_2 adducts, involving Fe centers, in aqueous electrolytes, has not been proven directly, but only inferred from careful analyses of kinetic and thermodynamic data.

Although its relevance to aqueous electrochemistry may be questioned, interesting information regarding the nature of porphyrin dioxygen adducts has been gained from solution phase experiments performed in non-aqueous electrolytes. For example, exposure to air of the face-to-face porphyrin Co–Co 5 (similar to L in Scheme 3.2, with five as opposed to four atoms in the strut) solutions of 1-triphenyl methyl imidazole in toluene–dichloromethane mixtures yielded a diamagnetic species, ascribed to a μ-peroxo adduct, as evidenced by the lack of a signal in the electron spin resonance (ESR) spectra. However, subsequent oxidation by I_2 yielded a 15 line ESR spectrum attributed to the trans-μ-superoxo form of the adduct (see insert in Figure 3.6), a species believed to be an important intermediate in the reduction of dioxygen [29].

Much of the effort toward furthering our understanding of interactions between O_2 and iron macrocycles has focused on the studies of model structural systems that mimic the active site of enzymes involved in the transport and storage of dioxygen in living organisms [30]. In particular, theoretical calculations strongly suggest that the most favorable configuration of dioxygen adducts of hemoglobin, Hb, and myoglobin, Mb, two heme-based oxygen carriers, and also of simple Fe(II) porphyrin models, can be best represented as a superoxide ion bonded to the metal site, forming an angle with the macrocyclic ring in a bent-type geometry [31, 32]. In contrast, heme-based cytochrome-c oxidases do not form stable adducts, but, as mentioned earlier, reduce oxygen to water in a complex multielectron transfer process that may involve hydrogen bonding of the adduct with neighboring amino acid residues. More specific examples of O_2–macrocycle interactions will be given later in this chapter.

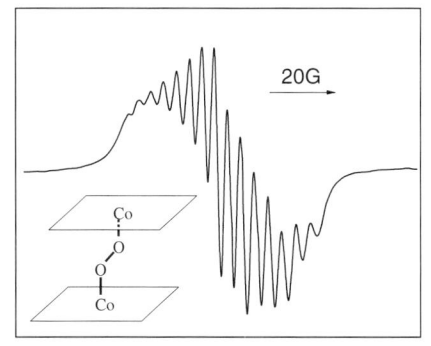

Figure 3.6 Electron spin resonance spectrum of the μ-superoxo complex of an axially coordinated imidazole derivative of a Co–Co face-to-face (Co–Co 5) porphyrin in toluene–dichloromethane solutions recorded at room temperature. A simplified schematic structure of the metal microenvironment of the proposed *trans*-type adduct is shown in the insert.

Systematic studies involving transition metal porphyrins have unveiled intriguing structure–activity correlations. For example, the onset potentials for O_2 reduction in solutions containing Fe(III) 5,10,15,20-tetrakis(*N*-methyl-4-pyridyl)porphyrin (FeTM-PyP) coincide precisely with the corresponding values found for the reduction of the macrocycle in solutions devoid of O_2 over a wide pH range [33]. This behavior is unlike that reported for the closely related Fe(II) 5,10,15,20-tetraphenylporphyrin (FeTPP) immobilized on carbon surfaces in the form of coatings of molecular dimensions, for which O_2 reduction commences ahead, about 100 mV more positive, of the onset of the voltammetric peak associated with its intrinsic reduction [22]. Yet, in stark contrast, the intrinsic redox peaks of some Co porphyrins occur at potentials positive of the onset of O_2 reduction [34]. Indeed, much work remains to be performed to identify the factors responsible for these trends and thereby determine the often subtle factors that govern structure–function relationships both in biological systems (biomimetic chemistry) and in electrocatalysis.

It must be emphasized, in closing this introductory section, that the Co(III/II) couple displays the most positive redox potential of all monomeric metalloporphyrins with demonstrated affinities for oxygen, and, not surprisingly, much of the effort in the area of dioxygen reduction has focused on the synthesis and characterization of species involving Co centers [12]. Also attractive from this viewpoint are the much less studied ruthenium porphyrins, for which the intrinsic redox transitions and presumably their catalytic activity for dioxygen reduction would be significant at reasonably positive potentials. Unfortunately, metalloporphyrins incorporating metal sites with very positive, metal-based redox potentials, such as Ni(III/II) and Fe(IV/III), do not show any tendency to bind oxygen due to their low electron density.

The main sections to follow address various aspects of homogeneous (Section 3.2) and heterogeneous (Section 3.3) electrocatalysis, using examples reported in the literature, which, in our view, best illustrate the rich behavioral spectrum this unique type of macrocyclic compounds, including dimeric species, display toward oxygen reduction. Despite sharing many commonalities, the theoretical and experimental tools involved in the study of these two types of electrocatalytic phenomena are sufficiently different to warrant separate discussion. Because of their impact in controlling structure–activity correlations, some of the relevant intrinsic physical and electrochemical properties of macrocycles, including the most salient techniques involved in their characterization, will also be reviewed. Issues related to pyrolyzed macrocycles are described in the subsequent and final section of this chapter, Section 3.4).

3.2
Homogeneous Electrocatalysis

Among the plethora of solution phase, electrocatalytically active macrocycles for dioxygen reduction, only very few have been found to display no affinity for electrode surfaces. Attention in this Section will be mostly focused on the intrinsic electrochemical and electrocatalytic properties of two of such unique materials:

1. iron(III) tetrakis(N-methyl-4-pyridyl)porphyrin, Fe(III)TMPyP (see A, Scheme 3.4), which is rendered water soluble via the positively charged methylpyridyl groups bound to each of the *meso*-positions of the porphine ring [35–37];

2. the *trans*-form of the Co(III) complex of 1,4,8,11-tetraazacyclotetradecane, *trans*-[Co([14]aneN$_4$)(OH$_2$)$_2$]$^{3+}$, to be denoted hereafter as Co(III)(cyclam) [38], which is shown in one of its forms in B, Scheme 3.4.

3.2.1
Intrinsic Properties of Solution Phase Transition Metal Macrocycles

3.2.1.1 Formal Redox Potentials and Diffusion Coefficients
Electrochemical techniques can effectively be employed for determining the formal potentials, $E^{o\prime}$, and the number of electrons transferred, n, of redox reactions, in

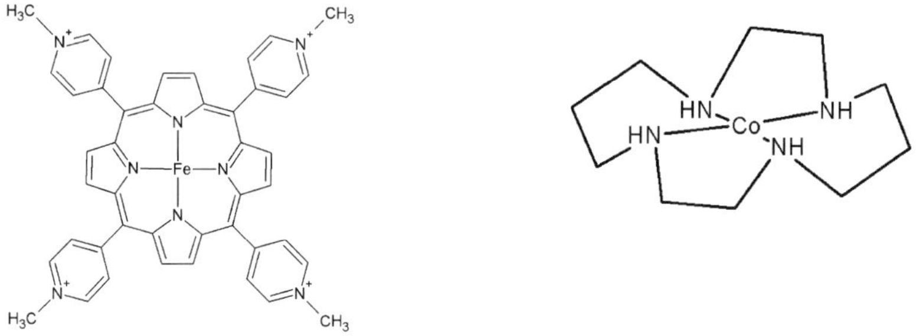

(A)   (B)

Scheme 3.4 Molecular structures of iron(III) tetrakis(N-methyl-4-pyridyl)porphyrin, Fe(III)TMPyP (A) and *trans*-[Co([14]aneN$_4$)(OH$_2$)$_2$]$^{3+}$, Co(III)(cyclam) (B).

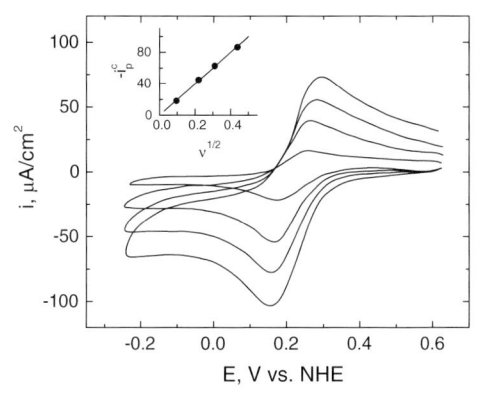

Figure 3.7 Cyclic voltammograms of 6×10^{-4} M Fe(III)TMPyP in deaerated 0.1 N H_2SO_4 aqueous solutions recorded using a glassy carbon (GC) electrode for $v = 0.01, 0.05, 0.10$ and $0.20\, V\, s^{-1}$ (in order of increasing peak currents). Insert: plot of cathodic peak currents, $-i_p^c$ ($\mu A\, cm^{-2}$) versus $v^{1/2}$ ($V^{1/2}\, s^{-1/2}$).

addition to the diffusion coefficients, D, of electroactive species. Shown in Figure 3.7 are cyclic voltammetric curves obtained for 6×10^{-4} M Fe(III)TMPyP in 0.1 N H_2SO_4 aqueous solutions for scan rates, v, in the range $0.01 \leq v \leq 0.2\, V\, s^{-1}$ using highly polished glassy carbon (GC) electrodes [36]. The separation between the peak potentials of the well defined features in the scans in the positive, E_p^a, and negative directions, E_p^c, denoted as $\Delta E_p = E_p^a - E_p^c$, was, for the lower scan rates, about 59 mV [39] and thus consistent with a kinetically reversible, that is, very fast, one-electron transfer process. Values of $E^{\circ\prime}$ can be obtained from:

$$E^{\circ\prime} = E_{av} - (RT/nF)\ln(D_R/D_O)^{1/2} \qquad (3.3)$$

where

$$E_{av} = \frac{1}{2}(E_p^a + E_p^c) \qquad (3.4)$$

provided the ratio of D_R and D_O, the diffusion coefficients of the reduced and oxidized forms of the redox species, respectively, is known. In many instances, however, $D_R \approx D_O$; hence, $E^{\circ\prime} \sim E_{av}$.

Plots of the peak current densities observed during the scan toward negative potentials, $-i_p^c$ versus $v^{1/2}$, extracted from the cyclic voltammetric data (see insert, Figure 3.7) were found to be linear with a zero intercept. This behavior is characteristic of a strictly diffusion controlled process for which

$$i_p = I_p/A = (2.69 \times 10^5)n^{2/3}D_O^{1/2}C_O^\infty v^{1/2} \qquad (3.5)$$

where C_O^∞ is the bulk concentration of the reactant O, I_p the current and A is the area of the electrode [39]. As will be discussed in Section 3.2.1.3, variations in ΔE_p with v, such as those observed at the higher scan rates in Figure 3.7, are often indicative of partial kinetic control and the analysis of their functional dependence can be used to estimate heterogeneous electron transfer rate constants.

Provided C_O^∞ is known, n can be determined directly using thin layer coulometry, a technique that affords a measure of the total charge required to effect a complete reduction or oxidation of stable electroactive species confined within the layer. Application of this methodology to Fe(III)TMPyP yielded $n = 1$ over the pH range 1–13 [36]. Values of D can also be found from experiments performed in a conventional (non-thin layer) cell in which the current is monitored as a function of time following a potential step to a value, either sufficiently positive or negative, to bring the concentration of the reactant at the interface to zero, or potential step chronoamperometry. Under such diffusion controlled conditions, and in the absence of other mechanistic complications, the transient charge determined by the integration of chronoamperometric curves, $Q(t)$, can be shown to be given by:

$$Q(t) = 2nFAD_O^{1/2}\pi^{-1/2}C_O^\infty t^{1/2} \tag{3.6}$$

where F is Faraday's constant. Hence, a plot of Q versus $t^{1/2}$, or chronocoulometric curve, should be linear with a zero intercept [39], allowing D to be extracted from the slope. Not surprisingly, and in line with its high charge and rather large molecular size, about 1.5×1.5 nm^2, analysis of chronocoulometric plots for Fe(III)TMPyP (see Figure 3.8) yielded a value for $D_{Fe(III)TMPyP}$ of around 2.5×10^{-6} cm^2 s^{-1}, which is about an order of magnitude smaller than D_{O_2} in the same media, around 2×10^{-5} cm^2 s^{-1} [36]. This disparity introduces complications into the analysis of electrocatalytic data, as will be discussed below.

Another means of determining D, provided n is known, relies on the use of the rotating disk electrode (RDE). Based on the pioneering work of Levich [39], the steady state diffusion limiting current, I_{disk}^{lim}, measured at an RDE is given by:

$$I_{disk}^{lim} = 0.62nFAC_O^\infty D_O^{2/3}\upsilon^{-1/6}\omega^{1/2} \tag{3.7}$$

where υ is the kinematic viscosity of the media, about 0.01 cm^2 s^{-1} for aqueous solutions, and ω is the rotation rate of the disk (in radians s^{-1}). In agreement with this formalism, a plot of I_{disk}^{lim} versus $\omega^{1/2}$ (see B, Figure 3.9) for the one-electron reduction of Co(III)(cyclam) extracted from the slow linear potential scan curves (see upper panel, A, Figure 3.9), recorded in 1.5 mM Co(III)(cyclam) using the pyrolytic graphite (PG) disk electrode of a Pt|PG RRDE in 0.5 M HClO$_4$ aqueous solutions, was found to be linear with a zero intercept, yielding $D_{Co(III)(cyclam)} = 5 \times 10^{-6}$ cm^2 s^{-1} [4]. Although strictly point by point measurements would be required to meet the theoretical steady state assumption, acquisition of data in the dynamic polarization mode at slow scan rates does not lead to serious errors, as capacitive contributions are minimized and the concentration profiles can relax sufficiently. Also depicted in the lower panel A in this figure are the currents measured at the concentric Pt ring electrode, I_{ring}, of the RRDE polarized at $E_{ring} = 0.6$ V versus SSCE, a potential sufficiently positive for the diffusion limited reoxidation of $[Co([14]aneN_4)(OH_2)_2]^{2+}$ to ensue, yielding back the original material.

An independent determination of $E^{o\prime}$ and n can be made in some instances, by coupling a thin layer electrochemical cell with UV–visible spectroscopy, as described by Kuwana and Heinemann [40]. In its most common embodiment, such an optically

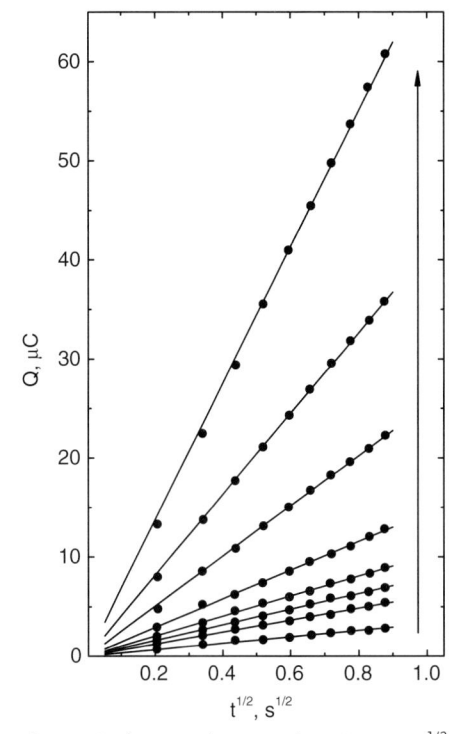

Figure 3.8 Chronocoulometric plots, Q versus $t^{1/2}$, obtained for Fe(III)TMPyP in 0.1 N H_2SO_4 aqueous solutions using a GC electrode ($A = 0.48$ cm^2) for various macrocycle concentrations: 2.5×10^{-5}, 5.0×10^{-5}, 7.5×10^{-5}, 9.9×10^{-5}, 1.5×10^{-4}, 2.9×10^{-4}, 4.6×10^{-4} and 8.4×10^{-4} M shown in the sequence specified by the arrow.

transparent thin layer electrode (OTTLE) assembly incorporates an Au mini-grid of a mesh large enough for light to pass through without significant attenuation. For these experiments, the thin layer cell is placed normal to the path of the beam, making it possible to record spectra as a function of the potential applied to the grid. The thickness of the cell is usually of the order of 0.02 cm, allowing full equilibration to be achieved within a few minutes. Transition metal macrocycles are especially suited for this type of investigation, as the molar absorptivities of their most prominent bands can reach values as high as 10^5 M^{-1} cm^{-1}, enabling detection of species at very small concentrations.

Shown in Figure 3.10A is a series of spectra recorded during incremental reduction of a solution of 10 μM Fe(III)TMPyP in aqueous 0.1 N H_2SO_4 using an Au mini-grid OTTLE. Clearly evident in this plot is the presence of at least three isosbestic points consistent with the quantitative conversion of a single species into another [36]. Also shown in B in this figure are a series of difference spectra following injection of a given amount of charge using the spectrum of the fully reduced species as a reference. This representation is useful in that it emphasizes the spectral changes

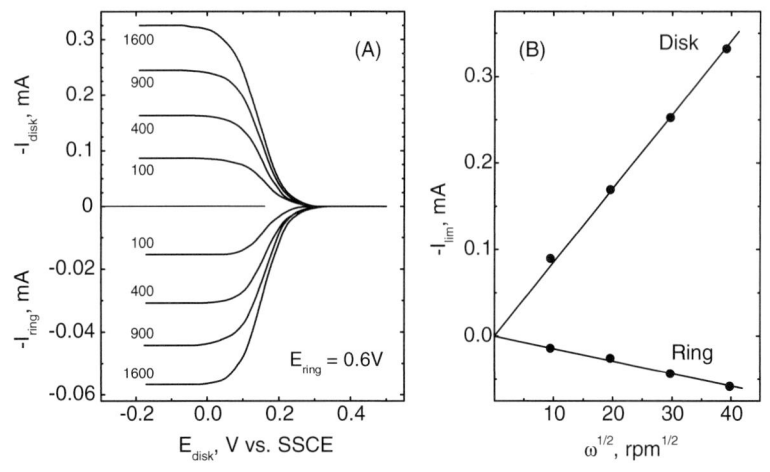

Figure 3.9 (A) Dynamic polarization curves obtained using the pyrolytic graphite (PG) disk ($A = 0.46$ cm^2) of a Pt|PG RRDE recorded in 1.5 mM Co(III)(cyclam) in 0.5 M HClO$_4$ aqueous solutions, for $\omega = 100, 400, 900$ and 1600 rpm, where SSCE stands for a sodium chloride saturated calomel electrode. The corresponding ring currents obtained for $E_{ring} = 0.6$ V are shown in the lower panel. (B) Plots of I_{disk}^{lim} and I_{ring} versus $\omega^{1/2}$.

induced by the redox transition. The OTTLE method allows for n and $E^{o\prime}$ to be determined directly from the Nernst equation expressed in terms of spectroscopic observables, that is:

$$E_{app} = E^{o\prime} + \frac{RT}{nF} \log \frac{[O]}{[R]} = E^{o\prime} + \frac{RT}{nF} \log \frac{A_2 - A_1}{A_3 - A_1} \tag{3.8}$$

where E_{app} is the potential applied to the electrode, A_3 and A_1 are absorbances evaluated at the absorption peak maximum (λ_{max}) of the fully oxidized species, for potentials at which the species is present in the fully oxidized and fully reduced forms, respectively, and A_2 is the corresponding value for mixtures of the two species. It thus follows from Equation 3.8 that a plot of E_{app} versus $\log[(A_2 - A_1)/(A_3 - A_1)]$ will be linear with an intercept at $E^{o\prime}$ and a slope inversely proportional to n. Analysis of data collected for 2.04×10^{-4} M Co(III)TMPyP in dearated 0.5 M NaNO$_3$/0.01 M HNO$_3$ aqueous solutions employing the OTTLE technique at $\lambda_{max} = 434$ nm (see A, Figure 3.11), yielded values of $E^{o\prime}$ of around $+0.175$ V versus SCE and $n = 1$ (see B in this figure) [41].

Spectroelectrochemical experiments in which the potential sequence is applied in both directions in a repetitive fashion allow an assessment of the stability of the electrogenerated species, which was indeed confirmed for both the Fe(III)TMPyP and Co(III)TMPyP systems. Yet another virtue of this methodology is the expedient preparation and spectroscopic characterization of reduced and oxidized forms of materials without the need for synthesizing often highly reactive and thus difficult to handle species.

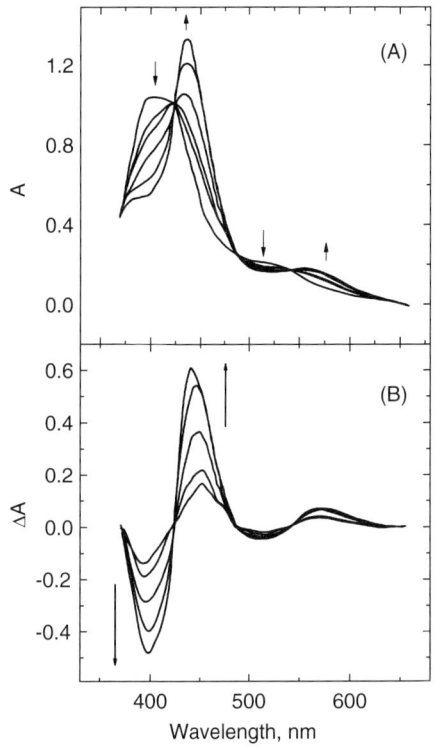

Figure 3.10 (A) Spectra obtained during incremental reduction of 10^{-5} M Fe(III)TMPyP in 0.1 N H_2SO_4 aqueous solutions recorded with an Au mini-grid OTTLE. (B) Difference spectra using the spectrum of the fully reduced species as a reference. The arrows indicate the direction of the absorbance (A) or difference absorbance changes (B).

3.2.1.2 Molecular Speciation

Shifts in the redox potentials induced by the pH can also afford additional details into acid–base aspects associated with the species involved. For Fe(III)TMPyP, plots of E_{av} versus pH were found to be independent of pH in acidic solutions, but yielded a straight line with a slope of 60 mV per unit of pH in the alkaline regime, that is, $9 \leq$ pH 13 [35, 36]. (see Figure 3.12). Two reactions consistent with these observations are given in Equations 3.9 and 3.10, where the latter, in particular, accounts explicitly for the transfer of one electron per hydronium ion:

$$[\text{Fe(III)TMPyP}(H_2O)]^{5+} + e^- \rightarrow [\text{Fe(II)TMPyP}(H_2O)]^{4+} \quad \text{acid} \quad (3.9)$$

$$[\text{Fe(III)TMPyP}(H_2O)(OH)]^{4+} + e^- + H^+ \rightarrow [\text{Fe(II)TMPyP}(H_2O)_2]^{4+} \quad \text{base} \quad (3.10)$$

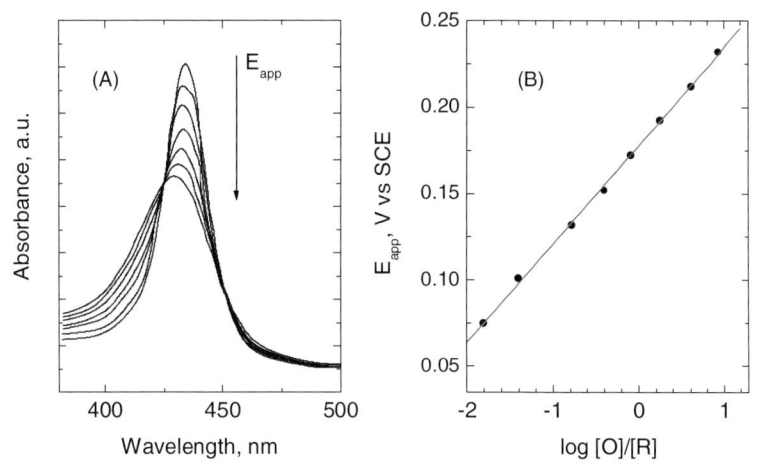

Figure 3.11 (A) UV–visible spectra of 2.04 × 10⁻⁴ M Co(III)TMPyP in 0.5 M NaNO₃/ 0.01 M HNO₃ aqueous solutions recorded in sequence (see arrow) in an OTTLE for E_{app} ranging from values at which the material is fully oxidized (open circuit, upper curve) to values at which the material is fully reduced, that is, $E_{app} = -0.050$ V versus SCE, lower curve. (B) Plot of E_{app} versus $\log\{[O]/[R]\}$ [$= \log\{(A_2 - A_1)/(A_3 - A_1)\}$] based on the data in (A), where all A_i values were measured at $\lambda_{max} = 434$ nm.

In addition, conventional spectral measurements in the UV region as a function of pH yielded for Fe(III)TMPyP, pK_a values of 4.7 and 6.5, as well as evidence for the formation of the μ-oxo type dimer (see insert Figure 3.14 below, and also G in Scheme 3.2) in which a single oxygen atom forms a bridge between the iron centers of two rings. This species is believed to be responsible for the presence of an

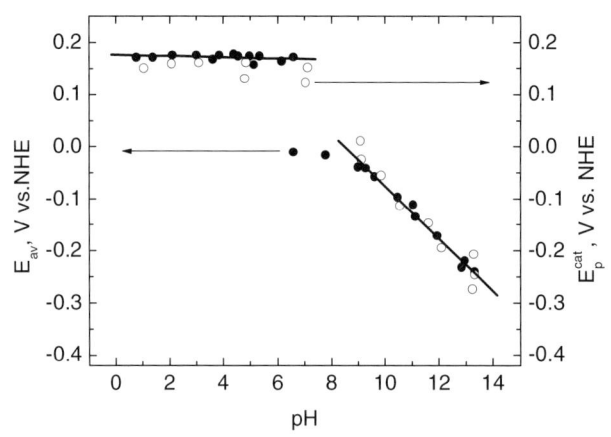

Figure 3.12 Plot of $E_{av} = \frac{1}{2}(E_p^a + E_p^c)$ versus pH for the reduction of Fe(III)TMPyP in deaerated solutions (solid circles, left ordinate) based on voltammetric measurements performed at $v = 0.1$ V s⁻¹. Also included in this figure are the peak potentials observed in the presence of dioxygen, (open circles, right ordinate) otherwise under the same conditions [35].

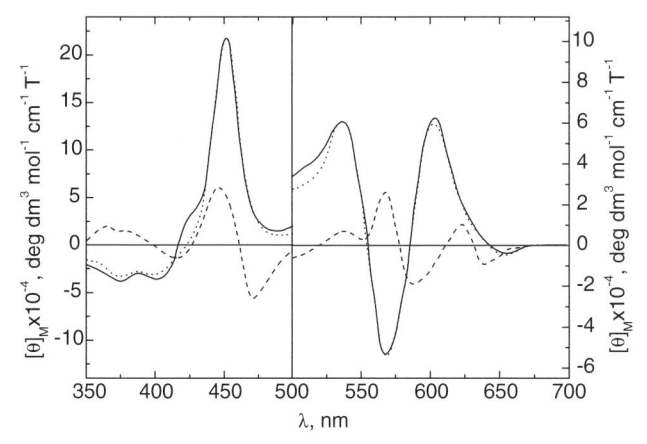

Figure 3.13 MCD spectra of Fe(II)TMPyP acquired in an OTTLE (path length = 0.29 mm) for [Fe(II)TMPyP] = 1.82 × 10^{-4} M in 0.05 H$_2$SO$_4$ (pH 1.0, dotted line), 0.1 M NaHCO$_3$ (pH 8.3, solid line) solutions and [Fe(II)TMPyP] = 7.83 × 10^{-4} M in 1 M NaOH solutions (pH 14, dashed line). Magnetic field 1.15 T; $E_{app} = -0.7$ V versus SCE.

additional pH independent reduction peak in alkaline solutions (not shown in Figure 3.12) at a potential more negative than that associated with the redox peak discussed above [36].

Further insight into the geometry and spin state of macrocycles complementary to that provided by UV–visible spectra, can be obtained from magnetic circular dichroism (MCD) [42]. Shown in Figure 3.13 are MCD spectra of Fe(II)TMPyP recorded at values of 1.0, 8.3 and 14 (see caption for details), where the magnitude of the MCD is expressed in terms of molar ellipticity per tesla, T (1 T = 10 000 G), [θ]$_M$ 10^4 deg mol^{-1} dm^3 cm^{-1} T^{-1}, collected *in situ* in an OTTLE at −0.7 V versus SCE. Analysis of these results suggests that Fe(II)TMPyP is a penta-coordinated, high spin complex, in which the Fe center is away from the plane of the porphyrin.

Also valuable are measurements of the paramagnetic susceptibility, χ_M, as a function of pH of the type shown in Figure 3.14 for Fe(III)TMPyP (solid circles) and iron(III)meso tetrakis(4-sulfonatophenyl)porphyrin, Fe(III)TPPPS$_4$ (open circles), yet another water soluble macrocycle [43]. As clearly indicated, χ_M undergoes a large decrease at about pH = 3, an effect associated with the formation of the μ-oxo dimer a species that displays antiferromagnetic coupling.

3.2.1.3 Rates of Heterogeneous Electron Transfer Reactions

Studies of the dependence of ΔE_p on v, as shown originally by Nicholson and Shain [44, 45], afford means of determining the rate of heterogeneous electron transfer processes, k_s. Application of these principles, assuming the diffusion coefficients of the reduced and oxidized forms to be the same, yielded for the reduction of Fe(III)TMPyP an average value of $k_s = 5.8 \times 10^{-3}$ cm s^{-1}. Alternatively, although not as yet applied to the study of solution-phase macrocycles, plots of $1/I_{disk}$

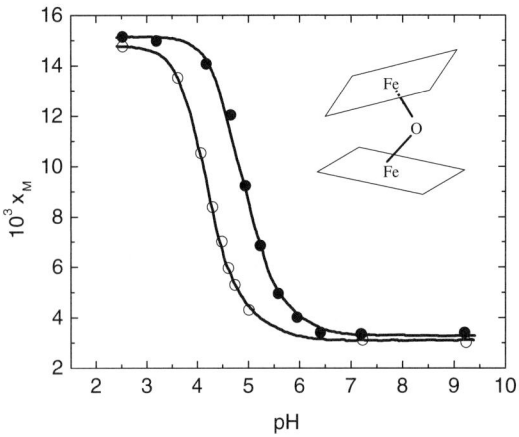

Figure 3.14 Plot of molar paramagnetic susceptibilities, χ_M, as a function of pH for 0.01 M Fe(III)TMPyP (solid circles) and 0.01 M Fe(III)TPPS$_4$ (open circles) aqueous solutions. Actual compositions can be found in the original literature. Insert: schematic representation of a dimeric Fe μ-oxo porphyrin (see also G in Scheme 3.2).

versus $\omega^{-1/2}$ at constant overpotential recorded with an RDE [45] yield linear plots from which k_s can be determined from the slopes and intercepts, and then used to extract values of exchange current densities and transfer coefficients.

3.2.2
Macrocyclic-Mediated Reduction of Dioxygen in Aqueous Electrolytes

A key step in the macrocyclic-mediated homogeneous electrocatalytic reduction of dioxygen involves encounters between the two species within the electrolyte region immediately adjacent to the electrode surface (see step a in Scheme 3.3). Hence, the course of the process will depend on the relative concentrations, or, more precisely, the fluxes, to be denoted hereafter as $f_{M(II)P/O_2}$, of the mediator and dioxygen.

Solutions to the governing differential equations that account for the large disparity between the diffusion coefficients of the macrocycles and dioxygen can be obtained using numerical methods, as illustrated by Bowers et al. for the RDE [46].

3.2.2.1 Model Systems
The macrocycles Co(III)(cyclam) and Fe(III)TMPyP display high activity for dioxygen reduction and negligible affinity for carbonaceous surfaces providing close to ideal conditions to warrant analyses of electrochemical data within the strict homogeneous electrocatalysis framework. Their most salient features are summarized in the two sub-sections to follow.

3.2.2.1.1 *trans*-[Co([14]aneN$_4$)(OH$_2$)$_2$]$^{3+}$, Co(III)(cyclam)
Various aspects of the chemistry of Co(III)(cyclam) [38], and, in particular, the reactivity of its reduced

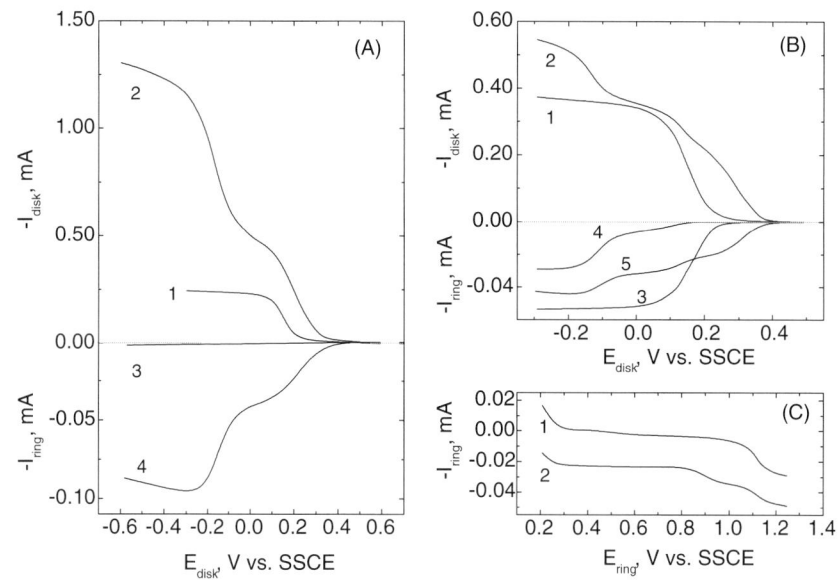

Figure 3.15 (A) Dynamic polarization curves obtained with a carbon disk ($A = 0.46 \text{ cm}^2$, top) of a Pt|carbon RRDE ($N = 0.175$) and corresponding Pt ring (bottom) currents recorded in 1.5 mM Co(III)(cyclam) in 0.5 M HClO$_4$ aqueous solutions in the absence (curve 1) and in the presence of O$_2$ at 900 rpm (curve 2), that is, $f_{\text{Co(cyclam)/O}_2}$ small, for $E_{\text{ring}} = 0.6$ (curve 3) and 1.0 V versus SSCE (curve 4) both in the presence of O$_2$. (B) Corresponding dynamic polarization curves recorded with the same RRDE assembly for 2.2 mM Co(III) (cyclam) in air-saturated (curves 2 and 4), that is, $f_{\text{Co(cyclam)/O}_2}$ large, and the flux of cyclam to the surface is 3.4-times larger than that of O$_2$, and deaerated (curves 1 and 3) in 0.5 M HClO$_4$ aqueous solutions for $E_{\text{ring}} = 0.6$ V. Curve 5 in this panel was obtained for $E_{\text{ring}} = 1.2$ V. (C) Current versus potential curves for the Pt ring upon polarizing the carbon disk at 0.1 V (curve 1) and -0.4 V (curve 2).

form, Co(II)(cyclam), toward O$_2$ have been extensively studied using conventional solution phase methods, particularly by Endicott and coworkers [47]. As reported by these workers, Co(II)(cyclam) forms adducts with O$_2$ stable enough to allow full spectroscopic characterization, at rates readily measurable by stopped flow techniques. Indeed, the information derived from these experiments has shed significant light into the electrocatalytic behavior of solution phase Co(III)(cyclam) in the presence of dioxygen. Clear evidence that the specific homogeneous electrocatalytic pathway for O$_2$ reduction is determined by the values of $f_{\text{Co(cyclam)/O}_2}$ was obtained from the analysis of dynamic polarization curves obtained in dioxygen-saturated 0.5 M HClO$_4$ using a Pt|carbon RRDE under conditions in which the Co(III)(cyclam)/O$_2$ is large and small (see, respectively, curves 2 in the upper subpanels, Panels A and B, Figure 3.15).

Also displayed in these subpanels, are curves recorded in solutions of Co(III)(cyclam) devoid of O$_2$ (see curves 1). As is evident from these data, the onset potential for reduction on the disk in the presence of O$_2$ is about 0.1 V positive than that observed in the absence of O$_2$, regardless of the value of $f_{\text{Co(cyclam)/O}_2}$. This behavior is

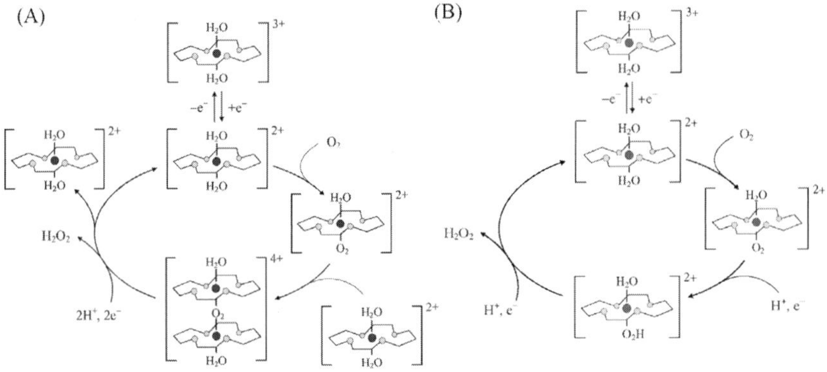

Scheme 3.5 Pathways for the reduction of dioxygen mediated by solution phase Co(cyclam) for large (A) and small (B) values of $f_{Co(cyclam)/O_2}$ (see text for details).

consistent with the rapid consumption of Co(II)(cyclam) by a chemical reaction that follows the electrochemical step (see below). Polarization of the ring at $E_{ring} = 0.6$ V, a potential which is positive to the oxidation of Co(II)(cyclam), but negative to the oxidation of hydrogen peroxide, failed to elicit a response over the potential range associated with the first wave at the disk. This indicates that the products generated in this potential range are not oxidizable at this potential.

In fact, when excess O_2 is present (Figure 3.15A), no current is observed at the ring over the entire range examined, signaling that Co(II)(cyclam) is being completely sequestered by the reaction with O_2 while in transit to the ring. This is not so for large $f_{Co(cyclam)/O_2}$ (Figure 3.15B), where a fraction of the Co(II)(cyclam) can evade capture and be detected by the ring (see curve 4, Panel B in the region $-0.1 < E_{disk} < 0.1$ V).

Two mechanistic pathways consistent with these observations are shown in Scheme 3.5 for large (A) and small (B) values of $f_{Co(cyclam)/O_2}$, which, as indicated, require, as an obligatory common first step, the reduction of Co(III)(cyclam) to yield the electrocatalytic active form of the macrocycle, that is, Co(II)(cyclam). For large $f_{Co(cyclam)/O_2}$, this latter species reacts with O_2 and, subsequently with a second reduced Co(II)(cyclam), to form a binuclear μ-peroxo bridged Co(III) complex (A, Scheme 3.5), which is then reduced electrochemically to yield solution phase hydrogen peroxide and Co(II)(cyclam). For small $f_{Co(cyclam)/O_2}$ (see B, Scheme 3.5), however, the species formed is believed to be a mononuclear hydroperoxide complex, which undergoes two sequential one-electron transfer steps regenerating Co(II)(cyclam). Regardless of the values of $f_{Co(cyclam)/O_2}$, the last stage in both pathways involves generation of solution phase hydrogen peroxide.

Further support for the mechanism proposed in B, Scheme 3.5, was provided by the results in Figure 3.15. In particular, for $E_{disk} < -0.1$ V in the presence of O_2, I_{disk} displays a new wave associated with the one-electron reduction of the adduct to generate hydrogen peroxide (see curve 2, upper panel B in this figure). This overall process releases Co(II)(cyclam), which is clearly detected at the ring for $E_{ring} = 0.6$ V (curve 3, Figure 3.15B). Polarization of the ring at $E_{ring} = 1.0$ V (curve 4, in A) and

1.2 V (curve 5, in B) was found to be sufficient to oxidize all species generated at the disk, namely, the dioxygen adducts, Co(II)(cyclam) and also hydrogen peroxide, that is, the most negative plateau in both curve 4 in A, and curve 5 in B. These assignments were confirmed by experiments in which E_{disk} was fixed, while scanning E_{ring} over a wide potential range (see Panel C, Figure 3.15). More specifically, for $E_{disk} = 0.1$ V, I_{ring} is negligible in the range $1.0 \leq E_{ring} \leq 0.3$ V, that is, no hydrogen peroxide is produced at the disk (curve 1, Figure 3.15C). The increase in I_{ring} at $E_{ring} = 1.0$ V can be attributed to the oxidation of the adduct and also of free Co(II)cyclam, whereas that at $E_{ring} = 0.3$ V to the reduction of Co(III)cyclam. For $E_{disk} = -0.4$ V, I_{ring} shows a clear shoulder at E_{ring} of about 0.9 V, consistent with the oxidation of hydrogen peroxide generated at the disk (see curve 2, C). Formation of a binuclear μ-peroxo bridged Co(III)species was supported by comparison of its electrochemical properties with those of the species synthesized independently [38].

Plots of I_{lim}^{disk} versus $\omega^{1/2}$ recorded in solutions containing Co(II)(cyclam) at submillimolar concentrations were found to bend downward within the region defined by the one-electron and two-electron Levich lines (see straight lines in Figure 3.16) both for $f_{Co(cyclam)/O_2} < 1$ and, $f_{Co(cyclam)/O_2} > 1$ (see open and solid circles, in the same figure, respectively).

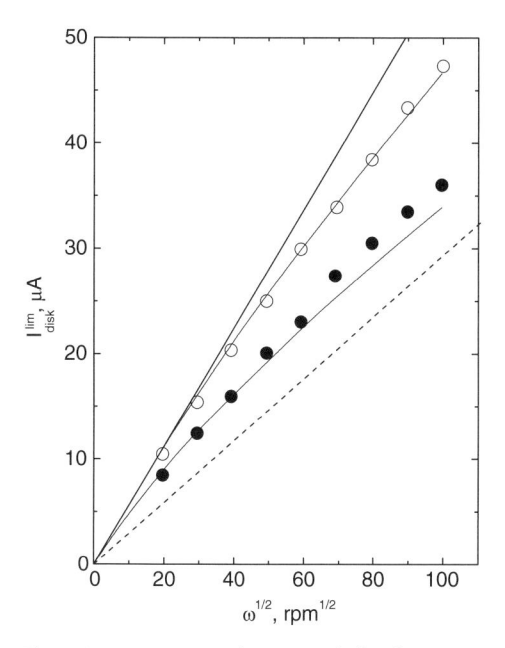

Figure 3.16 Comparison between calculated (lines) and experimental (symbols) plateau currents for $k_1 = 2 \times 10^6 \, M^{-1} s^{-1}$ (other parameters can be found in the original literature) for Co(III)(cyclam) = 74 μM for $[O_2] = 28$ μM (solid circles) and $[O_2] = 280$ μM (open circles). The straight dashed line corresponds to the best fit to the one-electron reduction of Co(III)(cyclam) in the absence of O_2, whereas the slope of the solid line is twice that of the dashed line [48].

Numerical simulations of these RDE experiments were carried out based on the sequence of reactions shown in Scheme 3.5, using values of the rate constants, k_{-1}, k_2 and k_{-2} in Equations 3.11 and 3.12 (which correspond to a and c in Scheme 3.3) determined by Endicott and coworkers by classical stop-flow techniques [47]:

$$Co(II)(cyclam) + O_2 \underset{k_{-1}}{\overset{k_1}{\rightleftharpoons}} Co(II)(cyclam)\text{-}O_2 \qquad (3.11)$$

$$Co(II)(cyclam)\text{-}O_2 + Co(II)(cyclam) \underset{k_{-2}}{\overset{k_2}{\rightleftharpoons}} [Co(II)(cyclam)]_2 O_2 \qquad (3.12)$$

The results obtained yielded average values for k_1, the rate constant for the formation of the Co(II)(cyclam)–O_2 adduct, of about $3 \times 10^6 \, M^{-1} \, s^{-1}$ [48].

Based on purely geometric arguments, it might be expected that addition of relatively bulky groups to the periphery of cyclam could prevent formation of a dimeric μ-peroxo species due to steric crowding and thus simplify the reduction pathway compared with Co(cyclam). Indeed the Co(II) complex of C-meso, 5,7,7,12,14,14-hexamethyl-1,4,8,11-tetraazacyclotetradecane (hmc) was found to form only a monomeric dioxygen adduct, a species that underwent a rapid one-electron reduction to yield (hmc)CoOOH^{2+} at the potentials at which (hmc)Co(III) was reduced to the corresponding Co(II) analogue. At more negative potentials the adduct was irreversibly reduced generating hydrogen peroxide and (hmc)Co(II).

The ability to fairly unambiguously identify macrocyclic–O_2 adducts for Co(cyclam) may be regarded as exceptional, as for most macrocycles, particularly those involving iron centers, the formation of adducts has only been inferred based on indirect evidence.

3.2.2.1.2 Iron tetrakis(N-methyl-4-pyridyl)porphyrin, Fe(III)TMPyP

The electrocatalytic properties of FeTMPyP [36] toward O_2 reduction in aqueous electrolytes are eloquently illustrated by comparing the voltammetric response obtained both before and after introducing the mediator into an O_2-containing solution (see Figure 3.17). Whereas on a non-catalytic electrode, such as glassy carbon (GC), the onset for O_2 reduction occurs at a very negative potential, that is, −0.15 V versus NHE (see solid curve, a, Figure 3.17), addition of FeTMPyP brings about a large increase in the current at much more positive potentials, that is, +0.3 V versus NHE (dotted curve) [50]. Careful inspection of the voltammetric scans obtained in solutions containing FeTMPyP in the absence (see curve b in this figure) of O_2 reveals that the onset potential for the reduction of the macrocycle, $E(I_p)$, appears to coincide with the onset potential for the FeTMPyP-mediated reduction of O_2, $E(I_p^{cat})$. In fact, as shown in Figure 3.12, excellent agreement was found between average peak potentials for the macrocycle in the absence of O_2, E_{av}, and the peak potentials for reduction obtained in the presence of O_2, over the entire range of pH values examined.

As the concentration of Fe(III)TMPyP is increased, while keeping the bulk concentration of O_2 constant, $E(I_p^{cat})$ shifts positively, reaching values which actually exceed those of $E(I_p)$ (see Figure 3.18). This behavior is consistent with an

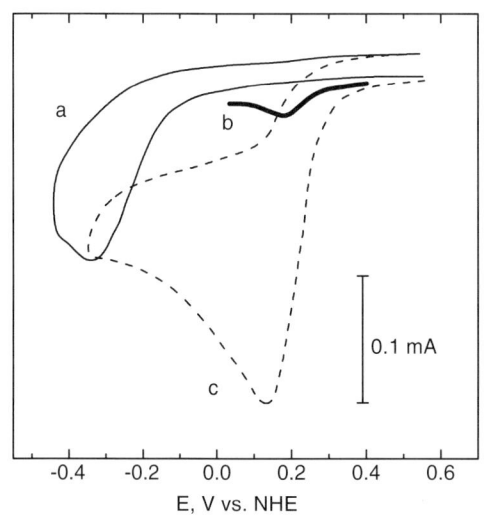

Figure 3.17 Cyclic voltammetry curves for O_2 reduction on a GC electrode in 0.05 M H_2SO_4 aqueous solutions containing O_2 at a concentration of 0.24 mM, in the absence (solid line a), and in the presence of 0.25 mM Fe(III)TMPyP (dashed line c). The single scan curve (b, thick line) was obtained for the macrocyclic solution in the absence of O_2 [50].

electrochemical–chemical (EC) cycle of the type shown in Scheme 3.6 in which both of the steps are very fast.

Insight into the nature of the reaction products generated by this electrocatalytic process was gained from plots of the peak current in the presence of dioxygen, I_p^{cat}, normalized by the corresponding peak currents for a one-electron reduction process,

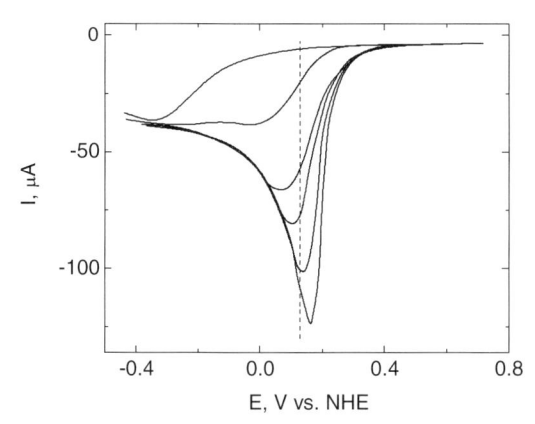

Figure 3.18 Cyclic voltammetry curves (only scans in a single direction are shown) for the reduction of dioxygen by FeTMPyP in 0.05 M H_2SO_4 containing O_2 at a concentration of 0.2 mM, for [Fe(III)TMPyP] = 0.0, 0.022, 0.13, 0.25, 0.51 and 0.99 mM (top to bottom). Scan rate: 50 mV s^{-1}. Electrode Area: 0.39 cm^2.

Electrochemical Step

$$Fe(III)TMPyP + e^- \xrightarrow{k_e} Fe(II)TMPyP$$

Chemical Steps

$$2Fe(II)TMPyP + O_2 + 2H^+ \xrightarrow{k_{f1}} Fe(III)TMPyP + H_2O_2$$

$$2Fe(II)TMPyP + H_2O_2 + 2H^+ \xrightarrow{k_{f2}} 2Fe(III)TMPyP + H_2O$$

Scheme 3.6 Mechanism for the FeTMPyP-mediated electrocatalytic reduction of dioxygen.

I_p, as a function of the concentration of Fe(III)TMPyP (see Figure 3.19). The open circles in this figure are the actual measured values and the solid circles were obtained after subtracting from these values contributions due to the one-electron reduction of the mediator.

It becomes evident from these data that as [Fe(III)TMPyP] increases, the net number of electrons transferred per O_2 molecule approaches four, and is thus consistent with the reduction of O_2 beyond the peroxide step to yield water as a product. In other words, as the concentration of catalytic species is increased, the frequency of encounters with the substrate and its reaction intermediates is enhanced leading to higher rates of O_2 conversion into water. Support for this so-called 2 + 2 mechanism was obtained from RRDE experiments in which the ring was polarized at a potential sufficiently positive for the oxidation of hydrogen peroxide to proceed under diffusion controlled conditions [51].

Despite the good agreement between theory and experiments, there are some rather subtle issues that require close attention. In particular, the redox mediator can store only a single electron for delivery to the reactant; hence, a reasonable initial step would involve a reaction between Fe(II)TMPyP and dioxygen to yield superoxide and Fe(III)TMPyP, that is:

$$Fe(II)TMPyP + O_2 \leftrightarrow Fe(III)TMPyP + O_2^- \qquad (3.13)$$

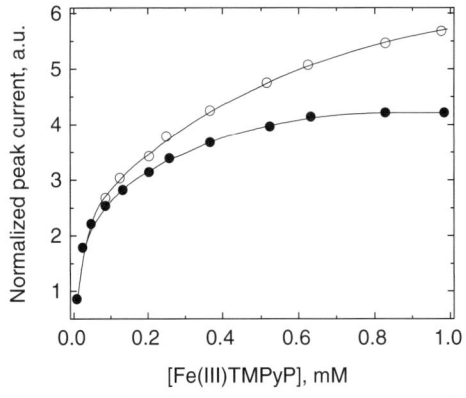

Figure 3.19 Plots of normalized peak current ($n = 1$) for dioxygen reduction, using the Randles–Sevcik equation, [39] (open circles) and corrected normalized current (solid circles) versus Fe(III)TMPyP concentration in solutions containing O_2 at 0.24 mM for $v = 0.05\,V\,s^{-1}$.

However, based strictly on the standard reduction potentials of the two redox couples involved, the equilibrium constant of the reaction as written would be exceedingly small. To further complicate matters, the rates of disappearance of O_2 in the presence of Fe(II)TMPyP, as monitored by RRDE techniques, and the rates of disappearance of O_2^- in the presence of Fe(III)TMPyP, as measured by spectrophotometric methods [52], yielded very similar values. The most likely resolution of this seeming quandary invokes formation of a macrocycle–dioxygen adduct as a short-lived, albeit yet to be detected, intermediate.

As the electrocatalytic behavior of Fe(III)TMPyP indicates, an increase in the concentration of the mediator leads to a corresponding increase in the net number of electrons transferred to dioxygen. Hence, sizable gains in electrocatalytic efficiency could, in principle, be achieved by immobilizing electrocatalysts on electrode surfaces, as will be discussed in detail in the next section.

3.3
Heterogeneous Electrocatalysis

Confinement of electrocatalysts to electrode surfaces makes it possible to increase their local concentration to levels well above solubility limits. Furthermore, metal centers in close proximity can open possible new pathways for interactions with species that can accept or release multiple electrons not readily available with individual molecules. In particular, binding of O_2 to two metal centers could facilitate bond activation and, as highly desirable, expedite delivery of four electrons before intermediates, such as peroxide, can escape into the bulk solution. It is also conceivable that chemical interactions between adsorbed macrocycles and the electrode surface may promote formation of O_2 adducts with redox properties very different from those of their solution phase counterparts and thus lead to changes in both the mechanism and overall rates of O_2 reduction. An interesting illustration of some of these effects is provided by $[(tim)Co]^{2+/3+}$, tim = 2,3,9,10-tetramethyl-1,4,8,11-tetraazacyclotetradeca-1,3,8,10-tetraene (see insert Figure 3.20), a material that is soluble in aqueous acidic solutions, and also displays affinity for carbon surfaces. As shown by Espenson et al. [53], in acidic solutions, $[(tim)Co]^{2+}$ catalyzes O_2 reduction to water very slowly, yielding no evidence for peroxide being formed. However, upon adsorption on roughened graphite, as reported much later by Bhugun and Anson, it becomes a rather powerful strict two-electron electrocatalyst that generates hydrogen peroxide quantitatively [54].

This section describes various strategies for the immobilization of macrocycles on electrode surfaces and their characterization by both electrochemical and in situ spectroscopic techniques in solutions devoid of dioxygen. It also provides theoretical foundations involved in the analysis of the mechanisms of oxygen reduction at such interfaces based on measurements performed under forced convection. Studies involving a number of carefully selected phthalocyanines, and porphyrins, will be presented and discussed, which in our view best illustrate the nuances of the rich behavior this class of adsorbed electrocatalysts can exhibit. These examples serve to

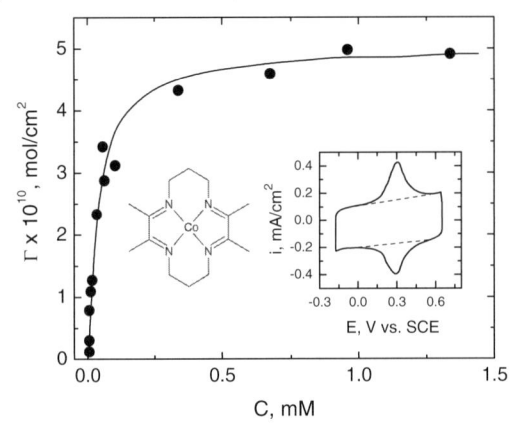

Figure 3.20 Plot of the surface concentration Γ (mol cm^{-2}) of [(tim)Co]$^{3+}$ (see insert) adsorbed on a roughened graphite electrode as a function of the concentration of [(tim)Co]$^{3+}$ in 0.2 M CF$_3$COONa|0.2 M CF$_3$COOH solutions determined from a coulometric analysis of background corrected voltammetric data ($v = 0.30$ V s^{-1}, see insert) such as that shown in the insert for measurements performed in 0.12 mM [(tim)Co]$^{3+}$ after extensive cycling. The dotted lines in this insert represent contributions to the capacitive current ascribed to the underlying substrate (see text for details).

underscore the need for a better understanding of the factors that control structure–activity relationships, including those associated with adsorption, as a necessary step toward the rational design of electrocatalytic interfaces displaying optimum performance.

3.3.1
Adsorption Isotherms

Mathematical expressions that relate the coverage of adsorbed species and their concentration in solution at constant temperature are defined as adsorption isotherms. Such correlations can be established with relative ease for redox active molecules, such as the vast majority of macrocycles displaying activity for O$_2$ reduction, by exploiting the differences in the voltammetric behavior between solution phase and adsorbed species. More specifically, the voltammetric peak currents, I_p, due to solution phase species (see Section 3.2) are proportional to the bulk concentration, to the square root of the scan rate, $v^{1/2}$, and to the cross-sectional area of the electrode, whereas those due to adsorbed species are proportional to v and to the total number of adsorbates on the surface (see below). Hence, at high enough v, the overall voltammetric behavior will be chiefly dominated by the adsorbed species. As a means of illustration, the cyclic voltammogram of a roughened graphite electrode in a solution 0.12 M [(tim)Co]$^{3+}$ in 0.2 M CF$_3$COONa and 0.2 M CF$_3$COOH yielded characteristic mirror image peaks associated with the redox transition of the adsorbed species (see insert, Figure 3.20). An absolute measure of the number of such species can be obtained from a coulometric analysis of the voltammetric peaks provided the number of electrons transferred, n, is known, and the featureless background currents derived from the purely capacitive behavior of the interface (see dotted lines in the insert, Figure 3.20) are properly subtracted out. Shown in Figure 3.20 is a plot of the surface concentration, Γ (in mol cm^{-2} of a cross-sectional area of electrode) of [(tim)Co]$^{3+}$, as a function of its bulk concentration, C, determined by this method in the same supporting electrolyte. As indicated, Γ reaches saturation at about 5×10^{-10} mol cm^{-2} [53].

Attention in what follows, however, will be focused on systems in which the adsorbate is irreversibly confined to the electrode surface (at least within a time scale longer than that required for electrocatalytic studies), even in the absence of solution phase material. Various aspects of the preparation and characterization of such chemically modified electrodes will be presented next.

3.3.2
Chemically Modified Electrodes

The general concept of chemical modification [55, 56] of electrode surfaces can be attributed by and large to the seminal work of Hubbard and coworkers [57] in the early 1970s. Since then, electrode surfaces modified by a wide variety of adsorbed species, including polymers and other types of films have received wide attention not only in electrocatalysis, but also in sensor technology and other applications.

3.3.2.1 Preparation and Electrochemical Characterization
A variety of tactics for the surface confinement of transition metal macrocycles (and also other species) have been developed, particularly by the groups of Anson at Caltech [58], Kuwana at Kansas State [59], Yeager at Case Western Reserve University [60] and Murray at the University of North Carolina [61]. These are described in some detail below.

3.3.2.1.1 **Spontaneous Adsorption** The affinity of certain macrocycles for surfaces such as carbon (for which functional groups present on the surface may in some instances be responsible for the adsorption) promotes spontaneous formation of strongly adsorbed layers even for species that are soluble in the adjacent media. A good illustration of this phenomenon is provided by iron(III) protoporphyrin IX chloride (or hemin, Hm) adsorbed on the basal plane of highly oriented pyrolytic graphite HOPG(bp) from borate buffer aqueous solutions. As shown in panel A, Figure 3.21, Hm|HOPG(bp) interfaces yield, in the absence of Hm in the electrolyte, very well defined redox peaks consistent with the presence of a single equivalent monolayer of Hm on the surface. *In situ* scanning tunneling microscopy (STM) images of Hm|HOPG(bp) collected in the same electrolyte (see Figure 3.22) [62], revealed a very regular pattern akin to closed packed tiles of about 1.4 nm per side (C), and thus very close to the molecular dimensions of Hm determined by other techniques. Also included in panels B and D in this figure, respectively, are the cyclic voltammetry and *in situ* STM image obtained for a mixture of self-assembled

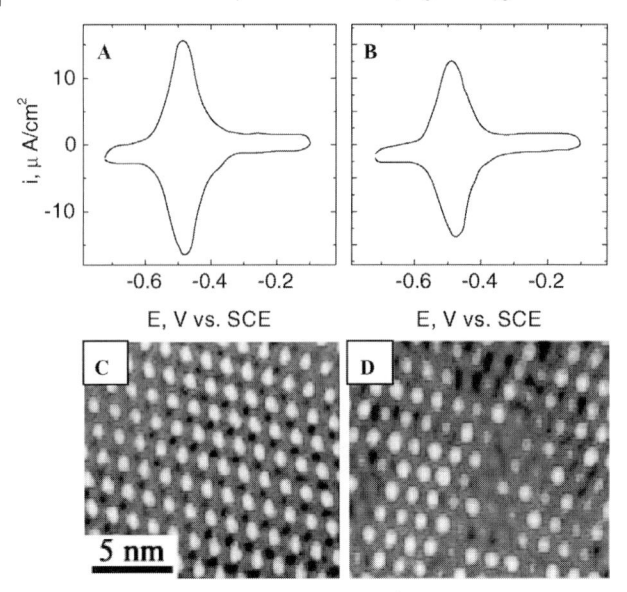

Figure 3.21 Cyclic voltammetry ($v = 0.2\,V\,s^{-1}$) of a closed packed monolayer of Hm (A), and of a mixture (4:1) of Hm and its metal free analogue (B), irreversibly adsorbed on HOPG(bp) in neat 0.2 M $Na_2B_4O_7$ aqueous solutions (see text). Panels C and D display scanning tunneling microscopy (STM) images of the interfaces specified in (A) and (B), respectively, where the spots of lower intensity in (D) are attributed to metal free Hm.

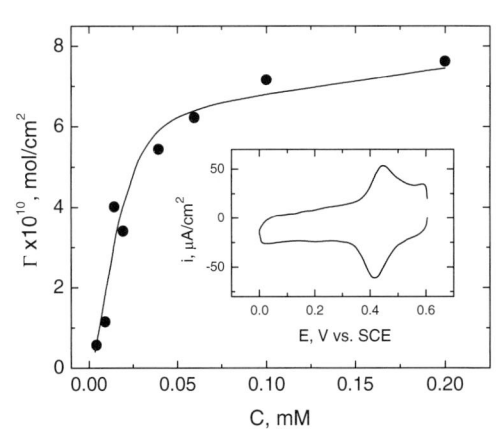

Figure 3.22 Plot of the surface concentration Γ (mol cm^{-2}) of CoOEP spontaneously adsorbed on an EPG electrode surface from $CHCl_3$ solutions as a function of their bulk concentration, based on a coulometric analysis of voltammetric data collected in aqueous 1 M $HClO_4$ recorded at $v = 50\,mV\,s^{-1}$ of the type shown in the insert (see text for details).

Hm and its metal free analogue, which, under the tunneling conditions selected for these experiments, yields spots of lower intensity compared with Hm.

Spontaneously adsorbed molecular layers can also be formed using solvents other than water. In particular, exposure of edge plane pyrolytic graphite (EPG) surfaces to solutions of CoOEP (the Co complex of F in Scheme 3.2) in CHCl$_3$, followed by rinsing with neat solvent to remove non-adsorbed material and subsequent transfer to aqueous 1 M HClO$_4$, yielded voltammograms displaying a single peak centered at about 0.45 V versus SCE (see insert, Figure 3.22). Experiments of this type, in which the concentration of CoOEP in the CHCl$_3$ solution was systematically varied, yielded saturation surface coverages, Γ^*, of about 0.7 nmol cm^{-2} [63].

Generally, adsorbed layers of macrocycles that are soluble in the adjacent media undergo slow desorption over time, a process that can be prevented in some instances by adding very small amounts of the material to the solution. This specific approach was implemented to obtain highly detailed voltammetric curves for spontaneously adsorbed water-soluble iron 4,4′,4″,4‴-tetrasulfophthalocyanine (FeTsPc) on OPG in deaerated aqueous solutions pH = 10.7 containing 10 µM FeTsPc (see Figure 3.23). As expected, plots of i_p, after subtraction of the background capacitive current, as a function of v, yielded a straight line, as illustrated for the peak centered at around 0.1 V in the insert [64]. Stabilization against desorption can also be achieved by adding a counterion to the supporting electrolyte with which the macrocycle forms an insoluble salt. Examples of this tactic include, perchlorate, ClO$_4^-$ for FeTPyP in acidic

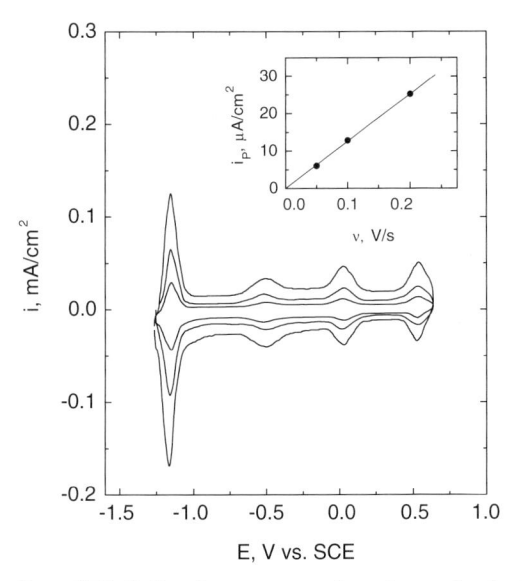

Figure 3.23 Cyclic voltammograms of spontaneously adsorbed FeTsPc on OPG in N$_2$-saturated aqueous buffered solutions pH = 10.7 containing 10 µM FeTsPc for v = 50, 100 and 200 mV s^{-1} yielding, in sequence, peaks of increasing amplitude. Shown in the insert is a plot of i_p versus v for the peak centered at around 0.1 V versus SCE.

electrolytes [58], and hexafluorophosphate, PF_6^-, for ruthenated complexes of *meso*-tetrakis(4-pyridyl)porphyrin, $CoP(py)_4$ [65], to be discussed later in this chapter.

3.3.2.1.2 Precipitation
For macrocycles that are insoluble in aqueous solutions, chemical modification can also be achieved by delivering a solution of the species in a high vapor pressure organic solvent, such as chloroform, or dichloroethane, through a micropipette onto the surface of the electrode. Once the solvent evaporates, and depending on the concentration of the macrocycle in the dispensing solution, amounts equivalent to a fraction, up to several monolayers, can be left on the surface. In the latter case, however, the coatings may become very compact, preventing ions from migrating within the structure to counterbalance the charge associated with redox transitions. In fact, the electrochemical, and thus the electrocatalytic activity of such multilayer deposits is often restricted to the outermost layer.

3.3.2.1.3 Covalent Attachment
A more elegant and better controlled methodology involves the binding of judiciously functionalized macrocycles to chemically modified electrode surfaces. Two such examples, involving the anchoring of an amino porphyrin to acyl-derivatized carbon and Pt surfaces are shown in Figure 3.24, where, without loss of generality, a two-end binding has been assumed. Evidence for the

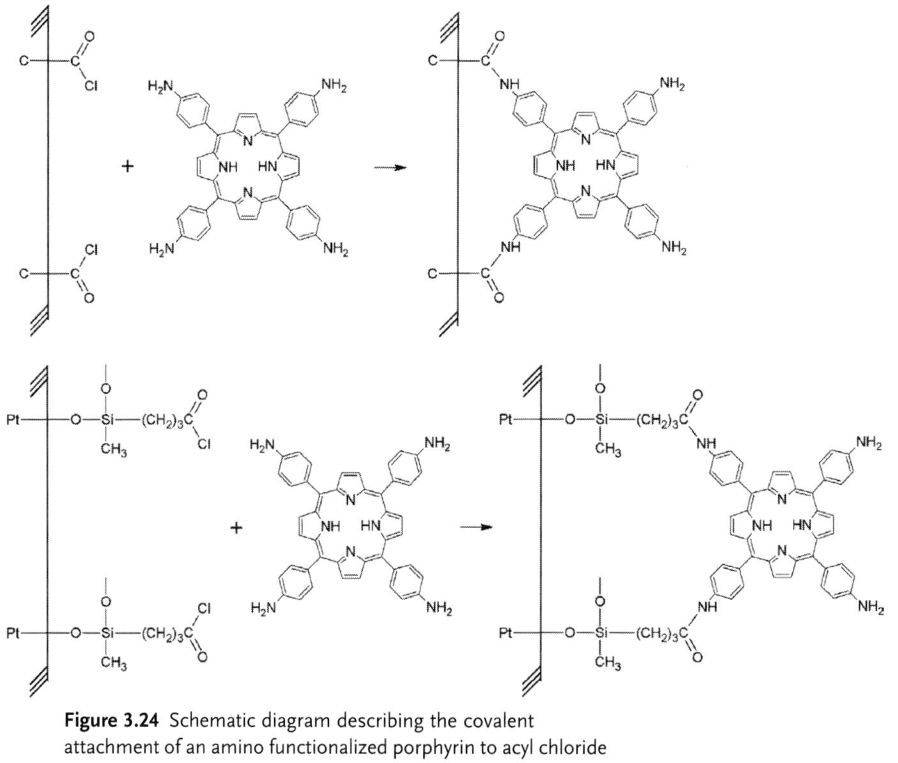

Figure 3.24 Schematic diagram describing the covalent attachment of an amino functionalized porphyrin to acyl chloride derivatized carbon (top) and Pt surfaces (bottom).

covalent attachment of the porphyrin to the functional groups on the surface was obtained from X-ray photoelectron spectroscopy (XPS). Prior to this analysis the surface was further derivatized by converting $-NH_2$ moieties not bound to the surface into nitro groups, for which the N(1s) binding energies are very different. Yet another form of covalent attachment involves the use of surfactants bound to the electrode surface bearing functional groups that can axially coordinate to the metal center in the macrocycle [55].

Both of these techniques render interfaces incorporating up to one monolayer of material. Voltammograms of covalently bonded mixtures of unmetallated and tetra (p-aminophenol)porphyrin, $T(p-NH_2)PP$ and $Co-T(p-NH_2)PP$ on carbon prepared by the method shown in Figure 3.24 in dimethyl formamide, DMF, with added pyridine (see upper curve, left ordinate, Figure 3.25) yielded, in addition to the well defined peak centered at about -0.9 V versus SSCE, two small and rather broad features at -0.2 and -1.2 V [66]. Significant improvements in peak definition can be obtained using differential pulse voltammetry [39] (see caption for specific conditions), as illustrated by the data in the lower curve (right ordinate) in this figure.

3.3.2.1.4 Other Methods
Macrocyclic-modified electrode surfaces can also be prepared by physical vapor deposition under reduced pressure, provided the species can be sublimed without undergoing decomposition [67], or by electropolymerization of monomeric macrocycles bearing suitable functional groups [68]. Another interesting tactic involves incorporation of species into thin polymer films supported on electrode substrates [69]. Shown in Figure 3.26 are cyclic voltammetry curves obtained for spin-coated films of Nafion®, a perfluorinated polymer with pendant sulfonic groups, containing Co meso-tetraphenyl porphyrin (CoTPP), on OPG

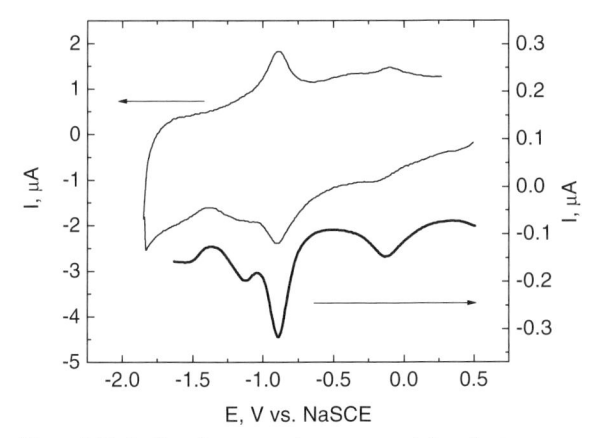

Figure 3.25 Cyclic voltammetry (upper curve, left ordinate, $v = 0.1\,V\,s^{-1}$) and differential pulse voltammetry (lower curve, right ordinate, $v = 2\,mV\,s^{-1}$, pulse rate 125 ms per pulse, 10 mV peak amplitude) for a mixture of H_2-, and $CoT(p-NH_2)PP$ chemically bonded to an acyl-functionalized carbon surfaces in DMF with added pyridine.

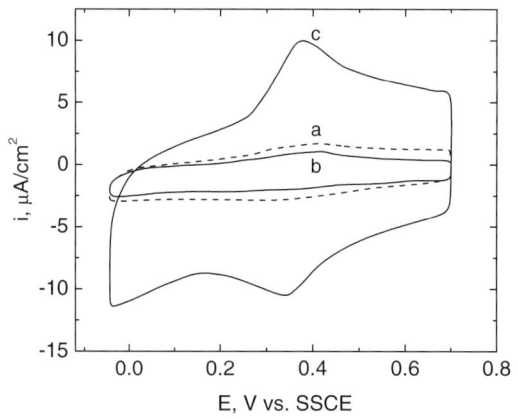

Figure 3.26 Cyclic voltammograms ($v = 50\,mV\,s^{-1}$) of a graphite electrode coated with a layer of Nafion™ (about $3.7 \times 10^{-8}\,mol\,cm^{-2}$) incorporating H_2TPP, before (curve a) and after conversion into Co(III)TPP (curve b, see text) in Ar-purged 0.5 M H_2SO_4. Also shown for comparison (curve c) is the response of about $10^{-8}\,mol\,cm^{-2}$ CoTPP adsorbed directly on the carbon surface, that is, no Nafion™ film.

surfaces. For these experiments, the supported Nafion™ films were first exposed to H_2TPP then to a solution of a Co salt to form CoTPP and finally to oxygen to oxidize the metal center. As shown in Figure 3.26, the broad voltammetric peak observed before (curve a) and after metallation (curve b) were virtually identical [69] and thus unrelated to the presence of the macrocycle. For comparison, CoTPP bound directly to the OPG surface exhibits a very well defined peak (curve c), for which the potential happens to coincide, rather fortuitously, with that of the other two Nafion™-modified surfaces. The lack of a redox response for CoTPP in the Nafion™ film is due in part to its very low mobility within this medium, but mostly to the fact that the polymer matrix is electronically insulating. As demonstrated by Buttry and Anson [69], electrons can be relayed from the carbon electrode to the redox active species using a highly mobile species as a shuttle. This is clearly shown in Figure 3.27, which displays cyclic voltammetry curves for Co(III)TPP|Nafion™ films following incorporation of very small amounts of $Ru(NH_3)_6^{3+}$, where the redox feature centered at about 0.3 V can be ascribed to the reduction of the Co(III) center to yield Co(II)TPP. In fact, the difference between the cathodic and anodic peaks affords a very good measure of the total amount of CoTPP in the film.

3.3.2.2 *In situ* Spectroscopic Characterization

Despite their sensitivity, electrochemical methods lack the required specificity to allow structural and electronic aspects of adsorbed species on electrode surfaces to be examined. This important complementary information can be obtained by employing spectroscopic techniques *in situ*, that is, with the electrode immersed in the solution under potential control. As discussed above, the number of species that can be accommodated on surfaces is of the order of only a fraction of a nmol cm^{-2}, posing

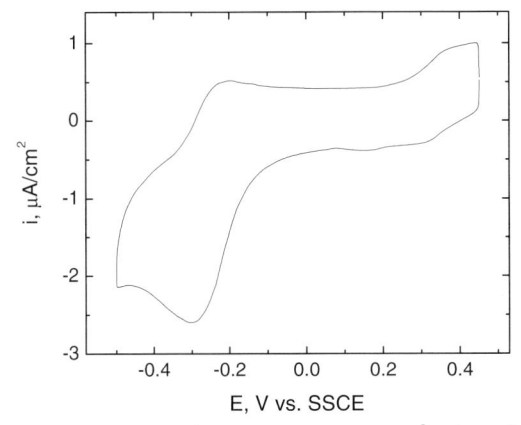

Figure 3.27 Cyclic voltammogram ($v = 5$ mV s^{-1}) of a Nafion™ film (about 2.4×10^{-7} mol cm^{-2}) incorporating CoTPP recorded in a 0.5 M H$_2$SO$_4$ solution containing 1×10^{-8} M Ru(NH$_3$)$_6^{3+}$ (see text for details).

challenging, and, at times, rather frustrating experimental problems for their detection and study.

At least three factors have greatly contributed to advances in this field: implementation of modulation and averaging schemes in reflectance spectroscopy to offset the very small optical changes induced by adsorption of species on electrode surfaces and/or by changes in oxidation state; the discovery of surface enhanced Raman spectroscopy on a few metal surfaces, prominently Ag and Au; and, more recently, the use of high-area carbon supports, which offer a high degree of transparency to X-rays, allowing synchrotron-based X-ray absorption methods to be implemented.

Rather serendipitously, transition metal macrocycles display electronic transitions in the UV and visible spectral regions with very large cross sections. These conditions do not only lead to increased sensitivity in reflectance measurements in this energy range, but also contribute to enhance Raman signals through electronic resonances. This sub-section addresses salient aspects of *in situ* spectroscopy as applied to the study of macrocyclic modified surfaces of relevance to the electrochemical reduction of dioxygen.

3.3.2.2.1 **Electronic Properties** The spectral properties of solution phase species in the transmission mode can be readily determined using the absorbance spectra of the pure solvent (or electrolyte) as a reference (or normalizing factor). In stark contrast, classical optics predicts that the reflection spectrum of an interface comprised of a film in direct contact with the substrate, normalized by the corresponding spectrum of the bare substrate, namely:

$$\Delta R/R = (R_{ref} - R_{ads})/R_{ads} \quad (3.14)$$

where R_{ref} and R_{ads} are proportional to the detected light intensity for the bare and modified surfaces, respectively, does not, in general, yield the desired 'pure' spectrum of the adsorbed layer [70]. In other words, the substrate surface cannot be regarded as a simple mirror, as the reflectance spectra 'mixes' the optical properties of the adsorbate and the substrate in a highly non-linear fashion (see also Ref. [71]). Moreover, physical and chemical interactions between the adsorbate and the surface can evoke changes in the optical properties of both phases, and thus in the reflectance spectrum, a factor that further complicates theoretical analyses. Even if these effects were not significant, the experimental determination of $\Delta R/R$ has proven to be far from routine, owing to the extraordinary difficulties involved in reliably measuring minute changes (of the order of a fraction of a percent) in light intensity.

Far higher success has been obtained using reflection absorption wavelength modulation spectroscopy (RAWMS), a technique that relies on periodic changes in the wavelength of the incident beam, generated, for example, by vibrating the optical grating about an axis normal to the plane of incidence, using lock-in techniques to detect correlated changes in the intensity of the reflected beam. An illustration of this novel tactic is shown in Figure 3.28 [72], which displays *in situ* p-polarized RAWMS spectra of HOPG(bp) in 0.05 M H_2SO_4 before (dashed curve, panel A) and after (solid

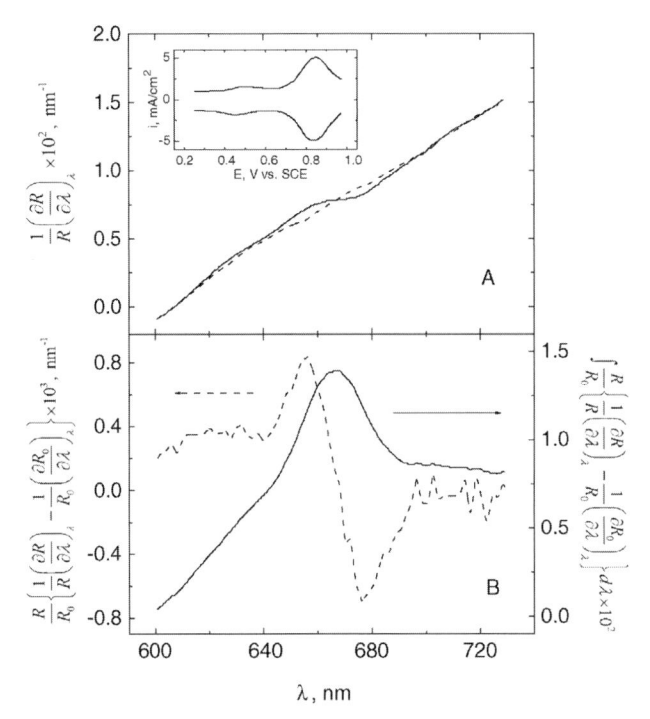

Figure 3.28 (A) *In situ* p-polarized RAWMS spectra of HOPG(bp) in 0.05 M H_2SO_4 before (dotted line) and after (solid line) adsorption of about 0.018 nmol cm^{-2} of CoTsPc with the electrode polarized at 0.5 V versus SCE (see text for details). Insert: cyclic voltammogram ($v = 100$ mV s^{-1}) of CoTsPc|HOPG(bp) in 0.05 M H_2SO_4. (B) Normalized difference RAWMS spectrum, (dotted line, left ordinate) and the corresponding integrated RAWMS spectrum (solid line, right ordinate) for the CoTsPc|HOPG(bp) interface based on the data shown (A).

curve in the same panel) adsorption of a single monolayer, about 0.018 nmol cm^{-2} of CoTsPc, with the electrode polarized at 0.5 V versus SCE. HOPG(bp) is a highly non-interacting substrate that helps mitigate 'chemical' effects associated with formation of the interface.

The cyclic voltammogram of the CoTsPc|HOPG(bp) recorded at $v = 100$ mV s^{-1} in the base electrolyte is given in the insert, panel A. Also shown in panel B in this figure are the normalized difference (dotted line, left ordinate) based on the results shown in A, and the resulting integrated RAWMS spectrum (solid line, right ordinate), which exhibits a peak at about the same position as CoTsPc dissolved in the aqueous solution. Potential modulation affords yet another means of exploring the dynamic aspects of chemically modified interfaces, as will be illustrated later in this section.

3.3.2.2.2 Vibrational Properties
Although generally restricted to metals with suitable optical properties, principally Ag, Cu and Au, and requiring, in addition, a certain degree of surface roughening [73], surface enhanced Raman scattering (SERS) has emerged as a powerful tool for examining vibrational spectra of adsorbed species. The extraordinary sensitivity of this method is illustrated in Figure 3.29, which displays the SERS spectra of a single monolayer of Hm (curve a) and its reduced counterpart (curve b) adsorbed on a roughened Ag electrode obtained *in situ* in a pH = 3 solution [74].

For these experiments, the excitation wavelength, $\lambda_{exc} = 532$ nm, was close to the energy associated with one of the electronic transitions of Hm in the UV–visible range, a factor that contributes to further enhancements in the SERS due to resonance. Cursory inspection of these data shows marked differences between the spectral features of Hm and its reduced form. In particular, the most prominent SERS peaks found for adsorbed Hm in the range $0.0 < E < 0.25$ V were found to be in very good agreement with those of Hm in powder form using the same excitation

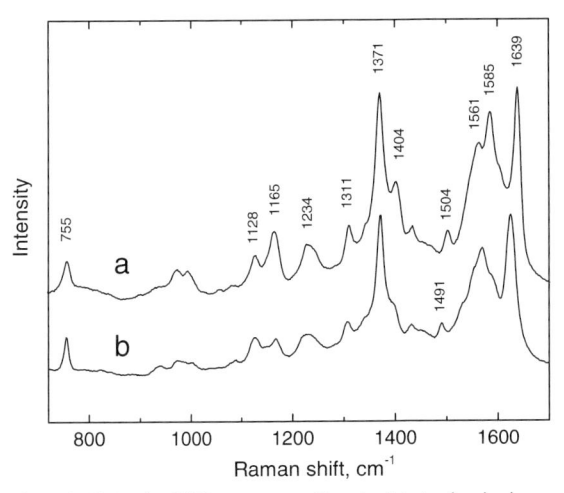

Figure 3.29 *In situ* SERS spectrum of hemin (Hm) adsorbed on a roughened Ag electrode (curve a, −0.50 V versus SCE) and its reduced counterpart (curve b, 0.1 V) recorded in an aqueous solution pH = 3, $\lambda_{exc} = 532$ nm.

wavelength. However, those recorded at $E = -0.50$ V versus SCE are characteristic of the intermediate spin Fe(II) (or reduced) form of Hm. Applications of quantitative *in situ* SERS to studies of the Hm|Ag interface will be described later in this section.

3.3.2.2.3 Structural Properties
Synchrotron radiation provides a source of tunable X-rays allowing both electronic and structural aspects of materials to be examined in great detail [75, 76]. The use of high-area carbon supports, which are virtually transparent to X-rays of sufficiently high energy, has made it possible to monitor *in situ* changes in the local environment surrounding the metal center of adsorbed Fe porphyrins and phthalocyanines as a function of the applied potential [77]. Because of their high energy, X-rays can induce electronic transitions from orbitals deep in the core subject to the same selection rules as transitions in the UV–visible spectral range.

Shown in the insert in Figure 3.30 is the cyclic voltammogram of 40% w/w (FeTMPP)$_2$O adsorbed on Black Pearls. a high-area carbon of about 1000 m^2 g^{-1}, in 0.1 M NaOH, which displays clearly defined voltammetric peaks centered at -0.75 V versus SCE. The normalized fluorescence Fe K-edge X-ray absorption near edge structure (XANES), recorded *in situ* at 0.43 V, that is, positive to the voltammetric wave (see curve b), was found to yield a peak in the so-called pre-edge region, centered at 7112 eV, attributed to transitions from the 1s orbital to localized empty orbitals with p-character about the Fermi level. The same feature was also observed in the

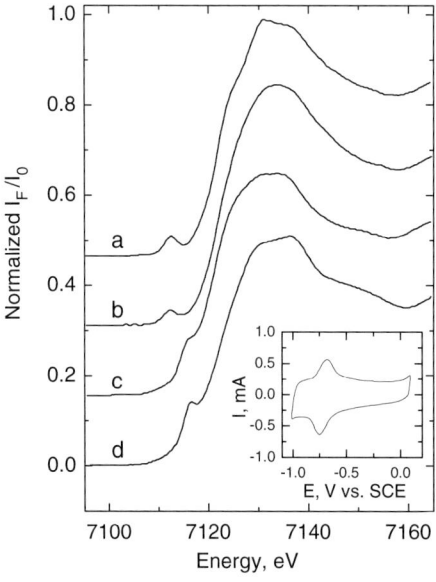

Figure 3.30 Normalized Fe K-edge fluorescence XANES of (a) microcrystalline (FeTMPP)$_2$O, (b) and (c) 40% w/w (FeTMPP)$_2$O on Black Pearls (BP) high-area carbon in 0.1 M NaOH in the original oxidized state, 0.43 V and in the reduced state, -0.90 V versus SCE, respectively, recorded *in situ*; (d) microcrystalline FePc. Insert: cyclic voltammetry (5 mV s^{-1}) of 40% (FeTMPP)$_2$O|BP in the same solution.

corresponding XANES of pure, microcrystalline (FeTMPP)$_2$O (see curve a), which strongly suggests that the local environment of the ferric centers, and presumably the structure of (FeTMPP)$_2$O as a whole, is not affected significantly following adsorption on the carbon surface. Furthermore, the rather high intensity of this pre-edge feature is consistent with Fe being in a low symmetry environment, as would be expected for the Fe(III) centers being displaced away from the plane of the porphyrin. Polarization of the electrode at −0.9 V, that is, negative to the redox peaks, led to the disappearance of the pre-edge feature yielding, instead, a shoulder at 7115 eV (see curve c), which is very similar to that found in the *ex situ* Fe-edge XANES of pure microcrystalline Fe(II)Pc powder shown in curve d in this figure. It can thus be surmised, that reduction of adsorbed (FeTMPP)$_2$O leads, as expected, to its dissociation to render two monomeric Fe(II)TMPP species.

At energies above the absorption edge, the presence of neighboring atoms leads to oscillations in the absorption coefficient known as extended X-ray absorption fine structure or EXAFS. Individual components of this oscillatory signal can be extracted by Fourier transform techniques resulting in shells that represent interactions between the metal and neighbors of the same type. Each of these shells can then be individually analyzed to yield the number of and distance to equivalent neighboring atoms. Shown in Figure 3.31 are the *in situ* reduced Fe K-edge EXAFS functions, $k^3\chi(k)$ versus k (panel A), where k is the wavevector, recorded in the same solution as specified in Figure 3.30 for (FeTMPP)$_2$O (curve a) and its reduced counterpart (curve b). Also shown are their corresponding Fourier transforms (see B in Figure 3.32) [78], where the peak at r' at about 1.5 Å represents the Fe to nearest N shell, Fe–N. Analyses of these data were consistent with a decrease in the average distance between the iron center and the neighboring nitrogen atoms in the ring, d(Fe–N), upon reduction, from 2.08 ± 0.01 Å for the dimer, down to d(Fe–N) = 2.02 ± 0.01 Å, for the monomer, as would be expected for the metal center moving into the ring plane. In strongly acidic

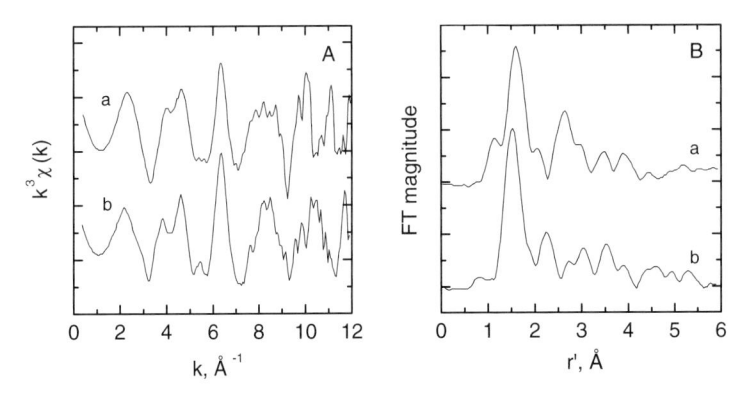

Figure 3.31 *In situ* reduced Fe K-edge EXAFS functions (A) and their corresponding Fourier transforms (B) for (FeTMPP)$_2$O adsorbed on Black Pearls high-area carbon recorded in aqueous solutions, pH = 13 polarized at a potential at which the adsorbate is present in its oxidized (dimer) (a) and reduced (b) (monomer) forms.

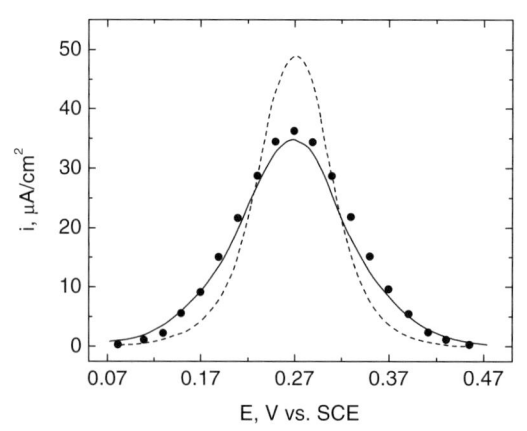

Figure 3.32 Background-corrected voltammetric feature of Co_2FTF4 adsorbed on carbon centered at $E = 0.27$ V versus SCE (solid line) and the calculated classical Nerstian response (dashed line, see text for details). The scattered points represent best fits to the data using the model proposed by Anson for a non-ideality parameter, $r\Gamma^* = -0.7$ [82].

solutions, the dimer spontaneously dissociates to yield monomeric ferric species in which the metal sits in the porphyrin ring plane. Nor surprisingly, $d(Fe-N)$ for the two redox species extracted from the *in situ* EXAFS were found in this instance to be virtually identical, that is, around 2.03 Å.

3.3.3
Redox Active Chemically Modified Electrodes

3.3.3.1 Thermodynamic Aspects

3.3.3.1.1 **Electrochemical Measurements** Under conditions in which the interactions among adsorbates can be neglected, and assuming that all adsorption sites are strictly identical (Langmuir isotherm), and that the electron transfer rates are very fast, the peak potential, E_p, and peak current density, i_p, observed in cyclic voltammetry are given, respectively, by:

$$E_p = E^{\circ\prime} - \left(\frac{RT}{nF}\right)\ln\frac{\beta_O \Gamma_O}{\beta_R \Gamma_R} \tag{3.15}$$

and

$$i_p = \frac{n^2 F^2}{4RT} v \Gamma^* \tag{3.16}$$

In these equations $E^{\circ\prime}$ is the formal potential of the adsorbed, redox active species, β_O/β_R is related to the difference in the standard Gibbs free energies of adsorption for O and R, Γ_i ($i =$ O or R) is the surface concentration of the adsorbed species, and Γ^* is the total surface concentration of the species regardless of oxidation state [39]. For most irreversibly adsorbed macrocycles, however, the shape of the redox peaks has been found to deviate from this classical behavior. An illustration of this effect is shown in Figure 3.32, which displays the background corrected voltammetric peak for a dimeric Co porphyrin, Co_2FTF4 [79], to be discussed later in this chapter (see solid line), and the predicted curve based on the classical Nernstian response assuming E_p to be the potential at the observed current maximum (dotted line). Several explanations have been offered to account for this effect, including site heterogeneity [80] and molecular reorientation induced by the change in oxidation state [81].

In particular, Anson has proposed the use of a modified Nernst equation by introducing a parameter r that accounts for interactions among the adsorbed species [82]. Application of this model yielded very good agreement with the experimental data for $r\Gamma^* = -0.7$ as shown by the scattered points in Figure 3.32.

3.3.3.1.2 Spectroscopic Measurements

UV–Visible Reflectance Correlations between the composition of a redox active irreversibly adsorbed layer and the applied potential can be obtained in some instances from the analysis of reflectance measurements. As shown in panel A, Figure 3.33 for Hm|HOPG(bp), the UV–visible reflectance (right ordinate) as a function of potential undergoes a large increase upon reduction and, correspondingly, a large decrease as the species is subsequently oxidized [83]. Plots of Q, obtained from coulometric analyses, versus $\Delta R/R$ were found to be linear (see panel C, Figure 3.33); hence, the potential at which $\Delta R/R$ is half of the total change corresponds to $E°$, or equivalently E_p.

This strategy allows for $E°$ to be extracted in cases in which the voltammetric features are ill defined, as illustrated for CoPc|GC in neat 0.2 M NaOH recorded at

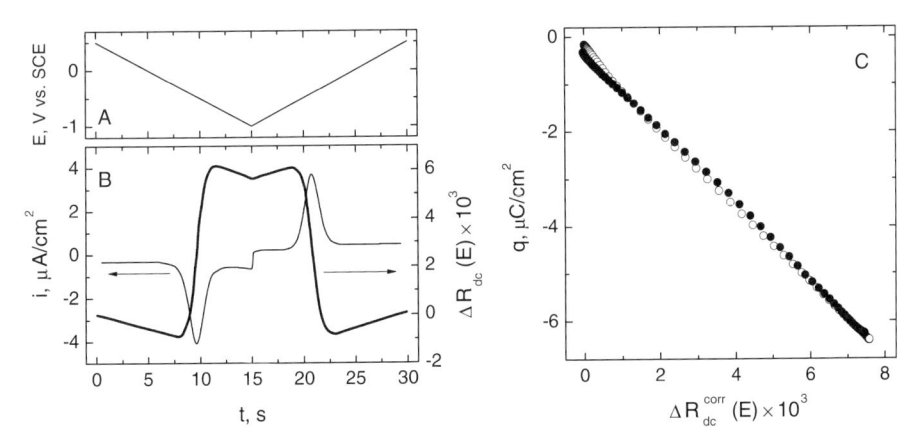

Figure 3.33 Simultaneous cyclic voltammetry ($v = 100\,mV\,s^{-1}$, left ordinate) and reflectance ($\lambda = 440\,nm$, smoothed, right ordinate) measurements displayed in the time domain for Hm|HOPG(bp) in 0.1 M $Na_2B_4O_7$ aqueous solutions (B). The potential program for the cyclic voltammetry is shown in (A). (C) Plots of charge determined from a coulometric analysis of the voltammetric peak as a function of the background corrected $\Delta R/R$ based on the optical data in (B). Solid and empty circles represent data extracted from the anodic and cathodic peaks, respectively.

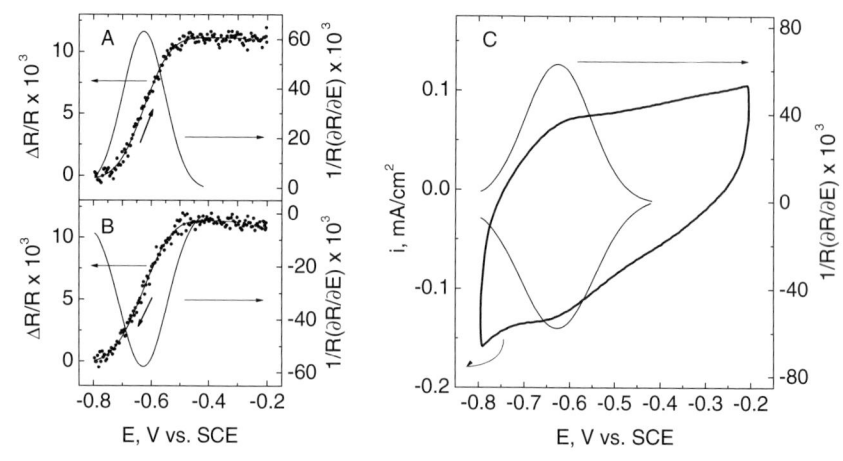

Figure 3.34 Plots of $\Delta R/R$ versus E (scattered points and best fit) for CoPc adsorbed on a GC electrode in neat 0.2 M NaOH recorded at $v = 200\,mV\,s^{-1}$ for the scans in the positive and negative directions, (A) and (B), respectively (left ordinates). The lines in these panels (right ordinates) are the derivatives of the fitted data shown along the scattered points. Shown in (C) is the cyclic voltammogram (left ordinate) and the normalized current contribution due to the redox activity of the adsorbed species (right ordinate) deduced from the optical data in the other two panels.

$v = 200\,mV\,s^{-1}$ (see thick line, left ordinate, Figure 3.34C). Also shown in thin lines in the left panels (right ordinates) in this figure are the derivatives of the fitted $\Delta R/R$ versus E functions for the scans in the positive and negative directions, which yielded curves proportional to the component of the current in the voltammogram that accounts for the oxidation or reduction of the adsorbed species. The contribution to the optical signal due to the bare substrate (not shown) is negligible [84].

Surface Enhanced Raman Scattering The extent of redox conversion as a function of the applied potential, E, can also be deduced from changes in the vibrational spectra as monitored by *in situ* SERS. Shown in Figure 3.35 are *in situ* SERS spectra recorded for Hm irreversibly adsorbed on a roughened Ag electrode in an aqueous electrolyte (pH = 3) in the range $0.25 \leq E \leq -0.55$ V versus SCE in 50 mV steps in ascending order in the regions 1140–1180 cm^{-1} (A) and 1470–1545 cm^{1} (B) [74]. Also provided in the insert in this figure is the rather ill-defined voltammogram of the Hm|Ag interface in the same electrolyte.

The spectral contributions of Hm and its reduced counterpart as a function of E were extracted from a statistical analysis of these data using as standards, normalized spectra collected at 0.25 V for Hm, and at −0.55 V for its reduced form. Specifically, the peaks in the range of from 1140 to 1180 cm^{-1}, and that at around 1504 cm^{-1}, shown in panels A and B in Figure 3.35 were selected for constructing plots of the coverage as a function of potential. To this end, the integrated areas under each of these features were assumed to be proportional to the actual amounts of adsorbed Hm present in their respective oxidation states. The results of this procedure are illustrated for the feature at 1504 cm^{-1} in Figure 3.36, which shows plots of the coverage of the reduced form of Hm versus E (see open circles). Also

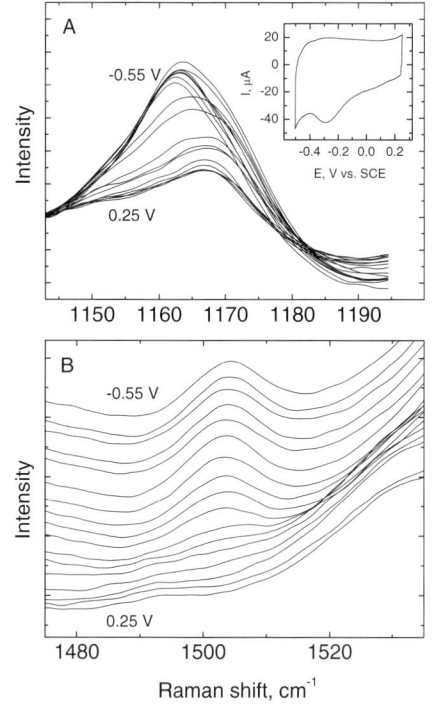

Figure 3.35 Spectra of hemin adsorbed on a roughened Ag electrode as a function of the applied potential in the potential range $0.25 \leq E \leq -0.55$ V versus SCE, in the 1140–1180 cm^{-1} (A) and 1470–1545 cm^{-1} (B) regions recorded in pH = 3 aqueous solutions, in ascending order. The cyclic voltammogram of Hm|Ag in this electrolyte is shown in the insert.

displayed in this figure is a statistical fit to the experimental data (thin solid line), which upon differentiation produced a peak-shaped curve analogous to the expected, strictly redox-based, voltammetric response (thick solid line). The sum of the contributions due to both species was constrained to one. Correlations between coverage and potential obtained using this procedure (see full circles, Figure 3.36) and the corresponding fit and derivative (dashed lines in the same figure), were found to be in excellent agreement with those derived from the 1504 cm^{-1} feature, both in terms of its shape and peak position, around −0.11 V versus SCE, lending strong support to the validity of the overall method. It becomes evident from this analysis that the peak positions of the derivative curves may be ascribed to the $E°$ for the adsorbed Hm at this pH.

3.3.3.2 Redox Speciation

In direct analogy with the behavior found in solution phase, redox active irreversibly adsorbed macrocycles involving iron and cobalt centers also display pH dependent voltammetric features, as shown in Figure 3.37 for FeTsPc (solid circles) and CoTsPc (open circles) [64].

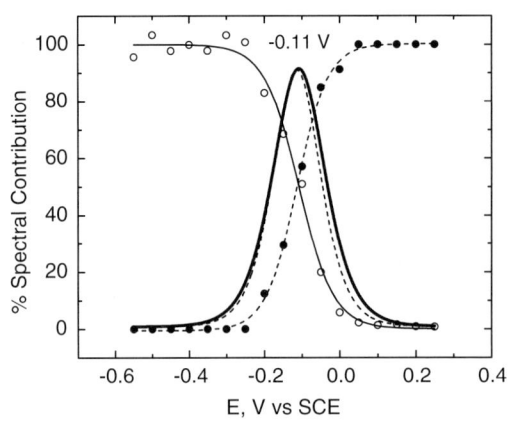

Figure 3.36 Plots of the percentage spectral contributions of Hm (filled circles) and its reduced counterpart (empty circles) as a function of potential derived from a linear combination analysis and the integrated intensities of the peaks around 1165 and 1504 cm^{-1}, respectively. The peak-shaped curves in this figure represent the derivatives of the best fits to the experimental data for Hm (solid line) and its reduced counterpart (dotted line).

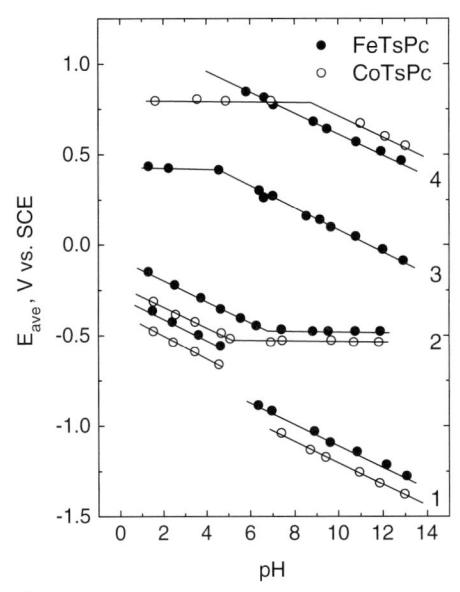

Figure 3.37 Dependence of the voltammetric peak potentials for CoTsPc|OPG (solid circles) and FeTsPc|OPG (in 10 µM FeTsPc, open circles) in N_2-saturated aqueous solutions as a function of pH. The labels 1–4 refer to the voltammetric features for FeTsPc in Figure 3.23 from negative to positive potential values.

Evidence in support of peak 3 for FeTsPc|OPG involving a metal-based orbital was obtained from the Fe K-edge XANES recorded *in situ* for the closely related FePc adsorbed on Ketchen Black (KB) high-area carbon in 0.5 M H_2SO_4 [85]. As shown in Figure 3.38, the absorption edge of Fe(II)Pc (recorded at the potential labeled as a in the insert) shifted significantly toward higher energies upon (a one-electron) oxidation (b) to yield formally Fe(III)Pc. These changes are very similar to those found for FeTMPP discussed above. In contrast, only very minor differences could be discerned between the *in situ* Fe K-edge XANES recorded at potentials a and c in the insert, which strongly suggests that the microenvironment around the metal center remains unaffected by the change in oxidation state, and, therefore, that the corresponding peak 2 is ring-, rather than metal-based.

Yet another example of the nuances involved in the assignment of redox peaks is provided by face-to-face porphyrins incorporating Co centers (see, for example, L, in Scheme 3.2). Differential pulse voltammetry curves for Co_2FTF4 (see Section 3.4) adsorbed on a graphite electrode are characterized by two redox features centered at around 0.2 (peak 1) and 0.6 V (peak 2) (see curve a, Figure 3.39). Whereas peak 1 displays a

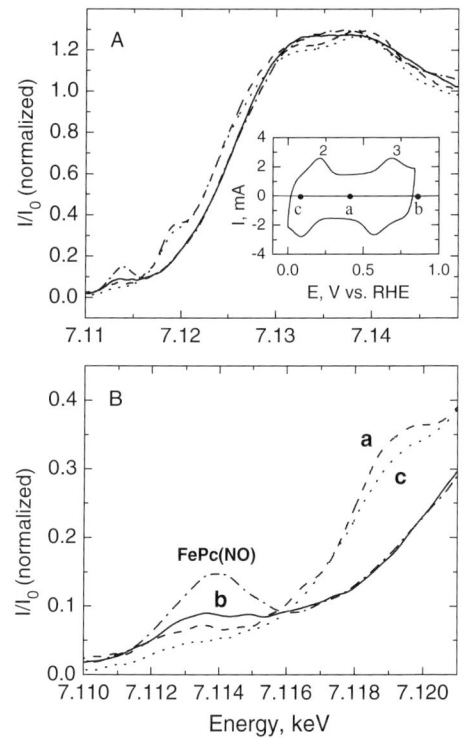

Figure 3.38 (A) *In situ* Fe K-edge XANES of FePc|KB in 0.5 M H_2SO_4 at 0.4 V (a, dashed line), 0.84 V (b, solid line), and 0.08 V (c, dotted line). (B) Same as (A) in an expanded scale. Insert: cyclic voltammogram of FePc/KB recorded at 10 mV s^{-1} in 0.5 M H_2SO_4, where a, b and c indicate the potentials at which XANES spectra were recorded.

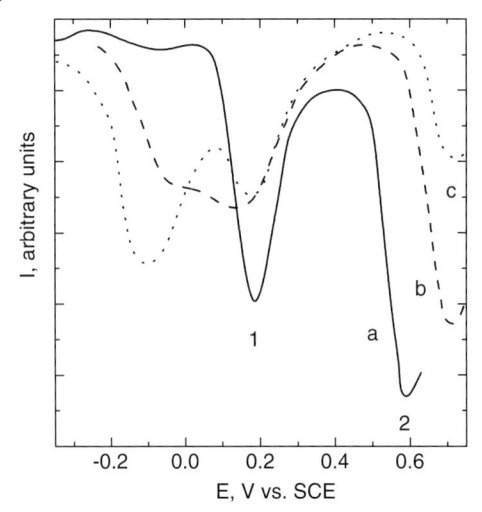

Figure 3.39 Differential pulse voltammograms of Co_2FTF4 adsorbed on a graphite electrode in aqueous solutions pH = 8 in the absence (solid curve a) and in the presence of 1 mM N-methyl imidazole (NMI) (dashed curve b) and 10 mM NMI (dotted curve c).

60 mV per pH slope, peak B remains invariant as the pH was changed [86] (not shown in the figure). Addition of 1 mM N-methyl imidazole (NMI), an N-based strong Lewis base, led to the decrease in the magnitude of peak 1 and to the emergence of a new feature at more negative potentials (curve b), which became more pronounced as the concentration of NMI was increased to 10 mM (curve c). This effect is consistent with the axial coordination of NMI to the metal center and provides clear evidence that the orbitals involved have predominantly metal character. In contrast, the changes in the peak potential of peak 2 following NMI additions were significantly smaller, which suggests that the orbitals involved are mainly localized on the ring.

3.3.3.3 Redox Dynamics

The dependence of the rate of elementary electron transfer reactions on the applied potential, E, is governed by the Butler–Volmer equation. For irreversibly adsorbed redox active species, this rate can be expressed, without loss of generality, in terms of the surface concentration of the reduced form of the species, Γ_{red}, as follows:

$$\frac{d\Gamma_{red}}{dt} = k_s[(\Gamma^* - \Gamma_{red})\exp\{-\alpha n_a F(E-E^{\circ\prime})/RT\} \\ - \Gamma_{red}\exp\{(1-\alpha)n_a F(E-E^{\circ\prime})/RT\}] \quad (3.17)$$

where k_s is the heterogeneous electron transfer rate constant in s^{-1}, α the transfer coefficient and n_a is the apparent number of electrons transferred, a factor that accounts for the non-Langmuirian character of the adsorbate (see above). Although various methods have been described for determining k_s, the optical properties of macrocycles render measurements based on absorption of light particularly useful.

As illustrated earlier in this section, the intensity of the reflected light can be used as a measure of the extent of conversion of one irreversibly adsorbed redox species into another. Advantage has been taken of this effect to monitor the rates of electron transfer using potential modulation reflectance techniques. As described by Feng and Sagara [87], a sinusoidal voltage perturbation of small amplitude, $\Delta E_{ac} \sin \omega t$, where ω is the frequency, is applied to the cell, generating, in turn, both an electrical and an optical response of the same frequency, for which the amplitude and phase are related to parameters intrinsic to the cell, including the interface of interest. The electrical response is often represented in terms of the real (Re) and imaginary (Im) components of a complex function that depends on ω, known as the impedance, Z, or its inverse, the admittance, Y. More specifically, for an applied potential:

$$E = E_{DC} + \Delta E_{AC} \sin \omega t \tag{3.18}$$

where E_{DC} is the DC electrode potential, the intensity of light reflecting off the electrode surface, R_t, can also be expressed in terms of DC and AC components, namely:

$$R_t = R_{DC} + \Delta R_{AC} \sin(\omega t - \varphi) \tag{3.19}$$

where φ is the phase shift. It thus follows that $\varphi \neq 0$, unless the process is infinitely fast. In practice, analysis of the experimental data relies on an equivalent circuit representation of the cell. For the system in question, this circuit is shown Figure 3.40, where, R_s and C_d are the resistance due to the electrolyte and the double layer capacity,

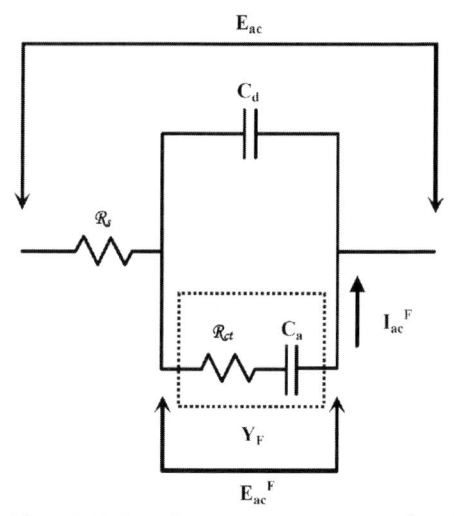

Figure 3.40 Equivalent circuit representation of an electrode incorporating an adsorbed redox active species (see text for notation). The components responsible for the ER (electroreflectance) response are shown within the dotted frame, where Y_F is the interfacial admittance, and the subscript F refer to the current and potential associated with the faradaic or interfacial modulation.

respectively, and R_{ct} and C_a represent the charge transfer resistance and capacitance associated with the adsorbed redox active species, respectively, namely:

$$R_{ct} \frac{2RT}{nn_a F^2 k_s \Gamma^*} \tag{3.20}$$

$$C_a = \frac{nn_a F^2 k_s \Gamma^*}{4RT} \tag{3.21}$$

where R is the gas constant (not to be confused with the reflectance). Based on these definitions:

$$k_s = (2R_{ct}C_a)^{-1} \tag{3.22}$$

Theoretical analysis of this equivalent circuit predicts that:

$$-\omega \cot \varphi = \frac{1-\omega^2 R_s R_{ct} C_a C_d}{R_{ct}C_a + R_s C_a + R_s C_d} \tag{3.23}$$

where $\cot \varphi$ is the ratio of the imaginary and real parts of the reciprocal of the *optical* impedance E_{ac}/R_{ac}.

Selection of the wavelength that provides optimized conditions for achieving the highest optical sensitivity is made based on plots of the difference in the intensity of the reflected light induced by a sinusoidal potential perturbation of small amplitude centered about the formal potential for the reduction of the adsorbed species, ΔR, normalized by the intensity of the reflected light, R, as a function of the wavelength. This quantity, also denoted as $\Delta R/R$, represents the difference between the interfacial spectra of the reduced and oxidized species. In the case of Hm adsorbed on GC recorded in 0.5 M NaF and 13 mM phosphate buffer at pH = 6.85, for which $E^{\circ\prime} = -0.31$ V versus Ag/AgCl, $\Delta R/R$ reached the desired maximum at $\lambda = 433$ nm (see Figure 3.41).

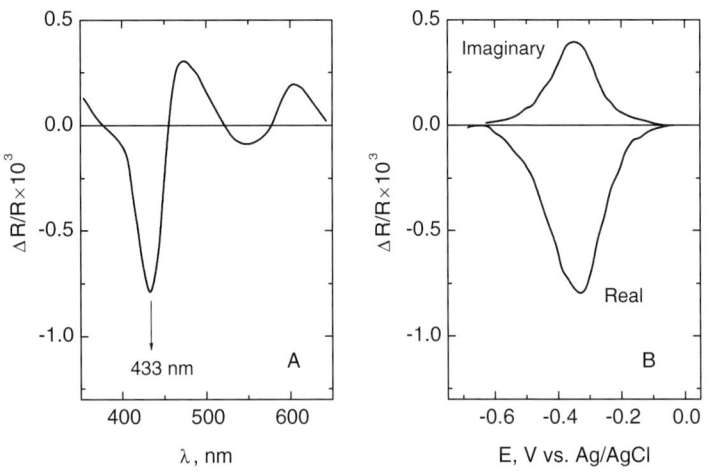

Figure 3.41 ER spectrum [(A) $E_{dc} = -0.31$ V] and ER voltammogram [(B) $v = 2$ mV s^{-1}; $\lambda = 433$ nm] for an Hm modified GC electrode in 0.5 M HF + 30 mM phosphate buffer (pH = 6.85) for $\omega = 14.27$ Hz and $\Delta E_{ac} = 14.5$ mV.

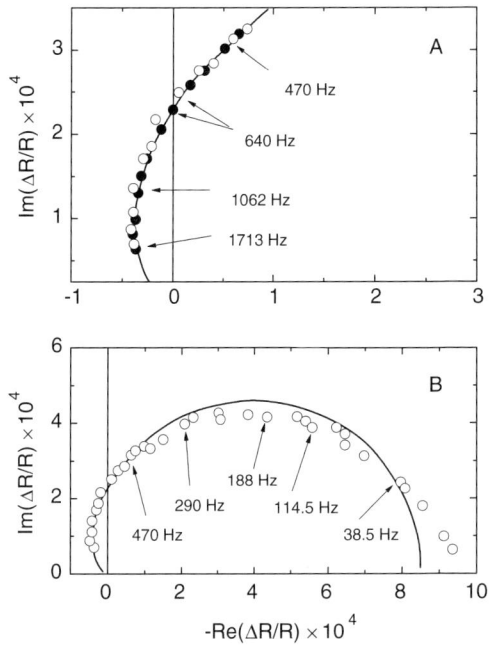

Figure 3.42 Complex plane plots of the ER response of the Hm|GC electrode in the solution specified in Figure 3.42 in the high frequency (A) and wide frequency (B) regions for $E_{dc} = -0.325$ V, $\Delta E_{ac} = 10.25$ mV and $\lambda = 433$ nm. Open circles are experimental values and solid circles in (A) and full lines in (A) and (B) represent best fits to the model.

Shown in Figure 3.42, are plots of the Im versus the Re components of $\Delta R/R$ (where R refers in this case to the reflectance) or Nyquist plots, recorded at $\lambda = 433$ nm, where the values specified identify the frequencies at which the data were collected. The solid line in this figure represents the best fit to the data based on the equivalent circuit proposed. As predicted by Equation 3.23, a plot of $-\omega \cot \varphi$ versus ω^2 was linear (see Figure 3.43 on page 242), yielding for $\mathcal{R}_s = 10.9\ \Omega\ \text{cm}^2$ and $C_d = 55.2\ \mu\text{F cm}^{-2}$, a value for k_s of $4.9 \times 10^3\ \text{s}^{-1}$.

3.3.4
Electrocatalytic Aspects of Dioxygen Reduction

3.3.4.1 Theoretical Considerations

The vast majority of irreversibly adsorbed macrocycles promote dioxygen reduction via the sequence of steps specified in Scheme 3.7, known as electrochemical–chemical–electrochemical, or ECE mechanism.

One of the primary aims from an experimental viewpoint is to identify the rate determining step and implement methodologies that can afford values for the kinetic rate constants involved. The mathematical analysis of data collected with the RDE and

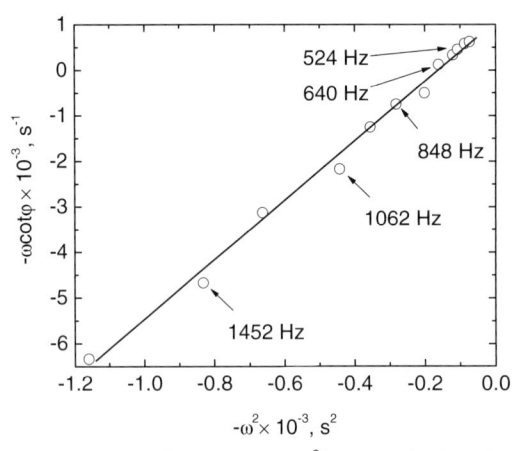

Figure 3.43 Plot of $-\omega\cot\varphi$ versus ω^2 based on the data shown in Figure 3.42. Open circles are experimental values, and the full line represents the best fit to the data.

RRDE is significantly simpler than with conventional cyclic voltammetry data in quiescent solutions [88, 89]. As such, these forced convection systems have been widely used in the study of electrocatalysis in general. Of special interest are situations where the rate determining step is chemical (a) or electrochemical (B) (Scheme 3.7) [60]. In particular, for an RDE at steady state, the rate at which the reactant is depleted at the interface must be equal to the rate at which it is replenished from the solution via convective mass transport. For a reaction first order in dioxygen this relationship reads:

$$k_2[O_2^*] = \frac{D_{O_2}}{\delta}\{[O_2^\infty]-[O_2^*]\} \tag{3.24}$$

where $[O_2^*]$ represents the concentration of dioxygen at the interface. Upon rearrangement and further substitution, Equation 3.24 leads to the so-called Koutecky–Levich (KL) equation:

$$\frac{1}{i} = \frac{1}{i_{Lev}} + \frac{1}{nFk_2[O_2^\infty]} \tag{3.25}$$

$M(III)P^+ + e^- \rightarrow M(II)P$ A
$M(II)P + O_2 \rightarrow M(II)PO_2$ a
$M(II)PO_2 + e^- \rightarrow M(II)PO_2^-$ B
$M(II)PO_2^- + ne^- \rightarrow$ products   C

Scheme 3.7 Electrochemical–chemical–electrochemical (ECE) mechanistic pathway for the reduction of O_2 promoted by an irreversibly adsorbed macrocycle.

where i_{Lev} is the diffusion limited current as defined in Equation 3.7, Section 3.2.1.1, and the term $nFk_2[O_2^\infty]$, often defined as i_{kin}, represents the current density that would be measured if $[O_2^*] = [O_2^\infty]$, that is, no mass transport limitations.

3.3.4.1.1 Case I. Electrochemical Rate-determining Step
For situations in which the rate determining step in Scheme 3.7 is the second electrochemical process (B), that is:

$$M(II)PO_2 + e^- \xrightarrow{k_3} M(II)PO_2^- \qquad (3.26)$$

where k_3 is the standard rate constant, i is given by:

$$i = nFk_3[M(II)PO_2]\exp\left[-\frac{\alpha F}{RT}(E-E°)\right] \qquad (3.27)$$

where n represents the total number of electrons transferred. Implied in this analysis is the fact that the overpotential is large enough for the second term in the corresponding Butler–Volmer expression to be negligible. Assuming K to be the equilibrium constant for the formation of the surface confined adduct, and replacing Equation 3.27 into Equation 3.25 may be shown to yield:

$$i = nFKk_3[M(II)P][O_2^\infty]\left[\frac{i_{Lev}-i}{i_{Lev}}\right]\exp\left[-\frac{\alpha F}{RT}(E-E°)\right] \qquad (3.28)$$

or, upon rearrangement:

$$\ln\left[\frac{i-i_{Lev}}{i}\right] = \ln\frac{i_{Lev}}{nFKk_3[M(II)P][O_2^\infty]} + \frac{\alpha F}{RT}(E-E°) \qquad (3.29)$$

Hence, at sufficiently negative potentials, so that the macrocycle will be present solely in its active reduced form, a plot of $\ln[(i-i_{Lev})/i]$ versus E (or Tafel plots) would be linear with a slope of $\alpha F/RT$, yielding, for $\alpha = 0.5$, a value of about 120 mV per decade.

3.3.4.1.2 Case II. Chemical Rate-determining Step
A second possible reaction pathway involves formation of $M(II)PO_2$ (see a in Scheme 3.7), that is:

$$M(II)P + O_2 \xrightarrow{k_2} M(II)PO_2 \quad \text{chem} \qquad (3.30)$$

as the rate limiting process. Although its rate constant, k_2, would not, in general, be potential dependent, the overall process will be a function of the applied potential through the surface concentration of $M(II)P$, as prescribed by the classical or modified Nernst equation. Provided the formation of the adsorbed adduct is first order both in the surface concentration of $M(II)P$ and in the concentration of O_2 at the interface, $[O_2^*]$, the corresponding Koutecky–Levich equation will read:

$$\frac{1}{i} = \frac{1}{i_{Lev}} + \frac{1}{nFk_2[M(II)P][O_2^\infty]} \qquad (3.31)$$

It thus follows that at sufficiently negative potentials, the kinetic current, i_{kin} and, hence, the measured current, i, will be a maximum, to be denoted as i_{kin}^{max} and i^{max}, respectively. If it is further assumed that (a) the kinetics associated with the redox process involving the bound catalyst are very fast so that the relative concentrations of its oxidized (inactive) and reduced (active) forms will be prescribed solely by the applied potential E through the generalized Nernst equation above and (b) the reaction is first order in the surface concentration of the active form of the electrocatalyst, the following can thus be shown:

$$\frac{i_{kin}}{i_{kin}^{max}} = \frac{\Gamma_{M(II)P}(E)}{\Gamma^*} = \theta_{M(II)P} \tag{3.32}$$

where $\Gamma_{Me(II)P}(E)$ represents the surface concentration of the adsorbed, redox active species M(II)P, and Γ^* is the surface concentration of the adsorbate regardless of its oxidation state, and, therefore, the ratio is, by definition, the coverage of M(II)P, $\theta_{M(II)P}$. Hence, assuming classical Nernstian behavior for the adsorbate:

$$E = E^{o\prime}_{surf} + \frac{RT}{nF} \ln\left(\frac{\Gamma^* - \Gamma_{M(II)P}}{\Gamma_{M(II)P}}\right) = E^{o\prime}_{surf} + \frac{RT}{nF} \ln\left(1 + \frac{i_{kin}^{max}}{i_{Lev}}\right)\left(\frac{i_{max}}{i} - 1\right) \tag{3.33}$$

and the potential observed at $i = i^{max}/2$, defined as $E^{O_2}_{1/2}$, or half-wave potential, will be given by:

$$E^{O_2}_{1/2} = E^{o\prime}_{surf} + \frac{RT}{nF} \ln\left(1 + \frac{i_{kin}^{max}}{i_{Lev}}\right) \tag{3.34}$$

It becomes evident from this last expression that $E^{O_2}_{1/2}$ will shift toward more positive values as i_{kin}^{max}, or equivalently, the magnitude of the rate constant of association k_2 is increased. It is conceivable that the half-wave potential for the oxygen reduction reaction can be shifted by as much as 100 mV from $E^{o\prime}_{surf}$; however, larger shifts are not likely, as a subsequent electron transfer step may become rate-limiting.

It is important to note that a single electrocatalytic interface may promote O_2 reduction predominantly via Case 1 in one potential range and via Case 2 in another. Indeed, illustrations of this effect will be given in the sections to follow.

3.3.4.2 Model Systems

Systematic studies of the role of such factors as the nature of the metal center and the detailed structure of the chelating ring, particularly its peripheral functionalization, can afford valuable information toward unveiling structure–activity relationships for macrocycles as electrocatalysts for oxygen reduction. The following sub-sections describe some of the most salient aspects of a selected number of transition metal phthalocyanines and porphyrins, including the effects of redox and non-redox active substituents on the properties of Co porphyrins.

3.3.4.2.1 Transition Metal Phthalocyanines

Water-soluble tetrasulfonated phthalocyanines incorporating transition metals of the first row, particularly, FeTsPc and CoTsPc, may be regarded as among the first adsorbed macrocycles for which the reaction mechanisms for oxygen reduction in aqueous electrolytes have been studied in depth using RDE and RRDE techniques [60, 90].

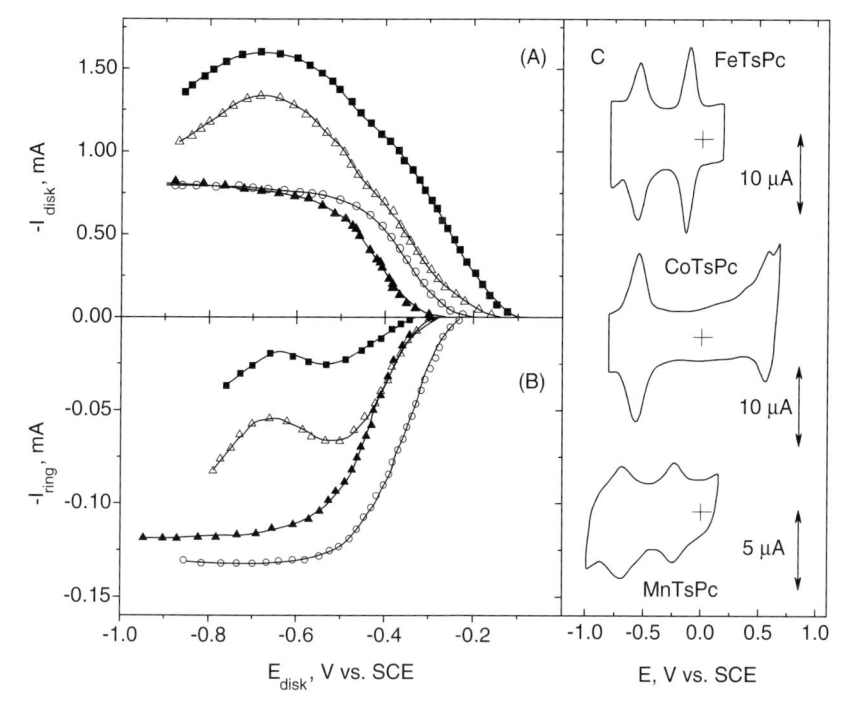

Figure 3.44 (A) Dynamic polarization curves ($v = 2\,\text{mV s}^{-1}$, $\omega = 1000\,\text{rpm}$) recorded in O_2-saturated 0.1 M NaOH, with a Pt|OPG RRDE ($A_{disk} = 0.5\,\text{cm}^2$, $N = 0.146$) with the OPG disk modified by a spontaneously adsorbed layer of FeTsPc (solid squares), MnTsPc (empty triangles), CoTsPc (empty circles) and NiTsPc (solid triangles). (B) Pt ring currents, I_{ring}, recorded simultaneously with the ring polarized at a potential positive enough to oxidize hydrogen peroxide generated at the disk under diffusion limited control. (C) Cyclic voltammetry curves ($v = 0.2\,\text{V s}^{-1}$) for FeTsPc|OPG (upper, the more positive peak is not shown), CoTsPc|OPG (middle) and MnTsPc|OPG electrodes (lower) recorded in N_2-saturated 0.1 M NaOH under quiescent conditions.

Shown in panel A, Figure 3.44 are dynamic polarization curves ($v = 2\,\text{mV s}^{-1}$) for spontaneously adsorbed FeTsPc (solid squares), CoTsPc (empty circles), MnTsPc, (empty triangles) and NiTsPc (solid triangles) on the OPG disk of a Pt|OPG RRDE recorded in O_2-saturated 0.1 M NaOH at $\omega = 1000\,\text{rpm}$. The corresponding symbols in panel B in this figure represent plots of I_{ring}, the current recorded at the Pt ring polarized at a potential positive enough to oxidize hydrogen peroxide generated at the disk under diffusion limited control, as a function of E_{disk}. Also included as inserts in panel C in this figure are cyclic voltammetry curves ($v = 0.2\,\text{V s}^{-1}$) for FeTsPc-, CoTsPc- and MnTsPc-modified OPG electrodes recorded in quiescent, N_2-saturated 0.1 M NaOH [90].

Several important observations can be gleaned from a cursory inspection of these data. Firstly, the onset potentials for O_2 reduction, $E^{O_2}_{onset}$, in addition to the magnitudes of i_{disk} for low to medium overpotentials, increase in the sequence Ni-, Co-, Mn- and FeTsPc|OPG. In particular, $E^{O_2}_{onset}$ for FeTsPc is about 0.3 V more positive than that found for OPG (not shown). Secondly, the values of $E^{O_2}_{onset}$ are close to the onset

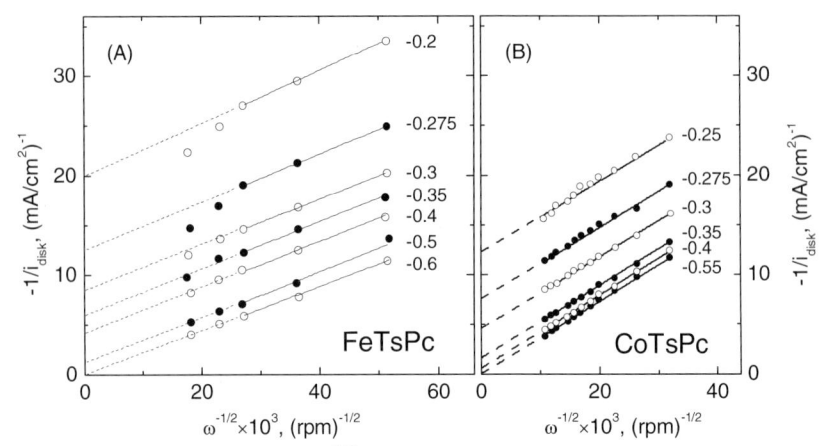

Figure 3.45 Plots of $1/i_{disk}$ versus $\omega^{-1/2}$ for FeTsPc|HOPG(bp) (A) and CoTsPc|OPG (B) RDE based on data collected in O_2-saturated 0.1 M NaOH at the specified potential values (V versus SCE). A 1 μM CoTsPc solution was added during acquisition of the data in (B) to prevent material from undergoing desorption.

potential for the reduction of the metal center (see the more positive peak in the voltammograms) for FeTsPc|OPG and MnTsPc|OPG, but much more negative (about 0.6 V) than that for CoTsPc|OPG. This behavior is generally, although not universally, found when comparing single-ring Fe and Co porphyrins (see below). Lastly, no peroxide can be detected at the Pt ring for FeTsPc|OPG and MnTsPc|OPG for small overpotentials, consistent with the reaction proceeding via a direct four-electron mechanism. At larger overpotentials, however, I_{disk} decreases and I_{ring} increases, signaling generation of solution phase peroxide. In contrast, CoTsPc|OPG and NiTsPc|OPG were found to yield large amounts of peroxide over the entire potential range examined, as is evident by the magnitudes of I_{ring}. In fact, plots of $1/i_{disk}$ versus $\omega^{-1/2}$ for $E > \approx -0.6$ V for FeTsPc|OPG and CoTsPc|OPG (see A, Figure 3.45), yielded, respectively, slopes very close to those predicted for a four- and two-electron process, using accepted values for D_{O_2} and the solubility of O_2 in this media.

In the case of FeTsPc|OPG, corresponding $1/i_{disk}$ versus $\omega^{-1/2}$ plots for $E_{disk} <$ 0.6 V, (not shown here) yielded higher slopes and substantial ring currents. As noted by Zagal et al. [60], this change in mechanism occurs close to the potential at which Fe(II)TsPc is further reduced (most negative peak in the upper insert, Figure 3.44) to yield a less efficient catalyst for the four-electron reduction of dioxygen compared with Fe(II)TsPc.

Further insight into mechanistic aspects was obtained from the Tafel plots (see Figure 3.46) as a function of pH. For low polarization ($E_{disk} > \approx 0$ V) a regime for which the reaction order in [OH$^-$] was -1, the Tafel slopes for FeTsPc|OPG increased from about 30 mV per decade in alkaline solutions to about 60 mV per decade for weakly acidic media (pH around 4.4). At higher polarization, $\log[i/(i_{Lev} - i)]$ reached a limiting value of about -0.1, independent of pH, pointing to a chemical process with a potential independent kinetics as rate controlling. At even higher polarization, the

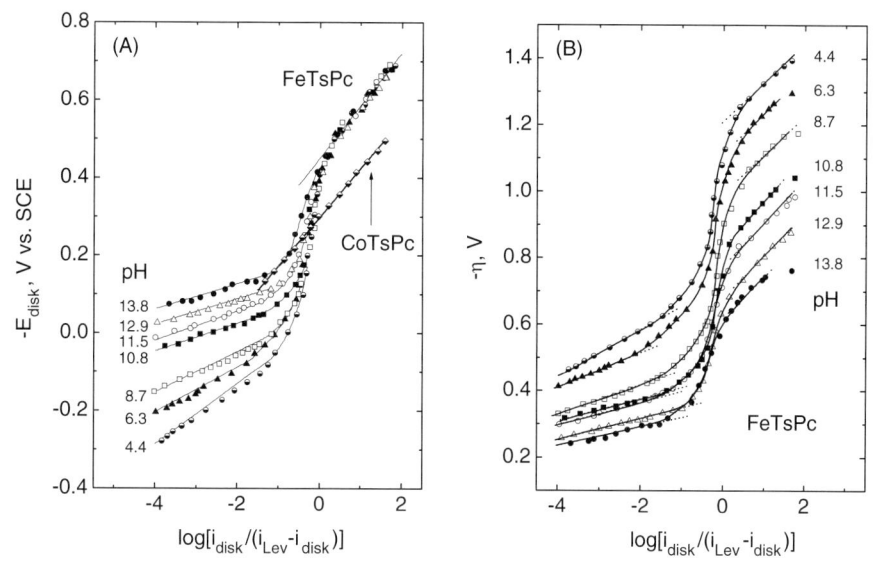

Figure 3.46 Tafel plots for (A) E_{disk} versus $\log[i_{disk}/(i_{Lev} - i_{disk})]$ and (B) η versus $\log[i_{disk}/(i_{Lev} - i_{disk})]$ for O_2 reduction on an FeTsPc|OPG RDE ($\omega = 3370$ rpm) at the specified pH values. Also shown in (A) is the corresponding Tafel plot for a CoTsPc|HOPG (bp) RDE at $\omega = 2920$ rpm in 0.1 M NaOH.

Tafel slopes (both for acid and base) yielded values of about 120 mV per decade consistent with a one-electron process as rate determining ($\alpha = 0.5$). In fact, the Tafel slope for CoTsPc|HOPG(bp) over the potential range specified in A, Figure 3.46, yielded the same value, indicating a common mechanistic pathway. It is worth noting that for FeTsPc|OPG, E_{disk}, for constant i_{disk} (and ω), becomes more negative as the pH increases in the low polarization region, $E_{disk} > -0.2$ V. However, the same data plotted as overpotential, η, versus $\log[i/(i_{Lev} - i)]$ (B, Figure 3.46), underscores the fact that the overall reduction of O_2 becomes more difficult as the pH decreases. The essential mechanistic aspects of this process in both acid and in basic media for high and low polarization are summarized in Table 3.2.

3.3.4.2.2 **Porphyrins** Much of the information regarding the electrocatalytic activity of metal porphyrins for O_2 reduction in aqueous electrolytes originates from the group of Anson at Caltech, who examined, fairly systematically, the behavior of a number of *meso*- and β-pyrrole substituted species. Shown in Figure 3.47 are cyclic voltammograms ($v = 20$ mV s^{-1}, lower panels) and steady state RRDE polarization curves (upper panels) for coatings of FeTPP (A), FePPIX (B) and FeTPyP (C) on OPG in deaerated 0.1 M HClO$_4$ [58]. Corresponding data for CoTPP|OPG (A), CoPPIX|OPG (B) and CoTPyP|OPG (C) in deaerated 1.0 M HClO$_4$ + 0.1 M NaClO$_4$ solutions [63] are given in Figure 3.48 (see captions for details). To ease the notation, and unless otherwise noted, the macrocycle|substrate interface will be represented by the formula ascribed to the macrocycle, for example, CoTPyP|OPG ≡ CoTPyP.

Table 3.2 Proposed mechanism and Tafel slopes for the reduction of dioxygen mediated by adsorbed FeTsPc on carbon for low and high polarization as a function of pH. For simplicity the acid-base properties of hydrogen peroxide were not considered.

Polarization	pH	Mechanism	Tafel slope (mV per decade)
Low	Base	$Fe(III)TsPcOH + e^- \rightleftarrows Fe(II)TsPc + OH^-$ *fast* $Fe(II)TsPc + O_2 \xrightarrow{k_2} Fe(III)TsPcO_2^-$ *fast* $Fe(III)TsPcO_2^- + e^- \rightarrow Fe(II)TsPcO_2^-$ *slow* $Fe(II)TsPcO_2^- + ne^- \rightarrow$ products *fast*	35
	Acid	$Fe(III)TsPc + e^- \rightleftarrows Fe(II)TsPc$ *fast* $Fe(II)TsPc + O_2 \xrightarrow{k_2} Fe(III)TsPcO_2^-$ *slow* $Fe(III)TsPcO_2^- + H^+ + e^- \rightarrow$ intermediate *fast*	60
High	All	$Fe(II)TsPc + O_2 \xrightarrow{k_2} Fe(III)TsPcO_2^-$ *fast* $Fe(III)TsPcO_2^- + e^- \rightarrow Fe(II)TsPcO_2^-$ *slow* $Fe(II)TsPcO_2^- + ne^- + H^+ \rightarrow Fe(II)TsPc + O_2H^-$ *fast*	120

As evident from a comparison between the voltammetric data in the lower panels in Figures 3.47 and 3.48, the Fe porphyrins display better defined metal-based redox peaks compared with their Co counterparts. Furthermore, and in analogy with FeTsPc, $E_{onset}^{O_2}$ for all Fe porphyrins (solid circles in Figure 3.47) was slightly more

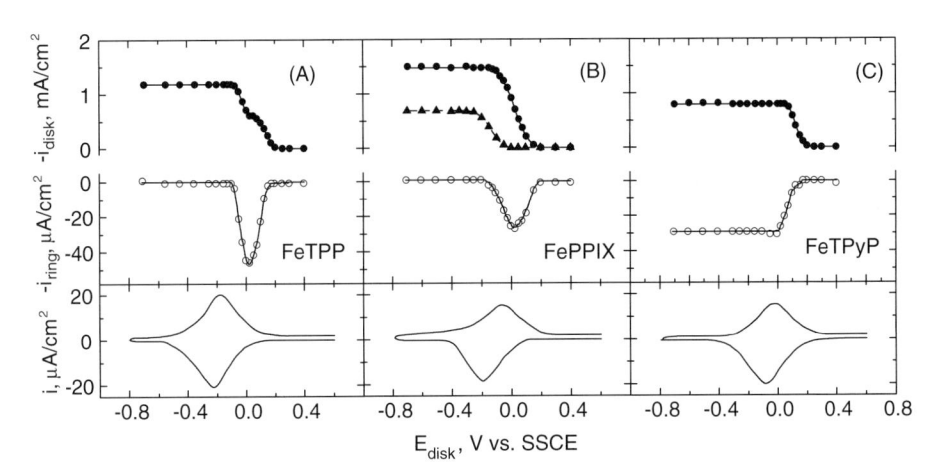

Figure 3.47 Upper panels: steady state polarization curves recorded with a Pt|OPG RRDE ($\omega = 100$ rpm) for coatings (6×10^{-8} mol cm^{-2}) of FeTPP (A), FePPIX (B) and FeTPyP (C) (see solid circles) deposited on the OPG disk in aqueous O_2-saturated 0.1 M $HClO_4 + 0.1$ M $NaClO_4$, where each of the points was collected for a freshly prepared coating. The data in empty circles are ring currents recorded for $E_{ring} = 1.2$ V versus SSCE. The triangles in (B) represent steady state polarization data acquired with a Co-porphyrin modified carbon paste (ring)|FePPIX coated OPG disk RRDE ($\omega = 100$ rpm) in a 1 mM H_2O_2 solution in the same electrolyte specified above. Lower panels: cyclic voltammograms ($v = 20$ mV s^{-1}) for OPG disk electrodes modified with the same coatings in the deaerated electrolytes.

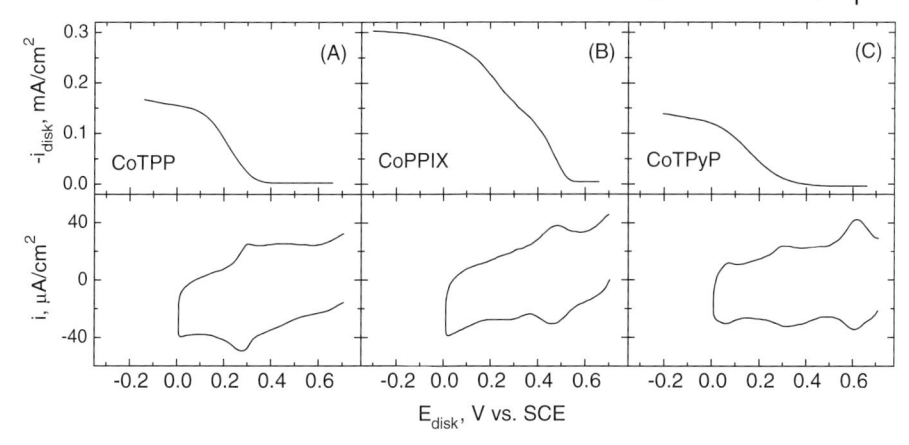

Figure 3.48 Upper panels: dynamic polarization curves ($v = 5\,\text{mV s}^{-1}$) recorded with an edge plane PG RDE ($\omega = 100\,\text{rpm}$) electrode coated with CoTPP (A), CoPPIX (B) and CoTPyP (C) in air-saturated 1 M $HClO_4$ + 0.1 M $NaClO_4$ aqueous solutions. Lower panels: cyclic voltammograms ($v = 50\,\text{mV s}^{-1}$) for the same modified PG disk electrodes in the same deaerated electrolyte.

positive than, or coincident with, the onset of the metal reduction, and vastly more positive than that found for bare OPG (not shown here). This behavior is similar to that observed for the β-pyrrole substituted CoPPIX (Figure 3.48B) for which $E^{O_2}_{\text{onset}}$ was more positive, about 0.5 V versus SCE, than for FePPIX. As was the case with CoTsPc, $E^{O_2}_{\text{onset}}$ values for the *meso*-substituted Co porphyrins are negative to the onset of the Co(III)|Co(II) transition. Also worth noting is the fact that for FeTPP and CoPPIX, O_2 reduction occurs in two rather distinct steps, whereas for all other supported porphyrins only one step is apparent.

Measurements performed with FePPIX as a function of pH yielded similar trends to those found for FeTsPc, that is, $E^{O_2}_{\text{onset}}$ shifted to more negative values as the solution became more alkaline. Plots of i_{lim} as a function of the coverage of FePPIX, Γ_{FePPIX}, recorded at two different values of ω (see Figure 3.49) were found to be close to linear for $\Gamma_{\text{FePPIX}} < \Gamma^*_{\text{FePPIX}}$ (about $0.7\,\text{nmol cm}^{-2}$) and reached a plateau for $\Gamma_{\text{FePPIX}} \geq \Gamma^*_{\text{FePPIX}}$, which indicates that only the outermost layer of the catalyst is electrocatalytically active.

Plots of i_{lim} (or second current plateau when applicable) versus $\omega^{1/2}$ for all three Fe porphyrins in Figure 3.47, (see Figure 3.50A), and also of the first plateau for CoPPIX (empty circles, Figure 3.50C) were found to bend downward at high ω, that is, away from the otherwise linear character predicted by the Levich equation. Koutecky–Levich (KL) plots, that is, $1/i_{\text{lim}}$ versus $\omega^{-1/2}$, (see panels on the right, Figure 3.50) for FeTPP and FePPIX (see triangles and solid circles, respectively, in B), and the β-pyrrole substituted CoPPIX (see solid circles, D) based on the data in A in this figure, were linear with slopes consistent with the four-electron reduction ($n = 4$) of O_2. For FeTPyP (squares, B) and the second plateau for CoPPIX (solid circles, D), however, n was found to be slightly larger than 2, that is, the reduction proceeds

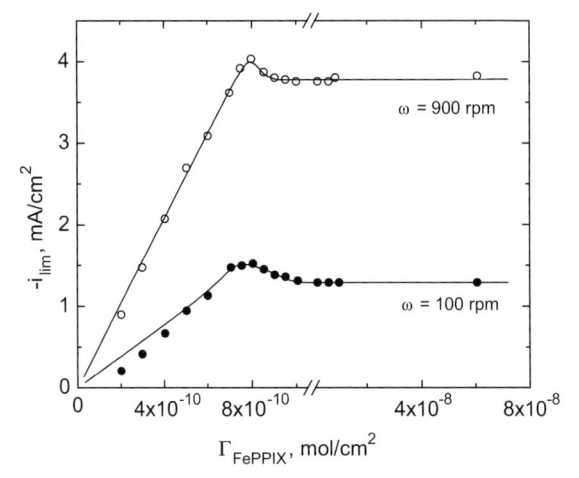

250 | *3 Transition Metal Macrocycles as Electrocatalysts for Dioxygen Reduction*

Figure 3.49 Plots of the limiting current, i_{lim}, for O_2 reduction at an FePPIX|carbon RDE in O_2-saturated 0.1 M $HClO_4$ in 0.1 M $NaClO_4$ aqueous solutions as a function of Γ_{FePPIX} for two rotation rates.

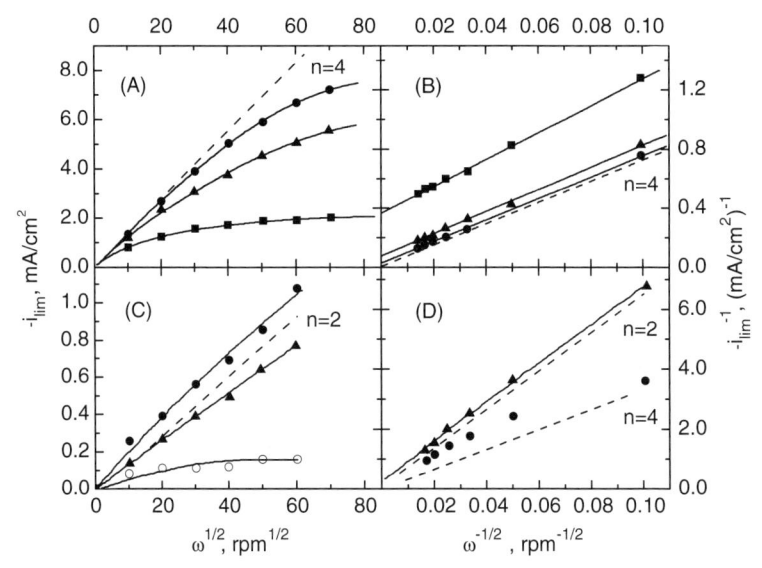

Figure 3.50 Levich (left panels) and Koutecky–Levich (KL) plots (right panels) for the reduction of O_2 (second plateau, see text) on graphite electrodes modified by Fe (upper panels) and Co (lower panels) porphyrin coatings (solid circles, PPIX; triangles, TPP; squares, TPyP; empty circles, first current plateau for CoPPIX) specified in the captions to Figures 3.47 and 3.48, respectively.

Table 3.3 Dependence on pH of the reduction of dioxyen and hydrogen peroxide mediated by selected iron porphyrin coatings on carbon electrodes.[a]

Porphyrin	pH	E^f (V vs. SSCE)	O$_2$ reduction			H$_2$O$_2$ reduction		
			$E_{1/2}$ (V vs. SSCE)	i_{kin} (mA cm^{-2})	n_{app}	$E_{1/2}$ (V vs. SSCE)	i_{kin} (mA cm^{-2})	n_{app}
FePPIX	12	−0.68	−0.48	b	4	−0.38	b	2
	9.8	−0.54	−0.38	b	4	−0.36	b	2
	7.0	−0.37	−0.29	b	3.9	−0.32	b	1.9
	5.0	−0.32	−0.25	39	3.8	−0.26	17	2
	3.2	−0.17	−0.05	39	3.8	−0.11	17	1.9
FeTPP	12	−0.65	−0.37	27	4	−0.39	12	1.9
FeTPyP	12	−0.65	−0.37	9	4	−0.39	7	2

[a] Other details may be found in the original reference.
[b] Intercepts of the KL plots were too small to measure [58].

beyond the two-electron step. In contrast, the same analysis for Co porphyrins functionalized by bulky *meso*-substituents, such as CoTPP (see triangles, D, Figure 3.50) and CoTPyP (no data shown here), yielded $n = 2$, that is, solution phase peroxide is the only product generated. A compilation of formal redox potentials for FePPIX, and half-wave potentials, kinetic currents and values of n determined from the intercepts and slopes of the corresponding KL plots for O$_2$ reduction are given in Table 3.3. Also listed therein are the corresponding parameters for FeTPP and FeTPyP for pH = 12.

Further insight into mechanistic aspects of O$_2$ reduction mediated by these Fe porphyrins was provided via the analysis of i_{ring}. As shown in Figure 3.47 (upper panels, open circles), peroxide is generated over the first wave, that is, small η, for FeTPP (A) and FePPIX (B). This behavior suggests that the two clearly defined plateaus found for FeTPP merge into a single feature for FePPIX. It can therefore be surmised that the bent character of the Levich plots for these two macrocycles are not the result of a change in the reaction stoichiometry, but are due to a chemical step that precedes the electrochemical process. In contrast, the closely related FeTPyP (C, Figure 3.47) yielded peroxide over the entire potential range in which O$_2$ was reduced; nevertheless, the fact that n for this macrocycle was found to be 3.2 (see Ref. [58]) indicates that the two-electron reduction is not quantitative, that is, it proceeds via a mixed reaction pathway. In fact, dioxygen was reduced via a two- and a four-electron pathway in about equal proportions over the entire range of ω examined. It may be concluded for these results, that peripheral modifications to the ring structure can induce marked changes in electrocatalytic activity.

Yet another important aspect that can shed light on the mechanism of dioxygen reduction is the ability of the adsorbed catalyst to reduce and/or chemically decompose hydrogen peroxide. For example, CoTPP and CoTPyP (and certainly bare OPG) display no activity for hydrogen peroxide reduction ($n = 2$) in the potential region of relevance, whereas CoPPIX is only slightly active [63]. Hence, the large disk current observed with CoPPIX (Figure 3.48B) must be derived from the reduction of peroxide before it is desorbed. In contrast, the activity of all Fe porphyrins described so far in this section, for the electrochemical reduction of solution phase peroxide is indeed significant, as illustrated for FePPIX in Figure 3.47 (see triangles, B), a material that exhibits no activity for hydrogen peroxide decomposition at this pH (see Figure 3.5). The fact that the onset for peroxide reduction is more negative than $E^{O_2}_{onset}$ is in all likelihood responsible for the humps observed in i_{ring} in Figure 3.47. A few values of $E_{1/2}$, i_{kin} and n for the three Fe macrocycles in Figure 3.47 extracted from data collected in 1 mM H_2O_2 solutions devoid of O_2 are compiled in Table 3.3.

A reaction mechanism responsible for the results obtained for the Fe porphyrins is the same as that shown in Table 3.2 for FeTsPc in acid at low polarization. It must be emphasized, however, that in analogy with solution phase Fe(III)TMPyP (Section 3.2.2.1), no evidence has been found for the formation of adsorbed $FePO_2$ adducts, where P represents either TsPc or a porphyrin in aqueous electrolytes under ambient conditions. It may be argued that the fact that $E^{O_2}_{onset}$ is more positive than the onset of the redox transition at all pH values, may reflect the rapid reduction of the adduct to yield products, as theory would predict (see Equation 3.34); however, this would require the rate of the chemical step to be close to diffusion control values, and thus much higher than those determined based on the KL plots.

Among all the Fe and Co electrocatalysts introduced so far in this section, the *meso*-unsubstituted CoPPIX not only displays the most positive $E^{O_2}_{onset}$, but is also capable of promoting the reaction at least partially by a four-electron process. One possible explanation for this behavior may be found in the ability of two Co centers belonging to two porphyrins devoid of *meso*-substituents to face each other at a closer distance than porphyrins bearing a fairly bulky *meso*-group, such as phenyl or pyridyl. This lack of steric hindrance facilitates possible formation of Co—O—O—Co bridges, which may lead to the further activation and ultimate scission of the O—O bond, a concept that will be developed thoroughly in the sections to follow.

It should be noted at the outset that Ir(OEP) promotes the direct four-electron reduction of dioxygen in acid media with a value of $E^{O_2}_{onset} = 0.72$ V versus NHE in 0.1 M CF_3COOH, only 70 mV more negative than the corresponding value reported for Pt in the same electrolyte. Issues still remain to be resolved, however, regarding whether the electrocatalytically active species is a monomer or a dimer [91–93].

3.3.4.2.3 Systematic Studies of the Effect of Ring Functionalization on the Activity of Co Porphyrins for the Reduction of Dioxygen
The building platform for all porphyrins is a ring structure known as porphine shown in B, Scheme 3.2, and the study of its properties was expected to afford a baseline for systematic comparisons with its functionalized derivatives. Rather surprisingly, the electrochemical behavior of Co porphine, CoPn, in O_2-saturated aqueous 1 M $HClO_4$ was found to be unique among all single-ring Co porphyrins [94]. Specifically, its $E^{O_2}_{1/2}$ was the most positive of all species of this type, that is, 0.53 V versus SCE, with average n values ranging from 3.8 to 4.0 for low and high overpotentials, respectively. Small structural alterations, such as addition of methyl substituents in the *meso*-positions of CoPn to yield

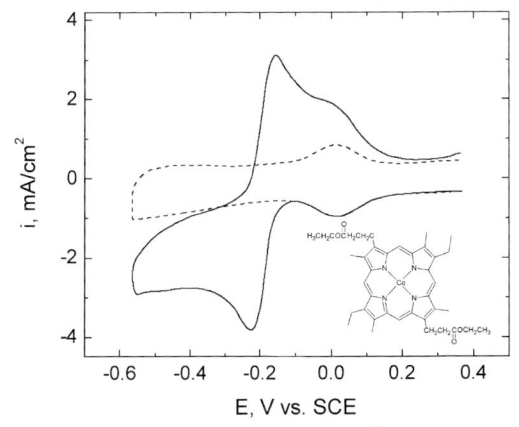

Figure 3.51 Cyclic voltammetry ($v = 1\,V\,s^{-1}$) of CoPI (see insert) modified graphite surface in neat (dotted line) and O_2-saturated (solid line) 1 M NaOH aqueous solutions. The surface concentration of CoPI as determined from coulometry was about $4 \times 10^{-10}\,mol\,cm^{-2}$.

5,10,15,20-tetramethyl porphyrin (CoTMP) led to shifts in $E_{1/2}^{O_2}$ of more than 100 mV in the negative direction, that is, a loss in performance, and to a drop in n down to 3.3 at low overpotentials [63]. Large decreases in n were also found for β-pyrrole substituted analogues, such as CoOEP and CoPPIX described in the previous section, for which the $E_{1/2}^{O_2}$ values were intermediate between those of CoPn and CoTMP. These observations provide additional support to steric factors as being essential to promoting formation of O_2 activating Co−O−O−Co bridges.

Among all the single Co porphyrins examined, a β-pyrrole substituted species bearing two ester groups, mentioned briefly in Section 3.1.2, (see insert, Figure 3.51) was found to display a rather unique behavior upon adsorption on carbon surfaces. This macrocycle, to be denoted as CoPI, is a precursor to the synthesis of the first face-to-face porphyrin examined in an electrochemical environment [95]; hence, a discussion of its properties serves to underscore the effects associated with changes in the electrocatalytic microenvironment following covalent dimerization (see below).

CoPI [95] Modification of graphite surfaces with CoPI, using the precipitation technique described in Section 3.3.2.1, yielded electrodes exhibiting fairly well defined voltammetric peaks over a wide pH range (see, for example, dashed lines in Figure 3.51 at pH = 14). A plot of the average redox potentials, E_{av}, versus pH (see Figure 3.52), revealed two clearly defined linear regions, with slopes consistent with the following redox processes:

$$Co(III)PI-OH_2 + e^- \leftrightarrow Co(II)PI-OH_2 \quad \text{for} \quad 0 < pH < 6 \tag{3.35}$$

$$Co(III)PI-OH + e^- + H^+ \leftrightarrow Co(II)PI-OH_2 \quad \text{for} \quad 6 < pH < 14 \tag{3.36}$$

The cyclic voltammogram of CoPI|OPG in O_2-saturated 1 M NaOH, shown as solid lines in Figure 3.51, was characterized by the occurrence of a well defined reversible

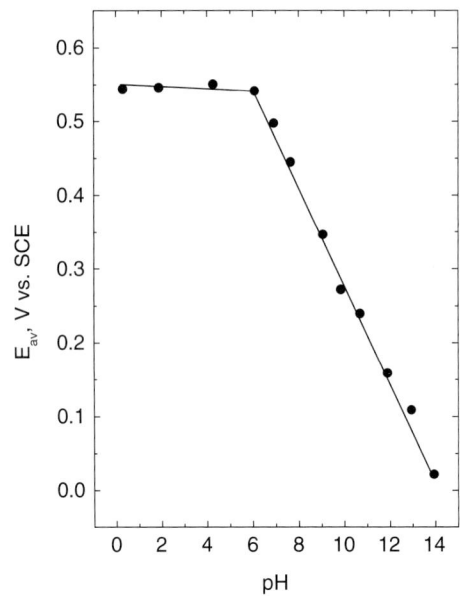

Figure 3.52 Plot of the average redox peak potential, E_{av}, versus pH for CoPI adsorbed on OPG.

redox peak with E_{av} at about -0.2 V, with $E_{onset}^{O_2}$ significantly more negative than the onset potential for the reduction of CoPI. Also worth noting is the fact that the reduction peak of the macrocycle was unaffected by the addition of O_2. These new voltammetric features can be attributed to the O_2–HO_2^- redox couple, for which the peak potential separation is close to the Nernstian (reversible) value of about 30 mV. In brief, adsorption of CoPI converts the otherwise fairly inactive bare OPG surface into an electrode reversible to the O_2–HO_2^- couple in alkaline solutions, a behavior similar to that displayed by bare Hg under the same conditions.

In analogy with the results found for the Fe porphyrins above, a plot of i_{lim} versus $\omega^{1/2}$ (see solid circles, left ordinate, lower abscissa, Figure 3.54B) based on the dynamic polarization curves collected with CoPI|OPG RDE (see Figure 3.54A), was found to bend downward, away from a strict two-electron Levich behavior (see dotted line in this figure). This behavior is consistent with the overall reaction rate being controlled by a chemical step that precedes the electron transfer process. Analyses of the linear KL plots (see open circles, right ordinate, upper abscissa, Figure 3.53B) yielded values for the rate constant for the reaction between O_2 and the adsorbed macrocycle in the range 1×10^5–2×10^5 $M^{-1} s^{-1}$, that is, very similar to the rates of coordination of dioxygen to a series of Co(II) complexes in aqueous solutions (see Section 3.2.2.1), and a slope with a magnitude in line with a 2e$^-$ reduction of dioxygen to yield peroxide as the product.

On this basis, the most likely mechanism for this reaction at high pH is given by the sequence of reactions in Scheme 3.8, where the formation of Co(II)PI–O_2 is the rate determining step. The fact that the reduction of O_2 occurs at potentials more negative

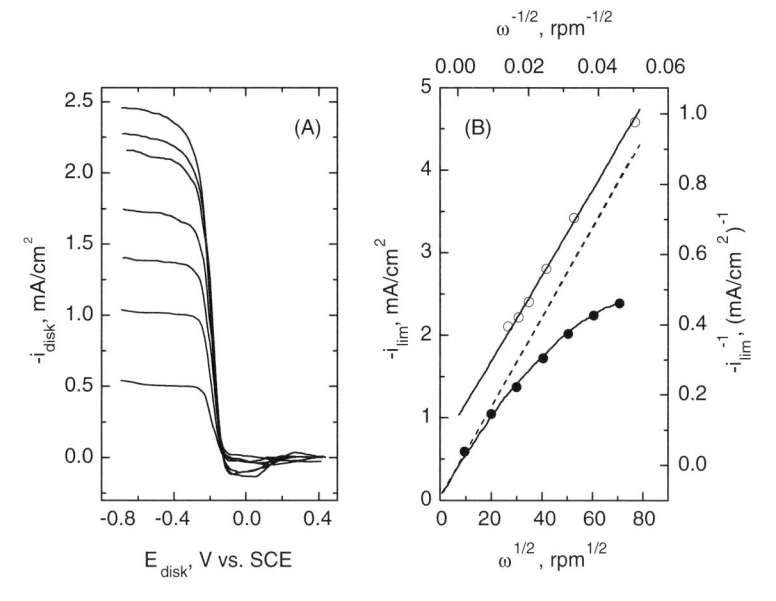

Figure 3.53 (A) Dynamic polarization curves ($v = 8.3$ mV s^{-1}) for a CoPI|OPG RDE in 1 M NaOH for $\omega = 100$, 400 (i_{lim} was adjusted to conform to the original corresponding value on the right panel), 900, 1600, 2500, 3600 and 4900 rpm in ascending order. (B) Levich (left ordinate, lower abscissa, solid circles) and Koutecky–Levich (right ordinate, upper abscissa, open circles) plots based on i_{lim} values from the polarization curves. The dashed line represent i_{lim} versus $\omega^{1/2}$ values calculated from the mass transport limited two-electron reduction of O_2.

than the reduction of adsorbed CoPI clearly shows that adduct formation is a necessary, but not sufficient condition for the catalytic cycle to be initiated.

As the pH is lowered, the overall rates decrease, that is, the reaction becomes less reversible, as evidenced by an increase in the peak separation in voltammetric curves recorded in quiescent solutions. Close inspection of the data (not shown here) indicates that this effect is due to a shift in the peak ascribed to hydrogen peroxide oxidation toward more positive values and not to a shift in $E^{O_2}_{onset}$.

Kinetic data extracted from the foot of RDE dynamic polarization curves for O_2 reduction yielded for pH < 11 linear $\log[i/(i_{lim} - i)]$ versus E, or Tafel plots, with a slope of around 120 mV per decade, and thus consistent with the first electron transfer as being rate determining for the reduction of the CoPI–O_2 adduct. As expected for a reversible (Nerstian) two-electron redox couple, the Tafel slope at pH = 14, decreased to 30 mV per decade. It is interesting to note that an oxidized form of the closely related CoOEP displays extraordinary reversibility for the O_2–H_2O_2 couple in solutions of pH < 1 [63].

Two approaches have been developed and implemented toward a rational optimization of Co porphyrin based electrocatalysts for the reduction of dioxygen. The first relies on the incorporation of metal complexes to either *meso-* or β-pyrrole positions of

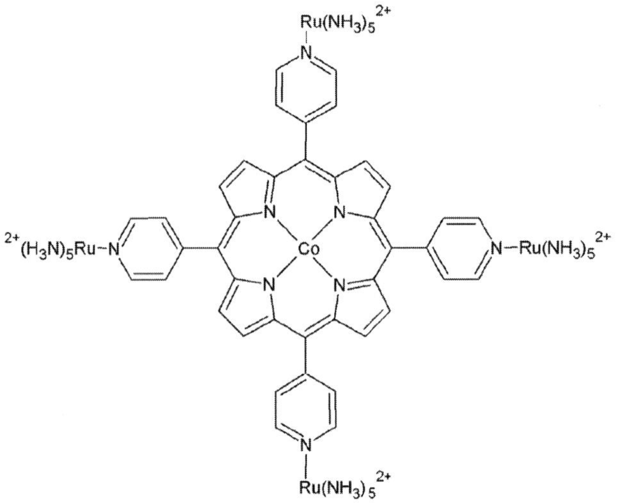

Figure 3.54 Tetra-Ru(NH$_3$)$_5$ N-substituted CoTPyP denoted as Ru(4)CoTPyP.

single ring Co porphyrins, whereas the second seeks to place metal centers of two different rings facing each other (see, e.g. L, Scheme 3.2) so as to better control the mode of bonding of dioxygen in bridge-type configurations. These will be discussed in some detail next.

Incorporation of Electron-donating Groups at Peripheral Sites of Single-ring Co Porphyrin
The electronic density at the Co(II) center can be increased by further functionalization of the phenyl or pyridyl moieties with groups that can donate electrons. This tactic appears to have been first proposed by Meyer and coworkers at the University of North Carolina [96], who attempted injection of multiple electrons into the coordinating metal center microenvironment, via reduction of redox groups attached to the periphery of the ring.

The most systematic study of the effects of addition of redox groups to the periphery of porphyrin rings were reported by Anson and coworkers [97]. Their approach involved coordination of [Ru(NH$_3$)$_5$(OH$_2$)]$^{2+}$, an electron donor species with a relatively positive redox potential (see below), to each of the pyridyl groups in CoTPyP (see Figure 3.55), a species denoted as Ru(4)CoTPyP, to allow, in principle, injection of four electrons *intramolecularly* into O$_2$ bound to the Co center to generate water as the product [98].

$$\text{Co(II)PI} + \text{O}_2 \rightleftharpoons \text{Co(II)PI-O}_2$$
$$\text{Co(II)PI-O}_2 + 2e^- \rightleftharpoons \text{Co(II)PI-O}_2^{2-}$$
$$\text{Co(II)PI-O}_2^{2-} + \text{H}^+ \rightleftharpoons \text{Co(II)PI-O}_2\text{H}^-$$
$$\text{Co(II)PI-O}_2\text{H}^- \rightleftharpoons \text{Co(II)PI} + \text{O}_2\text{H}^-$$

Scheme 3.8 Proposed mechanism for dioxygen reduction mediated by CoPI/OPG in alkaline solutions.

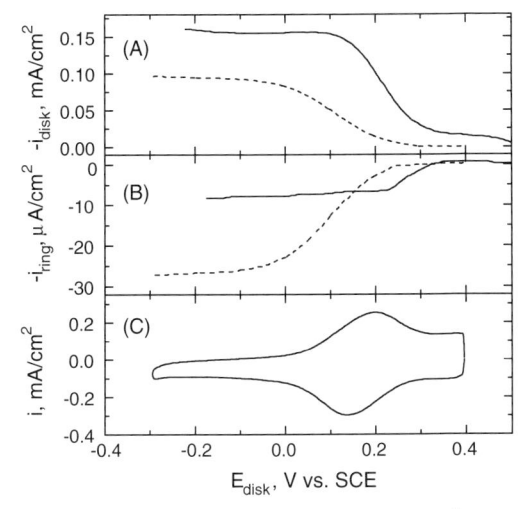

Figure 3.55 (A) Dynamic polarization ($v = 2\,\text{mV s}^{-1}$, $\omega = 100\,\text{rpm}$) curves for a CoTPyP-coated disk of a Pt|OPG RRDE ($N = 0.39$), in air-saturated 0.5 M NH_4PF_6 in 0.5 M $HClO_4$, before (dotted lines) and after (solid lines) ruthenation of the macrocycle. The corresponding ring currents ($E_{ring} = +1.0\,\text{V}$ versus SCE) are shown in (B). (C) Cyclic voltammogram of Ru(4)CoTPyP|OPG in 0.5 M NH_4PF_6 in 0.5 M $HClO_4$ ($v = 50\,\text{mV s}^{-1}$).

Shown in the Figure 3.55C is the cyclic voltammogram of Ru(4)CoTPyP adsorbed on the graphite disk of a Pt|OPG RRDE recorded under quiescent conditions in Ar-saturated 0.5 M NH_4PF_6 in 0.5 M $HClO_4$. The use of PF_6^- was required in this instance, in order to prevent dissolution of the complex. As indicated, Ru(4)CoTPyP|OPG displays a single well defined peak centered at 0.2 V versus SCE with a charge equivalent to four times the number of CoTPyP species in the coating [97]. The presence of a single peak implies that all redox sites are identical and behave independently.

As evident from the i_{disk} and i_{ring} data collected with a CoTPyP|Pt RRDE in the same electrolyte following air saturation (see dashed lines in A and B, Figure 3.55), ruthenation of the macrocycle (solid lines) brings about an increase in i_{disk}, a sizable decrease in i_{ring}, and an overall shift in $E_{onset}^{O_2}$ toward more positive potentials [97]. Of particular note is the fact that the onset for the reduction of the redox groups appears to coincide with $E_{onset}^{O_2}$ for Ru(4)CoTPyP|OPG. In analogy with the behavior found for FeTPP and FePPIX in the previous section, the Levich plots were bent and the corresponding KL plots yielded $n = 4$ (not shown here). Furthermore, a reduction in the number of ruthenium complexes bound to the ring led to a loss in electrocatalytic activity, from $n = 4$, for three and four $[Ru(NH_3)_5(OH_2)]^{2+}$ groups, to $n = 2$, for one and two $[Ru(NH_3)_5(OH_2)]^{2+}$ groups, despite the rapid electron transfer between the electrode and the Ru complex in the porphyrin [99]. Although appealing, because of its simplicity, and also consistent with the results presented above, other lines of evidence proved the original idea not only to be incorrect, but unveiling its true origin opened new prospects for the search of other types of functionalization capable of accomplishing the same goals. The string of events that led to what is currently viewed as the correct interpretation of the effects observed constitutes a fascinating tale that underscores the subtleties of electrocatalysis as a scientific discipline and the insightfulness and skill of its practitioners, which certainly warrants some digression.

The first sign that the mode of action proposed above might not be correct was provided by experiments employing the equally fast, but much more powerful reducing agent Ru(II)(edta)(OH$_2$), that is, about 0.2 V more negative than [Ru(NH$_3$)$_5$(OH$_2$)]$^{2+}$. As evidenced by the results obtained, the corresponding tetrasubstituted CoP(py)$_4$ was found to be ineffective toward imparting the porphyrin core four-electron reducing capabilities [99], leading these workers to seek other factors as being responsible for the effects observed.

Bonding between metals and ligands, such as carbon monoxide, has been described in terms two types of orbital interactions: a σ-bond between corresponding orbitals in carbon and the metal, in which the electronic density is highest along the internuclear axis leading to a net charge transfer from C to the metal M; and a π-bond involving filled metal orbitals of the correct geometry and empty π* orbitals localized on the CO, which lead to transfer of charge back to the CO, also known as π backbonding. In the case of porphyrins involving pyridine substituents in the *meso*-position, the nitrogen in pyridine plays the role of C in CO, allowing electronic density from the metal to become delocalized into the pyridine–porphyrin ring framework displacing, at the same time, electronic density from the ligand to the Co center. Support for this model was obtained from experiments involving judicious functionalization with other types of groups capable of enhancing or depressing the degree of back-bonding. In one such study, porphyrins bearing 4-cyano-phenyl [100] and 1-methyl pyridinium-4-yl substituents in the four *meso*-positions of CoP were exposed to [Ru(NH$_3$)$_5$(OH$_2$)]$^{2+}$. Although E^f was found to be virtually identical for all species examined, as expected, n varied from 2 to 3.9 and $E_{1/2}^{O_2}$ shifted steadily from 0.18 to 0.30 V, as the number of Ru coordinated 4-cyano-phenyl groups increased from 1 to 3. Evidence for the subtle character of this effect was provided by the better performance found for the doubly substituted Ru species involving adjacent, as opposed to opposite, *meso*-ring positions. Furthermore, the 3-cyano-phenyl substituted porphyrin, which is predicted to be less effective toward increasing the electronic density of the carbon bound to the porphyrin ring, could only promote dioxygen reduction via a two-electron route.

Yet another proof in support of the back-bonding hypothesis was obtained by introducing methyl groups in the 2- and 6-position of the phenyl ring, a modification that forces the 4-cyano-phenyl *meso*-substituent to rotate forming a larger angle with respect to the porphyrin plane. This geometry would decrease the extent of conjugation rendering the material a two-electron reducer [101]. It should be emphasized that incorporation of simple electron donors into the phenyl rings in CoTPP, such as OH groups in the *para*-position [102], failed to catalyze the direct reduction of O$_2$ to yield water, although it led in some instances to small positive shifts in $E_{1/2}^{O_2}$.

The demands for spin, ligand and solvent shell reorganization associated with an efficient four electron reduction of dioxygen to yield water are difficult to meet by macrocycles involving a single metal center, except perhaps the Ir porphyrin mentioned earlier. This was one of the major motivations for Collman et al. [103] to prepare ligands capable of holding two metal centers in a suitable geometry so that they may jointly bind a dioxygen molecule and, each metal center transferring two electrons, carry out the four-electron process (see below).

3.3.4.2.4 **Face-to-face Porphyrins** The presence of two redox centers in close proximity may afford, in principle, optimized conditions for the two-end binding of dioxygen and the subsequent delivery of multiple electrons, while preventing the undesirable release of intermediates, particularly peroxide, into the solution. Configurations of this type have been suggested to form upon increasing the concentration of monomeric macrocycles, for example, by surface confinement discussed above. A much more elegant approach, pioneered independently by the groups of Collman [103] and Chang [104], exploits the substitutional versatility of porphyrins for the synthesis of dimers in which the two rings are placed parallel to one another (face-to-face, FTF, geometry), using molecular struts or bridges linking either the *meso-* or β-pyrrole sites of the two porphyrins. For example, two identical, opposing struts of varying lengths, allow the metal-to-metal distance to be fixed fairly rigidly (see L, Scheme 3.2), whereas a single rigid hinge, or pillar, joining the two porphyrins (see K, Scheme 3.2) enables a flapping motion of the two opposing rings. Such tactics make it possible for the metal–metal distance to be varied either, statically or dynamically, to values believed to optimize their interaction with dioxygen, that is, 0.4–0.7 nm. In some instances, however, the rings and thus the metal centers were found to shift or slip in the dimer, such as for example in the amide linked FTF, [105, 106] an effect that almost invariably leads to losses in activity.

In one of the most extraordinary successes of targeted molecular design in electrocatalysis, Collman and Anson discovered that layers of a dicobalt FTF porphyrin, denoted as Co_2FTF 4–2, 1N–H (see Figure 3.56), adsorbed on graphite

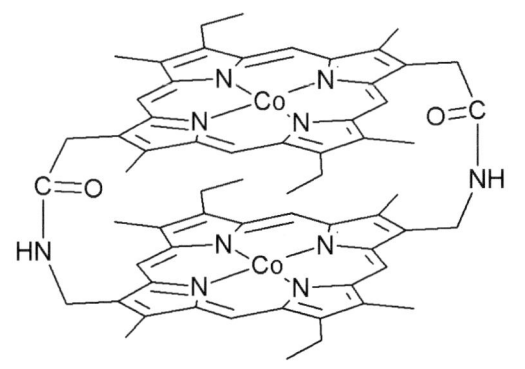

Figure 3.56 Molecular structure of the *syn* diastereomer of the face-to-face porphyrin Co_2FTF 4–2, 1N–H, where the 4-atom struts are linked to the β-pyrrole positions of the rings. The ordering of the groups in the struts is different to that in L in Scheme 3.2.

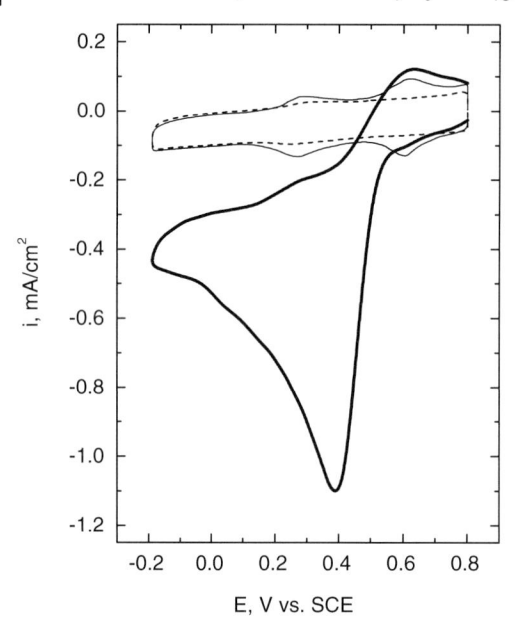

Figure 3.57 Cyclic voltammograms ($v = 0.1\,V\,s^{-1}$) of a Co_2FTF 4–2, 1N–H coated graphite electrode in deaerated (thin line) and O_2-saturated 1 M CF_3COOH (thick line).

(G) electrodes could reduce dioxygen to water in 1 M CF_3COOH at remarkably high potentials [103, 107]. As shown by the voltammetric data in thick lines in Figure 3.57, $E_{onset}^{O_2}$ is more than 200 mV more negative than the onset potential for the redox peak centered at about 0.7 V versus SCE, ascribed to the one-electron reduction of one of the Co centers in Co_2FTF 4–2, 1N–H (see thin line in this figure). Dynamic polarization curves collected with a disk coated with Co_2FTF 4–2, 1N–H of a Pt|carbon RRDE (see thick curve, upper panel, Figure 3.58) [107], yielded at small overpotentials, $0.6 < E < 0.8$ V versus NHE, i_{disk} values approaching i_{disk}^{lim} for the four-electron reduction of O_2 (upper panel, Figure 3.59), without generation of any detectable solution phase peroxide [102], that is, virtually negligible i_{ring} (see thick line, lower panel, in this figure).

In analogy with the behavior found for the single-metal porphyrins in the previous sub-section, i_{disk}^{lim} was found to display bent Levich behavior, pointing to the binding of dioxygen to the porphyrin as the rate limiting step (see Figure 3.59A). In fact, assuming the (potential independent) process to be first order in O_2, the rate constants in acidic solutions extracted from Koutecky–Levich plots (see Figure 3.59B), were found to be in the range $2 \times 10^5 – 3 \times 10^5\,M^{-1}\,s^{-1}$, which translates into turnover numbers of around $300\,s^{-1}$. These values compare well with those estimated for more conventional Pt-based catalysts, but are still much lower than those found for enzymes in biological systems, for example cytochrome c oxidase [108].

Additional studies revealed that the reduction of O_2 on Co_2FTF 4–2, 1N–H modified carbon surfaces is strongly influenced by pH, leading to changes both in

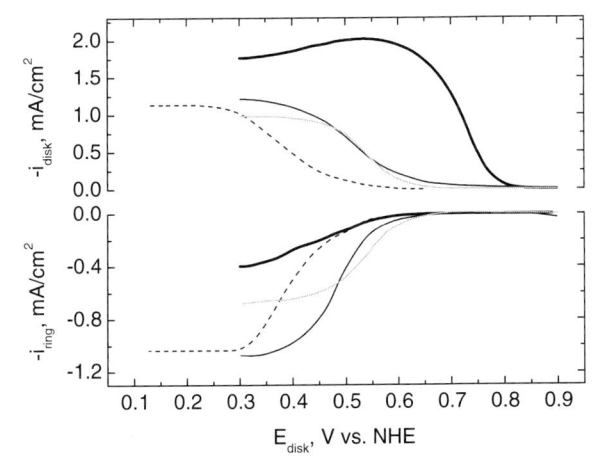

Figure 3.58 Dynamic polarization curves obtained for a Pt|G RRDE with the graphite disk coated with a layer of the four-atom strut Co_2FTF 4–2, 1N-H (thick black curves) and the six=atom strut Co_2FTF 6–3, 2N-H (thin black curves) in O_2-saturated 0.5 M CF_3COOH at $\omega = 250$ rpm (upper curves). The corresponding ring currents recorded for $E_{ring} = 1.4$ V versus NHE are shown in the lower curves in this figure. The dashed and grey curves were obtained for coatings of CoPI and the mixed metal PdCoFTF 4–2, 1N-H, respectively.

$E_{1/2}^{O_2}$ (see left ordinate, solid circles, Figure 3.60) and in the magnitudes of i_{disk}^{lim} (see right ordinate, open circles, Figure 3.60) [107]. More specifically, in unbuffered, close to neutral media, the potential range over which no hydrogen peroxide was detected became narrower compared with that found in strongly acidic media. In fact, at very high pH, O_2 was reduced to peroxide both reversibly and quantitatively, a behavior not unlike that found for CoPI discussed earlier.

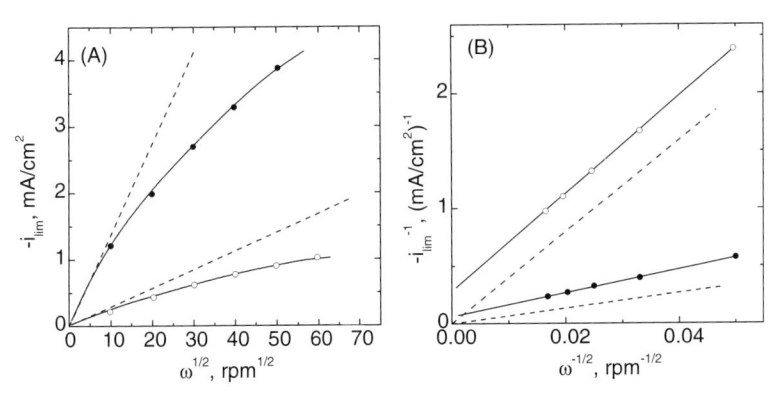

Figure 3.59 Levich (A) and Koutecky–Levich (B) plots obtained from the Co_2FTF 4–2, 1N-H coated graphite electrode of the RRDE in the caption to Figure 3.58 in air- (open circles), and O_2-saturated (solid circles) 1 M CF_3COOH solutions. The dashed lines in this figure represent theoretical values for i_{lim} for the four-electron reduction of O_2 based on the experimental conditions specified.

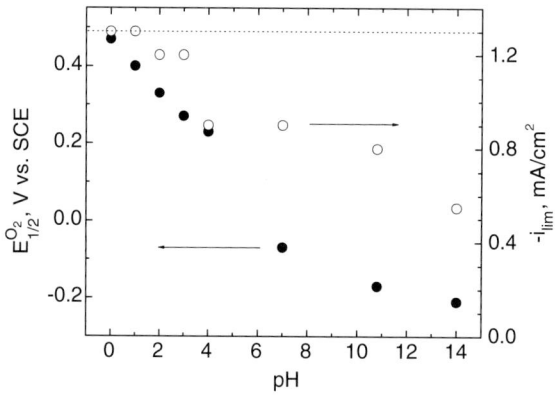

Figure 3.60 Plots of $E^{O_2}_{1/2}$ (left ordinate, solid circles) and i_{lim} (right ordinate, open circles) versus pH for O_2 reduction obtained with a Co_2FTF 4–2, 1N–H coated rotating graphite disk electrode in buffered O_2-saturated solutions as a function of pH (see original reference for actual compositions). The dashed line at $i_{lim} = 1.3$ mA cm^{-2} represents the theoretical value for the four-electron reduction of O_2.

In analogy with the results found for many Co and Fe monomeric porphyrins, RRDE experiments in deaerated solutions containing peroxide showed that Co_2FTF 4–2, 1N–H, in addition to analogue species with shorter and longer struts, were not active for peroxide reduction nor for peroxide decomposition.

The rather unique electrocatalytic properties of Co_2FTF 4–2, 1N–H and those of Co–Co4, a similar face-to-face porphyrin synthesized by Chang and coworkers [104], became evident upon modifying various aspects of their structure. In particular, extending the length of the struts to five and six atoms, such as Co_2FTF 6–3, 2N–H, yielded a catalyst capable of reducing O_2 at least partially via the two-electron route, generating detectable peroxide at the ring electrode (see thin lines, Figure 3.58). In fact, its overall performance was only slightly better than that of CoPl shown in dashed lines in the same figure. Furthermore, replacement of one of the two Co centers by Pd, a metal devoid of redox activity and known to display no affinity for dioxygen, that is, $PdCoFTF$ 4–2, 1N–H, was found to be detrimental, yielding once again a response similar to the Co monomer (see grey line, Figure 3.58). Several other metal centers and combinations thereof yielded species displaying by and large poor activities compared with Co_2FTF 4–2,1N–H. For example, the behavior of Fe_2FTF 4–2, 1N–H [103](not shown in this figure) was not much different than that of its monomeric counterpart. Rather surprisingly, the single metal CoH_2FTF 4–2,1N–H and $AlCoFTF$ 4–2, 1N–H yielded values of i^{lim}_{disk} consistent with a net $n = 3$, that is intermediate between two- and four-electron reduction, which suggests involvement of a Lewis acid site as a key factor in controlling the mechanistic pathways.

Further extension of the amide struts in Co_2FTF led to lateral ring slippage, which rendered the relative positions of the two metal centers displaced with respect to the axis normal to the ring and thus in changes in the overall FTF architecture of the

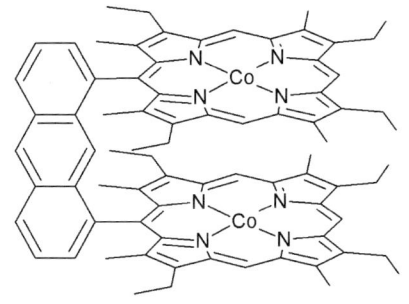

Figure 3.61 Co$_2$ FTF 1,8-anthryldiporphyrin.

electrocatalytic microenvironment. Prompted in part by these limitations, Chang et al. prepared Co–Co face-to-face porphyrins incorporating a single rigid *meso*-bound linkage, more specifically, 1,8-anthryldiporphyrin (see Figure 3.61) and 1,8-biphenylenediporphyrin linkages (see K, Scheme 3.2) [109], that greatly impeded slippage, allowing at the same time for the Co–Co separation to vary in a dynamic way. This 'chomping' action was perceived as beneficial as it would facilitate access of dioxygen and its further binding to the two metal centers, 'trapping' the μ-peroxo species believed to be required (albeit not sufficient) to effect the full four-electron reduction of dioxygen [109].

The electrochemical behavior of both these species in the form of coatings was very similar to that found for the original most active dicobalt porphyrins. This is illustrated in Figure 3.62A [110], which compares 1,8-anthryldiporphyrin (solid lines) and Co–Co4 (dashed lines). In particular, the values of $E_{1/2}$ were found to be in between those observed for Co–Co4 and Co$_2$FTF 4–2, 1N–H and the amount of hydrogen peroxide detected at the ring was about 8%.

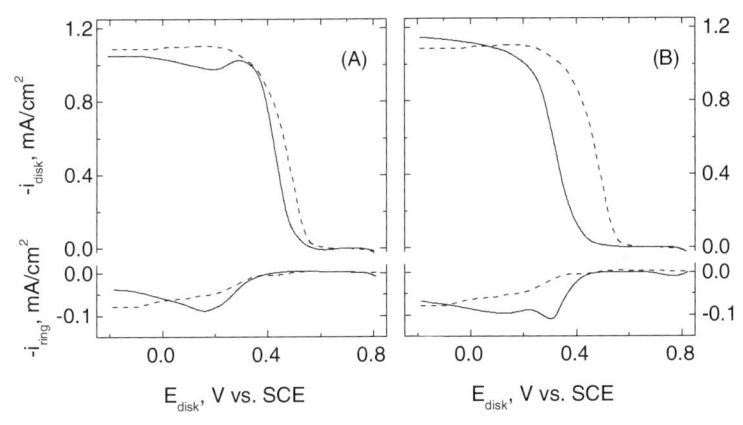

Figure 3.62 Dynamic polarization curves ($v = 10$ mV s^{-1}) for dioxygen reduction at a Pt|PG RRDE with the disk coated with 1.2×10^{-9} nmol cm^{-2} of the dicobalt 1,8-anthryldiporphyrin (A) and its corresponding monocobalt analogue (B) collected at 100 rpm in O$_2$-saturated 1 M CF$_3$COOH solutions. $E_{ring} = 0.9$ V. The dashed lines are corresponding curves recorded under identical conditions for coatings of the more active Co–Co4 on the PG disk.

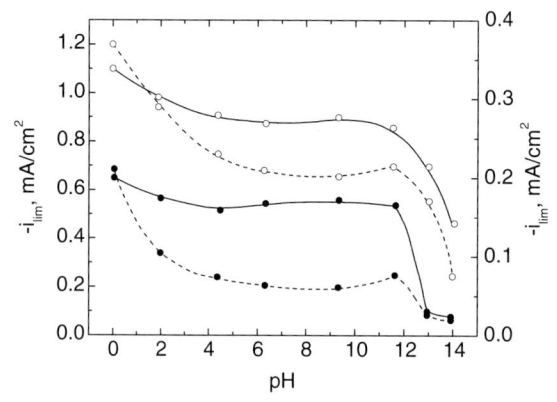

Figure 3.63 Limiting currents, i_{lim}, for O_2 reduction as a function of pH in air-saturated 1 M CF_3COOH (left ordinate, open circles, $\omega = 100$ rpm) and peroxide reduction in deaerated 1 mM H_2O_2, 1 M CF_3COOH (right ordinate, solid circles, $\omega = 400$ rprn) solutions at a graphite RDE coated with 1.2×10^{-9} mol cm^{-2} of dicobalt 1,8-anthryldiporphyrin (solid lines) and its corresponding monocobalt analogue (dotted lines).

As may have been predicted, however, both these materials were found to follow a similar mechanistic pathway to that of Co–Co4, that is, a chemical step was rate determining, as judged by the bent Levich plots. A more detailed kinetic analysis based on the Koutecky–Levich intercepts was in this instance complicated by the mixed reaction pathway.

Unlike other cofacial Co–Co porphyrins, these pillared materials were also found to be active for the reduction of hydrogen peroxide, where again, based on a full RDE analysis, a chemical step, in all likelihood the binding of hydrogen peroxide to the active site, was found to be rate limiting. It should be stressed, however, that none of these materials was active for peroxide decomposition based on experiments of the type described in Section 3.1.2. The activity of these materials for the reduction of both oxygen and hydrogen peroxide was found to be sensitive to the availability of protons, as evidenced from the values of i_{lim} as a function of pH shown in Figure 3.63.

Rather surprisingly, the monocobalt 1,8-anthryldiporphyrin material yielded values for i_{disk}^{lim} and i_{ring} (see solid lines, Figure 3.62B) very similar to those found for the dicobalt counterpart (see solid lines Figure 3.62A), and consistent with the reaction proceeding mostly through a four-electron pathway; however, the onset potential for the reduction was about 0.2 V more negative for the mono-, compared with the dicobalt derivative.

Further evidence of the subtleties associated with the unique behavior of Co_2FTF 4–2, 1N–H was provided by the low activities of a very similar dicobalt porphyrin in which the order of the groups in the linker was changed from CH_2–NH–CO–CH_2 (Figure 3.56) to CO–NH–CH_2CH_2 (see L, Scheme 3.2) in one instance [107], and by changing the point of attachment of the struts from the β-pyrrole to the *meso*-position (not shown here) in the other [103]. Both such modifications yielded hydrogen peroxide as the major product regardless of the length of the struts. This loss of activity has been attributed in the first instance to the lack of coaxiality between the two metal centers [105], which precludes formation of the μ-peroxocobalt species, believed to be responsible for the rapid four electron reduction of dioxygen found for Co$_2$FTF 4–2, 1N–H. In fact, structural studies involving electron spin resonance (ESR) have shown that in non-aqueous media incorporating a nitrogenous axial base, dioxygen reacts with the tightly bound Co$_2$FTF 4–2, 1N–H to yield an ESR silent species consistent with the formation of a diamagnetic μ-peroxocobalt dimer, a behavior not found for the corresponding *meso*-linked analogues under similar conditions. In fact, the strong pH dependence of the reduction process appears to be consistent with a protonation of the peroxo-bridge leading to the formation of a cationic complex that could facilitate electron transfer to the peroxo-unit at more positive potentials compared with its unprotonated counterpart, leading ultimately to the cleavage of the oxygen–oxygen bond. However, attempts to incorporate protons in the close vicinity of the cavity involving, for example, reduction of the amide to amine in the linkers, yielded dicobalt derivatives displaying activities not unlike those found with their monomeric Co(II) counterparts [103].

In contrast, and to underscore the exquisite sensitivity of geometrical and electronic aspects of the cavity microenvironment the *meso*-linked dicobalt dimer is a highly effective catalyst for electrochemical reduction of the closely related nitric oxide, whereas the corresponding β-linked species exhibits only very modest activities for the same reaction [111].

Shown in Scheme 3.9 is a mechanism consistent with the most salient features found experimentally for the reduction of O$_2$ in acidic electrolytes catalyzed by adsorbed dicobalt cofacial macrocycles [112]. As indicated therein, the active form of

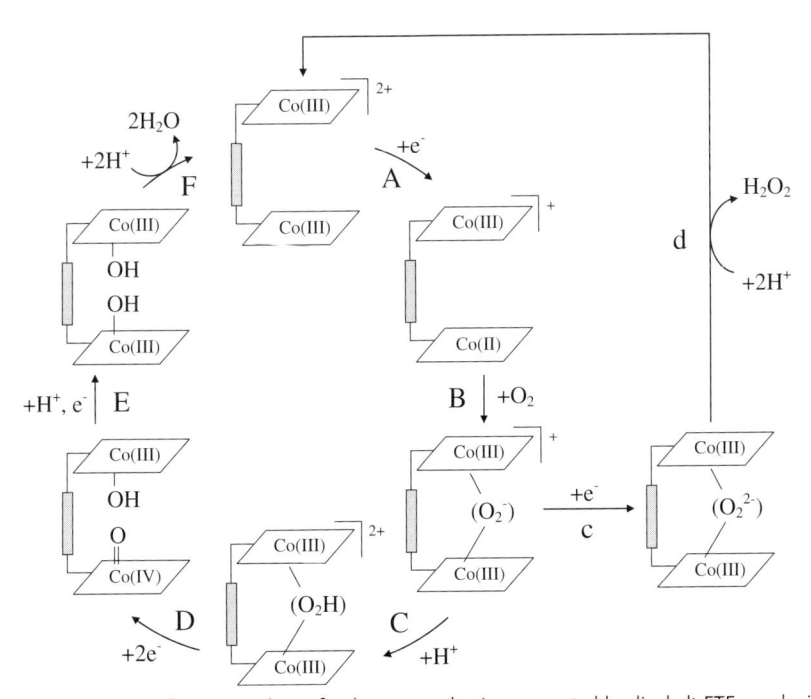

Scheme 3.9 Mechanistic pathway for dioxygen reduction promoted by dicobalt FTF porphyrins.

the macrocycle is a mixed valence species generated by the one-electron reduction of the fully oxidized form of the material (step A), which can form an adduct with O_2 to generate a superoxo-type species (step B). The fate of this adduct, and, hence, the ultimate mechanistic pathway, will be dictated by its relative reactivity toward either electron or proton transfer. The first path (c) leads, following a one-electron transfer, to the formation of a µ-peroxo-complex, which, upon further protonation, generates a labile adduct releasing peroxide into the solution. The second path involves protonation (step C) and subsequently a stepwise injection of three more electrons and protons (steps D, E and F) to ultimately yield the desired water product. This mechanism accounts for the shape of the current–potential curve, the potentials and pH dependence of the voltammetric features observed both in the presence and absence of O_2, the increase in the relative amounts of hydrogen peroxide to water produced as the pH of the solutions is increased and the potential-independent step in the reaction mechanism found upon increasing the *flux* of dioxygen. In addition it invokes formation of a superoxo-adduct whose existence has been inferred from ESR data.

It follows from Scheme 3.9, that the ability of the µ-superoxo-adduct (step C) to undergo protonation will ultimately determine whether the reaction will produce water or peroxide. Such properties will be controlled by the Lewis base character of the superoxide moiety in the adduct. To test this hypothesis, Nocera and coworkers synthesized two Pacman materials incorporating different hinges leading to different bite widths and thus to different Co–Co distances [113]. These are shown in A (DPX) and B (DPD) in Figure 3.64, where the DPX hinge is xanthene and that of DPD is dibenzofuran.

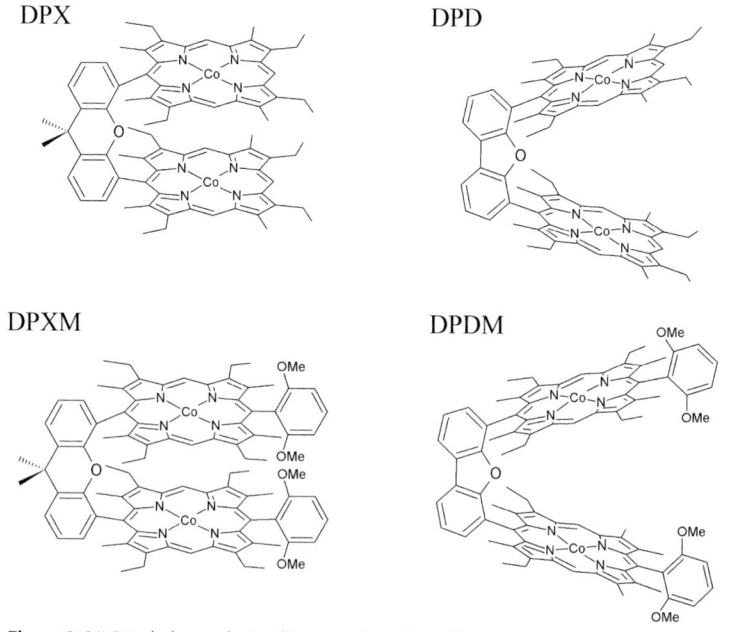

Figure 3.64 Dicobalt porphyrin dimers with wide and narrow bites.

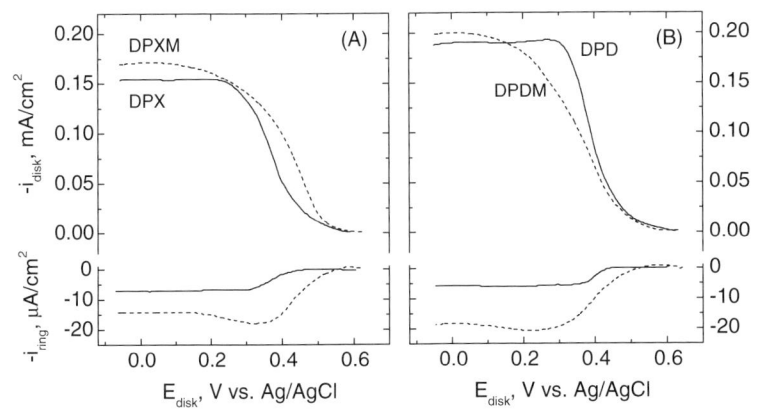

Figure 3.65 Dynamic polarization curves for O_2 reduction at a PG disk of a Pt|PG RRDE coated with DPX and DPXM (solid and dashed lines, respectively, A), and DPD and DPDM (solid and dashed lines, respectively, B) collected at 100 rpm in air-saturated solutions of 0.5 M $HClO_4$–1.5 M CF_3COOH for DPX and DPD, and 2 M $HClO_4$ for DPXM and DPDM. $E_{ring} = 1.0\,V$.

The behavior of DPX and DPD was compared with that of their methoxyaryl *meso*-substituted analogues *trans* to the hinge, DPXM and DPDM in C and D, Figure 3.64. Solution phase voltammograms collected in non-aqueous electrolytes (not shown here) yielded for narrow bite materials, DPX and DPXM, two clearly defined peaks attributed to electronic communication between the two sites, which collapsed into a single peak when the bite was made wider, that is, DPD and DPDM. In addition, the *meso*-substitution yielded only minor changes in the redox potentials. The electrocatalytic activity of these materials for O_2 reduction in aqueous electrolytes was examined using coatings deposited on PG electrodes. The results obtained revealed only small differences between the narrow and wide bite materials for a fixed ring structure (see Figure 3.65). However, introduction of the *trans*-substituents was found to significantly reduce the specificity toward the four-electron reduction, regardless of the width of the bite (Figure 3.66). In particular, at relatively small overpotentials, 0.35–0.40 V, the fraction of water produced by DPD and DPX is about double that found for DPDM and DPXM.

Close inspection of the solution phase redox peaks for the four species under analysis obtained in a thin layer cell in deaerated non-aqueous media (not shown here) yielded values for the onset of the reduction of the first Co site very similar to one another. Hence, it can only be surmised that the differences in specificity are related to the presence of the peripheral substituent and seemingly unrelated to the redox properties of the metal sites. Evidence that the differences in the specificity between the *meso*-substituted and non-*meso*-substituted materials are due to subtle electronic effects was provided by quantum mechanical calculations. These showed that upon addition of the *meso*-substituents, the electronic charge density associated with the HOMO (highest occupied molecular orbital) localized on the dioxygen moiety is markedly reduced, thereby weakening its Lewis acid character and hence its ability to coordinate a proton (see Figure 3.67).

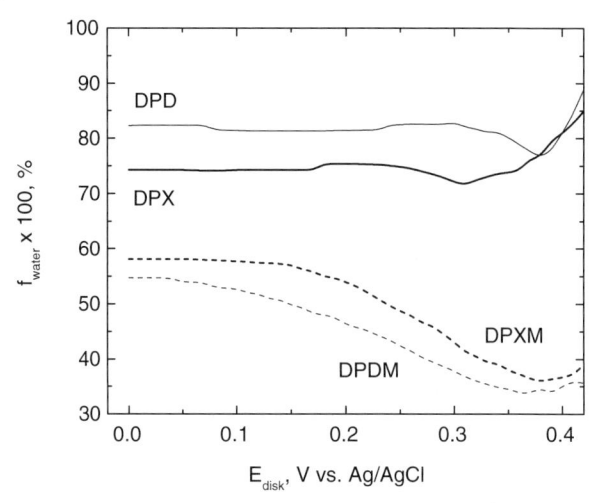

Figure 3.66 Plot of percentage water generated at the surface of a macrocyclic-modified graphite disk electrode of an RRDE based on the data in Figure 3.65, calculated using $f_{water} = (N - i_{ring}/i_{disk})/(N + i_{ring}/i_{disk})$.

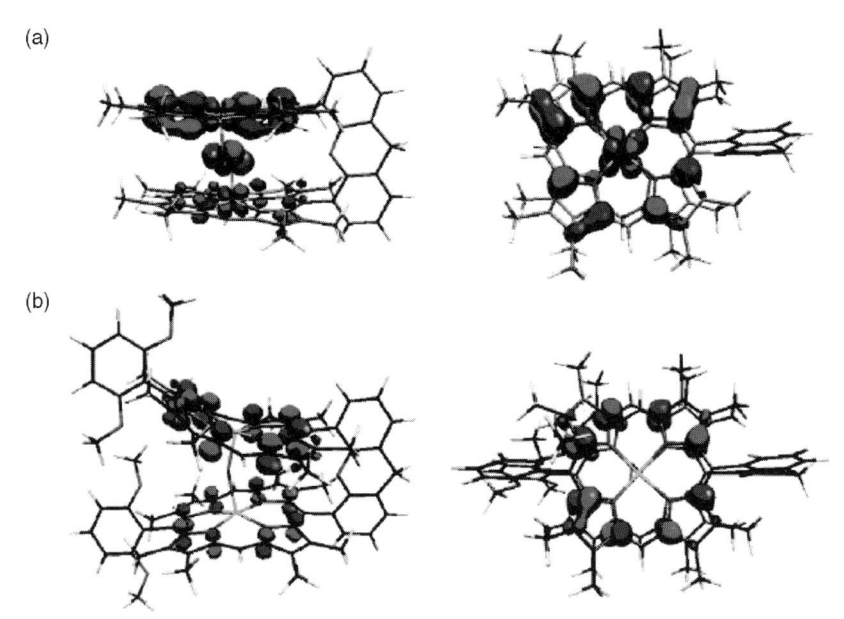

Figure 3.67 Singly occupied HOMOs of (a) $[Co_2(DPX)(O_2)]^+$ and (b) $[Co_2(DPXM)(O_2)]^+$. The structures on the left show a side view, normal to the bridge plane, whereas those on the right display a top view, perpendicular to the porphyrin planes (Reproduced with permission by ACS.)

3.4
Thermal Activation of Transition Metal Macrocycles

3.4.1
Brief Introduction

A number of transition metal macrocycles, either adsorbed on, and/or simply dispersed as small microcrystals in high surface area carbons, have been found to enhance the rates of O_2 reduction following exposure to temperatures in the range 700–900 °C in an inert atmosphere [114–117]. In addition, for some supported macrocycles, which exhibit electrocatalytic activity for O_2 reduction in their pristine states, the thermal treatment has also been shown to improve rather substantially the long term stability of operating O_2-fed cathodes. Since they were first reported by Alt et al. [114], these effects have sparked much interest in both the scientific and technical communities [118–125], because of the possibility of replacing platinum as a catalyst in fuel cells and other applications by materials based on more abundant and less expensive elements. This general area of research has been reviewed recently in the literature by Dodelet [117]; hence, only structural and spectroscopic studies aimed at unveiling the factors responsible for the electrocatalytic activity of this remarkable class of materials we regard as most significant will be highlighted in this survey. It must be stressed at the outset, that all of the experimental evidence collected to date is consistent with changes in the structure of supported macrocycles induced by the heat treatment. However, no consensus appears to have been reached regarding detailed aspects of the thermal decomposition pathways, and, particularly, the identity of the resulting electrocatalytic sites for dioxygen reduction. Because of their promising performance, attention in what follows will be focused mostly on macrocycles containing Fe and Co centers.

3.4.2
Electrochemical Characterization

Among the most expedient tactics developed for the electrochemical and electrocatalytic characterization of materials in high-area form is the mixing of the powder with a binder to render a paste, which is then applied to the surface of a slightly receded carbon disk of an RDE or RRDE assembly. This thin porous electrode technique has been found to best mimic the results of fuel cell tests and to allow for the electrocatalytic activity of different materials to be assessed with high degree of reliability. However, owing to its intrinsic three-dimensional nature, the analysis of polarization curves as a function of rotation rate precludes determination of kinetic rate constants employing the Koutecky–Levich formalism of compact solid electrodes described in earlier sections. Although only a few macrocycles have been the subject of detailed studies, the conclusions emerging from the data collected appear to be general enough to account for the behavior of a much wider class of materials. The sub-sections to follow present the results of a systematic investigation involving Co-, Fe- and H_2TMPP supported on Vulcan XC-72, a fairly pure high-area carbon with a specific area of about $250 \, m^2 \, g^{-1}$ [126].

3.4.2.1 Cyclic Voltammetry

Thin porous coating electrodes incorporating pristine CoTMPP and (FeTMPP)$_2$O supported on Vulcan XC-72 yielded cyclic voltammograms both in 0.1 M NaOH and 0.05 M H$_2$SO$_4$ (see upper curves, B and C, in Figures 3.68 and 3.69, respectively) displaying reversible peaks at potentials in good agreement with those of other Co and Fe porphyrins described earlier in this chapter. Well defined features were also observed for H$_2$TMPP|XC-72 in the alkaline electrolyte (see upper curve, Figure 3.68A). Coulometric analyses based on these voltammetric curves using as baselines the dotted lines, however, could account for not more than about 20% of the macrocycle present in the coating. This effect may be related to encapsulation of the active material by the binder during the electrode preparation and/or mass transport hindrances within the convoluted structure of the electrode, which would impair equilibrium conditions to be achieved within the time scale of the voltammetric experiments.

Following heat treatment at a temperature $T_p = 800\,°C$ (see lower curves, Figures 3.68 and 3.69), the voltammetric peaks either disappeared or gave rise to new features at potentials well removed from those observed for the untreated specimens. In particular, close inspection of the results obtained for heat-treated Fe(TMPP)$_2$O|XC-72 in 0.1 M NaOH (see lower curve, Figure 3.68C) revealed virtually

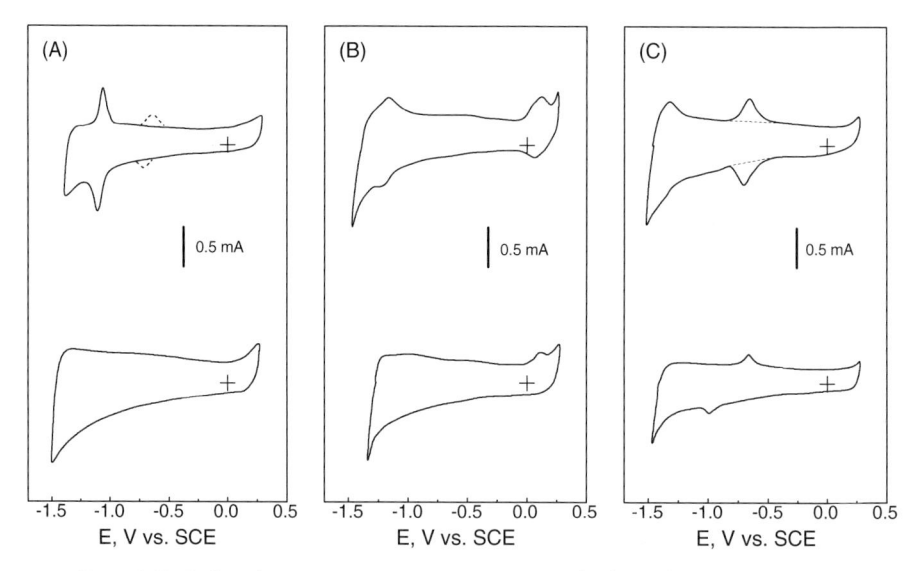

Figure 3.68 Cyclic voltammograms ($v = 20\,\text{mV s}^{-1}$) of thin porous coating electrodes containing 4.4% w/w H$_2$TMPP/XC-72 (A), 4.7% w/w CoTMPP|XC-72 (B) and 4.8% w/w (FeTMPP)$_2$O|XC-72 (C) in 0.1 M NaOH. The upper and lower curves in each of the panels were obtained with the materials before and after heat treatment at $T_p = 800\,°C$ for 2 h in Ar, respectively. Electrode cross-sectional area, 0.196 cm^2. The voltammetric features at about $-0.7\,V$ in the upper curve of (A) were obtained after the electrode had been cycled, or polarized at potentials negative to the redox peak at $-1.1\,V$. The dotted line in the voltammogram in Panel C was used as a baseline for the coulometric analysis.

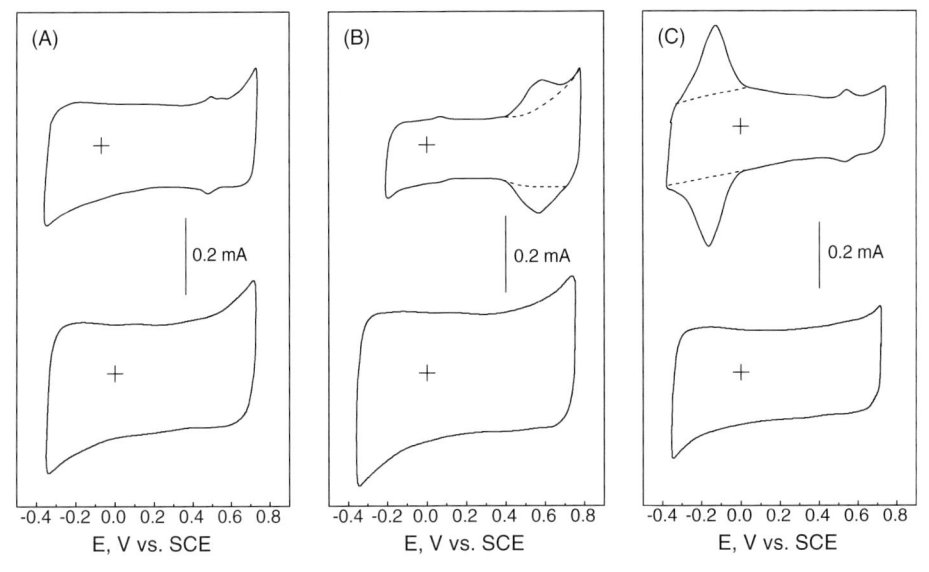

Figure 3.69 Cyclic voltammograms ($v = 20\,mV\,s^{-1}$) of thin porous coating electrodes containing 4.4% w/w H$_2$TMPP|XC-72 (A), 4.7% w/w CoTMPP|XC–72 (B) and 4.8% w/w (FeTMPP)$_2$O|XC-72 (C) in 0.05 M H$_2$SO$_4$. The upper and lower curves in each of the panels were obtained with the materials before and after heat treatment at $T_p = 800\,°C$ for 2 h in Ar, respectively. Electrode cross-sectional area, 0.196 cm^2. The dotted lines in the voltammograms in Panels B and C were used as baselines for the coulometric analyses.

identical peaks to those observed for iron hydroxide. In fact, evidence obtained from in situ ^{57}Fe Mossbauer effect spectroscopy (MES) measurements of FeOOH precipitated onto a high-area carbon support in alkaline electrolytes was consistent with the one-electron reduction of amorphous FeOOH to yield crystalline Fe(OH)$_2$. In analogy, the peak found at approximately 0.1 V versus SCE for pyrolyzed CoTMPP|XC-72 in the scan in the positive direction (see lower curve, Figure 3.68B), which often lacks a complementary reduction peak in the subsequent scan in the negative direction, has also been observed for Co(OH)$_2$ precipitated onto XC-72 in the same electrolyte [128]. Consistent with the formation of metal oxide or hydroxide particles are the featureless voltammograms recorded in acid electrolytes (see lower curves Figure 3.69), a media in which such species would be expected to dissolve.

3.4.2.2 Oxygen Reduction Polarization Curves

The effects of pyrolysis on the electrocatalytic properties of the same three macrocycles dispersed in Vulcan XC-72 for O$_2$ reduction were examined using thin porous electrodes applied to the carbon disk of an RRDE assembly. Steady state polarization curves obtained at $\omega = 2500$ rpm in 0.1 M NaOH and 0.05 M H$_2$SO$_4$ are shown in Figures 3.70 and 3.71. As is evident from the results presented therein, the activity of H$_2$TMPP in both media (see solid squares) before and after pyrolysis was much lower than that of the metallated counterparts and fairly close to that of

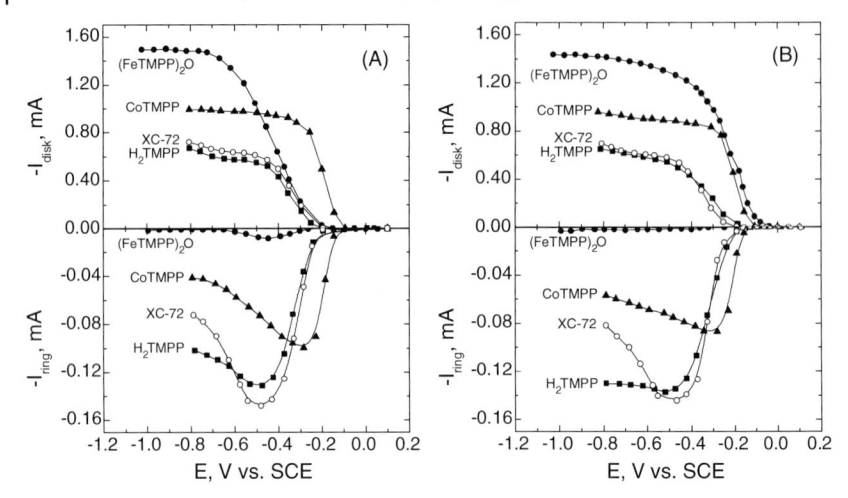

Figure 3.70 Steady state ring–disk polarization curves for O_2 reduction on a thin porous coating electrode containing 4.7% w/w CoTMPP|XC-72 (solid triangles), 4.8% w/w (FeTMPP)$_2$O|XC-72 (solid circles), 4.4% w/w H$_2$TMPP|XC-72 (solid squares) and Vulcan XC-72 carbon (open circles) in 0.1 M NaOH before (A) and after pyrolysis 800 °C, 2 h in Ar (B). Electrode cross-sectional area, 0.196 cm^2, $\omega = 2500$ rpm, room temperature, Au $E_{ring} = +0.1$ V versus SCE, $N = 0.38$.

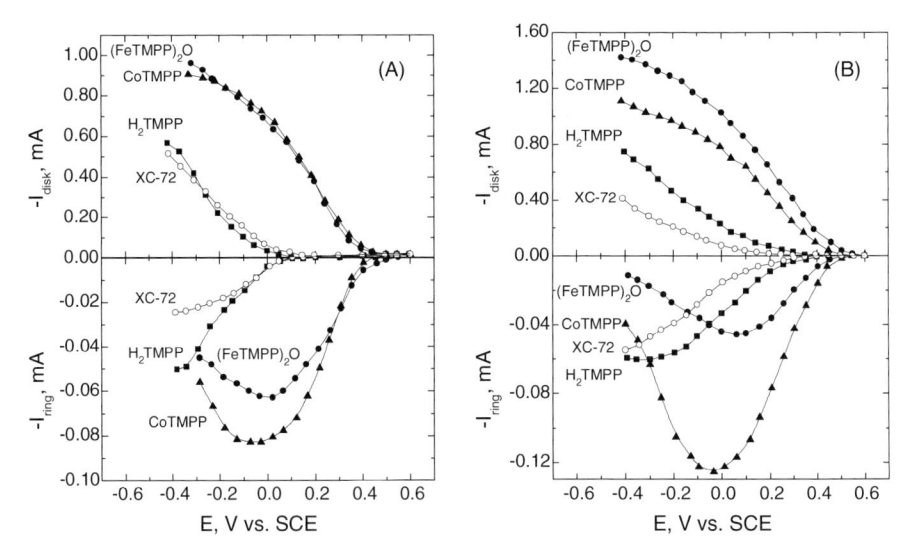

Figure 3.71 Steady state ring–disk polarization curves for O_2 reduction on a thin porous coating electrode containing 4.7% w/w CoTMPP|XC-72 (solid triangles), 4.8% w/w (FeTMPP)$_2$O|XC-72 (solid circles), 4.4% w/w H$_2$TMPP|XC-72 (solid squares) and Vulcan XC-72 carbon (open circles), in 0.05 M H$_2$SO$_4$ before (A) and after pyrolysis at $T_p = 800$ °C, 2 h in Ar (B). Electrode cross-sectional area, 0.196 cm^2, $\omega = 2500$ rpm, room temperature, Pt $E_{ring} = +1.1$ V versus SCE, $N = 0.38$.

pure XC-72 carbon, also shown in the figure for comparison (see open circles). It may thus be concluded that the presence of the metal is essential to impart the materials with electrocatalytic activity. Whereas $E_{onset}^{O_2}$ for CoTMPP|XC-72 in 0.1 M NaOH (see solid triangles, Figure 3.70) was about 0.15 V more positive than for (FeTMPP)$_2$O|XC-72 (see solid circles, Figure 3.70), the latter showed the largest i_{lim}, and the smallest i_{ring}, as measured with an Au ring polarized at $E_{ring} = 0.1$ V versus SCE. This behavior is consistent with the generation of very small amounts of solution phase peroxide, and contrasts that found for the other two catalysts, for which i_{ring} was significant.

Insight into possible changes in the reaction mechanism induced by the heat treatment was gained from measurements of peroxide decomposition. In particular, CoTMPP|XC-72 in 0.1 M NaOH at 25 °C displayed relatively high activity, increasing by a factor of four after heat treatment at $T_p = 800$ °C. Large enhancements in activity were also observed for the otherwise inactive (FeTMPP)$_2$O|XC-72 following heat treatment, which might account for the much improved performance for O$_2$ reduction of the pyrolyzed (see solid circles, Figure 3.71) compared with its intact counterpart (see solid circles, Figure 3.70).

Corresponding data for the same thin porous electrodes in 0.05 M H$_2$SO$_4$ revealed that the heat treatment had relatively little effect on the performance of CoTMPP|XC-72 (see solid triangles, Figure 3.71). Unexpectedly, it adversely affected that of (FeTMPP)$_2$O|XC-72 (see solid circles, Figure 3.71), the best catalyst in this media in the intact state (see solid circles, Figure 3.70), leading to shifts in $E_{onset}^{O_2}$ toward less positive potentials, a decrease in i_{lim} and an increase in solution phase H$_2$O$_2$ detected by a Pt ring electrode.

More recently, Faubert et al. [129] studied in a more systematic fashion the effect of T_p on both the activity and the stability of FeTPP (presumably in its μ-oxo form) and CoTPP dispersed on XC-72 carbon incorporated into a gas diffusion electrode in actual fuel cells. Shown in Figure 3.72A and 3.72B are polarization curves recorded at 50 °C for FeTPP|XC-72 and CoTPP|XC-72, respectively, pyrolyzed at the specified temperatures, in a Nafion™-based fuel cell configuration for which the performance of specimen in the range $700 \leq T_p \leq 900$ °C was fairly comparable to that of 2% w/w Pt supported on high-area carbon. These same materials also displayed good stability up to about 10 h compared to samples treated at higher and lower T_p under the same conditions (see Figure 3.73A and 3.73B). Also shown for comparison are the results obtained for 2% w/w Pt dispersed in the same carbon.

A better understanding of the structure of these pyrolyzed materials has been gained using an array of techniques, as will be illustrated in the following sections.

3.4.3
Spectroscopic and Structural Characterization

3.4.3.1 Pyrolysis-Mass Spectrometry
Information regarding the chemical identity of gases evolved from specimens as a function of the heating temperature can be obtained from pyrolysis–mass

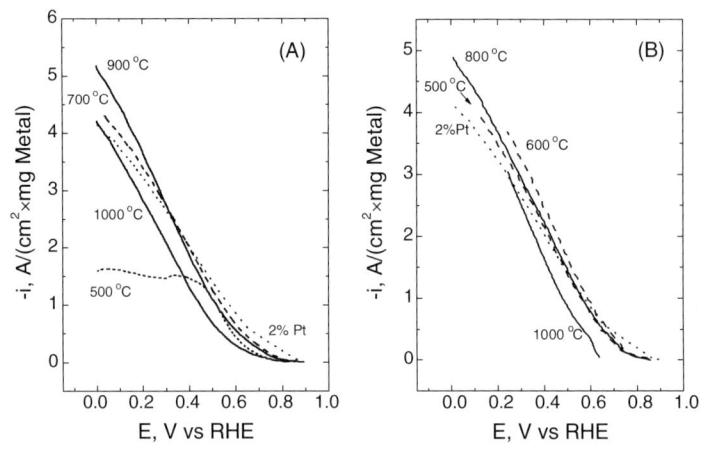

Figure 3.72 Polarization curves recorded at 50 °C for FeTPP|XC-72 (A) and CoTPP|XC-72 (B) pyrolyzed at the specified temperatures in a fuel cell configuration. Also shown in dotted lines for comparison are data collected under the same conditions for 2% w/w Pt, supported on high-area carbon (see original reference for details).

spectrometry (Py–MS) and its variant Py–GC–MS, where GC stands for gas chromatography [130]. These techniques were employed to examine the thermal decomposition patterns of ultrapurified Co-, Fe- and H_2-TMPP both in the form of bulk crystals and also dispersed in Vulcan XC-72 carbon. For these measurements,

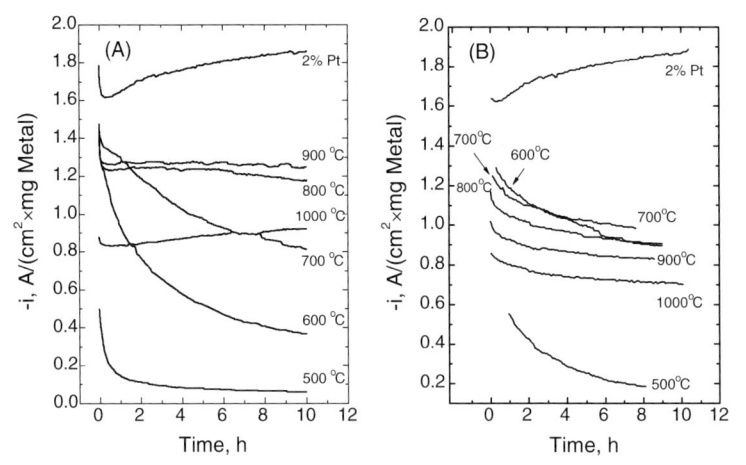

Figure 3.73 Current versus time curves for FeTPP|XC-72 (A) and CoTPP|XC-72 (B) pyrolyzed at the specified temperatures for electrodes polarized at 0.5 V in a fuel cell configuration. Corresponding data for 2% w/w Pt supported on high-area carbon are also shown for comparison (see original reference for details).

3.4 Thermal Activation of Transition Metal Macrocycles | 275

the materials were heated under 1 atm (=101.325 kPa) pressure of a flowing inert gas up to temperatures of about 600 °C. Of particular interest was to determine the identity of the molecular fragments released by specimens as a function of the pyrolysis temperature (T_p). This information can afford insight into the pyrolysis mechanism, and also into the effects induced by the adsorption of the macrocycles on the carbon support, on the overall thermal degradation patterns.

Shown in Figure 3.74 are plots of the normalized ion current as a function of temperature (see below) obtained for slow Py–MS for the three porphyrins supported on XC-72 carbon, for the specified values of m/z. The temperature in this instance was stepped from 150 °C to about 600 °C in increments of 50 °C and held at the intermediate values for 2 min. All of the features for $T < 400$ °C may be traced to residual impurities, including solvents used during the preparation of the samples,

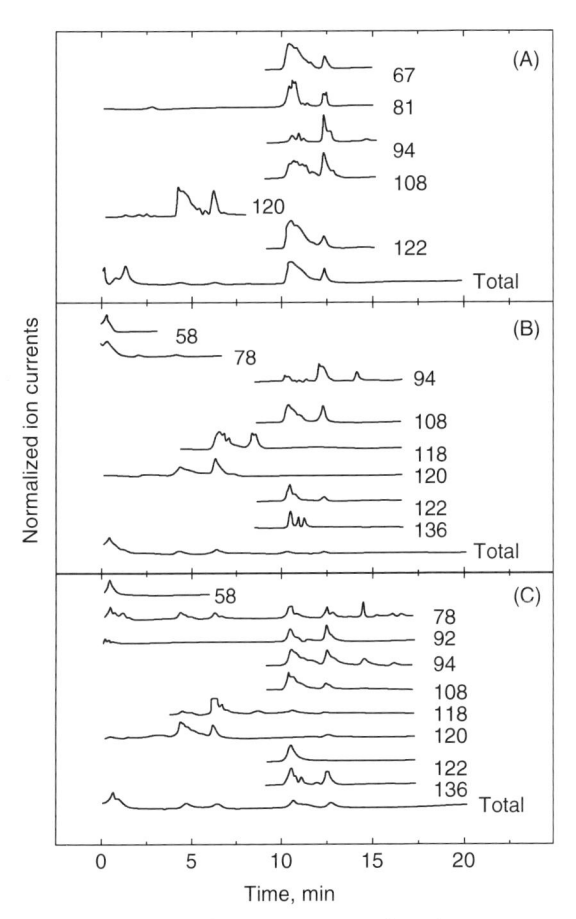

Figure 3.74 Slow pyrolysis–mass spectra of 4.4% w/w H_2TMPP|XC-72 (A), 4.7% w/w CoTMPP|XC-72 (B) and 4.8% w/w (FeTMPP)$_2$O|XC-72 (C) from room temperature up to about 600 °C.

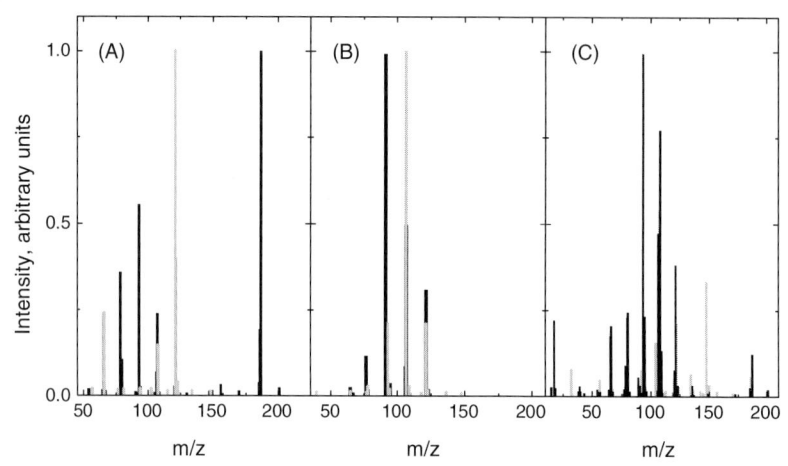

Figure 3.75 Integrated pyrolysis–mass spectra of (A) 4.4% w/w H$_2$TMPP|XC-72, (B) 4.7% w/w CoTMPP|XC-72 and (C) 4.8% w/w (FeTMPP)$_2$O|XC-72 (see grey bars). The data in black bars were obtained using the same methodology for the macrocycles in crystalline unsupported form.

which are completely removed before the onset of thermal decomposition of the macrocycles at about 400 °C.

Integrated pyrolysis–mass spectra for the thermal decomposition products for both the unsupported (thick black bars) and supported porphyrins (thin grey bars) over the temperature range 20–600 °C are shown in Figure 3.75. As indicated therein, phenol (m/z 94), methoxybenzene and/or methylphenol (m/z 108) and ethylphenol and/or methyl methoxybenzene (m/z 122) (originating from the *meso*-substitutents of the macrocycle), were detected for all specimens in varying amounts at temperatures in the range 400–450 °C. The pyrolysis fragments with m/z values between 130 and 200 (with the exception of 134 and 136, which correspond to isomeric mixtures of Me and COMe and C-3 phenols) consist mostly of a pyrrole group linked by an aliphatic chain to a functionalized benzene moiety. The amount of fragments released was much higher for H$_2$TMPP compared with (FeTMPP)$_2$O, but below the detection limit for CoTMPP.

The most important conclusions emerging from these studies may be summarized as follows:

1. The magnitude of the peak associated with the methoxy group ($m/z = 31$) was negligible as compared with that of benzene or its derivatives. Hence, the major thermal decomposition pathway up to T_p of about 500 °C involves the cleavage of the complete *meso*-substituent and not just the outer fringe of the chelate as reported originally by van Veen et al. [131]

2. The fraction of volatile nitrogen-, to non-nitrogen-containing species generated during thermal treatment is much higher for H$_2$TMPP|XC-72 than for CoTMPP|XC-72 and (FeTMPP)$_2$O|XC-72. These results are consistent with the

microanalysis of the same supported macrocycles for which the amount of nitrogen found following heat treatment at 800 °C was higher for the metallated than for the metal-free species. This effect was particularly pronounced upon dispersion of the macrocycles in the high-area carbon support. In other words, the metal center stabilizes the macrocyclic ring, and/or catalyzes pyrolytic processes that lead to the binding (or retention) of nitrogen to (or by) the carbon.

3. The high molecular weight volatile fragments detected for H_2TMPP and to a much lesser extent for $(FeTMPP)_2O$ are ascribed to species involving both pyrrole and substituted phenyl moieties.

3.4.3.2 Mossbauer Effect Spectroscopy

Insight into the fate of the metal center and its microenvironment for $(FeTMPP)_2O$ and CoTMPP brought about by the thermal treatment was gained from Mossbauer effect spectroscopy (MES) in the absorption (^{57}Fe) [126] and emission (^{57}Co) modes, [132] using isotopically enriched materials. The ^{57}Fe MES absorption spectrum of $(^{57}FeTMPP)_2O$ supported on XC-72 carbon at 5.2 K (see upper curve Figure 3.76) was found to be very similar to that recorded for pristine unsupported microcrystalline $(^{57}FeTMPP)_2O$ in bulk form at 4.2 K (not shown in this figure), affording evidence that the dispersion process does not alter the macrocyclic structure. After pyrolysis, however, the spectral features observed for the supported specimens were indeed significantly different (see lower curve in Figure 3.76),

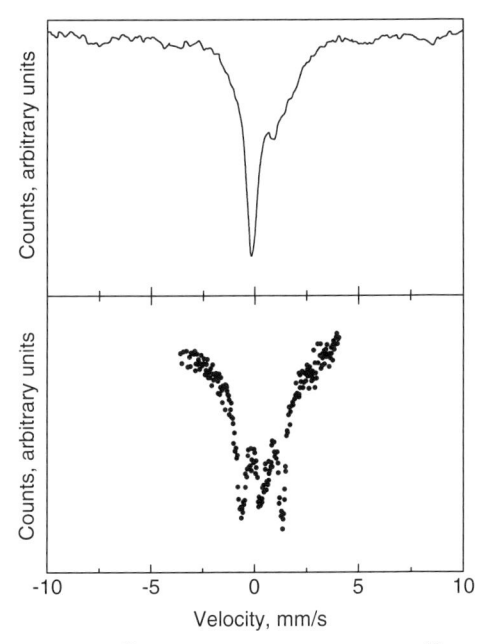

Figure 3.76 ^{57}Fe MES absorption spectrum of $(^{57}FeTMPP)_2O$ supported on XC-72 carbon at 5.2 K before (upper curve) and after pyrolysis at 800 °C (lower curve).

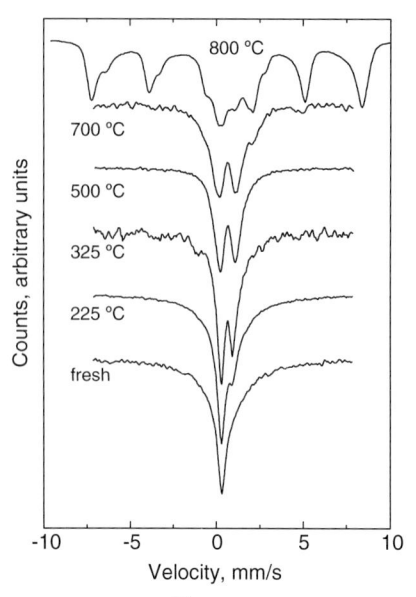

Figure 3.77 Ex situ ^{57}Fe MES spectra of isotopically enriched ^{57}FeTPPCl dispersed on Vulcan XC-72 for specimens pyrolyzed at the specified temperatures recorded at room temperature.

consistent with major changes in the environment surrounding the Fe site including the metal itself.

In more recent studies, Bouwkamp-Wijnoltz et al. [133] examined, in a systematic fashion, changes in the *ex situ* ^{57}Fe MES of isotopically enriched ^{57}FeTPPCl dispersed in Vulcan XC-72 as a function of T_p. As shown in Figure 3.77, the single peak found for the intact specimen gradually converted into a doublet and then into a very complex spectrum consisting of a multiplicity of lines at $T_p = 800\,°C$. Many of these spectral features decreased significantly in intensity upon washing with aqueous H_2SO_4, to yield a prominent doublet characteristic of specimens pyrolyzed at lower T_p (see Figure 3.78). The most likely explanation for this effect is the dissolution of iron oxide formed by exposure of the heat treated sample to the atmosphere, in agreement with information derived from cyclic voltammetry above. As the electrocatalytic activity in 0.5 M H_2SO_4 was found to be the highest for FeTPPCl|XC-72 pyrolyzed at $T_p = 700\,°C$, it seems reasonable to conclude that the acid insoluble material responsible for the doublet, is in fact the electrocatalytic species.

3.4.3.3 X-ray Absorption Fine Structure

Additional insight into the nature of the catalytic site was provided by the results of Fe K-edge X-ray absorption near edge structure (XANES) measurements of FeTMPPCl|BP [134]. As shown in Figure 3.79, the material pyrolyzed at 800 °C, as a dry powder, displays a very pronounced shoulder at energies very similar to those found for metallic Fe, which suggest that the heat treatment leads at least partially to

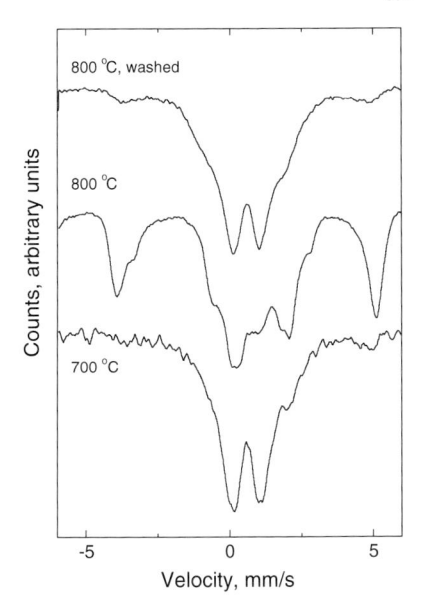

Figure 3.78 *Ex situ* ^{57}Fe MES of isotopically enriched ^{57}FeTPPCl dispersed on Vulcan XC-72 for specimens pyrolyzed at 700 °C (lower curve) and 800 °C (middle curve) following washing with an acidic solution. Also shown for comparison is the corresponding spectrum of the specimen pyrolyzed at 800 °C before washing.

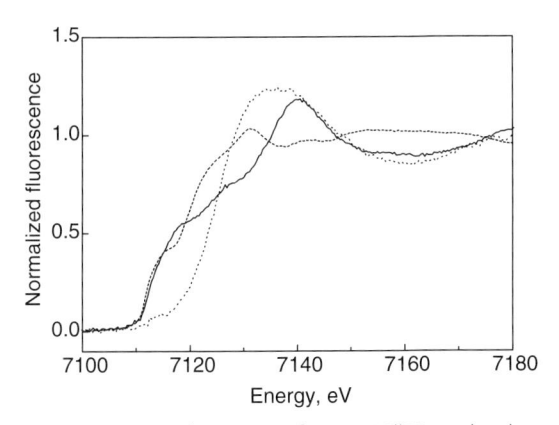

Figure 3.79 Fe K-edge XANES of FeTMPPCl|BP pyrolyzed at 800 °C in powder form (solid line) and following incorporation into a Teflon-bonded electrode immersed in 0.1 M H_3PO_4 at open circuit (about 0.6 V versus DHE, *in situ*, dotted line). Also shown in this figure is the *ex situ* XANES of an Fe foil for comparison (dashed line).

the reduction of the metal center to its elemental state. Marked changes in the XANES were observed for the same material in the form of a Teflon-bonded electrode immersed in 0.1 M H_3PO_4 (see dotted line) without potential control. Most prominent is the shift in the absorption edge to higher energies to yield values not unlike those found for the intact macrocycle (not shown here). These striking differences provide unambiguous evidence that conclusions regarding the nature of the electrocatalytic site drawn purely on the basis of *ex situ* data of dry materials must be made with extreme caution. Additional information on the closely related FeTPP adsorbed on Vulcan XC-72 carbon was obtained from the analysis of the *ex situ* Fe K-edge EXAFS, which showed a decrease in the Fe nearest neighbor distance (assuming it to be nitrogen for the pyrolyzed material) from 2.047 down to 1.925 Å, without changing the coordination number $N = 4$ [133]. This marked change provides yet another piece of the puzzle that in time will help unveil the identity of the species responsible for the enhanced electrocatalytic activity and stability under operation.

In yet another application of this powerful spectroscopic and structural technique, correlations were sought between changes induced by the thermal treatment and electrocatalytic activity for cobalt phthalocyanine (CoPc) [135]. Shown in Figure 3.80A are a series of ex situ Co K-edge XANES for CoPc dispersed in Vulcan XC-72, treated at various temperatures. Also included in that figure are the corresponding spectra for pristine unsupported CoPc (lower curve) As clearly evident from the data, a set of features quite different than those of the intact CoPc either in supported or unsupported form begin to emerge for the specimen heated

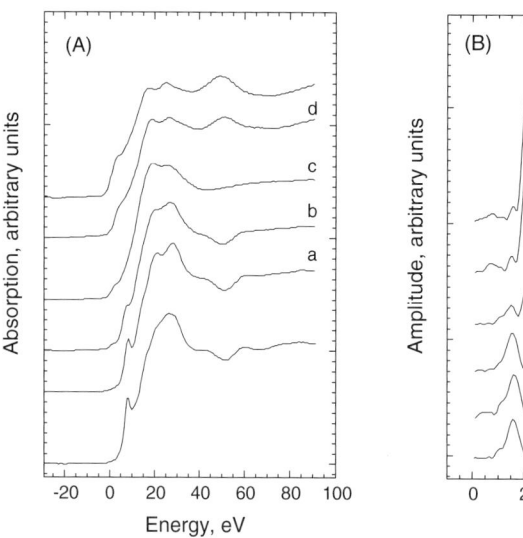

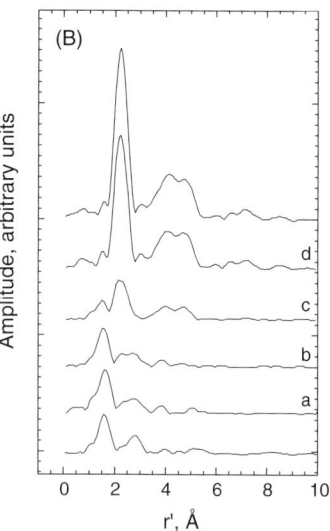

Figure 3.80 A. *Ex situ* Co K-edge XANES for CoPc adsorbed on Vulcan XC-72 not thermally activated (curve a) and heat treated at $T_p = 700$ (curve b), 800 (c) and 1000 °C (d). The lower and upper curves are, respectively, the corresponding XANES for the crystalline unsupported CoPc and of metallic Co for comparison. B. Fourier transforms of the Co K-edge EXAFS functions of the same specimens in A in this figure. The zero in the enegy axis is at 7709 eV.

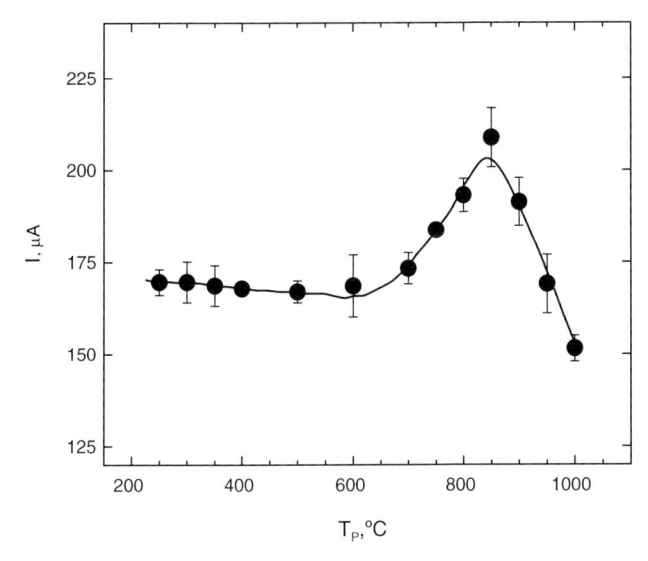

Figure 3.81 Current associated with dioxygen reduction in H_2SO_4 (pH = 0.5) for CoPc|XC-72 measured at −0.15 V vs SCE as a function of T_p. The catalyst loading was 6.4 mg/cm^2.

at $T_p = 700\,°C$ and become dominant for higher temperatures. The same overall behavior can be gleaned from the Fourier transform of the EXAFS function in Figure 3.80B, where the high temperature specimens are characterized by a shell at a distance longer than that for the pristine counterparts. In fact, the spectral features observed for the specimens treated at the highest temperature are quite similar to those found for metallic Co shown in the upper curves in Figures 3.80A and B. These observations are consistent with drastic changes in the metal microenvironment and more specifically with the loss of the MeN4 moeities. The electrocatalytic activity of these specimens in sulfuric acid (see Figure 3.81 and its caption for details), however, was found to peak for $T_p = 850\,°C$, i.e. a value at which no remnants of the original macrocyclic are detected by XAS.

3.4.3.4 X-ray Photoelectron Spectroscopy

Although not very useful in terms of detection and identification of metal sites due to their rather low cross-sections and overall dilute character of the specimens, X-ray photoelectron spectroscopy (XPS) studies involving FeTPP|XC-72, showed a single N 1s peak at about 398 eV for the dispersed intact macrocycle. As T_p was raised beyond 700 °C, a second feature at 401.3 eV was found to emerge at the expense of the low energy peak [129].

3.4.4
In Situ and Quasi *In Situ* Spectroscopic Characterization

Perhaps the most significant information regarding the redox properties of pyrolyzed macrocycles has been obtained from *in situ* Fe K-edge XANES measurements for

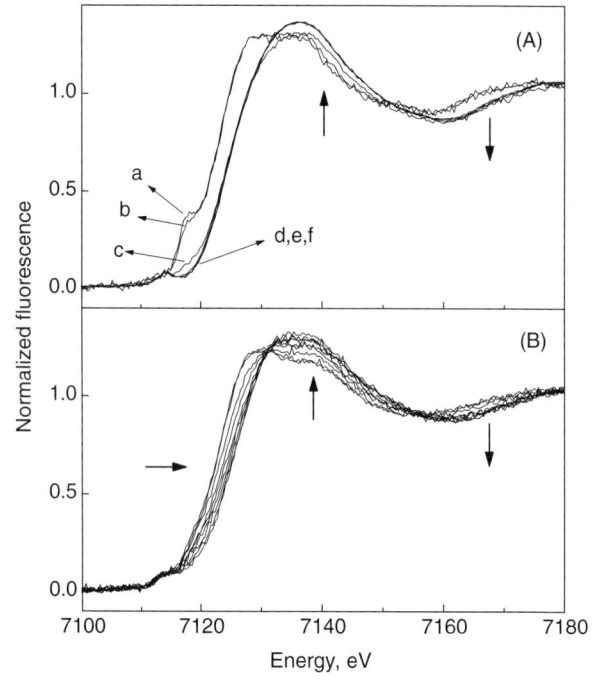

Figure 3.82 Series of *in situ* Fe K-edge XANES for intact (A) and heat treated FeTMPPCl|BP (B) as a function of the applied potential in 0.1 M H$_3$PO$_4$. The arrows point in the direction of increasing potentials for data collected at −0.20, 0.00, 0.20, 0.30, 0.50, 0.60, 0.80, 0.90 and 1.15 V versus DHE.

intact (A) and heat treated FeTMPPCl|BP (B) as a function of the applied potential in 0.1 M H$_3$PO$_4$ (see Figure 3.82 and caption for details) [134]. The large differences that exist between the *in situ* XANES of these two specimens are better illustrated in A and B of Figure 3.83, for electrodes polarized at −0.2 and 0.9 V, respectively. Plots of ΔE_{edge} versus E extracted from data shown in Figure 3.82 are given in Figure 3.84 for the intact (A) and heat-treated FeTMPPCl|BP (B), over the ranges $0.35 < E < 0.90$ and $-0.40 < E < 1.20$ V versus a dynamic hydrogen reference electrode (DHE), where the solid and open circles and triangles in B are based on results for the heat-treated FeTMPPCl|BP in three completely independent runs collected at two different synchrotron radiation facilities. Within the uncertainty of these measurements, the dependence of ΔE_{edge} on E in the range $0.0 < E < 1.2$ V for the heat treated specimens is almost linear (2 eV V^{-1}) and fairly reproducible. This behavior is unlike that observed for the intact form of FeTMPPCl|BP, which displays a sharp transition, within about 0.2 V, between the two oxidation states (see A). However, the overall shift in ΔE_{edge}, about 2.5 eV, is almost identical for both the intact and heat-treated species and consistent with a net oxidation state change of a single electron per iron site. The rather broad potential range over which the shift occurs for the heat treated specimen is most likely due to the presence of an ensemble of sites with similar electronic

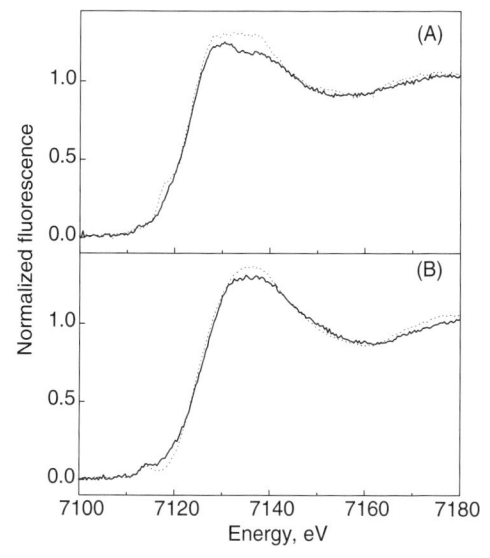

Figure 3.83 Comparison between the *in situ* Fe K-edge XANES of FeTMPPCl|BP before (dotted line) and after (solid line) heat treatment at −0.20 (A) and 0.90 V (B) based on the data in Figure 3.82.

properties and, thus, redox potentials. This interpretation is in harmony with the voltammetry, which showed an extremely broadened set of peaks.

Drastic changes in the nature of the Fe sites as a function of the applied potential could also be discerned from quasi *in situ* ^{57}Fe MES measurements reported by Bouwkamp-Wijnoltz *et al.*, [133] who implemented the *in situ* freezing techniques developed by Fierro *et al.* in our laboratories [136]. This tactic relies on exposure of a polarized electrode to low temperatures while under potential control so as to prevent the interface from undergoing discharge. Rather surprisingly, however, the spectra of the material in the reduced state recorded in this fashion, yielded very similar features to those in the oxidized state, which suggests that the number of electroactive species may represent only a small fraction of the total present in the specimen.

3.4.5
Concluding Remarks

Although no agreement has yet been reached regarding the structural nature of the electrocatalytic sites generated by the heat treatment, the evidence so far collected points to the need for the metal, nitrogen and carbon as the key ingredients to impart specimens with such unique characteristics. Prompted in part by these observations, researchers are exploring the properties of materials prepared by incorporating such elements from sources other than transition metal macrocycles. Indeed, pyrolysis of a variety of lower cost precursors containing these elements in the range 700–1100 °C have been shown to yield materials displaying spectroscopic, electrochemical and electrocatalytic properties not unlike those observed with macrocycles. In particular,

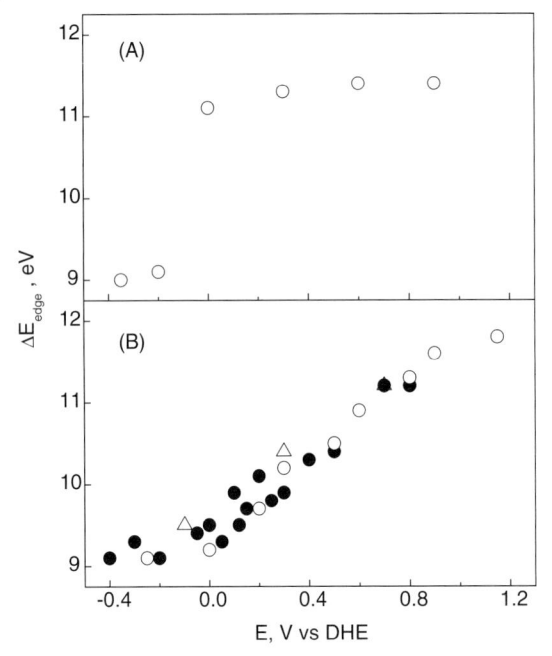

Figure 3.84 Plots of ΔE_{edge} versus E extracted from data shown in Figure 3.82 for the non-heat-treated (A) and heat-treated FeTMPPCl/BP (B), over the ranges $-0.35 < E < 0.90$ and $-0.40 < E < 1.20$ V versus DHE, where the solid and open circles and triangles in (B) are based on results for the heat-treated FeTMPPCl|BP in three completely independent runs collected in two different synchrotron radiation facilities.

some of the best catalysts in terms of long-term performance (10–300 h) and high activity have been produced by high-temperature treatment of iron(II) acetate adsorbed on microporous carbon in an NH_3–H_2–Ar atmosphere at 900 °C [137]. Perhaps the best reported catalyst of this class was obtained by pyrolysis of $RuCl_3$, propylene diamine and urea on Ketjen Black carbon at 700 °C [138]. Not only were the peroxide yields found to be very low, but the reported overvoltage was only 50 mV higher than for state-of-the-art Pt–C catalyst (20 µg cm^{-2} loading). Although the long-term stability of this pyrolyzed Ru–N–C material has not as yet been investigated, the loadings of Ru required to achieve optimum performance are fairly high (120 µg cm^{-2}) and, as such, can hardly be considered as an economically viable substitute for Pt.

Acknowledgements

The authors would like to express their gratitude to Prof. Auro Atsushi Tanaka, Department of Chemistry, Federal University of Maranhao, Sao Luis (City) - Maranhao (State), Brazil for making available some of the data in Section 3.4 in digital form, and to Dr. Eugene O'Sullivan from IBM Yorktown Heights, NY, US for a critical reading of a few sections of the manuscript. Support for this work was provided by a Grant from NSF.

References

1. Adzic, R.R. (1998) in: *Electrocatalysis*, (eds J. Lipkowski and P.N. Ross), Wiley, New York.
2. Jasinski, R.J. (1964) *Nature*, **201**, 1212.
3. Jasinski, R.J. (1965) *J. Electrochem. Soc.*, **112**, 526.
4. Ferguson-Miller, S. and Babcock, G.T. (1996) *Chem. Rev.*, **96**, 2889.
5. Iwata, S., Ostermeier, C., Ludwig, B., and Michel, H. (1995) *Nature*, **376**, 660.
6. McKeown, N.B. (1998) *Phthalocyanine Materials: Synthesis, Structure, and Function*, Cambridge University Press, Cambridge, UK, New York.
7. Cotton, F.A., Wilkinson, G., Murillo, C.A., and Bochmann, M. (1999) *Advanced Inorganic Chemistry*, 6th edn, Wiley, New York.
8. Falk, J.E. (1975) *Porphyrins and metalloporphyrins: a new edition based on the original volume* (ed. K.M. Smith), Elsevier Scientific, Amsterdam, New York.
9. Dolphin D. (ed.) (1978) *The Porphyrins*, Academic Press, New York.
10. Collman, J.P., Elliott, C.M., Halbert, T.R., and Tovrog, B.S. (1977) *Proc. Natl. Acad. Sci. USA*, **74**, 18.
11. Chang, C.K. (1977) *J. Am. Chem. Soc.*, **99**, 2819.
12. Collman, J.P., Denisevich, P., Konai, Y., Marrocco, M., Koval, C., and Anson, F.C. (1980) *J. Am. Chem. Soc.*, **102**, 6027.
13. Radyushkina, K.A. and Tarasevich, M.R. (1986) *Sov. Electrochem.*, **22**, 1087.
14. Lipkowski J. and Ross P.N. (eds) (1998) *Electrocatalysis*, Wiley VCH, New York.
15. Bielski, B.H.J. and Allen, A.O. (1977) *J. Phys. Chem.*, **81**, 1048.
16. Wroblowa, H.S., Pan, Y.-C. and Razumney, G. (1976) *J. Electroanal. Chem.*, **69**, 195.
17. Tanaka, A.A., Fierro, C., Scherson, D.A. and Yeager, E. (1989) *Mater. Chem. Phys.*, **22**, 431.
18. Adzic, R.R., Strbac, S., and Anastasijevic, N. (1989) *Mater. Chem. Phys.*, **22**, 349.
19. Lima, F.H.B., Zhang, J., Shao, M.H., Sasaki, K., Vukmirovic, M.B., Ticianelli, E.A. and Adzic, R.R. (2007) *J. Phys. Chem. C.*, **111**, 404.
20. Zagal, J., Bindra, P. and Yeager, E. (1980) *J. Electrochem. Soc.*, **127**, 1506.
21. Song, E.H., Shi, C.N. and Anson, F.C. (1998) *Langmuir*, **14**, 4315.
22. Shigehara, K. and Anson, F.C. (1982) *J. Phys. Chem.*, **86**, 2776.
23. Zhang, J.L., Vukmirovic, M.B., Sasaki, K., Nilekar, A.U., Mavrikakis, M. and Adzic, R.R. (2005) *J. Am. Chem. Soc.*, **127**, 12480.
24. Lever, A.B.P., Milaeva, E.R., and Speier, G. (1993) *Solution Phthalocyanines, Properties and Applications*, Vol. 3 (eds C.C. Leznoff and A.B.P. Lever), VCH, New York, p. 63.
25. Guilard, R. and Kadish, K.M. (1988) *Chem. Rev.*, **88**, 1121.
26. Jones, R.D., Summerville, D.A. and Basolo, F. (1979) *Chem. Rev.*, **79**, 139.
27. Munakata, M. and Endicott, J.F. (1984) *Inorg. Chem.*, **23**, 3693.
28. Wong, C.-L. and Endicott, J.F. (1981) *Inorg. Chem.*, **20**, 2233.
29. Chang, C.K. (1977) *J. Chem. Soc. Chem. Comm.*, 800.
30. Momenteau, M. and Reed, C.A. (1994) *Chem. Rev.*, **94**, 659.
31. Torrens, F. (2003) *Polyhedron*, **22**, 1091.
32. Nakashima, H., Hasegawa, J.Y. and Nakatsuji, H. (2006) *J. Comput. Chem.*, **27**, 426.
33. Forshey, P.A., Kuwana, T., Kobayashi, N. and Osa, T. (1982) *Electrochemical and Spectrochemical Studies of Biological Redox*

Components, vol. 201 (ed. K.M. Kadish), American Chemical Society, Washington, D.C., p. 601.
34. Durand, R.R. and Anson, F.C. (1982) *J. Electroanal. Chem.*, **134**, 273.
35. Forshey, P.A. and Kuwana, T. (1983) *Inorg. Chem.*, **22**, 699.
36. Forshey, P.A. and Kuwana, T. (1981) *Inorg. Chem.*, **20**, 693.
37. Forshey, P.A., Kuwana, T., Kobayashi, N. and Osa, T. (1982) *Advances In Chemistry Series*, 601.
38. Geiger, T. and Anson, F.C. (1981) *J. Am. Chem. Soc.*, **103**, 7489.
39. Bard, A.J. and Faulkner, L.R. (2001) *Electrochemical Methods: Fundamentals and Applications*, 2nd edn, John Wiley and Sons, New York.
40. Kuwana, T. and Heineman, W.R. (1976) *Acc. Chem. Res.*, **9**, 241.
41. Rohrbach, D.F., Deutsch, E., Heineman, W.R. and Pasternack, R.F. (1977) *Inorg. Chem.*, **16**, 2650.
42. Kobayashi, N., Koshiyama, M., Osa, T. and Kuwana, T. (1983) *Inorg. Chem.*, **22**, 3608.
43. Goff, H. and Morgan, L.O. (1976) *Inorg. Chem.*, **15**, 3180.
44. Nicholson, R.S. (1965) *Anal. Chem.*, **37**, 1351, and references therein.
45. Jahn, D. and Vielstich, W. (1962) *J. Electrochem. Soc.*, **109**, 849.
46. Bowers, M.L., Anson, F.C. and Feldberg, S.W. (1987) *J. Electroanal. Chem.*, **216**, 249.
47. Wong, C.L., Switzer, J.A., Balakrishnan, K.P. and Endicott, J.F. (1980) *J. Am. Chem. Soc.*, **102**, 5511.
48. Bowers, M.L. and Anson, F.C. (1984) *J. Electroanal. Chem.*, **171**, 269.
49. Kang, C. and Anson, F.C. (1995) *Inorg. Chem.*, **34**, 2771.
50. Kuwana, T., Fujihira, M., Sunakawa, K., and Osa, T. (1978) *J. Electroanal. Chem.*, **88**, 299.
51. Bettelheim, A. and Kuwana, T. (1979) *Anal. Chem.*, **51**, 2257.
52. Pasternack, R.F. and Halliwell, B. (1979) *J. Am. Chem. Soc.*, **101**, 1026.
53. Marchaj, A., Bakac, A., and Espenson, J.H. (1993) *Inorg. Chem.*, **32**, 2399.
54. Bhugun, I. and Anson, F.C. (1996) *Inorg. Chem.*, **35**, 7253.
55. Murray R.W. (ed.) (1992) *Molecular Design of Electrode Surfaces*, J. Wiley and Sons, New York.
56. Murray, R.W. (1980) *Acc. Chem. Res.*, **13**, 135.
57. Lane, R.F. and Hubbard, A.T. (1973) *J. Phys. Chem.*, **77**, 1401.
58. Shigehara, K. and Anson, F.C. (1982) *J. Phys. Chem.*, **86**, 2776.
59. Tse, D.C.-S. and Kuwana, T. (1978) *Anal. Chem.*, **50**, 1315.
60. Zagal, J., Bindra, P., and Yeager, E. (1980) *J. Electrochem. Soc.*, **127**, 1506.
61. Rocklin, R.D. and Murray, R.W. (1981) *J. Phys. Chem.*, **85**, 2104.
62. Tao, N.J. (1996) *Phys. Rev. Lett.*, **76**, 4066.
63. Song, E.H., Shi, C.N., and Anson, F.C. (1998) *Langmuir*, **14**, 4315.
64. Zecevic, S., Simic-Glavaski, B., Yeager, E., Lever, A.B.P., and Minor, P.C. (1985) *J. Electroanal. Chem.*, **196**, 339.
65. Shi, C.N. and Anson, F.C. (1992) *Inorg. Chem.*, **31**, 5078.
66. Lennox, J.C. and Murray, R.W. (1978) *J. Am. Chem. Soc.*, **100**, 3710.
67. Kahl, J.L., Faulkner, L.R., Dwarakanath, K. and Tachikawa, H. (1986) *J. Am. Chem. Soc.*, **108**, 5434.
68. Bettelheim, A., White, B.A., and Murray, R.W. (1987) *J. Electroanal. Chem.*, **217**, 271.
69. Buttry, D.A. and Anson, F.C. (1984) *J. Am. Chem. Soc.*, **106**, 59.
70. Kolb, D.M. (1988) *Spectroelectrochemistry: Theory and Practice* (ed. R.J. Gale), Plenum Press, New York.
71. Kim, S., Wang, Z.H., and Scherson, D.A. (1997) *J. Phys. Chem. B*, **101**, 2735.
72. Kim, S.H. and Scherson, D.A. (1992) *Anal. Chem.*, **64**, 3091.
73. Campion, A. and Kambhampati, P. (1998) *Chem. Soc. Rev.*, **27**, 241.
74. Cai, W.B., Stefan, I.C., and Scherson, D.A. (2002) *J. Electroanal. Chem.*, **524**, 36.
75. Gerson, A.R., Halfpenny, P.J., Pizzini, S., Ristic, R., Roberts, K.J., Sheen, D.B., and Sherwood, J.N. (1999) *X-ray Characterization of Materials* (ed. E. Lifshin),

Wiley-VCH Verlag GmbH, Weinheim, Germany, p. 105.
76 Zubavichus, Y.V. and Slovokhotov, Y.L. (2001) *Russ. Chem. Rev. (Engl. Transl.)*, **70**, 373.
77 Kim, S., Bae, I.T., Sandifer, M., Ross, P.N., Carr, R., Woicik, J., Antonio, M.R., and Scherson, D.A. (1991) *J. Am. Chem. Soc.*, **113**, 9063.
78 Kim, S.Y., Tryk, D.A., Bae, I.T., Sandifer, M., Carr, R., Antonio, M.R., and Scherson, D.A. (1995) *J. Phys. Chem.*, **99**, 10359.
79 Durand, R.R., Bencosme, C.S., Collman, J.P. and Anson, F.C. (1983) *J. Am. Chem. Soc.*, **105**, 2710.
80 Albery, W.J., Boutelle, M.G., Colby, P.J., and Hillman, A.R. (1982) *J. Electroanal. Chem.*, **133**, 135.
81 Gerischer, H. and Scherson, D.A. (1985) *J. Electroanal. Chem.*, **188**, 33.
82 Brown, A.P. and Anson, F.C. (1977) *Anal. Chem.*, **49**, 1589.
83 Sagara, T., Murase, H., Komatsu, M. and Nakashima, N. (2000) *Appl. Spectrosc.*, **54**, 316.
84 Shi, P., Geraldo, D., Fromondi, I., Zagal, J.H. and Scherson, D.A. (2005) *Anal. Chem.*, **77**, 6942.
85 Stefan, I.C., Mo, Y.B., Ha, S.Y., Kim, S., and Scherson, D.A. (2003) *Inorg. Chem.*, **42**, 4316.
86 Ngameni, E., Lemest, Y., Lher, M., Collman, J.P., Hendricks, N.H. and Kim, K. (1987) *J. Electroanal. Chem.*, **220**, 247.
87 Feng, Z.Q., Sagara, T. and Niki, K. (1995) *Anal. Chem.*, **67**, 3564.
88 Xie, Y.W. and Anson, F.C. (1996) *J. Electroanal. Chem.*, **414**, 91.
89 Xie, Y.W., Kang, C. and Anson, F.C. (1996) *J. Chem. Soc., Faraday Trans.*, **92**, 3917.
90 Zagal, J., Paez, M., Tanaka, A.A., Dossantos, J.R. and Linkous, C.A. (1992) *J. Electroanal. Chem.*, **339**, 13.
91 Collman, J.P. and Kim, K. (1986) *J. Am. Chem. Soc.*, **108**, 7847.
92 Collman, J.P., Chng, L.L. and Tyvoll, D.A. (1995) *Inorg. Chem.*, **34**, 1311.
93 Shi, C.N., Mak, K.W., Chan, K.S. and Anson, F.C. (1995) *J. Electroanal. Chem.*, **397**, 321.
94 Shi, C., Steiger, B., Yuasa, M. and Anson, F.C. (1997) *Inorg. Chem.*, **36**, 4294.
95 Durand, R.R. and Anson, F.C. (1982) *J. Electroanal. Chem.*, **134**, 273.
96 Barley, M.H., Rhodes, M.R. and Meyer, T.J. (1987) *Inorg. Chem.*, **26**, 1746.
97 Anson, F.C., Shi, C.N. and Steiger, B. (1997) *Acc. Chem. Res.*, **30**, 437.
98 Shi, C. and Anson, F.C. (1991) *J. Am. Chem. Soc.*, **113**, 9564.
99 Steiger, B., Shi, C., and Anson, F.C. (1993) *Inorg. Chem.*, **32**, 2107.
100 Steiger, B. and Anson, F.C. (1994) *Inorg. Chem.*, **33**, 5767.
101 Steiger, B. and Anson, F.C. (1997) *Inorg. Chem.*, **36**, 4138.
102 Yuasa, M., Steiger, B., and Anson, F.C. (1997) *J. Porph. Phthal.*, **1**, 181.
103 Collman, J.P., Denisevich, P., Konai, Y., Marrocco, M., Koval, C. and Anson, F.C. (1980) *J. Am. Chem. Soc.*, **102**, 6027.
104 Liu, H.Y., Weaver, M.J., Wang, C.B. and Chang, C.K. (1983) *J. Electroanal. Chem.*, **145**, 439.
105 Collman, J.P., Chong, A.O., Jameson, G.B., Oakley, R.T., Rose, E., Schmittou, E.R., and Ibers, J.A. (1981) *J. Am. Chem. Soc.*, **103**, 516.
106 Hatada, M.H., Tulinsky, A., and Chang, C.K. (1980) *J. Am. Chem. Soc.*, **102**, 7115.
107 Durand, R.R., Bencosme, C.S., Collman, J.P., and Anson, F.C. (1983) *J. Am. Chem. Soc.*, **105**, 2710.
108 Vik, S.B. and Capaldi, R.A. (1980) *Biochem. Biophys. Res. Commun.*, **94**, 348.
109 Chang, C.K., Liu, H.Y., and Abdalmuhdi, I. (1984) *J. Am. Chem. Soc.*, **106**, 2725.
110 Liu, H.Y., Abdalmuhdi, I., Chang, C.K., and Anson, F.C. (1985) *J. Phys. Chem.*, **89**, 665.
111 Collman, J.P., Marrocco, M., Elliott, C.M., and Lher, M. (1981) *J. Electroanal. Chem.*, **124**, 113.
112 Chang, C.J., Loh, Z.H., Shi, C.N., Anson, F.C., and Nocera, D.G. (2004) *J. Am. Chem. Soc.*, **126**, 10013.

113 Chang, C.J., Deng, Y.Q., Shi, C.N., Chang, C.K., Anson, F.C., and Nocera, D.G. (2000) *J. Chem. Soc., Chem. Commun.*, 1355.

114 Alt, H., Binder, H., and Sandstede, G. (1973) *J. Catal.*, **8**, 28.

115 Jahnke, H., Schonborn, M. and Zimmermann, G. (1976) *Fortschr. Chem. Forsch. (Topics in Current Chemistry)*, **61**, 133.

116 Radyushkina, K.A. and Tarasevich, M.R. (1987) *Elektrokhimiya*, **22**, 1155.

117 Dodelet, J.P. (2006) *N4-Macrocyclic Metal Complexes* (eds J.H. Zagal, F. Bedioui and J.P. Dodelet), Springer Science.

118 Radyushkina, K.A., Tarasevich, M.R., and Andruseva, S.I. (1975) *Elektrokhimiya*, **11**, 1079.

119 Tarasevich, M.R., Radyushkina, K.A., and Zhutaeva, G.V. (2004) *Russ. J. Electrochem.*, **40**, 1174.

120 Maldonado, S. and Stevenson, K.J. (2004) *J. Phys. Chem. B*, **108**, 11375.

121 Manzoli, A. and Boccuzzi, F. (2005) *J. Power Sources*, **145**, 161.

122 Matter, P.H., Wang, E., Arias, M., Biddinger, E.J., and Ozkan, U.S. (2006) *J. Phys. Chem. B*, **110**, 18374.

123 Sidik, R.A., Anderson, A.B., Subramanian, N.P., Kumaraguru, S.P., and Popov, B.N. (2006) *J. Phys. Chem. B*, **110**, 1787.

124 Zhang, L., Zhang, J., Wilkinson, D.P., and Wang, H. (2006) *J. Power Sources*, **156**, 171.

125 Easton, E.B., Yang, R.Z., Bonakdarpour, A., and Dahn, J.R. (2007) *Electrochem. Solid State Lett.*, **10**, B6.

126 Tanaka, A.A., Gupta, S.L., Tryk, D., Fierro, C., Yeager, E.B., and Scherson, D.A. (1992) *Structural Effects in Electrocatalysis and Oxygen Electrochemistry*, **vol. 99–11** (eds D. Scherson, D. Tryk, M. Daroux and X. Xing), The Electrocehmical Society, Inc., Pennington, pp. 555–572.

127 Fierro, C.A., Carbonio, R.E., Scherson, D., and Yeager, E.B. (1987) *J. Phys. Chem.*, **91**, 6579.

128 Totir, D., Mo, Y., Kim, S., Antonio, M.R., and Scherson, D.A. (2000) *J. Electrochem. Soc.*, **147**, 4594–4597.

129 Faubert, G., Lalande, G., Cote, R., Guay, D., Dodelet, J.P., Weng, L.T., Bertrand, P., and Denes, G. (1996) *Electrochim. Acta*, **41**, 1689.

130 Scherson, D.A. and Tanaka, A.A., Gupta, S.L., Tryk, D., Fierro, C., Holze, R., Yeager, E.B. and Lattimer, R.L. (1986) *Electrochim. Acta*, **31**, 1247.

131 van Veen, J.A.R., van Baar, J.F., and Kroese, K.J. (1981) *J. Chem. Soc., Faraday Trans. I*, **77**, 2827.

132 Scherson, D.A., Gupta, S.L., Fierro, C., Yeager, E.B., Kordesch, M.E., Eldridge, J., Hoffman, R.W., and Blue, J. (1983) *Electrochim. Acta*, **28**, 1205.

133 Bouwkamp-Wijnoltz, A.L., Visscher, W., van Veen, J.A.R., Boellaard, E., van der Kraan, A.M., and Tang, S.C. (2002) *J. Phys. Chem. B*, **106**, 12993.

134 Bae, I.T., Tryk, D.A., and Scherson, D.A. (1998) *J. Phys. Chem.*, **102**, 4114.

135 Alves, M.C.M., Dodelet, J.P., Guay, D., Ladouceur, M. and Tourillon, G. (1992) *J. Phys. Chem.*, **96**, 10898.

136 Fierro, C.A., Mohan, M., and Scherson, D.A. (1990) *Langmuir*, **6**, 1338.

137 Jaouen, F., Charreteur, F. and Dodelet, J.P. (2006) *J. Electrochem. Soc.*, **153**, A689.

138 Liu, L.Y., Kim, H., Lee, J.W., and Popov, B.N. (2007) *J. Electrochem. Soc.*, **154**, A123.

4
Multiscale Modeling and Design of Electrochemical Systems

Richard D. Braatz, Edmund G. Seebauer, and Richard C. Alkire

4.1
Introduction

Recent advances in computing technology have set the stage for a new era in computational modeling and simulation. With such capabilities, it will be possible to understand and predict the behavior of complex natural and engineered systems, in addition to extending dramatically the exploration of fundamental processes of nature. Computational modeling is regarded by many as an additional mode of scientific enquiry along with observation (experiment) and hypothesis (theory). Because complex systems typically involve orders-of-magnitude variation in time and length scales, and because different simulation methods are most effective at different scales, the overarching goal which motivates efforts to simulate multiscale systems is to couple multiple simulation methods that produce the best predictions at each scale. This chapter describes the algorithms, software and data analysis tools under development for multiscale simulation, and describes the use of these tools in emerging electrochemical applications for which increased predictability is of great importance.

While multiscale simulation activities have seen significant growth over the past decade, the field today may best be described as being separate groups of researchers in individual disciplines (e.g. solid mechanics, chemical engineering, materials, biophysics) and sub-disciplines (e.g. solid mechanics of crack propagation, plasma dynamics, polymers) than a single cohesive field within the area of computational science and engineering. For example, most books and the vast majority of review papers on multiscale simulation methods are focused on a particular discipline. Further, most research papers on multiscale simulation cite exclusively or nearly exclusively papers within the sub-discipline rather than papers on multiscale simulation that are outside of the sub-discipline. This results in much "re-inventing of the wheel" which is inefficient and slows progress in the field of multiscale simulation, as well as in the particular applications.

The significance of multiscale simulation to the future of science and engineering has been recognized by many groups that span a wide range of particular interests. Multiscale modeling and simulation is the number one challenge in advancing knowledge and understanding in medicine, homeland security, energy, environment, materials and industrial and defense applications, according to the NSF Blue Ribbon Panel report on Simulation-based Engineering Science (http://www.nsf.gov/pubs/reports/sbes_final_report.pdf). 'Integrating multiscale (in space and time) multi-disciplinary simulations' is listed as one of the *top strategic priorities* of the Networking and Information Technology Research and Development (NITRD) federal agencies in High End Computing Research and Development in the NITRD Supplement to the President's FY (fiscal year) 2007 Budget (http://www.nitrd.gov/pubs/2007supplement/). The NITRD federal agencies include the National Science Foundation (NSF), the National Institutes of Health (NIH), the Defense Advanced Research Projects Agency (DARPA), the National Aeronautics and Space Administration (NASA), the National Institute of Standards and Technology (NIST), the National Security Agency (NSA), the National Oceanic and Atmospheric Administration (NOAA), the Department of Defense (DOD), the Environmental Protection Agency (EPA), the Agency for Healthcare Research and Quality (AHRQ), the Department of Homeland Security (DHS), the Department of Energy National Nuclear Security Administration (DOE/NNSA), the DOE Mathematical, Information, and Computational Science Division (DOE/SC), and the National Archives and Records Administration (NARA; http://www.nitrd.gov/subcommittee/agency-websites.html). Better understanding of the 'relationship between human-built systems and natural systems *at all spatial scales*' is listed as one of the *Six Themes of Environmental Sustainability in the Science and Technology for Sustainability* Multi-Year Plan that specifies how the EPA plans to organize research for FY 2008–2012 (http://www.epa.gov/sustainability/releasepubcommt.html). The Department of Energy report *Multiscale Mathematics Initiative: A Roadmap* ([1]; http://www.si.umich.edu/InfrastructureWorkshop/documents/NSF_2004_CIMultiscaleMath.pdf) strongly argues the need for 'a mathematical framework and software infrastructure to integrate heterogeneous models and data over the wide range of scales that characterize most physical phenomena' and for 'fundamentally new mathematics and considerable development of computational methods and software ... to address the challenges of multiscale simulation'. The *Multiscale Modeling Initiative* (http://www.nibib.nih.gov/Research/MultiscaleModeling) supports grants across eight of the National Institutes of Health, three Directorates of the National Science Foundation and two offices in the DOE and NASA. The NSF *Petascale Computing Initiative* lists as motivation the enabling of researchers to 'perform simulations that are intrinsically multiscale' (http://www.nsf.gov/pubs/2005/nsf05625/nsf05625.pdf).

Electrochemical phenomena play a central role in the fabrication and the functional capabilities of a great many materials, processes and devices. A common feature of these systems is that their behavior is largely determined as a result of concerted interactions that extend over many length scales. During the past several decades, the electrochemical field has advanced rapidly based primarily on a suite of remarkable new tools that: provide the ability to create precisely characterized systems for fundamental study; monitor behavior at unprecedented levels of sensitivity, resolution and chemical specificity; and predict behavior with new theories and improved computational abilities.

These capabilities have revolutionized fundamental understanding, and also contributed to the present rapid pace of the discovery of novel materials and devices where product quality is determined at the molecular scale. However, the manufacturability of such devices requires precise, quantitative understanding at a magnitude, sophistication and completeness that is extraordinarily difficult to assemble today. To speed development, engineers follow the 'targeted design' approach, which builds on the foundation of curiosity-driven discoveries, but involves working backward from the desired function or product to perfect the underlying material and the process conditions by which it can be fabricated. These trends point the way toward future opportunities for *molecular engineering* which, as Carl Wagner observed over in the 1960s (Wagner, 1962), 'may be important in the future development of industrial electrochemical processes'.

4.2
Background and Motivation

The main objective of this work is to discuss recent developments in molecular simulation, multiscale simulation and multiscale systems engineering, and how these developments enable the targeted design of processes and products at the molecular scale. The control of events at the molecular scale is critical to product quality in many new applications in medicine, computers and manufacturing.

4.2.1
Multiscale Simulation

The anticipated benefits of multiscale simulation to society are substantial. For example, the NSF Blue Ribbon Panel report on *Simulation-based Engineering Science* (http://www.nsf.gov/pubs/reports/sbes_final_report.pdf) describes benefits of accurate predictions for multiscale systems which include:

- improved understanding of protein biology and disease leading to improved pharmaceuticals and other therapeutics for improving human health;

- improved predictions that lead to the design of energy-efficient fuel cells for electric vehicles;

- improved understanding of our environment leading to techniques to optimize physical and cyberinfrastructure to mitigate the effects of man-made and natural disasters such as hurricanes and earthquakes on human life and property;

- improved predictions for the behavior of electronic materials within nanoscale devices, enabling the design of nanoelectronic processors with order-of-magnitude increases in clock speed and integrated circuit density;

- development of biomedical and nanotechnology applications through predictive simulation at the nanoscale.

Potential new technological developments that may results from these discoveries include nanobiological devices, micromachines, nanoelectronic devices and protein microarrays and next-generation computer chips (Alkire and Braatz, 2005 [179]; Sematech, 2005; [187]). These examples represent only a few of many potential technologies on the horizon that we cannot hope to understand, develop, or utilize in a timely manner without simulation-based engineering methods.

On the other hand, well engineered manufacturing operations depend on the availability of manipulated variables for real-time feedback control. These variables usually operate at macroscopic length scales (e.g. the power to heat lamps above a wafer, the fractional opening of valves on flows into and out of a chemical reactor, the applied potential across electrodes in an electrochemical process). The combination of a need for product quality at the molecular scale with the economic necessity that feedback control systems utilize macroscopic manipulated variables motivates the creation of methods for the simulation, design and control of multiscale systems.

The focus of this Section 4.2 on Background and Motivation is to summarize techniques for composing and utilizing multiscale simulation models to perform systems engineering tasks, such as parameter estimation, design and control ([9–11]; and references cited therein). As multiscale simulation models are capable of directly and simultaneously addressing phenomena across length scales from the sub-atomic to macroscopic, the molecular and nanometer length scales are also covered. Any systems problems posed for processes at these length scales are just special cases of systems problems defined for multiscale systems. That is, a systems approach that is sufficiently general to handle multiscale systems can also address the problems posed by the various definitions of nanotechnology, molecular manufacturing, molecular nanotechnology and molecular engineering. The incorporation of models that couple molecular through macroscopic length scales within systems tools enables a systematic approach to the simultaneous optimization of all of the length scales, including the optimal control of events at the molecular scale. Such a multiscale systems framework would address the 'grand challenge' of nanotechnology: how to move nanoscale science and technology from art to an engineering discipline [12].

Multiscale simulation involves the use of distinct methods appropriate for different length and time scales that are applied simultaneously to achieve a comprehensive description of a system. The range of length scales in the applications addressed here range from electronic to atomic to molecular to nanoscale to microscale to macroscale. Figure 4.1 illustrates some of the computational methods that have been developed over many decades in order to deal with phenomena at different time and length scales to compute properties and model phenomena. These include *quantum mechanics* for accurate calculation of small and fast phenomena, *statistical mechanics* including molecular dynamics and Monte Carlo methods for mechanistic understanding, the *mesoscopic scale* where new methods have emerged and *continuum mechanics* for macroscopic reaction and transport modeling.

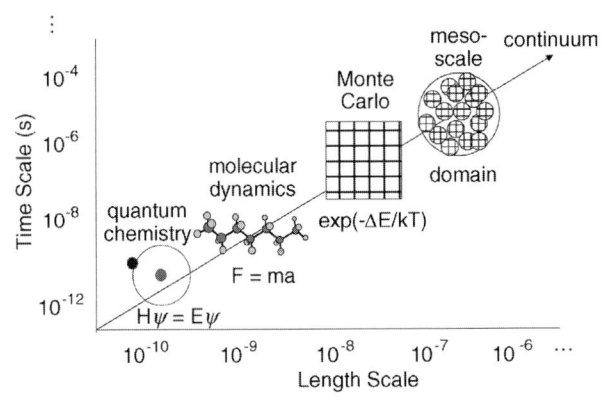

Figure 4.1 Various simulation methods and their suitability for multiple time and length scales.

4.2.2
Electrochemical Systems

Figure 4.2 shows the broad range of time and length scales, over 15 orders of magnitude, that are associated with phenomena that are involved in modeling electrochemical processing associated with one particular application: fabrication of on-chip interconnects for microelectronic devices. The figure is divided into three vertical bands in which different types of numerical simulation tools are used: noncontinuum, continuum and

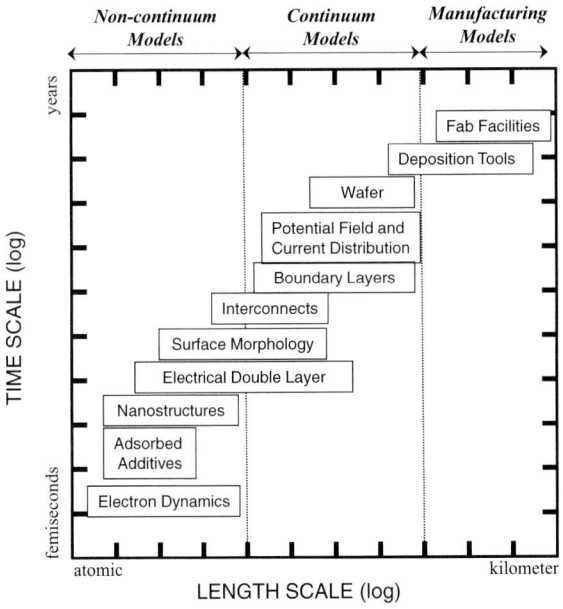

Figure 4.2 Range of time and length scales for electrochemical processing for chip manufacture [37].

manufacturing scale. The phenomena indicated in the boxes are placed in the approximate position to which they correspond. As an illustrative example of the multiple length scales involved in a microelectronics manufacturing process, consider the electrodeposition of fine copper which interconnect on-chip components. To be profitable, the integrated circuits are simultaneously manufactured on a 300 mm wafer, so each manufacturing process, including the copper electrodeposition process, requires spatial uniformity over the 30 cm across the wafer. On the other hand, control of length scales of a few nanometers is required to achieve uniform filling of trenches to form copper wires that serve to interconnect chip components.

In general, there is a now sophisticated understanding of the various phenomena indicated in Figure 4.1, each at its relevant scale. Multiscale simulation involves linking the pieces in order to understand interactions within an entire system [13, 14]. The challenges posed by composing multiscale systems serve to specify the requirements for development of new simulation algorithms and systems tools.

The range of possible applications of multiscale simulations methods to electrochemical systems is extensive. Electrochemical phenomena control the existence and movement of charged species in the bulk, and across interfaces between ionic, electronic, semiconductor, photonic and dielectric materials. The existing technology base of the electrochemical field is massive and of long-standing [15, 16]. The pervasive occurrence of these phenomena in technological devices and processes, and in natural systems, includes:

- materials include metals, alloys, ceramics, ionic solids, semiconductors, membranes, coatings, colloids, conducting polymers and biological materials including proteins and enzymes;

- phenomena that arise include conduction, potential field effects, electron or ion disorder, electroluminescence, ion exchange, passivity, membrane transport, double layers at boundaries between phases involving free charges, osmotic flow and electrokinetic phenomena;

- processes that depend critically on these phenomena include energy storage and conversion, corrosion, membrane separations, electrodeposition, etching, desalination, electrosynthesis of chemicals, refining of metals and many others;

- products that result include microelectronic devices, sensors, batteries, fuel cells, coatings, films, metals, gases, chemicals and ceramics.

The number of discoveries that have the potential for the development of new devices and processes in the future is too large to do more than mention a few examples:

- molecular electronic devices [17, 18] at the 1–2 nm scale;
- molecular electronic junctions based on carbon nanotubes [19];
- ion channels as nanoscale pathways for transport across membranes [20–22];
- organic light-emitting display devices based on electroluminescence [23, 24];
- anodized porous silicon [25] that lead to remarkable properties including photoluminescence, electroluminescence, and thermally induced ultrasonic emission;

- thin film magnets formed from electrochromic materials [26];
- charge transport in DNA elucidated from electrochemical rate measurements [27];
- switchable transitions in single-molecule surface layers that alter wetting, adhesion, friction and biocompatibility [28];
- nanostructures fabricated by electrodeposition [29, 30], including nanoclusters [31, 32], atomic layer epitaxy [33].

Future trends in electrochemical engineering will be influenced by the need to develop molecular-based discoveries into new and improved products and processes. What is needed is to develop a multiscale systems approach that builds upon the traditional base of continuum-scale mathematical models.

The traditional electrochemical technology base now represents a major market force with impact in three general areas: (a) as a major industry for materials and chemicals production, (b) as an enabling technology for other industries (such as batteries for portable electronic devices and corrosion control for oil production) and (c) as a means of promoting well-being (such as health care and environmental fields). Because large scale electrolytic processes are invariably driven to a transport-limited rate, the electrochemical engineering research literature of the past half century focused mainly on understanding how ohmic and mass transport processes, including the effect of hydrodynamic flow, influence the potential field between electrodes in addition to the current distribution, or rate of electrochemical reaction along a surface.

Continuum codes dominate the extensive modeling literature in electrochemical systems [34]. Sophisticated simulations of the current and potential distribution in electrochemical systems are now widely used to predict behavior and provide a rational basis for engineering design, scale-up, optimization and process control [35]. A wide variety of continuum-scale phenomena can be included with the result that models are widely used for sorting out competing effects, resolving experimental data, articulating scientific hypotheses of mechanism, measuring system parameters and predicting behavior.

Although continuum models can provide a rational basis for engineering design, optimization and control, they usually incorporate highly simplified characterization of interfacial processes, which appear as 'boundary conditions'. As a consequence, such models have a blind spot at the molecular scale [36, 37], where a great deal of rich scientific discovery is occurring, and where the quality control of next-generation technologies is determined. A multiscale approach is therefore needed for the design and control of next-generation systems based on electrochemical phenomena which, as illustrated in the examples discussed below, can span time and length scales in excess of ten orders of magnitude.

4.2.3
Microelectronic Applications

Beyond its increased focus on molecular simulations, the semiconductor industry has been moving steadily toward integration of simulation domains. For example, the following quotations are from the International Technology Roadmap for Semiconductors (Sematech, 2005 [187]), which describes the detailed technology requirements for semiconductor devices up to 2020, written by hundreds of company representatives from the semiconductor industry associations of the United States, Europe, Taiwan, Japan and Korea:

- 'The most important trend ... is the ever increasing need for improving integration between the various areas of simulation.'
- 'Different effects which could in the past be simulated separately will in future need to be treated simultaneously ...'

One of the most difficult challenges in modeling and simulation for the semiconductor industry is reported as being the:

- 'Integrated modeling of equipment, materials, feature scale processes, and influences on devices' (Table 121 of [38]).

This trend of the semiconductor industry towards multiscale simulation began to be explored in the late 1990s, and has been an active area of academic research (e.g. see [6, 13, 39–45]; and references cited therein).

More recently, techniques have been developed for utilizing multiscale simulation models to perform systems engineering tasks, such as parameter estimation, optimization and control (e.g. see reviews by [9, 10] and [11], and references cited therein). This incorporation of models that couple molecular through macroscopic length scales within systems engineering tools enables a systematic approach to the simultaneous optimization of all of the length scales of the process.

The expected future impact of multiscale simulation in the semiconductor industry is suggested by some of the issues associated with the most difficult challenges in modeling and simulation, as stated in the International Technology Roadmap for Semiconductors (Table 121 of [38]):

- 'modeling hierarchy from atomistic to continuum for dopants and defects in bulk and at interfaces';
- 'linked equipment/feature scale models';
- 'process modeling tools for the development of novel nanostructure devices (nanowires, carbon nanotubes, quantum dots, molecular electronics)';
- 'device modeling tools for analysis of nanoscale device operation'.

4.2.4
Nanoscale Science and Technology

While the first two issues highlighted in the previous paragraph are clearly associated with multiscale simulation, the second two are associated with nanotechnology and nanoelectronics, which are widely perceived as the future of the semiconductor industry. The relationship between multiscale simulation and nanoscale science and technology, however, depends on one's perspective as there are different definitions of the word 'nanotechnology':

- 'the processing of separation, consolidation and deformation of materials by one atom or one molecule [46];

- 'the ability to work at the atomic, molecular and supramolecular levels (on a scale of $\approx$1–100 nm) in order to understand, create and use material structures, devices and systems with fundamentally new properties and functions resulting from their small structure' [47];

- that which includes 'the ability to individually address, control and modify structures, materials and devices with nanometer precision, and the synthesis of such structures into systems of micro- and macroscopic dimensions such as MEMS-based devices' and 'encompasses the understanding of the fundamental physics, chemistry, biology and technology of nanometer-scale objects and how such objects can be used in the areas of computation, sensors, nanostructured materials and nano-biotechnology' [48].

To distinguish between the very broad and very narrow definitions, some researchers have introduced additional terminology, for example, Drexler [2] defined *molecular manufacturing* as 'the construction of objects to complex, atomic specifications using sequences of chemical reactions directed by non-biological molecular machinery' and *molecular nanotechnology* as comprising 'molecular manufacturing together with its techniques, its products and their design and analysis'. These terms contrast with *molecular engineering*, which is typically defined to be the design of a molecule to have desired properties [49–53].

As described earlier in Section 4.2, a systems approach that is sufficiently general to handle multiscale systems can also address the problems posed by the various definitions of nanotechnology, molecular manufacturing, molecular nanotechnology and molecular engineering. Such a multiscale systems framework would address the 'Grand Challenge' of nanotechnology: how to move nanoscale science and technology from art to an engineering discipline [12].

4.2.5
Other Electrochemical Applications

Chemically reactive surfaces play a central role in determining the behavior and the operating characteristics of an enormous variety of electrochemical systems. Precise control of surface processes is invariably required in order to form, manipulate and stabilize the assembly of such systems, from new materials to medical implants and microelectronic devices, including many that are envisioned for nanotechnology applications [2, 3, 5, 7, 8]. Nanotechnology will be used increasingly to assemble novel materials including functional materials (membranes and separators), hard materials (catalysts) and soft materials (additives and chemically modified surface films). By assembling nanostructured composites (involving semiconductors, conducting polymers, redox mediators, etc.), we will learn to manipulate pores with significant double-layer regions in order to achieve unique properties. Engineering methods for exploiting double-layer properties should grow in response to applications that use control at small scales such as in electrophoretic and osmotic flows, microfluidics in MEMS and colloidal and interfacial phenomena.

Biomedical and health care applications are deeply coupled to electrochemical phenomena, as are the very processes of life itself: action potentials, membrane and neurological phenomena, cell fusion, sensory and energy transduction, motility and reproduction. These phenomena are based on interactions between ions, polyelectrolytes (e.g. proteins) or charged membranes containing enzymes and ion-selective channels. Technological applications will grow as engineers merge qualitative biological insights with quantitative engineering methods of analysis, many of which have been developed in electrochemical applications during the past half century.

Many additional growth areas could also be mentioned, from energy storage and conversion with batteries and fuel cells, to environmental monitoring and waste treatment. Because the role of the electrical potential is ubiquitous at the small scale, where a great deal of scientific discovery and innovation is taking place, the electrochemical engineering community has a natural position of advantage for steering technology developments in the future.

4.3
Trend Toward Atomistic/Molecular Simulation

So that the trends can be recognized in the context of the applications that drive them, this section begins with examples from the area of microelectronic device fabrication. The discussion then addresses the simulation methods used at several different scales, as indicated in Figure 4.2. These include continuum methods, which are used for simulating macroscale to microscale lengths, followed by molecular simulation methods, which range from electronic to atomic to molecular to the nanoscale, and coarse-grained methods which bootstrap molecular simulation methods to handle longer time and length scales.

4.3.1
Integrated Circuit Example

Anyone familiar with the complexity of an integrated circuit knows that its design and manufacture would be impossible without the extensive application of simulation. In the semiconductor industry the simulation of carrier transport in electronic devices is referred to as *device simulation* and of the processing steps used to manufacture these devices are referred to as *process simulation*. Most of the early simulation codes for both device and process simulation were based on the simultaneous solution of conservation equations written as partial differential-algebraic (PDAE) equations. Device simulation was necessary for the design of transistors in the 1960s [54–56], when the semiconductor industry was just starting, whereas simulation was applied to many microelectronics manufacturing processes in the 1980s (Barnes et al., 1986 [57–60, 180]).

Process and device simulation becomes more challenging as physical dimensions shrink. This trend towards smaller length scales is well illustrated by the well known law by Intel co-founder Gordon E. Moore [61], who noted that the number of transistors per chip had doubled every year since the integrated circuit was invented, and predicted that the trend would continue into the future. For several decades, Moore's Law was achieved by packing more transistors into smaller dimensions, which requires shrinking the physical dimensions of features in electronic devices from micrometers to nanometers, with molecular dimensions under active investigation (Sematech, 2005 [187]). Although this pace of device innovation has somewhat slowed recently, advances in computer architecture (especially parallelism) and speedup in computational algorithms have sustained Moore's Law. The International Technology Roadmap for Semiconductors indicates that Moore's Law has a good chance of holding for at least another decade (Sematech, 2005 [187]).

To gain some appreciation for these length scales, integrated circuits on a 300 mm wafer exhibit a DRAM product half pitch of 65 nm, which is the standard *nominal feature size* for tracking trends. The nominal feature size has been reduced by many orders-of-magnitude since 1960, to a dimension of 65 nm in 2007. Many of the physical dimensions of features in a transistor are actually much smaller than the nominal feature size (see Table 4.1), and the spatial dimensions required to simulate the formation of these features is two orders-of-magnitude smaller. For example, consider the schematic of a metal oxide semiconductor field effect transistor (MOSFET) in Figure 4.3, which is a common type of transistor. The thin layer of oxide acts as an insulator to separate the channel from the gate. Applying a gate voltage causes the semiconductor in the channel to switch behavior from that of an insulator to a conductor, so that electrons flow from the source to the drain (the source and the drain are typically constructed from copper metal). The regions below the source and drain consist of doped silicon, with the *junction* defined as the interface between each doped silicon region and the silicon material below it. A junction depth in a modern MOSFET is typically about 28 nm. Simulation of the rapid thermal annealing process used to manufacture these junctions requires at least 100 grid cells, resulting a grid cell length of $28/100$ nm $= 0.28$ nm, which is the diameter of a silicon atom! Of course under such situations the continuum assumption breaks down, which motivates the modeling of this process using atomistic simulation (e.g. the DADOS Monte Carlo simulation package, [62–64]).

Table 4.1 Some physical dimensions of features in a high-volume microprocessor in 2004 (Sematech, 2004, Tables 1g, 47a and 81a [187]).

Feature	Length
Nominal feature size	90 nm
Physical gate length	32 nm
Barrier thickness for copper wiring	9 nm
Physical gate oxide thickness	1.1 nm

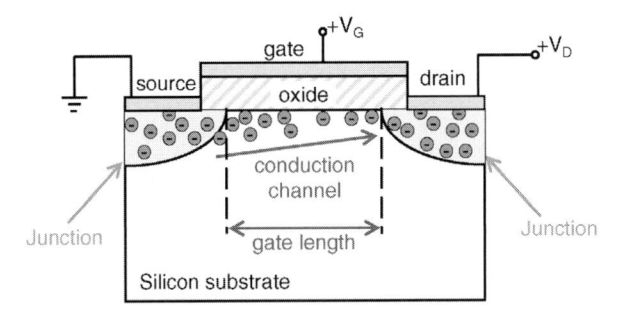

Figure 4.3 Schematic of a metal oxide semiconductor field effect transistor (courtesy of U. Ravaioli, University of Illinois at Urbana-Champaign).

4.3.2
Continuum Methods

Macroscopic phenomena are described by systems of integro-partial differential algebraic equations (IPDAEs) that are simulated by continuum methods such as finite difference, finite volume and finite element methods ([65] and references cited therein; [66, 67]). The commonality of these methods is their use of a mesh or grid over the spatial dimensions [68–71]. Such methods form the basis of many common software packages such as Fluent for simulating fluid dynamics and ABAQUS for simulating solid mechanics problems.

While most modern simulators for process and device simulation are still based on PDAE continuum models [72–76], a purely IPDAE representation becomes less applicable as physical dimensions shrink, which has resulted in molecular and mesoscale simulation becoming important, either as a substitute or as an augmentation to solving PDAE models [77–83]. Molecular simulation has been applied to the chemical vapor deposition of gallium arsenide, which is used in optoelectronics and certain high-speed integrated circuits [39, 43], the electrodeposition of copper to form interconnects [41, 42] and reactive ion etching, TiN sputtering and tungsten chemical vapor deposition to form contacts to connect transistor electrodes to wires [84]. For some device components quantum effects have become important, in which case quantum mechanical calculations must be incorporated [85–87].

4.3.3
Molecular Simulation Methods

Most molecular simulation techniques can be categorized as being among three main types: (1) quantum mechanics, (2) molecular dynamics (MD) and (3) kinetic Monte Carlo (KMC) simulation. Quantum mechanics methods, which include *ab initio*, semi-empirical and density functional techniques, are useful for understanding chemical mechanisms and estimating chemical kinetic parameters for gas-phase and solid-state systems [88, 89]. Density functional theory has been very heavily used to compute energy barriers in both diffusion and chemical reactions [90–92]. The potential energy surface computed by quantum mechanics can be incorporated into molecular dynamics methods, which solve Newton's equations of motion for large numbers of molecules to compute their velocities and positions over time. The forces in Newton's equation ($F = ma$) can be the strong forces due to bonds between atoms and/or weaker forces such as van der Waals or electrostatic forces. Molecular dynamics methods have been used to construct mechanisms and compute diffusion coefficients for many semiconductor processes (Catellani et al., 2004; [89, 93–95, 181]). A closely related method is Langevin dynamics (LD), in which two force terms are added to Newton's equation to approximate neglected degrees of freedom. One term represents the frictional force and the other term the stochastic effects due to the thermal motions of solvent molecules. This approach can provide similar accuracy but is much more computationally efficient than applying MD to all of the solvent molecules.

A typical molecular dynamics simulation may involve up to a million atoms and simulate a time period of several nanoseconds with time steps of femtoseconds (10^{-15} s), which is shorter than the time scales associated with most micro- and nanostructure formation. By restricting the *configuration* or *state* of the process to the identity and positions of atoms and/or molecules (see Figure 4.4), KMC methods can simulate structural properties of matter that cannot be represented by a continuum description while being able to simulate for much longer times (hundreds of seconds) than can be achieved using molecular dynamics. The chemical mechanism, energy barriers and diffusion coefficients in the KMC simulation can be obtained from molecular dynamics and quantum mechanics calculations [44]. The KMC simulation is inherently stochastic, which is consistent with the reality of molecular motion. Owing to the large number of degrees of freedom and the stochastic nature of molecular motion, the real process will follow a different state trajectory and arrive at a different state/configuration, each time an experiment is run. Similarly, the configuration resulting from a KMC simulation

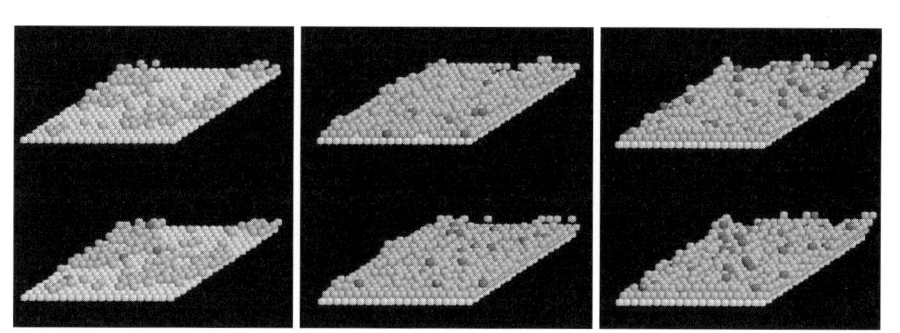

Figure 4.4 Six configurations computed during the KMC simulation of the electrodeposition of copper on a substrate (grey spheres) in the presence of additives (red and green spheres) to form a copper film 2–3 atoms thick [183, 184].

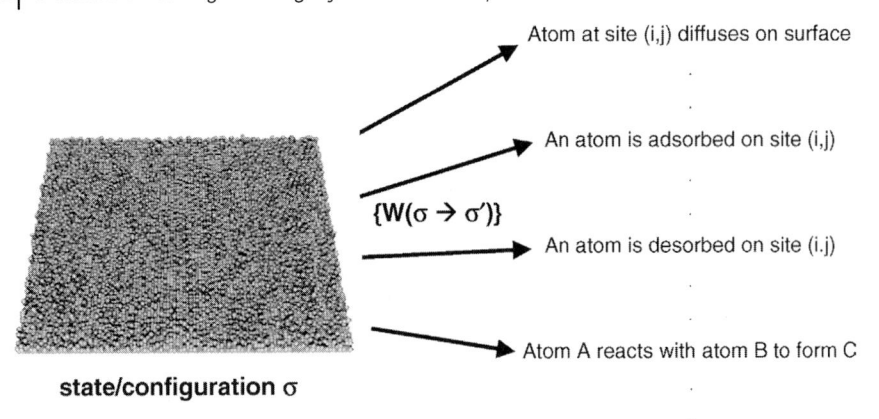

Figure 4.5 The set of transition probabilities from a specific configuration σ to a large number of alternative configurations σ′ involving a single reaction step (surface diffusion, adsorption, desorption, surface reaction) during the KMC simulation of the electrodeposition of copper.

will be different almost every time a KMC simulation is run. The probability per unit time, W(σ, σ′), that the process will undergo a transition from state σ to σ′ can be computed uniquely from the kinetic rates for the individual kinetic steps that can occur in the system (see Figure 4.5).

The probability distribution for each configuration is described by the Master equation [96]:

$$\frac{\partial P(\sigma, t)}{\partial t} = \sum_{\sigma'} W(\sigma', \sigma) P(\sigma', t) - \sum_{\sigma'} W(\sigma, \sigma') P(\sigma, t) \qquad (4.1)$$

where $P(\sigma, t)$ is the probability that the system is in state σ at time t. This is the conservation equation for the probability distribution for each configuration (accumulation = in − out), with the overall system described by writing Equation 4.1 for every possible state/configuration of the system. Because the number of possible states/configurations is too high to solve these state equations directly, KMC simulation follows a single realization of the Master equation by calling a random number generator to select among the possible transitions with probabilities defined by the kinetic rate laws for each allowed kinetic event (e.g. molecule A moves from one lattice site to the adjoining lattice site, molecule A reacts with adjacent molecule B to form molecule C, etc.). At most one kinetic step can be taken during each time step of the KMC algorithm, with the time step (typically of the order of 1 ns) selected so that the time simulated in the KMC algorithm corresponds to real time [96]. Although KMC simulation is much faster than solving the Master equation (Equation 1) exactly for each possible configuration, an efficiently implemented KMC simulation for a process of industrial importance (e.g. 100×100 lattice with 10^{11} time steps) typically takes in the order of a day to run.

4.3.4
Coarse-grained Simulation Methods

Many simulation methods have been developed which try to directly extend molecular simulation methods to longer time and length scales by using a coarser representation for the system. One of the big applications areas is to model the biological phenomena of living cells, which can only be understood through properties and processes that occur at longer time and length scales than molecular. There are very fast and localized processes, such as light absorption and chemical reactions, whereas other processes are spatially spread over a considerable volume of the cell and occur at much longer time scales, for example, signal transduction cascades. While technological advances in numerical algorithms and software have enabled increasingly large structures to be simulated with all-atom molecular dynamics, time and size limitations inherent to all-atom MD simulations do not permit a description of an average-sized cell, or even of common sub-cellular systems.

The coarse-grained approach utilizes a simplified system representation with fewer degrees of freedom, resulting in faster simulations but with reduced spatial and/or temporal resolution [97–99]. Different coarse-graining (CG) schemes have been devised to preserve the most relevant properties of the molecular system. Such methods can be applied to describe time scales that are far beyond the scope of all-atom MD or KMC simulations, and thus extend the scope of molecular simulation to the nanoscale. Some examples of successful application of CG methods are the simulation of the different phases of the lipid–water system, interactions of peptides and proteins with biological membranes, and the electrodeposition of copper to form nanowires, nanofilms and nanoclusters in kinetic-limited regimes [182].

A large class of currently used CG models employs what some researchers call 'particle-based' coarse-graining, in which a finite number of atoms of the molecular system are grouped into a single point mass (a CG particle, sometimes called a mesoparticle or a pseudo-particle). In the simulations of biomolecules, most commonly only a few heavy atoms (between 3 and 5) are replaced by a CG particle. In materials simulations it is common to have 10–100 molecules replaced by a CG particle. Depending on the system, researchers have derived the interaction of the CG particles from molecular theory considerations, from all-atom simulations, or by calibration to fit experimental data and results of all-atom simulations. In the coarse-grained molecular dynamics (CGMD) approach [100], the constitutive relations are derived directly from the interatomic potential by means of a statistical coarse-graining procedure [101]. Thus, the average thermodynamic effect of the atomic-scale quantities is retained in the coarse-scale motion. A coarse-grained Monte Carlo simulation approach has been derived that can represent mesoscopic length scales while correctly capturing atomistic information on intermolecular forces [102, 103]. Another approach is to use wavelets [104, 105] to track combinations of atom motions instead of individual atom motions. A closely related approach is the gap-tooth method [106], which tracks for some times and/or for some physical locations, and extrapolates to estimate the behavior for the atoms that are not simulated. The methods can be combined or integrated with other acceleration methods such as the tau-leaping [107–109] to further reduce the computational expense.

One class of CG models used in biomolecular simulations utilizes an elastic network in which a biomolecule is described as a network of nodes interacting through harmonic potentials. Such networks allow for computing the normal modes and principal components, thereby providing a basis for analyzing the large-scale motions of the system, although with the limitation that dynamics can be addressed only within linear response theory. CG modeling may be used to primarily describe shapes of macromolecules with as few point-like particles as possible. Such models can reach beyond the linear modes of motion when combined with non-harmonic potentials for interactions between the CG particles.

4.4
Multiscale Simulation

It is generally the case that advances in numerical algorithms, and their associated applications, have been greatly accelerated when general-use software becomes widely available for use by others. For example, the late John A. Pople was recognized by the Nobel Prize in Chemistry in 1998 for development of computational methods in quantum chemistry. It was his development of the software 'Gaussian' that implemented those computational methods that resulted in a significant research impact.

As another example, while molecular dynamics was developed in the 1950s by theoretical physicists [110], it was not until general-use software implementing these MD algorithms such as CHARMM became widely available that rapid growth occurred in numerical algorithms, usage and increased capabilities. In comparison, KMC simulation software are much less developed than quantum chemistry and molecular dynamics software, which is one of the reasons for much less usage of KMC (e.g. approximately 4000 journal papers on KMC and approximately 92 000 journal papers on MD have been published, according to ISI Web of Science) in addition to KMC typically having much less sophistication in terms of complexity of the systems that are studied. Similarly, general-use finite volume codes such as Fluent and finite element codes such as ABAQUS greatly accelerated applications by placing tools in the hands of applications researchers (instead of only researchers in computational science and engineering).

The numerical algorithm development for multiscale simulation is more complex than for numerical algorithms at the individual time and length scales. As a consequence, the development of multiscale simulation as a cohesive high-impact field within computational science and engineering requires significant development of general-use numerical algorithms and of general-use software. Advances in numerical algorithms and software for multiscale simulation will accelerate its applications in addition to accelerating the development of multiscale simulation as a field within computational science and engineering.

There are various levels of pursuit that may be taken simultaneously to address the challenges that arise in multiscale problems. One is to select one or more challenging exemplar problems, selected for their importance in specific high impact applications. A second is to abstract more general principles and procedures for multiscale simulation that may be used for other applications beyond those for which they were originally developed. Such procedures include error estimation and uncertainty quantification tools. At the same time, it is worth noting to keep in mind the need for appropriate computations level, as emphasized in the observation made recently by the late John Pople in the quote below ([14], page 34):

> *I have a comment about . . . the convergence with respect to basis sets in the computation or calculation of molecular energies. Probably the most notable feature of that is it became very expensive as you moved to the larger and larger basis sets but the last order of magnitude in computing that you in fact spent gave you almost no information, a result differing very little from the previous numbers and in fact it seems to me that one is in some danger of exaggerating the need for very large computation to achieve chemical accuracy.*
> *Many of those results we have been able to find, we have found can be achieved by the judicious use of small empirical corrections or simple extrapolation schemes.*   John A. Pople, Northwestern University

There is a limit to how much the coarse-grained methods described in the previous section can extend molecular simulation towards longer time and length scales. Different simulation methods are most computationally efficient at different length scales, a fact which motivates efforts to simulate overall multiscale systems by linking multiple simulation codes created by experts at each scale [111, 112]. There are numerous examples where length and time scales are coupled in a serial fashion, where the results from a simulation code is used in another simulation code (Broughton et al., 1999; [113–118]). For example, it is common to use quantum mechanics to compute force fields that are used in a molecular dynamics code. For quasi-static problems, the quasi-continuum method [119, 120] couples the atomistic and continuum scales by using a system-wide finite element mesh that is refined to the atomic dimensions where needed. Unlike conventional finite elements, the energy of each cell is computed from the underlying atomistic Hamiltonian. Many other papers propose iterative algorithms to converge codes at multiple length scales to a steady-state or quasi-steady-state solution [121–124].

A widely used type of multiscale simulation combines quantum mechanical and molecular mechanical (QM/MM) simulations. In this approach, the functional core of the molecular system, for example, the catalytic sites of an enzyme, is described at the electronic level (QM region), whereas the surrounding macromolecular system is treated using a classical description (MM region). Some of the biological applications for which QM/MM calculations have been widely utilized are chemical reactions in enzymes, proton transfer in proteins and optical excitations. In QM/MM

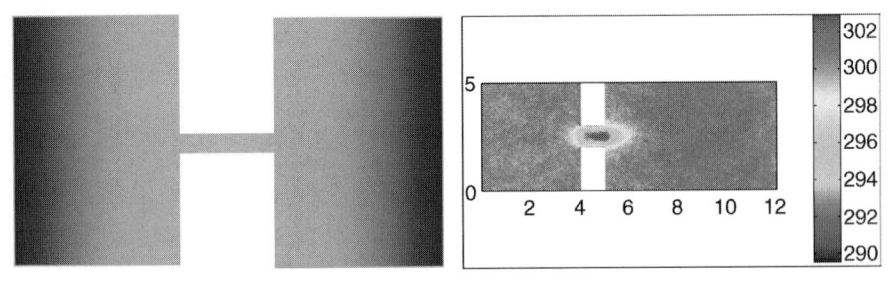

Figure 4.6 Left: a microfluidic filter example. The baths are simulated by using the continuum Stokes equations and the filter is simulated by using the direct simulation Monte Carlo (DSMC) method. Right: plot of temperature in the device obtained by using the multiscale method. Non-continuum effects such as temperature jump are captured by the multiscale method (courtesy of Narayana R. Aluru, University of Illinois at Urbana-Champaign).

calculations, the main focus is usually on the QM region, where the electronic structure and forces are calculated quantum mechanically in the presence of the effective electric field of the MM region. The QM region, in turn, can affect the MM region through both bonded and non-bonded interactions. Exchange of forces between the two sub-systems is adequate for a proper description of the system.

As another example of a multiscale simulation method, consider the combination of continuum Stokes equations with the direct simulation Monte Carlo (DSMC) method (see Figure 4.6). The continuum and DSMC regions were combined by an overlapped Schwarz alternating method with Dirichlet–Dirichlet type boundary conditions [125]. A scattered point interpolation scheme was developed to interpolate the solution between sub-domains. The convergence characteristics of the multiscale approach were investigated numerically. Specifically, the dependence of convergence on the overlap size, the DSMC noise and the number of time steps employed in the DSMC algorithm were studied. Their results indicated that, while the convergence depended weakly on the DSMC noise and the overlap size, the number of DSMC time steps simulated in each coupling iteration should be selected so that the total time steps simulated until convergence of the coupled process is close to the time constant of the DSMC subsystem. They applied this multiscale simulation method to investigate gas flow through micro/nano-pores, a membrane with an array of nanopores and an MEMS device with a nanopore filter array for particle trapping and sorting.

The need to simulate *dynamical* systems where a wide range of time and length scales are tightly coupled has motivated efforts to address the more challenging problem of *concurrent* multiscale simulation [13]. The FE/MD/TB hybrid approach [126] spatially divides the system into continuum, atomistic and handshake regions, and appropriate boundary conditions are shared among the various regions to obtain a self-consistent solution. This method has been implemented on supercomputers via parallel algorithms allowing the solution of large problems for short times (about 1 ns). An approach applicable to some reacting systems is the use of an effective reactivity to atomic and continuum scales [43, 45]. The direct numerical

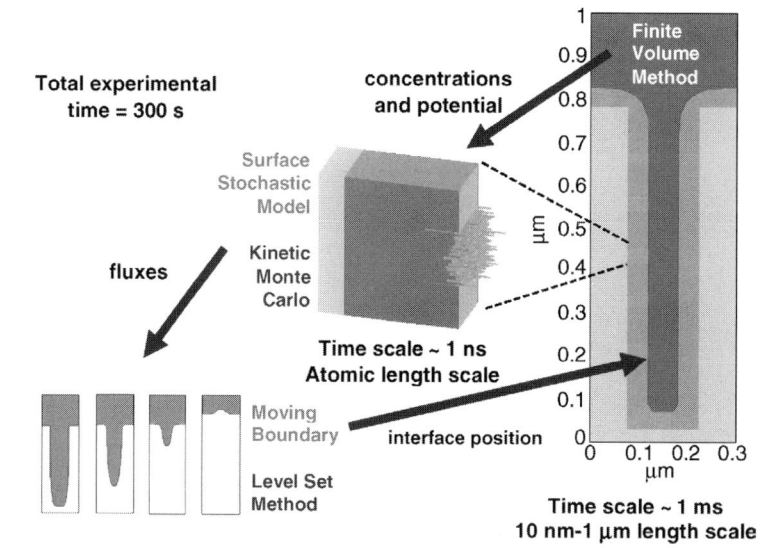

Figure 4.7 Schematic of the dynamically coupled multiscale simulation of the electrodeposition of copper into a trench to form a copper wire. A finite volume code that simulates the potential field and concentration fields of all chemical species in aqueous solution sends the solution concentrations and potential at the solid–liquid interface to a KMC code, which simulates adsorption, desorption and chemical and electrochemical reactions that occur on the surface. The KMC code computes fluxes that are sent to a level set code, which simulates the movement of the solid–liquid interface and sends its position to the finite volume code for the subsequent iteration. The different time scales are addressed by implementing many events in the KMC code for each time interval of the finite volume code. Higher accuracy can be obtained by iterating all three sets of codes to convergence at each time instance [131].

simulation approach involves running the simulation codes at each length scale simultaneously, with each code continually passing updated boundary conditions to the other codes [124, 127, 128]. A modified approach uses internal iterations to force convergence of the information passed between simulation codes [129, 130]. Another modification updates the continuum codes more slowly than the atomistic codes, in accord with their different time scales [42]. The last approach has been used to simulate the electrodeposition of copper into trenches to form interconnects in electronic devices (see Figure 4.7 from [131]).

Another important consideration when implementing a multiscale simulation code is that the coupling of simulation codes can induce numerical instabilities. While numerically stable codes are available for simulating each length scale, numerical instabilities can be induced in the coupling of such codes by temporal and spatial mismatches at the interfaces between the codes. For the coupling of continuum codes, one approach to numerically stabilize the coupling codes is by passing *both* boundary conditions and *associated sensitivity information* between the codes [39]. This approach cannot be applied to dynamic couplings that include noncontinuum codes; however, as the associated sensitivities cannot be computed to high enough accuracy for such codes. We have used nonlinear systems theory to design numerical coupling algorithms that modify the dynamic information passed between the simulation codes to numerically stabilize their coupling, and to increase the numerical accuracy of the simulation results [42, 132]. In this approach, the simulation codes are represented by deterministic or stochastic discrete-time nonlinear operators, with mismatches at the interfaces between simulation codes modeled as norm-bounded perturbations. Dynamic coupling algorithms are exactly modeled as additional operators inserted into the block diagram, which are designed by optimal control theory [132]. A constructive procedure for testing whether an arbitrary interconnection of simulation codes is well posed is provided [132], along with general conditions regarding the numerical stability and accuracy of dynamically coupled simulation codes. The dynamic coupling algorithms designed by this control theoretic approach have also been analyzed using the classical methods of numerical analysis (such as described in the textbook by [133]) and compared with more commonly used algorithms, such as direct coupling. Nonlinear systems theory and numerical analysis are complementary approaches for drawing clear and direct comparisons between the various dynamic coupling algorithms described in the literature (e.g. see Dollet *et al.*, 2004; [6, 11, 13, 40, 43, 134]; and references cited therein).

Application of these methods to analyze the numerical stability and accuracy of a wide variety of algorithms for coupling simulation codes has motivated the design of numerical coupling algorithms that increase the numerical accuracy of the overall simulation results by modifying the dynamic information passed between simulation codes or by introducing predictor–corrector iterations [185]. These analyses provide guidelines on how to best trade off numerical accuracy with computational expense when coupling simulation methods to simulate a particular multiscale system. For example, these analytical tools indicate that the direct simulation approach to coupling codes, which has the lowest computational expense, is restricted to first-order accuracy irrespective of the accuracy of the individual simulation codes. In contrast, a predictor–corrector algorithm can be designed to achieve second-order accuracy, but with higher computational cost. Applications of these analytical tools, experience in multiscale simulation codes and common sense have resulted in several guidelines to follow when constructing dynamically coupled simulation codes [186]:

1. Prefer numerical algorithms that satisfy conservation laws (such as finite volume methods) over algorithms that do not force the numerical solution to satisfy conservation laws (such as finite difference methods).
2. Select the order of passing of boundary conditions between codes based on physical insights.
3. Speed up and thoroughly validate individual simulation codes as much as possible before coupling with other codes.
4. Utilize accurate estimates from numerical analysis, when possible, to select the tradeoff between accuracy and computational expense.

5. If needed, filter or otherwise improve the estimation of outputs of stochastic simulation codes before passing to other simulation codes (whether stochastic or deterministic).

As an example application, consider the design of the dynamic coupling of simulation codes for the electrochemical process used to manufacture copper interconnects. A multiscale simulation model in Figure 4.7 couples multiple instances of a solid-on-solid KMC simulation code [41, 131] to an internally coupled moving boundary (MB) finite-volume/level set continuum simulation code [135–137] to simulate the filling of on-chip features (trenches) by electrodeposition in the presence of additives. The KMC and MB simulations dynamically pass interface conditions during the simulations. The MB code sends surface concentrations and the solution potential to each KMC code, which computes reaction rates from the simulation of the chemistry and physics that occur at the electrode surface. The KMC codes send species fluxes to the MB code that are used as surface boundary conditions. The MB code advances the copper–solution interface using the level-set method, and simulates the chemistry and physics in the electrolyte in and above the trench using the finite volume method. Nonlinear systems analysis indicates that this approach to numerically coupling the many simulation codes is well posed.

This implementation follows Guideline 1, in that the finite volume method exactly conserves mass, moles and current and the KMC code exactly conserves mass, moles and current as it tracks atoms. By exactly matching species fluxes at the solid–liquid interface, the coupled codes can exactly conserve mass, moles and current within the spatial resolution of the KMC codes along the interface. It is our experience that such conservation constraints tend to lead to more robust code linkages. The implementation follows Guideline 2, as the rates of the heterogeneous chemical and electrochemical reactions on the copper surface directly depend on the species concentrations (e.g. Cu^{2+}) in the electrolyte, and the fluxes directly depend on the rates of the reaction on the copper surface. The alternative coupling of passing fluxes from the electrolyte code to the KMC codes would be inconsistent with the strong dependence of the fluxes on the adsorbed species on the copper surface. Guideline 3 was followed to speed up and thoroughly validate codes before coupling [41]. Following Guideline 4, the fairly large amount of stochastic noise in the KMC codes indicated that the extra computational expense associated with a high-order dynamic coupling method would not significantly improve the overall accuracy of the multiscale simulation, so a first-order direct coupling with the modification of having more iterations in the KMC codes between the passing of information was used. As per Guideline 5, filtering was used to reduce the effect of stochastic KMC noise on the boundary conditions passed to the deterministic code.

In summary, even the most state-of-the-art multiscale simulation methods run the gamut of information passing between individual simulation methods from serial to parallel to modified parallel approaches. There remain important unresolved issues in multiscale simulation in which information must pass dynamically between the individual simulation methods: how to optimally design dynamic coupling algorithms for the full range of simulation codes of widely varying time and length scales.

4.5
Challenges and Requirements of Multiscale Modeling

A variety of challenges arise when building models for molecular and multiscale systems. For the systems that have been studied to date, the following have been identified in nearly every instance:

- uncertainties in physicochemical mechanisms;
- sparsity of on-line measurements at the molecular scale;
- sparsity of manipulated variables during processing,
- the need to dynamically couple model structures; and
- high computational costs for model simulation.

These challenges specify the requirements for next-generation tools needed for a systematic approach to the design and control of electrochemical systems *from molecules to devices*.

Although quantum mechanics and molecular dynamics calculations can reduce uncertainties in physicochemical mechanisms for some semiconductor processes, such as in solid-state systems [138], these techniques are not sufficiently developed for obtaining accurate results for other microelectronics processes. As an example, consider the electrodeposition of copper to form interconnects. In this process Cu^{2+} ions in aqueous solution diffuse and migrate to the surface in response to a potential applied between counter and working electrodes. Many chemical species, such as sulfuric acid, copper(II) chloride, polyethylene glycol, mercaptopropanesulfonic acid and 1-(2-hydroxyethyl)-2-imidazolidinethione, must be added to the aqueous solution to produce void-free copper deposits in sub-100 nm trenches [139, 140]. Although there is an extensive experimental literature in this area, the detailed chemical interactions of these additives with the copper surface are not well understood [141–147]. Examples of questions that must be answered to completely specify a hypothesized mechanism are:

- Which chemical species are formed and consumed at the surface?
- What is the physical configuration of each molecule on the surface (e.g. is the molecule sticking out into the solution or flat against the surface)?
- How many sites does each surface molecule cover?

Existing quantum mechanics and molecular dynamics methods have not yet advanced to the stage where water can be reliably modeled by itself [148–151], much less when involved in electrochemical reactions at a surface [152]. *An important requirement of any general multiscale systems framework is that it must be able to enable the resolution of the unknowns in complex heterogeneous mechanisms.*

Most molecular systems involve unknown complex heterogeneous chemical mechanisms with no sensors available for measuring concentrations at small length scales (<100 nm) at industrially relevant operating conditions. For example, the main sensors available for measuring length scales less than 10 microns for the electrodeposition process are scanning electron microscopy and atomic force microscopy, which measure surface morphology at the end of the process. Typical sensors measure macroscopic properties such as current, potential and temperature.

A challenge for such systems is how to construct an accurate chemical mechanism without in-process concentration measurements being available.

Most processes in which control of events at the molecular scale is important have only one or two manipulated variables available during processing. For example, the applied potential is the only manipulated variable during the electrodeposition of copper to form an interconnect. *A challenge for constructing predictive models for such systems is how to excite the dynamics during model identification experiments when so few manipulated variables are available.*

Another challenge when addressing molecular and multiscale systems is that the computational costs for model simulation are high, and the model structures may require dynamically coupling multiple simulation codes for the various time and length scales (e.g. see Figure 4.7). The variety of simulation codes used to model the various time and length scales, and the variety of ways in which these codes may be coupled, indicates that systems techniques developed for these systems should be designed to act directly on simulation inputs and outputs rather than being developed for every possible mathematical structure for the model equations. Input–output methods are also motivated by considering that the state dimension for the governing equations implemented in these simulation codes, such as in KMC simulation, can be too high for systems methods to be developed that act directly on the states. The high computational cost also indicates that multiscale systems tools must be very computationally efficient.

A challenge is to develop an advanced cyberinfrastructure needed to address multiscale systems. There are several aspects. Development of the numerical software and data analysis algorithms is needed to construct multiscale simulations that optimally utilize high end computing to make predictions and facilitate new applications. Software is needed to distribute data and computations among processors for problems with a wide range of time and length scales to compute efficiently on massively parallel machines. Cyberinfrastructure is needed that will efficiently manage data, communications and simulation components with low scalability (such as fluid dynamics codes) with those of high scalability as the number of processors increase, creating combinations of high and low scaling methods in conjunction to those simultaneously run, while exploiting the capability of next-generation supercomputers. In addition, there is a need for efficient methods that quantify the effects of uncertainties in model parameters and the predictions of simulation algorithms of smaller length scales (e.g. quantum) on the simulation predictions for the overall multiscale simulation that includes the larger length scales. Also important will be the development of algorithms to rapidly visualize high-dimensional data with orders-of-magnitude variations in time and length scales, typically over the internet.

4.6
Addressing the Challenges in Multiscale Modeling

This section summarizes our views on how to address the challenges posed in the previous section. Substantial unknowns in a complex heterogeneous mechanism can

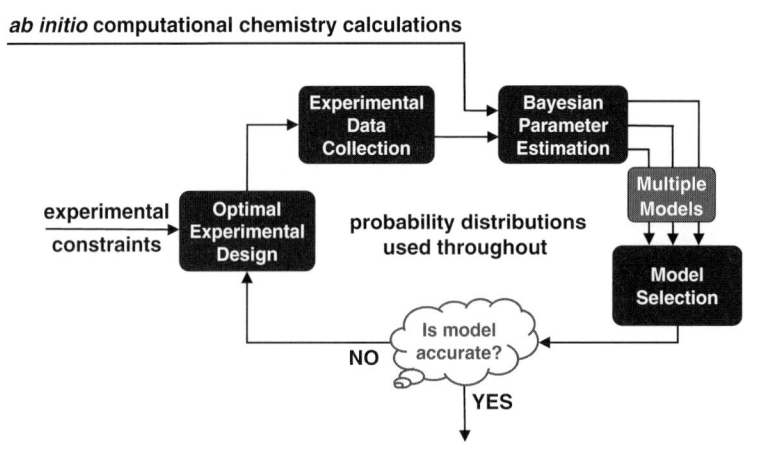

Figure 4.8 Model building using Bayesian estimation and model discrimination/selection.

be resolved by estimating the parameters in each hypothesized mechanism and applying existing criteria for model discrimination which select the mechanism that is most consistent with the experimental data (e.g. [153–155]; see Figure 4.8). The uncertainties in the parameters in each mechanism are quantified by probability distributions that are used to design each subsequent laboratory experiment, either to further reduce the model uncertainties or to maximize the ability to distinguish among the multiple hypothesized mechanisms [156]. Parameter estimates determined by quantum chemistry or molecular dynamics calculations can be incorporated using Bayesian estimation, which we have applied to identify the dominant chemical mechanisms during the formation of junctions in metal oxide semiconductor field effect transistors (Gunawan et al., 2003).

The questions of how to construct a chemical mechanism without having in-process concentration measurements and how to excite the dynamics during model identification while having few manipulated variables can be addressed by a combination of three methods:

- design and implement small-scale (e.g. micro- to millimeter scale) chemical systems so as to highly excite the experimental input space;
- extensively utilize scanning probe measurements; and
- use stop-and-repeat experiments, in which each batch experiment is stopped and analyzed for numerous time intervals to produce dynamic data.

In one application of this approach [157], a series of experiments was carried out in which each experiment deposited copper onto a flat copper substrate with the contact area being a circular disc with diameter of about 5 mm. To highly excite the input space, a D-optimal design with 36 experiments was implemented for the wide ranges of input variables shown in Table 4.2. Although the industrial interest is in depositing within trenches and vias to form copper wires and multilayer contacts, the experiments were designed for deposition on flat surfaces so that surface morphology can

Table 4.2 Lower bound, centerpoint, and upper bound for the inputs for a set of millimeter scale electrochemical experiments.

Perturbed experimental inputs	Experimental range
$CuSO_4$ concentration	0.3, 0.7, 0.8 M
H_2SO_4 concentration	5, 45, 175 g L^{-1}
SPS concentration	3, 26.5, 50 ppm
PEG concentration	0.1, 0.3, 3 g L^{-1}
Cl^- concentration	10, 55, 100 ppm
HIT concentration	5, 50, 200 ppb
Current density	3, 11.5, 20 mA cm^{-2}

be measured by using scanning probe measurements, in this case, atomic force microscopy which greatly expands the number of measurements obtained in each experiment.

Each of the experiments in the D-optimal design was repeated 2–3 times, with each experiment starting with a newly polished wafer. Eight surface locations were measured for each experiment, to characterize both experimental noise and biases. As the AFM images are only available at the end of each experiment, each experiment was repeated for about ten batch times, with times selected to capture the dynamics for each experimental condition (e.g. see Figure 4.9). The parameters in a KMC code were estimated by applying the methods in Figure 4.8 to the potential/current and root-mean-squared surface roughness curves (see Figures 4.9 and 4.10).

One question that arises with such an approach is how well the model parameters associated with surface diffusion and the chemical and electrochemical reactions can be extracted from the current, potential and *ex situ* surface morphology data, given the complex nature of the interactions of the additives with the surface (e.g. see Table 4.3). A key point is that current and potential curves and the surface morphology are very sensitive to changes in the experimental inputs (shown in Table 4.2), indicating that

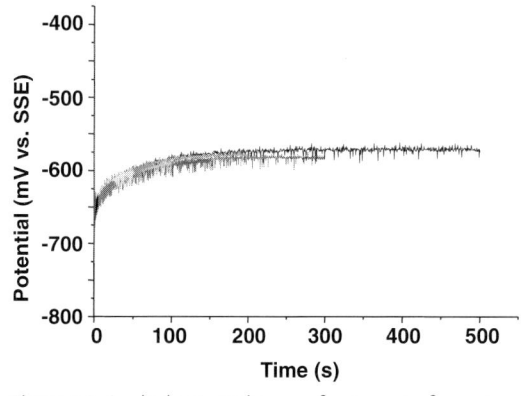

Figure 4.9 Applied potential curves for one set of experimental conditions in Table 4.2. Each color corresponds to a batch with a different batch time.

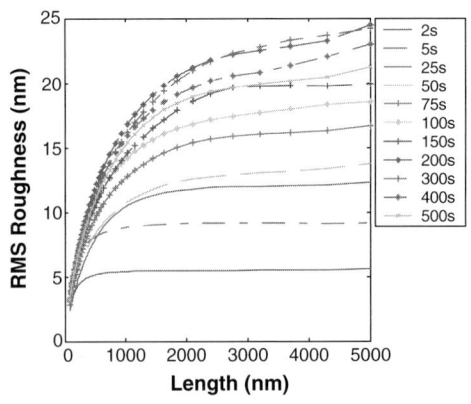

Figure 4.10 Root-mean-squared surface roughness as a function of length scale and time, averaged over 8 locations, at one set of experimental conditions given in Table 4.2.

the large quantity of experimental data collected for the wide range of experimental inputs contain substantial information about the underlying mechanism. Further, changes in the hypothesized mechanism or the sensitive model parameters being estimated have a very strong effect on the current, potential and *ex situ* surface morphology, which suggests that the data may be sufficient for the model parameters to be identifiable. The chemical–electrochemical mechanism that arose from applying the model building approach of Figure 4.8 to the experimental data is shown in Table 4.3, with the values of the parameters in the most sensitive reactions shown in Table 4.4. The small confidence intervals indicate that the model parameters were identifiable from the experimental data (readers interested in more details on the

Table 4.3 The hypothesized electrochemical–chemical mechanism for the electrodeposition of copper in the presence of additives resulting from application of the model building approach of Figure 4.8.

1. Cu^{2+} (aq) + e^- → Cu^+ (aq)
2. Cu^+ (aq) + e^- ↔ Cu (s)
3. Cu (s) → Cu (s) (surface diffusion)
4. Cu^+ (aq) + Cl^- (aq) ↔ CuCl (ads)
5. CuCl (ads) + e^- → Cu (s) + Cl^- (aq)
6. CuCl (ads) + PEG (aq) ↔ Cu(I)–Cl–PEG (ads)
7. SPS (aq) + $2e^-$ → 2thiolate$^-$ (aq)
8. Cu^+ (aq) + MPS (aq) ↔ Cu(I)thiolate (ads) + H^+ (aq)
9. Cu(I)thiolate (ads) + Cu^+ (aq) + e^- → Cu (s) + Cu(I)thiolate (ads)
10. Cu(I)thiolate (ads) + HIT (aq) → Cu(I)HIT (ads) + MPS (aq)
11. Cu(I)HIT (ads) + H^+ (aq) + e^- → Cu (s) + HIT (aq)
12. thiolate$^-$ (aq) + H^+ (aq) ↔ MPS (aq)
13. H_2SO_4 (aq) → H^+ (aq) + HSO_4^- (aq)
14. HSO_4^- (aq) ↔ H^+ (aq) + SO_4^{2-} (aq)

Table 4.4 Values of model parameters in the most sensitive reactions in the mechanism, with 95% confidence intervals and ($\rightarrow$) indicating that the sensitive model parameters are only in the forward reaction.

Reactions	k	α
1	$6.85 \pm 0.04 \times 10^{-2}$ m s^{-1}	0.20 ± 0.01
2($\rightarrow$)	$1.4 \pm 0.2 \times 10^{-3}$ m s^{-1}	0.511 ± 0.008
4($\rightarrow$)	14.94 ± 0.09 m^4 mol^{-1} s^{-1}	–
6($\rightarrow$)	100 ± 4 m^4 mol^{-1} s^{-1}	–
8($\rightarrow$)	$1.41 \pm 0.09 \times 10^8$ m^4 mol^{-1} s^{-1}	–

parameter estimation and discussion of results are referred to PhD theses by [158] and [159]).

Recently we have taken this approach of designing millimeter-scale chemical/electrochemical systems for high-throughput data collection a step further, by designing an apparatus in which the potential varies across the electrode. The apparatus was used to investigate the effect of additives on nucleation and growth of electrodeposited metal. In this manner, the experimental arrangement makes it possible to explore the effect of a wide range of potentials in a single batch experiment. That is, at the end of each experiment, the surface morphology is measured along the electrode (i.e. as a function of potential) by use of atomic force microscopy [159].

The implementation of the systems methods in Figure 4.8 is well established for deterministic simulation codes (e.g. see [10], and references cited therein), whereas the systems methods for stochastic simulation codes utilize multistep optimization (see Figure 4.11). The multistep optimization of Raimondeau et al. [160] utilizes sensitivity analysis to determine the key parameters, followed by response curve mapping to parameterize the responses of the simulation model as low-degree polynomials of the key parameters, and simulated annealing to optimize over the key parameters. Figure 4.11 revises this algorithm to utilize stochastic sensitivities [161] and optimal design methods to reduce the computational cost per iteration, and revises the structure of the low-order models, based on any known physics, to improve the quality of the low-order model identified in each iteration [10]. Either approach applies to multiscale simulation codes with arbitrary (but well posed) dynamic coupling of individual simulation codes.

4.7
Design Based on Multiscale Models

Once a model is constructed, it can be used for the optimization of all time and length scales, using a similarly constructed multistep algorithm (see Figure 4.11). The multistep algorithms in Figures 4.11 and 4.12 are the same numerically, but with

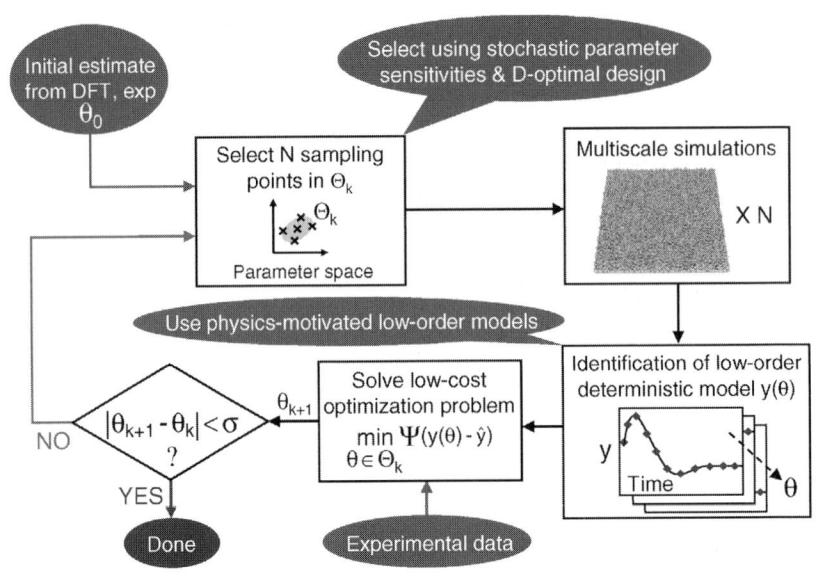

Figure 4.11 Multistep optimization for the estimation of parameters in stochastic simulation codes.

different optimization variables, objectives, and constraints. The optimization objective can be formulated to ensure robustness to the model uncertainties quantified in the parameter estimation procedure (e.g. see [162], and references cited therein). The optimization variables u can include operating variables (e.g. batch control trajectories), feedback controller parameters and parameters that define the mole-

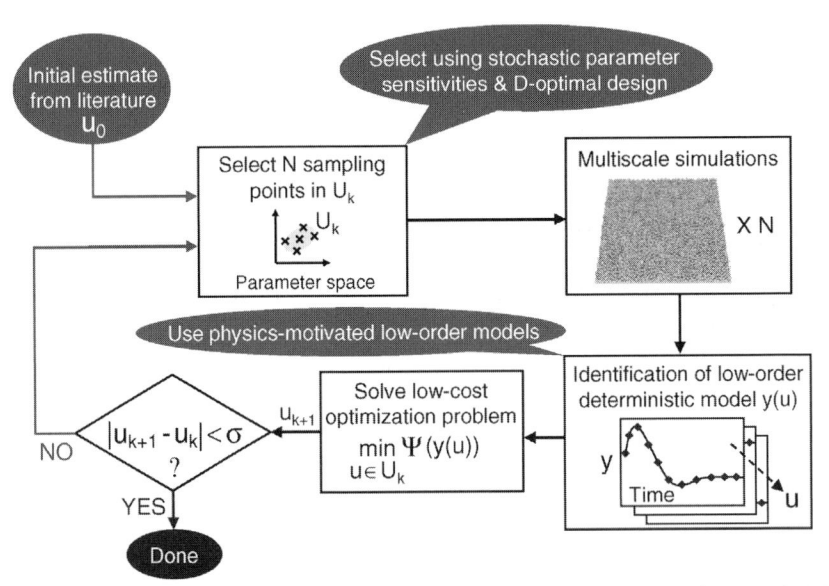

Figure 4.12 Multistep optimization for design and control using stochastic simulation codes [9].

cules in the system (e.g. polymer chain length, hydrophobicity, etc.). This provides a systematic methodology for the control of events at the molecular scale while simultaneously optimizing all length scales from the molecular to the macroscopic. By its inclusiveness of other time and length scales, the multiscale systems approach can address the modeling and design problems of nanotechnology, molecular nanotechnology and molecular manufacturing.

Although the multistep algorithms in Figures 4.11 and 4.12 were designed for use with stochastic simulation codes, these algorithms can also be applied for processes modeled by deterministic simulation codes, in which case the multistep algorithms are similar to successive quadratic programming. To illustrate such applications, consider the design of optimized processes for the formation of ultrashallow junctions in metal oxide semiconductor field effect transistors. The process for forming junctions involves ion implantation of dopants into a bulk semiconductor (see Figure 4.13). A particular example of intense interest for current digital logic technology is boron implanted into silicon. Although junctions can be made shallower by reducing the implant energy, the effectiveness of this approach has been limited by the need to anneal the resulting structure to over 1000 °C both to activate the dopant electrically and to eliminate implant-induced defects in the crystal

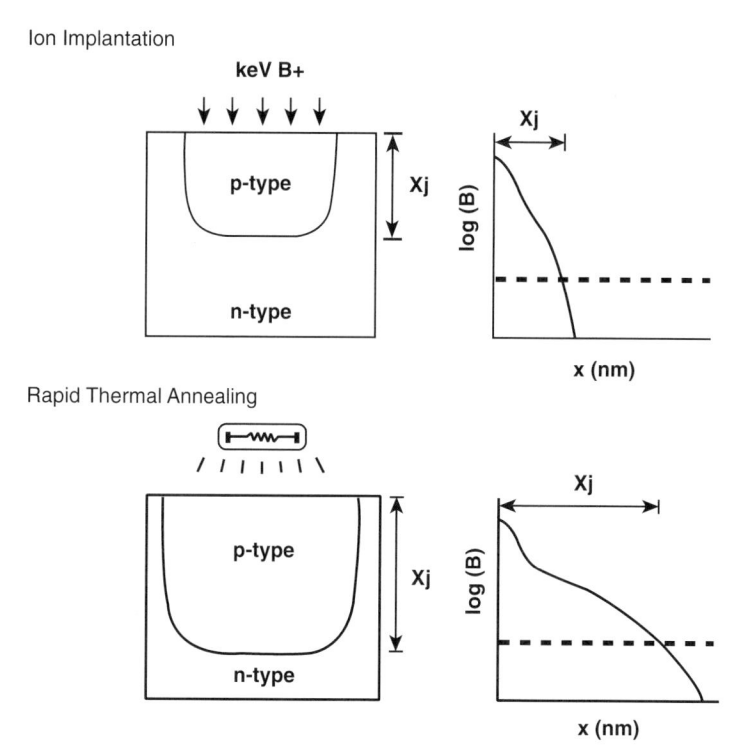

Figure 4.13 Schematic diagram of the increase in junction depth that takes place during rapid thermal annealing after ion implantation of dopant.

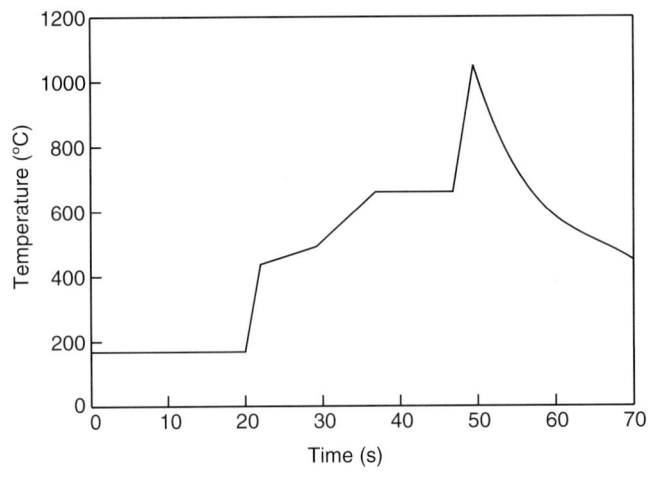

Figure 4.14 A typical rapid thermal anneal temperature program, which consists of a stabilization step and a spike-anneal (i.e. a fast linear heating step followed by a natural cool down step).

structure (see Figure 4.14 for an example of a temperature trajectory). Defects mediate unwanted diffusion of dopants during the anneal process, which often leads to a significant undesired increase in the junction depth. The aforementioned systems tools have been applied to the post-implant annealing process, to construct a simulation model and to minimize the junction deepening while maximizing dopant activation.

The simulation model includes the coupled reaction–diffusion–migration equations for interstitial atoms, interstitial clusters and related defects. These equations have the general form for species i:

$$\frac{\partial N_i}{\partial t} = -\frac{\partial J_i}{\partial x} + G_i \qquad (4.2)$$

where N_i denotes concentration and G_i is the net generation rate. The flux J_i incorporates terms due to diffusion and drift in response to electric fields. The net generation rates include the transfer of boron and silicon atoms between interstitial and substitutional positions and for the formation and dissolution of boron, silicon and mixed boron–silicon clusters. The model also includes Poisson's equation describing the electric field generated by spatial imbalance of the charge density. The simulation model consisted of about 25 partial differential equations, which were non-uniformly spatially discretized using between 200 and 800 points in the depth direction, resulting in up to 20 000 extremely stiff ordinary differential equations that were solved using the public domain software FLOOPS [163], which integrates the equations using a combination of the one-step trapezoidal rule and the backward differentiation formula [164].

The activation energies in the expressions for G_i and J_i were obtained by Bayesian parameter estimation, which incorporated information from density functional theory (DFT) calculations, past experimental studies and boron secondary ion mass spectroscopy data from the International Sematech consortium of semiconductor companies [165]. Model selection methods were used to select among competing chemical mechanisms and to simplify the reaction network [138, 166]. A combination of parameter sensitivity analysis and kinetic insights was used to select the physical mechanism [167], in which the most important part was the specification of the network of chemical reactions for the clusters. Parameter sensitivity analysis was a necessity in the construction of the physical mechanism, as the number of kinetic parameters was large, including 18 activation energies associated with the interstitial diffusion, cluster association and cluster dissociation reactions.

The systems tools in Figure 4.11 permitted the development of an understanding of the fundamental kinetic processes that govern diffusion and electrical activation of dopant [138, 168, 169]. The agreement between the simulated and experimental boron dopant profiles was within 2 nm for the entire junction (see Figure 4.15). To further validate the simulation model, the junction depth and sheet resistance (a measure of dopant activation) were computed for a wide range of temperature profiles, and compared with a large number of experimental values reported in the literature and to the 'Sematech curve', which summarizes additional experimental values (see Figure 4.16). The predictions of the simulation model were highly consistent with reported experimental values.

The simulation model was coupled with optimization (by using a similar algorithm as described in Figure 4.12) to compute an annealing temperature profile that minimized the junction depth while maintaining a desirable level of boron activation [170]. This analysis indicated that manipulating the rapid thermal annealing profile by itself is not able to simultaneously increase electrical activation and reduce junction depths to levels desired by industry. This limited actuation available at

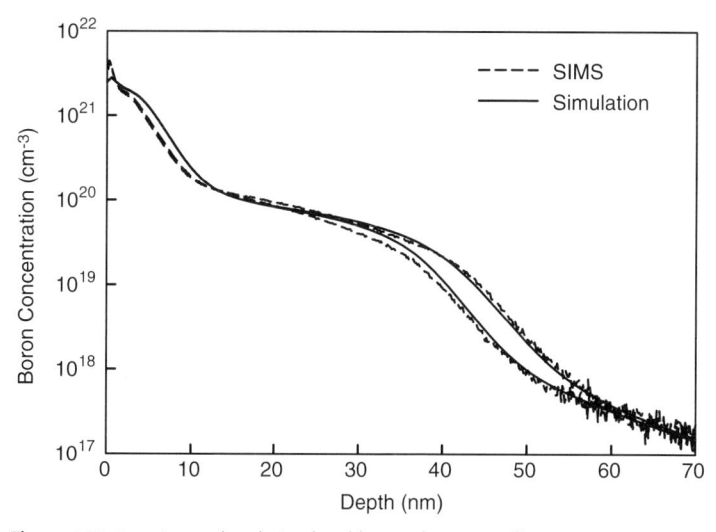

Figure 4.15 Experimental and simulated boron dopant profiles, for two batch operating recipes [165].

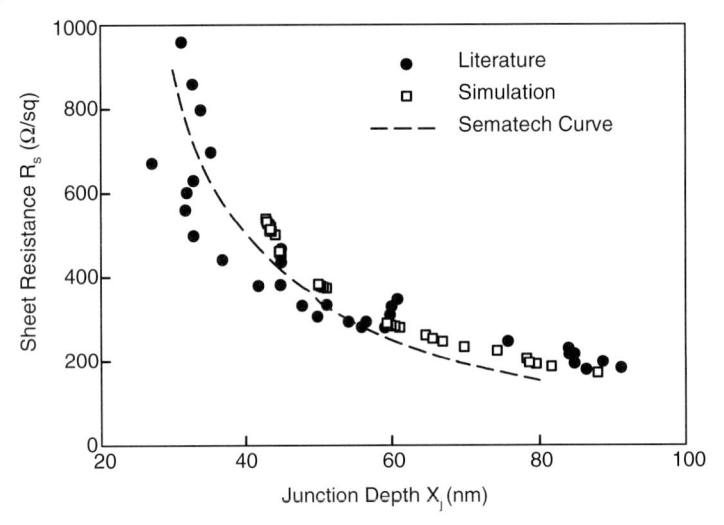

Figure 4.16 Comparison of junction depth–sheet resistance pairs from various published experimental papers and TED simulations employing various heating and cooling rates, and annealing temperatures. The Sematech curve summarizes the sheet resistance and junction depth data in experimental studies performed by International Sematech [170].

macroscopic length scales motivated the application of *molecular design*, and it was argued that the potential impact of molecular design optimization was much greater than the potential benefit of improved feedback control.

Indeed, the simulation model was crucial in helping to identify some important new mechanisms for accomplishing design optimization through 'defect engineering'. Various approaches to defect engineering intended to accomplish these purposes already exist, including millisecond-scale annealing methods such as laser and flash annealing, and co-implantation of elements such as C and F. Millisecond annealing carries problems of process integration, however, in addition to problems with the removal of implantation damage. Co-implantation can lead to undesired complications such as junction leakage due to tunneling. As junctions move closer to the surface, the possibility arises for using the surface itself for defect engineering. The simulation model was able to disaggregate the effects of two distinct mechanisms by which the surface influences the behavior of bulk defects created during implantation.

The first mechanism involves the reflection of charged interstitials from the surface due to electrically active surface defects that set up a repulsive electric field. The driving force for reflection is surface band bending that resulting from dangling bonds at a free Si surface (no oxide) or a damaged Si–oxide interface [171, 172]. For boron implantation, interstitial atoms of B and Si are positively charged in implanted material. The opposing field can transform the interface from a significant sink into a good reflector, which changes the average concentration of interstitials in the underlying bulk, which in turn influences the degree of dopant activation and diffusion [173]. The simulation model provided a precise indication of the magnitude of this effect under typical implantation conditions.

The second mechanism involves insertion of interstitials into dangling bonds at the surface. An atomically clean surface can annihilate interstitial atoms by simple addition of the interstitials to dangling bonds. Such a process resembles adding a hydrogen atom to a free radical in the gas phase; there is essentially no activation energy and the addition is very facile. However, if the same surface is saturated with a strongly bonded adsorbate, annihilation requires the insertion of interstitials into existing bonds. Such insertion should have a higher activation barrier and a much lower probability of occurrence. A schematic diagram of this idea appears in Figure 4.17. The simulation model enabled direct measurements of interstitial annihilation probabilities at surfaces [174].

As it is highly advantageous in terms of electronic device properties to restrict the chemistry to dopant and silicon molecules, the atomic species were kept unchanged, but the simulation model was used to change the bond structure at the silicon surface. It was shown that manipulating the structure of the silicon surface enables precise nanometer-scale control of the junction depth due to a

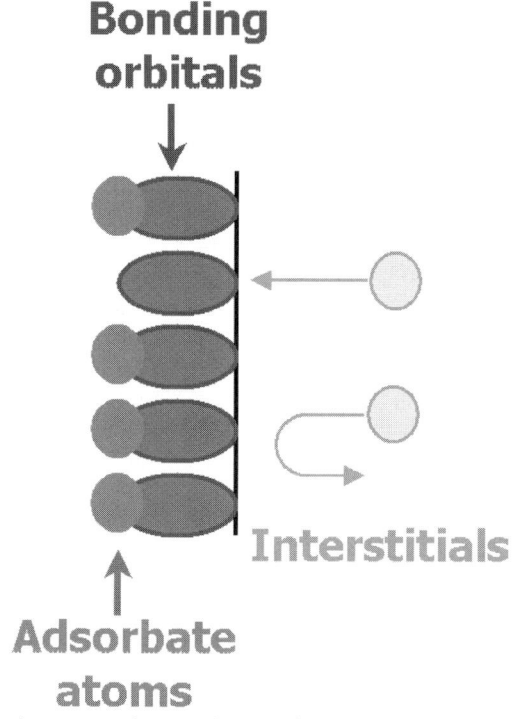

Figure 4.17 Schematic diagram of the bond insertion mechanism, showing how bulk interstitials should react relatively easily with surface dangling bond sites, but less easily with sites saturated by a strongly bonded adsorbate.

change in the effective surface boundary condition for interstitials, while providing high dopant activation. These simulation studies motivated discrete changes in processing conditions, which were shown to be effective in experiments [174–176] under conditions very close to actual manufacturing [177] and form the basis of a patent application [178].

4.8
Concluding Remarks

Future breakthroughs in science and engineering will likely require novel new combinations of experimental instrumentation and computational methods that extend our ability to make observations of smaller dimensions of length and time. In addition, there will be a steady stream of opportunities to move new understanding and innovative discoveries into well engineered products and processes. In general, there is now a sophisticated understanding of diverse electrochemical phenomena, each at its relevant time and length scale. Moreover, powerful computational methods have been developed at various time and length scales in order to simulate the behavior of such systems based on first-level physical–chemical principles. This chapter addresses the significant advancement of capability that is achieved by linking the pieces in order to understand interactions within an entire multiscale system.

The main objective of the chapter was to discuss recent developments in molecular simulation, multiscale simulation and multiscale systems engineering, and how these developments enable the targeted design at the molecular scale of processes and products based on electrochemical phenomena. The control of electrochemical phenomena at the molecular scale is critical to product quality in many new applications in medicine, computers and manufacturing. These applications include nanobiological devices, micromachines, nanoelectronic devices and protein microarrays and chips. On the other hand, for efficient operations the manipulated variables available for real-time feedback control operate at macroscopic length scales. This combination of a need for product quality at the molecular scale with the economic necessity of feedback control systems that utilize macroscopic manipulated variables motivates the creation of engineering methods for the simulation, design and control of *multiscale systems*.

Nowhere has the trend towards multiscale systems been more evident than in the microelectronics field, where multiscale simulation has been applied for nearly a decade. Subsequent efforts developed techniques for utilizing multiscale simulation models to perform systems engineering tasks, such as parameter estimation, optimization and control. This incorporation of models that couple molecular through macroscopic length scales within systems engineering tools enables a systematic approach to the simultaneous optimization of all of the length scales of the process, including the optimal control of events at the molecular scale. Such a multiscale systems framework addresses the 'grand challenge' of nanotechnology: how to move nanoscale science and technology from art to an engineering discipline.

4.8 Concluding Remarks

Although many of the trends discussed in this chapter have counterparts in other applications areas, the focus here is on electrochemical processes because of the many applications of molecular and multiscale simulation associated with these processes. A review of the progression of simulation for the design of electrochemical processes, including the increased importance of molecular and multiscale simulation, is followed by a discussion of the systems issues that arise when investigating multiscale systems. The efforts to address these issues to date serve to identify the challenges of multiscale systems, and how these challenges can be met by:

- design and implementation of high-throughput millimeter- and micrometer-scale chemical/electrochemical systems so as to highly excite the experimental input space;
- extensive utilization of scanning probe measurements;
- utilization of stop-and-repeat experiments;
- an iterative model-building procedure consisting of Bayesian estimation and mechanism selection; and
- multistep optimization.

The multiscale systems approach is directly applicable to problems in nanotechnology, molecular nanotechnology and molecular manufacturing. The key ideas have been illustrated with examples from two processes of importance to the semiconductor industry: the electrodeposition of copper to form on-chip interconnects and junction formation in metal oxide semiconductor field effect transistors.

While the modeling community has articulated the multiscale simulation vision for over a decade, the scale of computation required remained well beyond available capabilities. Dramatic recent increases in computation power have now enabled revolutionary work on multiscale systems. In the coming years, one may anticipate the development of new application algorithms, programming tools, libraries and application algorithms to allow science and engineering researchers without specific expertise in these multiscale technologies to take full advantage of them.

The most important direction for future research is the application of the multiscale systems approach to a broad range of additional non-trivial systems. There are a large number of such candidates, including many in which electrochemical phenomena play a significant role. The greatest number of electrochemical-based applications in the near term is likely to be in micro- and nanoelectronics, given the head-start in applications of multiscale simulation and the intense interest of the semiconductor industry, as cited earlier in this chapter. Additional applications are likely to arise in nanobiomedical sensors and other nanobiological devices, many of which are closely related to micro- and nanoelectronic processes in terms of chemistry, physics, materials and components. The pursuit of specific applications will also serve to improve the systems tools, as any nontrivial applications are apt to do.

Such applications would be accelerated by the development of a universal set of numerical software and data analysis algorithms for constructing multiscale simulations that optimally utilize high-end computing to make predictions and facilitate new applications. To address really challenging multiscale problems, these methods would be designed to distribute data and computations among processors to compute efficiently on *petascale machines*. Methods are desired to efficiently manage data, communications and simulation components with low scalability (such as fluid dynamics codes) with those of high scalability as the number of processors increase, creating combinations of high and low scaling methods in conjunction to run simultaneously while exploiting the capability of next-generation supercomputers.

Recall that the wide range in time and length scales, both internally in individual codes in multiscale linkages and in the updating of information passed between codes, creates temporal and spatial mismatches that must be tracked and controlled as the simulation progresses to ensure that the predictions of the overall multiscale simulation are accurate. Better numerical algorithms are needed to efficiently quantify the effects of stochastic uncertainties in model parameters and the predictions of stochastic simulation algorithms of smaller length scales (e.g. quantum) on the simulation predictions for the overall multiscale simulation that includes the larger length scales. This uncertainty quantification would enable the quantification of the degree of correspondence between a model and reality, and the determination of the propagation of stochastic uncertainties across scales and physical phenomena. Also of value would be multiscale extensions of error estimation methods for quantifying the effects of discretization, integration and basis set errors on model predictions, and how to use this information to best adaptively select the spatial and temporal resolution of individual simulation methods to achieve the highest accuracy in the outputs of the multiscale simulation code.

Acknowledgements

This material is based upon work supported by the National Science Foundation under Grant Nos. 9619019, 0086455, 0108053, 0135621, 0338215, 0426328, 0438356, and PACI NRAC-MCA 01S022. Any opinions, findings, and conclusions or recommendations expressed in this chapter are those of the authors and do not necessarily reflect the views of the National Science Foundation. The authors thank Narayana R. Aluru and Umberto Ravaioli for the use of some figures, Narayana R. Aluru for input on Section 4.4, and various students over the years who contributed directly or indirectly to the chapter.

References

1 Dolbow, J., Khaleel, M.A. and Mitchell, J. (2004) Multiscale Mathematics Initiative: A Roadmap. U.S. Department of Energy, December 2004 http://www.si.umich.edu/InfrastructureWorkshop/documents/NSF_2004_CIMultiscaleMath.pdf.

2 Drexler, K.E. (1992) *Nanosystems: Molecular Machinery, Manufacturing, and Computation*, Wiley Interscience, New York, NY.

3 Hoummady, M. and Fujita, H. (1999) Micromachines for Nanoscale Science

and Technology. *Nanotechnology*, **10**, 29–33.

4 Khanna, V.K. (2004) Emerging Trends in Ultra-miniaturized CMOS (Complementary Metal-Oxide-Semiconductor) Transistors, Single-Electron and Molecular-Scale Devices: A Comparative Analysis for High-Performance Computational Nanoelectronics. *J. Sci. Ind. Res.*, **63**, 795–806.

5 Lee, S.Y., Lee, S.J. and Jung, H.T. (2003) Protein Microarrays and Chips. *J. Ind. Eng. Chem.*, **9**, 9–15.

6 Nakano, A., Bachlechner, M.E., Kalia, R.K., Lidorikis, E., Vashishta, P., Voyiadjis, G.Z., Campbell, T.J., Ogata, S. and Shimojo, F. (2001) Multiscale Simulation of Nanosystems. *Comput. Sci. Eng.*, **3**, 56–66.

7 Prokop, A. (2001) Bioartificial Organs in the Twenty-First Century – Nanobiological Devices. Bioartificial Organs III: Tissue Sourcing, Immunoisolation, and Clinical Trials. *Ann. New York Acad. Sci.*, **944**, 472–490.

8 Tsukagoshi, K., Yoneya, N., Uryu, S., Aoyagi, Y., Kanda, A., Ootuka, Y. and Alphenaar, B.W. (2002) Carbon Nanotube Devices for Nanoelectronics. *Phys. B-Condens. Matt.*, **323**, 107–114.

9 Braatz, R.D., Alkire, R.C., Seebauer, E.G., Rusli, E., Gunawan, R., Drews, T.O. and He, Y. (2006a) Perspectives on the Design and Control of Multi-scale Systems. *J. Process Control*, **16**, 193–204.

10 Braatz, R.D., Alkire, R.C., Seebauer, E.G., Drews, T.O., Rusli, E., Karulkar, M., Xue, F., Qin, Y., Jung, M.Y.L. and Gunawan, R. (2006b) A Multiscale Systems Approach to Microelectronic Processes. *Comp. Chem. Eng.*, **30**, 1643–1656.

11 Vlachos, D.G. (2005) A Review of Multiscale Analysis: Examples from Systems Biology, Materials Engineering, and Other Fluid-Surface Interacting Systems. Technical report, University of Delaware, Newark, DE.

12 Stupp, S.I., Bawendi, M., Beebe, D., Car, R., Chiang, S., Gray, D., Heller, M., Hess, K., Iafrate, G., Jelinski, L., Jenks, T.S., Kuekes, P., Murray, C., Sohn, L., Sudarshan, T.S. and Theis, T.N. (2002) *Small Wonders, Endless Frontiers: A Review of the National Nanotechnology Initiative*, National Academies Press, Washington, D.C.

13 Maroudas, D. (2000) Multiscale Modeling of Hard Materials: Challenges and Opportunities for Chemical Engineering. *AIChE J.*, **46**, 878–882.

14 Alkire, R. and Ratner, M. (eds) (2003) *Beyond the Molecular Frontier: Challenges for the Chemical Sciences in the 21st Century*, National Research Council Report on Information, and Communications. National Academy Press Washington, DC.

15 Srinivasan, V. and Lipp, L. (2003) Report of the Electrolytic Industries for the Year 2002. *J. Electrochem. Soc.*, **150**, K15–K38.

16 National Materials Advisory Board (1986) *New Horizons in Electrochemical Science and Technology*, NMAB 438-1, National Academy Press, Washington, DC.

17 Aviram, A. and Ratner, M.S. (1974) Molecular Rectifiers. *Chem. Phys. Lett.*, **29**, 277–283.

18 Reed, M.A. and Lee, T. (eds) (2003) *Molecular Nanoelectronics*, American Scientific Publisher, Stevenson Ranch, CA.

19 Avouris, P. (2002) Molecular Electronics with Carbon Nanotubes. *Acc. Chem. Res.*, **35**, 1026–1034.

20 Cooper, K., Jakobsson, E. and Wolynes, P. (1985) The Theory of Ion Transport Through Membrane Channels. *Prog. Biophys. Mol. Biol.*, **46**, 51–96 (1985).

21 Hille, B. (2001) *Ion Channels of Excitable Membranes*, 3rd edn, Sinauer Associates, Sunderland, MA.

22 Jakobsson, E. and Chiu, S.-W. (1987) Stochastic Theory of Ion Movement in Channels with Single-Ion Occupancy. Application to Sodium Permeation of Gramicidin Channels. *Biophys. J.*, **52**, 33–45.

23 Forrest, S.R. (2000) Active Optoelectronics using Thin-film Organic Semiconductors. *IEEE J. Sel. Top. Quant. Electron.*, **6**, 1072–1083.

24 Friend, R.H., Gymer, R.W., Holmes, A.B., Burroughes, J.H., Marks, R.N., Taliani, C., Bradley, D.D.C., Dos Santo, D.A., Bredas, J.L., Logdlund, M. and Salaneck, W.R. (1999) Electroluminescence in Conjugated Polymers. *Nature*, **397**, 121–128.

25 Koshida, N. and Matsumoto, N. (2003) Fabrication and Quantum Properties of Nanostructured Silicon. *Mater. Sci. Eng. Res.*, **40**, 169–205.

26 Sato, O., Iyoda, T., Fujishima, A. and Hashimoto. K. (1996) Electrochemically Tunable Magnetic Phase Transition in a High-Tc Chromium Cyanide Thin Film. *Science*, **271**, 49–51.

27 Hartwich, G., Caruana, D.J., de Lumley-Woodyear, T., Wu, Y., Campbell, C.N. and Heller, A. (1999) Electrochemical Study of Electron Transport Through Thin DNA Films. *J. Am. Chem. Soc.*, **121**, 10803–10812.

28 Lahann, J., Mitragotri, S., Tran, T.N., Kaido, H., Sundaram, J., Choi, I.S., Hoffer, S., Somorjai, G.A. and Langer, R. (2003) A Reversibly Switching Surface. *Science*, **299**, 371–374.

29 Hansen, P.L., Wagner, J.B., Helveg, S., Rostrup-Nielsen, J.R., Clausen, B.S. and Topsøe, H. (2002) Atom-Resolved Imaging of Dynamic Shape Changes in Supported Copper Nanocrystals. *Science*, **295**, 2053–2055.

30 Kolb, D.M. (2002) The Initial Stages of Metal Deposition as Viewed by Scanning Tunneling Microscopy, in *Advances in Electrochemistry, Science and Engineering* (eds R.C. Alkire and D.M. Kolb), Wiley-VCH, Weinheim, vol. 7, pp. 107–150.

31 Engelmann, G.E., Ziegler, J.C. and Kolb, D.M. (1998) Electrochemical Fabrication of Large Arrays of Metal Nanoclusters. *Surf. Sci.*, **401**, 420–424.

32 Kolb, D.M., Engelmann, G.E. and Ziegler, J.C. (1999) Electrochemical Nanostructuring with a Scanning Tunneling Microscope, In *Ullmann's Encyclopedia of Industrial Chemistry*, 6th edn Wiley-VCH, Weinheim, chapter 3.7.

33 Stickney, J.L. (2002) *Electrochemical Atomic Layer Epitaxy: Nanoscale Control in the Electrodeposition of Compound Semiconductors. Advances in Electrochemisty Science and Engineering* (eds R.C. Alkire and D.M. Kolb), Wiley-VCH, Weinheim, vol. 7, pp. 1–106.

34 Alkire, R.C. and Chapman, T.W. (Winter 2003) Perspectives on the Evolution of Electrochemical Engineering. *Electrochem. Soc. Interface*, **12**, 47, 2003.

35 Newman, J. and Thomas-Alyea, K.E. (2004) *Electrochemical Systems*, 3rd edn John Wiley & Son, Hoboken, NJ.

36 Alkire, R. and Verhoff, M. (1995) Electrochemical Reaction Engineering in Materials Processing. *Chem. Eng. Sci.*, **49**, 4085–4093.

37 Alkire, R. and Verhoff, M. (1998) The Bridge from Nanoscale Phenomena to Macroscale Processes. *Electrochim. Acta*, **43**, 2733–2741.

38 Asada, S., Bakshi, V., Bork, I., Erdmann, A., Giles, M., Hall, E., Fujinaga, M., Heringa, A., Hwang, H.H., Jaouen, H., Le Carval, G., Lorenz, J., Merchant, T., Molzer, W., Meyyappan, M., Nakamura, M., Orlowski, M., Riccobene, C., Satoh, S., Schoenmaker, W., Singh, V., Szalapski, R., Trybula, W., Ventzek, P.L.G., Wada, T., Wang, T., Woltjer, R., Wu, J. and Yen, T. (2004) Modeling and Simulation. In *International Technology Roadmap for Semiconductors*. Sematech, http://public.itrs.net.

39 Cavallotti, C.M., Nemirovskaya, K. and Jensen, F. (2003) A Multiscale Study of the Selective MOVPE of $Al_xGa_{1-x}As$ in the Presence of HCl. *J. Crystal Growth*, **248**, 411–416.

40 Dollet, A. (2004) Multiscale Modelling of CVD Film Growth – A Review of Recent Works. *Surf. Coat. Technol.*, **177**, 245–251.

41 Drews, T.O., Braatz, R.D. and Alkire, R.C. (2004) Coarse-grained Kinetic Monte

Carlo Simulation of Copper Electrodeposition with Additives. *Int. J. Multiscale Comput. Eng.*, **2**, 313–327.

42 Drews, T.O., Krishnan, S., Alameda, J., Gannon, D., Braatz, R.D. and Alkire, R.C. (2005a) Multiscale Simulations of Copper Electrodeposition onto a Resistive Substrate. *IBM J. Res. Dev.*, **49**, 49–63.

43 Jensen, K.F., Rodgers, S.T. and Venkataramani, R. (1998) Multiscale Modeling of Thin Film Growth. *Curr. Opin. Solid State Mater. Sci.*, **3**, 562–569.

44 Nieminen, R.M. (2002) From Atomistic Simulation Towards Multiscale Modelling of Materials. *J. Phys.-Condens. Matt.*, **14**, 2859–2876.

45 Rodgers, S.T. and Jensen, K.F. (1998) Multiscale Modelling of Chemical Vapor Deposition. *J. Appl. Phys.*, **83**, 524–530.

46 Taniguchi, N., ed. (1974) On the Basic Concept of 'Nano-Technology.' *Proceedings of International Conference on Production Engineering*, Tokyo, Part II, Japan Society of Precision Engineering.

47 Roco, M.C. (2003) Nanotechnology: Convergence with Modern Biology and Medicine. *Curr. Opin. Biotechnol.*, **14**, 337–346.

48 Nanotechnology (2008). Scope of the journal *Nanotechnology*, as described on its website, Institute of Physics Publishing: Bristol, http://www.iop.org

49 Ghosh, T., Grade, S. and Garcia, A.E. (2003) Role of Backbone Hydration and Salt-Bridge Formation in Stability of Alpha-Helix in Solution. *Biophys. J.*, **85**, 3187–3193.

50 Green, D.F. and Tidor, B. (2004) Escherichia Coli Glutaminyl-tRNA Synthetase is Electrostatically Optimization for Binding of its Cognate Substrates. *J. Mol. Biol.*, **342**, 435–452.

51 Koberstein, J.T. (2004) Molecular Design of Functional Polymer Surfaces. *J. Polym. Sci. Part B-Polym. Phys.*, **42**, 2942–2956.

52 Larson, R.G. (2003) Molecular Engineering of Peptides. *Chem. Biol.*, **10**, 1005–1006.

53 Manstein, D.J. (2004) Molecular Engineering of Myosin. *Philos. Trans. R. Soc. London Ser. B-Biol. Sci*, **359**, 1907–1912.

54 Gummel, H.K. (1964) A Self-consistent Iterative Scheme for One-dimensional Steady State Transistor Calculations. *IEEE Trans. Electron Devices*, ED-11, 455–465.

55 Lee, C.M., Lomax, R.J. and Haddad, G.I. (1974) Semiconductor Device Simulation. *IEEE Trans. Microw. Theory Techn.*, MTT-22, 160–177.

56 Scharfetter, D.L. and Gummel, H.K. (1969) Large-Scale Analysis of a Silicon Read Diode Oscillator. *IEEE Trans. Electron Devices*, ED-16, 64–77.

57 Graves, D.B. (1987) Fluid Model Simulations of a 13.56-MHz rf Discharge – Time and Space Dependence of Rates of Electron-Impact Excitation. *J. Appl. Phys.*, **62**, 88–94.

58 Graves, D.B. and Jensen, K.F. (1986) Continuum Model of DC and rf Discharges. *IEEE Trans. Plasma Sci.*, PS-14, 78–91.

59 Park, S.K. and Economou, D.J. (1990) Analysis of Low-Pressure rf-Glow Discharges Using a Continuum Model. *J. Appl. Phys.*, **68**, 3904–3915.

60 Thompson, B.E. and Sawin, H.H. (1986) Comparison of Measured and Calculated SF_6 Breakdown in rf Electric-Fields. *J. Appl. Phys.*, **60**, 89–94.

61 Moore, G.E. (1965) Cramming More Components onto Integrated Circuits. *Electronics*, **38**, 8.

62 Martin-Bragado, I., Jaraiz, M., Castrillo, P., Pinacho, R., Rubio, J.E. and Barbolla, J. (2004) A Kinetic Monte Carlo Annealing Assessment of the Dominant Features from Ion Implant Simulations. *Mater. Sci. Eng. B-Solid State Mater. Adv. Technol.*, **114**, 345–348.

63 Pinacho, R., Jaraiz, M., Castrillo, P., Rubio, J.E., Martin-Bragado, I. and Barbolla, J. (2004) Comprehensive, Physically Based Modelling of As in Si. *Mater. Sci. Eng. B-Solid State Mater. Adv. Technol.*, **114**, 135–140.

64 Rubio, J.E., Jaraiz, M., Martin-Bragado, I., Pinacho, R., Castrillo, P. and Barbolla, J. (2004) Physically based Modelling of Damage, Amorphization, and Recrystallization for Predictive Device-Size Process Simulation. *Mater. Sci. Eng. B-Solid State Mater. Adv. Technol.*, **114**, 151–155.

65 Tezduyar, T.E. and Hughes, T.J.R. (eds) (1986) *Numerical Methods for Compressible Flows- Finite Difference, Element and Volume Techniques*, ASME Press, New York, AMD vol. 78.

66 Heath, M.T. (2002) *Scientific Computing: An Introductory Survey*, 2nd edn, McGraw-Hill, New York.

67 Karniadakis, G.E., Beskok, A. and Aluru, N.R. (2005) *Microflows and Nanoflows: Fundamentals and Simulation*, Springer Verlag, Berlin.

68 Jiao, X. and Heath, M.T. (2004a) Common-Refinement-Based Data Transfer between Non-matching Meshes in Multiphysics Simulations. *Int. J. Numer. l Meth. Eng.*, **61**, 2402–2427.

69 Jiao, X. and Heath, M.T. (2004b) Overlaying Surface Meshes. *Int. J. Comput. Geom. Appl.*, **14**, 379–419.

70 Jiao, X., Zheng, G., Alexander, P.A., Campbell, M.T., Lawlor, O.S., Norris, J., Haselbacher, A. and Heath, M.T. (2006) A System Integration Framework for Coupled Multiphysics Simulations. *Eng. Comput.*, **22**, 293–309.

71 van der Straaten, T.A., Tang, J.M., Ravaioli, U., Eisenberg, R.S. and Aluru, N.R. (2003) Simulating Ion Permeation Through the ompF Porin Ion Channel Using Three-dimensional Drift-Diffusion Theory. *J. Computat. Electron.*, **2**, 29–47.

72 Grasser, T. and Selberherr, S. (2002) Technology CAD: Device Simulation and Characterization. *J. Vac. Sci. Technol. B*, **20**, 407–413.

73 Kim, J., McMurray, J.S., Williams, C.C. and Slinkman, J. (1998) Two-step Dopant Diffusion Study Performed in Two Dimensions by Scanning Capacitance Microscopy and TSUPREM IV. *J. Appl. Phys.*, **84**, 1305–1309.

74 Pardhanani, A.L. and Carey, G.F. (2000) Multidimensional Semiconductor Device and Micro-scale Thermal Modeling Using the PROPHET Simulator with Dial-an-Operator Framework. *Comput. Model. Eng. Sci.*, **1**, 141–150.

75 Rafferty, C. and Smith, R.K. (2000) Making a PROPHET. *Comput. Model. Eng. Sci.*, **1**, 151–159.

76 Sibaja-Hernandez, A., Xu, M.W., Decoutere, S. and Maes, H. (2005) TSUPREM-4 Based Modeling of Boron and Carbon Diffusion in SiGeC Base Layers Under Rapid Thermal Annealing Conditions. *Mater. Sci. Semicon. Process.*, **8**, 115–120.

77 Cahill, D.G., Ford, W.K., Goodson, K.E., Mahan, G.D., Majumdar, A., Maris, H.J., Merlin, R. and Phillpot, S.R. (2003) Nanoscale Thermal Transport. *J. Appl. Phys.*, **93**, 793–818.

78 Grasser, T., Tang, T.W., Kosina, H. and Selberherr, S. (2003) A Review of Hydrodynamic and Energy-Transport Models for Semiconductor Device Simulation. *Proc. IEEE*, **91**, 251–274.

79 Kushner, M.J. (1985) Distribution of Ion Energies Incident on Electrodes in Capacitively Coupled rf Discharges. *J. Appl. Phys.*, **58**, 4024–4031.

80 Ravaioli, U. (1998) Hierarchy of Simulation Approaches for Hot Carrier Transport in Deep Submicron Devices. *Semiconduct. Sci. Technol.*, **13**, 1–10.

81 Saraniti, M., Tang, J., Goodnick, S.M. and Wigger, S.J. (2003) Numerical Challenges in Particle-based Approaches for the Simulation of Semiconductor Devices. *Math. Comput. Simulat.*, **62**, 501–508.

82 Schoenmaker, W. and Vankemmel, R. (1992) Simulation of Compound Semiconductor-Devices. *Microelectron. Eng.*, **19**, 31–38.

83 Sommerer, T.J. and Kushner, M.J. (1992) Numerical Investigation of the Kinetics and Chemistry of rf Glow-Discharge Plasmas Sustained in He, N_2, O_2, He/N_2/

O_2, $He/CF_4/O_2$, and SiH_4/NH_3 using a Monte-Carlo-Fluid Hybrid Model. *J. Appl. Phys.*, **71**, 1654–1673.

84 Takagi, S., Onoue, S., Iyanagi, K., Nishitani, K. and Shinmura, T. (2005) Topography Simulations for Contact Formation Involving Reactive Ion Etching, Sputtering and Chemical Vapor Deposition. *J. Vac. Sci. Technol. B*, **23**, 1076–1083.

85 Lake, R., Klimeck, G., Bowen, R.C. and Jovanovic, D. (1997) Single and Multiband Modeling of Quantum Electron Transport Through Layered Semiconductor Devices. *J. Appl. Phys.*, **81**, 7845–7869.

86 Sano, N., Hiroki, A. and Matsuzawa, K. (2002) Device Modeling and Simulations Toward Sub-10 nm Semiconductor Devices. *IEEE Trans. Nanotechnol.*, **1**, 63–71.

87 Vasileska, D. and Goodnick, S.M. (2002) Computational Electronics. *Mater. Sci. Eng. R-Rep.*, **38**, 181–236.

88 Dalpian, G.M., Janotti, A., Fazzio, A. and da Silva, A.J.R. (1999) Initial Stages of Ge Growth on Si(100): Ad-atoms, Ad-dimers, and Ad-trimers. *Physica B*, **274**, 589–592.

89 La Magna, A., Alippi, P., Colombo, L. and Strobel, M. (2003) Atomic Scale Computer Aided Design for Novel Semiconductor Devices. *Comput. Mater. Sci.*, **27**, 10–15.

90 Jeong, J.W., Lee, I.H., Oh, J.H. and Chang, K.J. (1998) First-Principles Study of the Equilibrium Structures of Si-N Clusters. *J. Phys.-Condens. Matt.*, **10**, 5851–5860.

91 Lin, L., Kirichenko, T., Banerjee, S.K. and Hwang, G.S. (2004) Boron Diffusion in Strained Si: A First-Principles Study. *J. Appl. Phys.*, **96**, 5543–5547.

92 Tuttle, B.R., McMahon, W. and Hess, K. (2000) Hydrogen and Hot Electron Defect Creation at the $Si(100)/SiO_2$ Interface of Metal-Oxide-Semiconductor Field Effect Transistors. *Superlatt. Microstruct.*, **27**, 229–233.

93 Goto, R., Shimojo, F., Munejiri, S. and Hoshino, K. (2004) Structural and Electronic Properties of Liquid Ge-Sn Alloys: Ab Initio Molecular-Dynamics Simulations. *J. Phys. Soc. Jpn.*, **73**, 2746–2752.

94 Ko, E., Jain, M. and Chelikowsky, J.R. (2002) First Principles Simulations of SiGe for the Liquid and Amorphous States. *J. Chem. Phys.*, **117**, 3476–3483.

95 Lee, S.M., Lee, S.H. and Scheffler, M. (2004) Adsorption and Diffusion of a Cl Adatom on the GaAs(001)-c(8×2) Zeta Surface. *Phys. Rev. B*, **69**, 125317.

96 Fichthorn, K.A. and Weinberg, W.H. (1991) Theoretical Foundations of Dynamical Monte Carlo Simulations. *J. Chem. Phys.*, **95**, 1090–1096.

97 Jönsson, M., Skepö, M., Tjerneld, F. and Linse, P. (2003) Effect of Spatially Distributed Surface Residues on Protein-Polymer Association. *J. Phys. Chem. B*, **107**, 5511–5518.

98 Lopez, C.F., Moore, P.B., Shelley, J.C., Shelley, M.Y. and Klein, M.L. (2002) Computer Simulation Studies of Biomembranes Using a Coarse Grain Model. *Comput. Phys. Commun.*, **147**, 1–6.

99 Shelley, J.C. and Shelley, M.Y. (2000) Computer Simulation of Surfactant Solutions. *Curr. Opin. Colloid. Surf. Sci.*, **5**, 101–110.

100 Rudd, R.E. and Broughton, J.Q. (1998) Coarse-grained Molecular Dynamics and the Atomic Limit of Finite Elements. *Phys. Rev. B*, **58**, R5893–R5896.

101 Rudd, R.E. and Broughton, J.Q. (2000) Concurrent Coupling of Length Scales in Solid State Systems. *Phys. Status Solidi B – Basic Res.*, **217**, 251–291.

102 Katsoulakis, M.A., Majda, A.J. and Vlachos, D.G. (2003) Course-grained Stochastic Processes and Monte Carlo Simulations in Lattice Systems. *J. Comput. Phys.*, **186**, 250–278.

103 Katsoulakis, M.A. and Vlachos, D.G. (2003) Coarse-grained Stochastic Processes and Kinetic Monte Carlo Simulation for the Diffusion of Interacting Particles. *J. Chem. Phys.*, **119**, 9412–9427.

104 Ismail, A.E., Rutledge, G.C. and Stephanopoulos, G. (2003a) Multiresolution Analysis in Statistical Mechanics – I. Using Wavelets to Calculate Thermodynamic Properties. *J. Chem. Phys.*, **118**, 4414–4423.

105 Ismail, A.E., Rutledge, G.C. and Stephanopoulos, G. (2003b) Multiresolution Analysis in Statistical Mechanics – II. The Wavelet Transform as a Basis for Monte Carlo Simulations on Lattices. *J. Chem. Phys.*, **118**, 4424–4431.

106 Gear, C.W., Li, J. and Kevrekidis, I.G. (2003) The Gap-Tooth Method in Particle Simulations. *Phys. Lett. A*, **316**, 190–195.

107 Gillespie, D.T. (2001) Approximate Accelerated Stochastic Simulation of Chemically Reacting Systems. *J. Chem. Phys.*, **115**, 1716–1733.

108 Gillespie, D.T. and Petzold, L.R. (2003) Improved Leap-size Selection for Accelerated Stochastic Simulation. *J. Chem. Phys.*, **119**, 8229–8234.

109 Rathinam, M., Petzold, L. and Gillespie, D. (2003) Stiffness in Stochastic Chemically Reacting Systems: The Implicit Tau-leaping Method. *J. Chem. Phys.*, **119**, 12784–12794.

110 Alder, B.J. and Wainwright, T.E. (1959) Studies in Molecular Dynamics. I. General Method. *J. Chem. Phys.*, **31**, 459–466.

111 Joseph, S. and Aluru, N.R. (2006) Hierarchical Multiscale Simulation of Electrokinetic Transport in Silica Nanochannels at the Point of Zero Charge. *Langmuir*, **22**, 9041–9051.

112 Qiao, R. and Aluru, N.R. (2004) Multiscale Simulation of Electroosmotic Transport Using Embedding Techniques. *Int. J. Multiscale Computat. Eng.*, **2**, 173–188.

113 Coronell, D.G., Hansen, D.E., Voter, A.F., Liu, C.-L., Liu, X.-Y. and Kress, J.D. (1998) Molecular Dynamics-based Ion-surface Interaction Modes for Ionized Physical Vapor Deposition Feature Scale Simulations. *Appl. Phys. Lett.*, **73**, 3860–3862.

114 Hansen, E. and Neurock, M. (2001) First-principles Based Kinetic Simulations of Acetic Acid Temperature Programmed Reaction on Pd(111). *J. Phys. Chem. B*, **105**, 9218–9229.

115 Hansen, U., Rodgers, S. and Jensen, K.F. (2000) Modeling of Metal Thin Film Growth: Linking Angstrom-scale Molecular Dynamics Results to Micron-scale Film Topographies. *Phys. Rev. B*, **62**, 2869–2878.

116 Maroudas, D. and Gungor, M.R. (2002) Continuum and Atomistic Modeling of Electromechanically-induced Failure of Ductile Metallic Thin Films. *Comput. Mater. Sci.*, **23**, 242–249.

117 Maroudas, D. and Shankar, S. (1996) Electronic Materials Process Modeling. *J. Comput.-Aided Mater. Des.*, **3**, 36–48.

118 Sinno, T. and Brown, R.A. (1999) Modeling Microdefect Formation in Czochralski Silicon. *J. Electrochem. Soc.*, **146**, 2300–2312.

119 Tadmor, E.B., Ortiz, M. and Phillips, R. (1996a) Quasicontinuum Analysis of Defects in Solids. *Philos. Mag. A – Phys. Cond. Matt. Struct. Defects, Mech. Prop.*, **73**, 1529–1563.

120 Tadmor, E.B., Phillips, R. and Ortiz, M. (1996b) Mixed Atomistic and Continuum Models of Deformation in Solids. *Langmuir*, **12**, 4529–4534.

121 Gobbert, M.K., Merchant, T.P., Borucki, L.J. and Cale, T.S. (1997) A Multiscale Simulator for Low Pressure Chemical Vapor Deposition. *J. Electrochem. Soc.*, **144**, 3945–3951.

122 Hadji, D., Marechal, Y. and Zimmerman, J. (1999) Finite Element and Monte Carlo Simulation of Submicrometer Silicon n-MOSFET's. *IEEE Trans. Magnetics*, **35**, 1809–1812.

123 Raimondeau, S. and Vlachos, D.G. (2002) Recent Developments on Multiscale, Hierarchical Modeling of Chemical Reactors. *Chem. Eng. J.*, **90**, 3–23.

124 Vlachos, D.G. (1997) Multiscale Integration Hybrid Algorithms for

Homogeneous-Heterogeneous Reactors. *AIChE J.*, **43**, 3031–3041.

125 Aktas, O. and Aluru, N.R. (2002) A Combined Continuum/DSMC Technique for Multiscale Analysis of Microfluidic Filters. *J. Comp. Phys.*, **178**, 342–372.

126 Broughton, J.Q., Abraham, F.F., Bernstein, N. and Kaxiras, E. (1999) Concurrent Coupling of Length Scales: Methodology and Application. *Phys. Rev. B*, **60**, 2391–2403.

127 Vlachos, D.G. (1999) Role of Macrotransport Phenomena in Film Microstructure during Epitaxial Growth. *Appl. Phys. Lett.*, **74**, 2797–2799.

128 Vlachos, D.G., Schmidt, L.D. and Aris, R. (1990) The Effects of Phase Transitions, Surface Diffusion, and Defects on Surface Catalyzed Reactions: Oscillations and Fluctuations. *J. Chem. Phys.*, **93**, 8306.

129 Merchant, T.P., Gobbert, M.K., Cale, T.S. and Borucki, L.J. (2000) Multiple Scale Integrated Modeling of Deposition Processes. *Thin Solid Films*, **365**, 368–375.

130 Cale, T.S., Bloomfield, M.O., Richards, D.F., Jansen, K.E. and Gobbert, M.K. (2002) Integrated Multiscale Process Simulation. *Comput. Mater. Sci.*, **23**, 3–14.

131 Li, X., Drews, T.O., Rusli, E., Xue, F., He, Y., Braatz, R. and Alkire, R. (2007) Effect of Additives on Shape Evolution during Electrodeposition. I. Multiscale Simulation with Dynamically Coupled Kinetic Monte Carlo and Moving-Boundary Finite-Volume Codes. *J. Electrochem. Soc.*, **154**, D230–D240.

132 Rusli, E., Drews, T.O. and Braatz, R.D. (2004) Systems Analysis and Design of Dynamically Coupled Multiscale Reactor Simulation Codes. *Chem. Eng. Sci.*, **59**, 5607–5613.

133 Ascher, U.M. and Petzold, L.R. (1998) *Computer Methods for Ordinary Differential Equations and Differential-Algebraic Equations*, SIAM Press, Philadelphia, PA.

134 Raimondeau, S. and Vlachos, D.G. (2000) Low-dimensional Approximations of Multiscale Epitaxial Growth Models for Microstructure Control of Materials. *J. Comput. Phys.*, **160**, 564–576.

135 Chang, Y.C., Hou, T.Y., Merriman, B. and Osher, S. (1996) A Level Set Formulation of Eulerian Interface Capturing Methods for Incompressible Fluid Flows. *J. Comp. Phys.*, **124**, 449–464.

136 Li, Z., Zhao, H. and Gao, H. (1999) A Numerical Study of Electro-migration Voiding by Evolving Level Set Functions on a Fixed Cartesian Grid. *J. Comput. Phys.*, **152**, 281–304.

137 Wheeler, D., Josell, D. and Moffat, T.P. (2003) Modeling Superconformal Electrodeposition using the Level Set Method. *J. Electrochem. Soc.*, **150**, C302–C310.

138 Jung, M.Y.L., Gunawan, R., Braatz, R.D. and Seebauer, E.G. (2004a) Pair Diffusion and Kick-Out: Contributions to Diffusion of Boron in Silicon. *AIChE J.*, **50**, 3248–3256.

139 Andricacos, P.C. (1999) Copper on-Chip Interconnections – A Breakthrough in Electrodeposition to Make Better Chips. *Electrochem. Soc. Interface*, **8**, 32–37.

140 Andricacos, P.C., Uzoh, C.J., Dukovic, O., Horkans, J. and Deligianni, H. (1998) Damascene Copper Electroplating for Chip Interconnections. *IBM J. Res. Dev.*, **42**, 567–574.

141 Datta, M. and Landolt, D. (2000) Fundamental Aspects and Applications of Electrochemical Microfabrication. *Electrochim. Acta*, **45**, 2535–2558.

142 Kondo, K., Matsumoto, T. and Watanabe, K. (2004) Role of Additives for Copper Damascene Electrodeposition: Experimental Study on Inhibition and Acceleration Effects. *J. Electrochem. Soc.*, **151**, C250–C255.

143 Moffat, T.P., Bonevich, J.E., Huber, W.H., Stanishevsky, A., Kelly, D.R., Stafford, G.R. and Josell, D. (2000) Superconformal Electrodeposition of Copper in 500-90 nm Features. *J. Electrochem. Soc.*, **147**, 4524–4535.

144 Moffat, T.P., Wheeler, D., Huber, W.H. and Josell, D. (2001) Superconformal Electrodeposition of Copper. *Electrochem. Solid State Lett.*, **4**, C26–C29.

145 Moffat, T.P., Wheeler, D. and Josell, D. (2004) Electrodeposition of Copper in the SPS-PEG-Cl Additive Sytem. *J. Electrochem. Soc.*, **151**, C262–C271.

146 Tan, M. and Harb, J.N. (2003) Additive Behavior During Copper Electrodeposition in Solutions Containing Cl$^-$, PEG, and SPS. *J. Electrochem. Soc.*, **150**, C420–C425.

147 West, A.C. (2000) Theory of Filling of High-Aspect Ratio Trenches and Vias in Presence of Additives. *J. Electrochem. Soc.*, **147**, 227–232.

148 Tu, Y. and Laaksonen, A. (2005) Quantum Chemistry Study of Monomer Electronic Properties in Water Clusters and Liquid Water and Methanol. *Int. J. Quantum Chem.*, **102**, 888–896.

149 Tuckerman, M.E., Marx, D. and Parrinello, M. (2002) The Nature and Transport Mechanism of Hydrated Hydroxide Ions in Aqueous Solution. *Nature*, **417**, 925–929.

150 Wernet, P., Nordlund, D., Bergmann, U., Cavalleri, M., Odelius, M., Ogasawara, H., Naslund, L.A., Hirsch, T.K., Ojamae, L., Glatzel, P., Pettersson, L.G.M. and Nilsson, A. (2004) The Structure of the First Coordination Shell in Liquid Water. *Science*, **304**, 995–999.

151 Xenides, D., Randolf, B.R. and Rode, B.M. (2005) Structure and Ultrafast Dynamics of Liquid Water: A Quantum Mechanics/Molecular Mechanics Molecular Dynamics Simulations Study. *J. Chem. Phys.*, **122**, 147506.

152 Onda, K., Li, B., Zhao, J., Jordan, K.D., Yang, J.L. and Petek, H. (2005) Wet Electrons at the H_2O/TiO_2(110) Surface. *Science*, **308**, 1154–1158.

153 Burke, A.L., Duever, T.A. and Penlidis, A. (1997) Choosing the Right Model: Case Studies on the Use of Statistical Model Discrimination Experiments. *Can. J. Chem. Eng.*, **75**, 422–436.

154 Gunawan, R., Ma, D.L., Fujiwara, M. and Braatz, R.D. (2002) Identification of Kinetic Parameters in a Multidimensional Crystallization Process. *Int. J. Mod. Phys. B*, **16**, 367–374.

155 Reilly, P.M. and Blau, G.E. (1974) The Use of Statistical Methods to Build Mathematical Models of Chemical Reacting Systems. *Can. J. Chem. Eng.*, **52**, 289–299.

156 Atkinson, A.C. and Donev, A.N. (1992) *Optimum Experimental Designs*, Clarenden Press, Oxford.

157 Rusli, E., Xue, F., Drews, T.O., Vereecken, P., Andracacos, P., Deligianni, H., Braatz, R.D. and Alkire, R.C. (2007) Effect of Additives on Shape Evolution during Electrodeposition. Part II: Parameter Estimation from Roughness Evolution Experiments. *J. Electrochem. Soc.*, **154**, D584–D597.

158 Rusli, E. (2006). PhD Thesis, University of Illinois, Urbana-Champaign.

159 Xue, F. (2006) PhD Thesis, University of Illinois, Urbana-Champaign.

160 Raimondeau, S., Aghalayam, P., Mhadeshwar, A.B. and Vlachos, D.G. (2003) Parameter Optimization of Molecular Models: Application to Surface Kinetics. *Ind. Eng. Chem. Res.*, **42**, 1174–1183.

161 Drews, T.O., Braatz, R.D. and Alkire, R.C. (2003) Parameter Sensitivity Analysis of Monte Carlo Simulations of Copper Electrodeposition with Multiple Additives. *J. Electrochem. Soc.*, **150**, C807–C812.

162 Nagy, Z.K. and Braatz, R.D. (2003) Robust Nonlinear Model Predictive Control of Batch Processes. *AIChE J.*, **49**, 1776–1786.

163 Law, M.E. and Cea, S.M. (1998) Continuum Based Modeling of Silicon Integrated Circuit Processing: An Object Oriented Approach. *Comput. Mater. Sci.*, **12**, 289–308.

164 Bank, R.E., Coughran, W.M. Jr., Fichtner, W., Grosse, E.H., Rose, D.J. and Smith, R.K. (1985) Transient Simulation of Silicon Devices and Circuits. *IEEE Trans.*

Computer-Aided Des. Integrated, Circuits Syst. **4**, 436–451.

165 Gunawan, R., Jung, M.Y.L., Seebauer, E.G. and Braatz, R.D. (2003a) Maximum A Posteriori Estimation of Transient Enhanced Diffusion Energetics. *AIChE J.*, **49**, 2114–2123.

166 Jung, M.Y.L., Kwok, C.T.M., Braatz, R.D. and Seebauer, E.G. (2005) Interstitial Charge States in Boron-implanted Silicon. *J. Appl. Phys.*, **97**, 063520.

167 Gunawan, R., Jung, M.Y.L., Seebauer, E.G. and Braatz, R.D. (2003b) Parameter Sensitivity Analysis of Boron Activation and Transient Enhanced Diffusion in Silicon. *J. Electrochem. Soc.*, **150**, G758–G765.

168 Jung, M.Y.L., Gunawan, R., Braatz, R.D. and Seebauer, E.G. (2003) Ramp-rate Effects in Transient Enhanced Diffusion and Dopant Activation. *J. Electrochem. Soc.*, **150**, G838–G842.

169 Jung, M.Y.L., Gunawan, R., Braatz, R.D. and Seebauer, E.G. (2004c) A Simplified Picture for Transient Enhanced Diffusion of Boron in Silicon. *J. Electrochem. Soc.*, **151**, G1–G7.

170 Gunawan, R., Jung, M.Y.L., Seebauer, E.G. and Braatz, R.D. (2004) Optimal Control of Transient Enhanced Diffusion in a Semiconductor Process. *J. Process Contr.*, **14**, 423–430.

171 Dev, K., Jung, M.Y.L., Gunawan, R., Braatz, R.D. and Seebauer, E.G. (2003) Mechanism for Coupling between Properties of Interfaces and Bulk Semiconductors. *Phys. Rev. B*, **68**, 195311 (1-6).

172 Dev, K. and Seebauer, E.G. (2004) Band Bending at the Si(111)-SiO$_2$ Interface Induced by Low-Energy Ion Bombardment. *Surf. Sci.*, **550**, 185–191.

173 Jung, M.Y.L., Gunawan, R., Braatz, R.D. and Seebauer, E.G. (2004b) Effect of Near-Surface Band Bending on Dopant Profiles in Ion-Implanted Silicon. *J. Appl. Phys.*, **95**, 1134–1139.

174 Seebauer, E.G., Dev, K., Jung, M.Y.L., Vaidyanathan, R., Kwok, C.T.M., Ager, J.W., Haller, E.E. and Braatz, R.D. (2006) Control of Defect Concentrations Within a Semiconductor Through Adsorption. *Phys. Rev. Lett.*, **97**, 055503.

175 Kwok, C.T.M., Dev, K., Braatz, R.D. and Seebauer, E.G. (2005) A Method for Quantifying Annihilation Rates of Bulk Point Defects at Surfaces. *J. Appl. Phys.*, **98**, 013524.

176 Zhang, X., Yu, M., Kwok, C.T.M., Vaidyanathan, R., Braatz, R.D. and Seebauer, E.G. (2006) Precursor Mechanism for Interaction of Bulk Interstitial Atoms with Si(100). *Phys. Rev. B*, **74**, 235301.

177 Yeong, S.H., Srinivasan, M.P., Colombeau, B., Chan, L., Akkipeddi, R., Kwok, C.T.M., Vaidyanathan, R. and Seebauer, E.G. (2007) Defect Engineering by Surface Chemical State in Boron-Doped Pre-amorphized Silcon. *J. Appl. Phys.*, **91**, 102112 (1–3).

178 Seebauer, E.G., Braatz, R.D., Jung, M.Y.L. and Gunawan, R. (2005) Methods for Controlling Dopant Concentration and Activation in Semiconductor Structures, US Application 20060024928.

179 Alkire, R.C. and Braatz, R.D. (2004) Electrochemical Engineering in an Age of Discovery and Innovation. *AIChE J.*, **50**, 2000–2007.

180 Barnes, M.S., Colter, T.J. and Elta, M.E. (1987) Large-Signal Time-Domain Modeling of Low-Pressure rf Glow-Discharges. *J. Appl. Phys.*, **61**, 81–89.

181 Catellani, A., Cicero, G., Righi, M.C. and Pignedoli, C.A. (2005) First Principles Simulations of SiC-based Interfaces. *Silicon Carbide Related Mater.*, **483**, 541–546.

182 Drews, T.O., Webb, E.G., Ma, D.L., Alameda, J., Braatz, R.D. and Alkire, R.C. (2005b) Coupled Mesoscale-Continuum Simulations of Copper Electrodeposition in a Trench. *AIChE J.*, **50**, 226–240.

183 Drews, T.O., Radisic, A., Erlebacher, J., Braatz, R.D., Searson, P.C. and Alkire, R.C. (2006) Stochastic Simulation

of the Early Stages of Kinetically Limited Electrodeposition. *J. Electrochem. Soc.*, **153**, C434–C441.

184 Drews, T.O., Braatz, R.D. and Alkire, R.C. (2007) Monte Carlo Simulation of Kinetically-limited Electrodeposition on a Surface with Metal Seed Clusters. *Z. Phys. Chem.*, **221**, 1287–1305.

185 He, Y., Gray, J.R., Alkire, R.C. and Braatz, R.D. (2004) In *Proceedings of the Topical Conference on Coupling Theory, Molecular Simulations and Computational Chemistry to the Physical World*, AIChE Annual Meeting, Austin, TX, November 7–12, paper 439b.

186 Karulkar, M., He, Y., Alkire, R.C. and Braatz, R.D. (2005) Guidelines for the Design of Multiscale Simulation Codes. In *Proceedings of the Topical Conference on Multiscale Analysis in Chemical, Materials and Biological Processes*, AIChE Annual Meeting, Cincinnati, OH, paper 503a.

187 Sematech (2004, 2005) *International Technology Roadmap for Semiconductors.* Sematech. http://public.itrs.net.

Index

a

ABAQUS 289, 304
accelerator coverage, curvature enhanced,
 see CEAC
activation
– barrier 143
– blocked electrodes 135–137
– thermal 269–283
active electronic devices 1–99
additive processes 113–146
adsorbates
– deactivation 121
– evolution 121–122
– impact on microstructure 122–125
– incorporated into growing deposit 119–121
– inhibition of metal deposition 125–130
– segregated 118–119
adsorption
– co-adsorption effects 134
– competitive 135–137, 141–146
– diffusion coefficient 117, 119–120
– isotherms 220–221
– kinetics 117
– potential-dependent 144–145
– spontaneous 221–223
– SPS 143–146, 164
– strong 118
Airy's formula 9
Al/Al$_2$O$_3$ system 53–54
algorithms 303–311, 315, 317, 323–324
– KMC, see KMC
– shape change 146, 149, 164
alkaline electrolytes 194
alternative anode materials 73–74
alternative dielectric materials 90–96
amino functionalized porphyrin 224
amino groups 138
amorphous films 2–4, 17

– Al/Al$_2$O$_3$ 54
– anodically formed oxides 48
– Beilby layer 40
– capacitor devices 67–70
– SiO$_2$ 75, 89
– Ta$_2$O$_5$ 53, 65
– TiO$_2$ 19, 23, 25, 40–41, 95
– ZrO$_2$ 42–46
anisotropic etching 86–89
anisotropic parameters 40
anisotropy
– optical 38, 48
– spectroscopic anisotropy micro-ellipsometry,
 see SAME
annealing
– rapid thermal 92–94, 299, 317–320
– simulated 315
– thermal 69
anode materials, alternative/modified 73–74
anode sinter body, thermal conductivity
 71–72
anodic ETR current density 35
anodic formation factor 3
anodic oxide formation 6, 29, 45, 51, 56, 64
– irreversible 55
– Nb/Nb$_2$O$_5$ system 49
anodic oxide growth 37
anodic passivation 76
anodic passive layer formation, single Ti
 grains 28–30
anodizing 64
applications
– electrochemical 297–298
– electrolytic capacitor manufacturing 57–74
– laser in electrochemistry 8–10
– microelectronic 295–296
– Nb/Nb$_2$O$_5$ system 53
– photoresist microelectrodes 28–35

– Si/SiO$_2$ system   77–90
– Zr/ZrO$_2$ system   44–46
aqueous electrolytes   193–199
– pH   197–199
aspect ratio   27
– DT Si etch   84–85
avalanche effects, electronic   67

b

band gap energy   3, 15
band structure   8, 28, 37, 52
barrier, activation   143
battery behavior   58
Bayesian estimation   312
Beer, Lambert–Beer's law   13–14
Beilby layer, isotropic   40
'Best'-cell DRAM   83
bit-line design, folded   81
Blue Ribbon Panel report   290–291
bond insertion mechanism   321
boron dopant profiles   319
bottle-shaped DT   85–88
boundary, moving   309
boundary layer
– diffusion   28
– hydrodynamic, see hydrodynamic boundary layer thickness
– thickness   154, 156, 161, 174
break down voltage   62
brightening   111–112
– catalyst-derived   161–176
bump formation, attenuation   177
Butler–Gärtner model   10–11
Butler–Volmer equation   238
Butler–Volmer kinetics   147

c

$C(U)$ curve, valve metal electrodes   7–8
Cabrera–Mott interface   6–7
capacitance
– density   80
– measurements   30–33
capacitors
– ceramic   62
– device types   62–65
– DRAM storage   82–90
– electrolytic, see electrolytes
– fundamentals   57–62
– manufacturing   57–74
– Nb/Nb$_2$O$_5$ 53
– polymer foil   62
– Ta   62–66
carbon electrode, glassy   205, 217
catalysis

– electro-, see electrocatalysis
– metal deposition   134
catalyst-derivatized electrodes   138–141
catalyst-derived brightening   111–112, 161–176
– stability analysis   173–176
catalyst-free electrolyte   165, 167
catalysts
– Cl   135
– competitive adsorption   135–137
– evolution   143
– function and consumption   139–143
– PEG–SPS–Cl   138, 141–144, 175
cathode material, conductive polymers   72–73
cathodic ETR current density   35
cation transfer coefficient   3
CEAC (curvature enhanced accelerator coverage)   111–113, 118, 161, 169–178
cell
– DRAM   80
– electrochemical   4
ceramic capacitors   62
– multi-layer   62
ceramic oxides, as dielectric films   1–99
charge transfer kinetics, site dependence   142–143
CHARMM   304
chelate structures   191, 244
chemical rate-determining step   243–244
chemical vapor deposition, see CVD
chemically modified electrodes   221–232
– redox active   232–241
chip manufacture, electrochemical processing   293
chronoamperometry   137–140, 206
chronocoulometric plots   206–207
CMOS (complementary metal oxide semiconductor)   77, 79
co-adsorption effects   130–134
co-deposition   125
Co porphyrins
– dioxygen reduction   252–259
– single-ring   256–259
$trans$-[Co([14]aneN$_4$)(OH$_2$)$_2$]$^{3+}$ 212–216
coarse-grained simulation methods   303–304
Co(III)(cyclam)   202, 204, 206, 212–216
competitive adsorption   135–137, 141–146
competitive interactions   122
complex dielectric constant   59
complex systems   289
conductive polymers, cathode material   72–73
conductivity, thermal   71–72
confidence intervals   315
consumption, catalysts   137–141

consumption processes 117–121
continuum methods 300
CoPI 253–255
– cyclic voltammogram 253
corrosion resistance 1
covalent attachment 223–225
critical wavenumber 175
crystallization nuclei 68
crystallographic orientation 17–18
current
– leakage, see leakage current
– limiting 206, 250, 264
– photo-, see photocurrent
– plateau 215, 249–251
– Poole–Frenkel 89
– threshold 160
current density
– diffusion-limited 143
– effect on microstructure 123
– ETR 35
curvature enhanced accelerator coverage, see CEAC
CVD (chemical vapor deposition) 2, 75, 89–96
cyberinfrastructure 311
cyclic voltammogram 20, 28–30
– Al/Al_2O_3 system 53–54
– anodic passivation 76
– capacitor devices 58–59
– conductive polymers 73
– CoPI 255
– Fe(III)TMPyP 205, 210–211, 217
– Hf/HfO_2 system 46–47
– Nafion 226–227
– Nb/Nb_2O_5 system 49
– thermally activated macrocycles 280
– Zr/ZrO_2 system 38, 43–44

d

D-optimal design 313
Damascene copper process 163
data collection, high-throughput 315
deactivation, adsorbates 121
Debye length 35, 57
deep trench, see DT
defect state density 38, 51, 57
defraction analysis, X-ray 53–54
density
– capacitance 80
– current, see current density
– defect state 38, 57
– donor 30–34, 36–37
deoxidization 68
depassivation, potential-dependent 131

deposit, growing 119–121
deposition
– co- 125
– electro-, see electrodeposition
– metals, see metal deposition
design
– D-optimal 313
– electrochemical systems 289–324
– molecular 259, 320
– multiscale model-based 315–322
desorption 117, 121–122, 302, 307
– hydrocarbon 93
– slow 223
destabilizing influences, superconformal film growth 108–110
device simulation 298–299
DHE (dynamic hydrogen reference electrode) 282, 284
dicobalt porphyrin dimers 266
dielectric constant 92–95
– Al/Al_2O_3 system 53
– capacitance measurements 30–31
– complex 59
– Hf/HfO_2 system 47–48
– Nb/Nb_2O_5 system 49, 52–53
– relative, see relative dielectric constant
– Ta/Ta_2O_5 system 56
– Zr/ZrO_2 system 44, 46
dielectric films
– experimental approaches 2–5
– for electronic devices 1–99
dielectric materials
– alternative 90–96
– Ta_2O_5 92–95
dielectrics, nano- 1–2
– nano- 97
differential pulse voltammetry 225, 238
diffuse double layer 7
diffusion
– boundary layer 28
– lateral hole 24
diffusion coefficient 28, 155, 158, 159, 301
– adsorption 117, 119–120
– electrocatalysis 204–205
– transition metal macrocycles 211–212
diffusion equation 148
diffusion-limited current density 143
dimers, dicobalt porphyrin 267
dioxygen reduction 191–284
– Co porphyrins 252–259
– electrocatalytic aspects 241–268
– in aqueous electrolytes 193–199, 212–219
– macrocyclic-mediated 212–219
– ring functionalization 252–259

dipping, single step 72
direct leakage current 65
direct tunneling 35, 81
disc electrode, rotating, see RDE
dislocation formation, Ta capacitors 67–70
dissipation factor 60
donor density 30–34, 36–37
dopant profiles, boron 319
doping 63, 70
– poly-Si 87, 96
– Si/SiO$_2$ system 75, 77, 83–84, 89
– sidewall 95
– Ti/TiO$_2$ system 36
– V- 51
double layer, diffuse 7
DRAM (dynamic random access memory)
– 'Best'-cell 83
– cell 80
– operation 81–82
– Si/SiO$_2$ system 80–82
– stack-capacitor 91
– storage capacitor 82–90
droplet method, nanoliter 25–28
DT (deep trench) 82–90
– bottle-shaped 85–88
– etching aspect ratio 84–85
dynamic hydrogen reference electrode (DHE) 282, 284
dynamic polarization curves 197, 198–208, 213, 245, 254–257
dynamically coupled multiscale simulation 307–309
dynamics
– Langevin 301
– molecular 300–302, 303–304
– redox 238–241

e
ECE (electrochemical–chemical–electrochemical) mechanism 241
ECR (electron cyclotron resonance) 92–93
electrical capacitance, see capacitance
electrical/electrical micro-methods 5
electrical method class, photocurrent measurements 10–15
electrocatalysis 192–193
– diffusion coefficient 211–212
– dioxygen reduction 241–268
– heterogeneous, see heterogeneous electrocatalysis
– homogeneous, see homogeneous electrocatalysis
electrocatalysts
– Co(III)(cyclam) 202, 204, 206, 212–216
– Fe(III)TMPyP 203–212, 216–219, 252
– metalloporphyrins 203
– phthalocyanines 191, 199, 244–247
– porphyrins, see porphyrins
– Pt-based 191, 196, 259
– transition metal macrocycles 191–284
electroceramic thin film dielectrics 46
electrochemical applications
– laser 8–10
– multiscale modeling 297–298
electrochemical cell 4
electrochemical–chemical–electrochemical (ECE) mechanism 241–243
electrochemical measurements
– modified electrodes 221–226
– redox active electrodes 232–233
– thermally activated macrocylces 280
electrochemical metrics 135
electrochemical oxide layer formation 5–7
electrochemical photocurrent measurements 10–15
electrochemical potential 1
electrochemical processing, chip manufacture 293
electrochemical rate-determining step 243
electrochemical systems, design 289–324
electrodeposition 108–116, 121–122
– multiscale modeling 310, 315
electrodes
– blocked 135–137
– catalyst-derivatized 138–141
– chemically modified 221–232
– DHE 282, 284
– glassy carbon 205, 217
– macroscopic single crystal 29
– OTTLE 232–241
– photoresist micro-, see photoresist microelectrodes
– redox active 232–241
– rotating disc, see RDE
– rotating ring-disk, see RRDE
– standard hydrogen, see SHE
– Teflon-bonded 279
– thin porous coating 270–271
electrohydrodynamic impedance spectroscopy 127
electrolytes
– alkaline 194
– anodic passivation 76
– aqueous, see aqueous electrolytes
– capacitance measurements 30, 46–47, 55
– catalyst-free 165, 167
– electrochemical measurements 16, 18, 25, 28

Index

– electrolyte capacitors   1, 37, 52–54, 97
– interfaces   6–7, 19, 36
– manufacturing of electrolyte capacitors   57–74
– PEG–SPS–Cl   108, 144, 164, 170, 178
– solid   64
– SPS adsorption   143–146
electron cyclotron resonance (ECR)   92–93
electron-donating groups   256–259
electron polarization   59
electron spin resonance spectra   203
electron transfer, heterogeneous   211–212
electron transfer reaction, see ETR
electronic applications
– Nb/Nb$_2$O$_5$ 53
– ZrO$_2$ applications   44–46
electronic avalanche effects   67
electronic devices, active/passive   1–99
electronic properties, modified electrodes   227–229
ellipsometry, spectroscopic anisotropy micro-, see SAME
energy
– band gap   3, 15
– mobility gap   19
engineering science, simulation-based   290–291
ESR (equivalent series resistance)   60–61, 73
estimation, Bayesian   312
etched foil technology   54, 64
etching   5
– anisotropic   85
– aspect ratio   84–85
– RIE   84–86
– Si/SiO$_2$ system   83, 85–89, 91, 95
etching aspect ratio, DT (deep trench)   84–85
ETR (electron transfer reaction)   34–35, 238–241
evolution
– adsorbates   121–122
– catalysts   143
– morphological   146–178

f

face-to-face porphyrins   259–268
feature filling
– catalyst-derived brightening   161–173
– geometric leveling   153–159
Fe(III)TMPyP   203–212, 216–219, 251
– cyclic voltammogram   205, 210, 217
Fermi level pinning   75
ferroelectrics   1, 90–91
field effect transistors, metal oxide semiconductor, see MOSFETs

filling   153–159
– simulation   148
– super-, see superfilling
film
– amorphous, see amorphous films
– dielectric   2–5
– HfO$_2$ 48
– monolayer   115
– superconformal growth, see superconformal film growth
– TiO$_2$ 19–21
– ultra-thin amorphous   11–15
film thickness   4, 10–11, 14–15, 18–22, 97
– effect on microstructure   123–124
– Hf/HfO$_2$ system   47
– Nb/Nb$_2$O$_5$ system   49–51
– Ta/Ta$_2$O$_5$ system   56–57
– variation   53
– Zr/ZrO$_2$ system   40, 44
– see also layer thickness
fluent   300
fluid dynamics codes   324
folded bit-line design   81
formal redox potential   204–209
formation factor   18, 42
– anodic   3
– HfO$_2$ 47
– Nb/Nb$_2$O$_5$ system   49
– oxide, see oxide formation factor
– Ta/Ta$_2$O$_5$ system   46
Fowler–Nordheim tunneling   81
Frenkel, Poole–Frenkel current   89
Frumkin isotherm   117
functional groups   225
functionalization, ring   252–259

g

galvanostatic oxide growth   7
gap energy
– mobility   19
– see also band gab energy
Gärtner, Butler–Gärtner model   10–11
gate length   77, 79–80
Gaussian temperature distribution   9
geometric leveling   110, 150–152
– stability analysis   160–161
glassy carbon electrode   205
grain orientation, Ti substrate   17–18
grain refinement   111–112
grains, hemispherical Si, see HSG
green body uniformity   72
groove
– hemi-cylindrical   162
– semicircular   152

– V-notch   151
groups
– amino   138
– electron-donating   256–259
– functional   225
– hydroxyl   7
– redox   256–257
growing deposit, adsorbates   119–121

h

hcp (hexagonal closed package)   39, 48
heat accumulation   66
Helmholtz layer   7, 19
hemi-cylindrical groove   162
hemin, see Hm
hemispherical Si grains, see HSG
heterogeneity, Ti/TiO$_2$ system   15
heterogeneous electrocatalysis   203, 219–268
– model systems   244–269
– pH   234–238, 246–255
heterogeneous electron transfer, reaction rates   211–212
heterogeneous oxygen reduction   194
hexagonal closed package (hcp)   39, 48
Hf/HfO$_2$ system   46–48
HfO$_2$, film formation   48
high-field equation   6
high-field model   6
high-throughput data collection   315
highest occupied molecular orbitals (HOMO)   201, 267–268
Hm (hemin)   221–224, 229, 233–236
hole diffusion, lateral   24
hole migration   19
HOMO (highest occupied molecular orbitals)   201, 267–268
homogeneous electrocatalysis   202–219
– model systems   212–219
HSGs (hemispherical Si grains)   85, 94
– formation   89–90
hydrodynamic boundary layer thickness   110, 117, 126, 128
hydroxyl groups   7
hysteretic voltammogram   136

i

ICs (integrated circuits)   77, 96–97, 291, 294
– molecular simulation   298–299
impedance   60
– optical   240
implantation, ion   317
impurity concentration   142
in situ microscopic sample monitoring   4
in situ spectroscopy   281–283

– modified electrodes   226–232
inhibition
– metal deposition   125–130
– PEG   130
inhibitor-based leveling   110–111, 152–161
inhibitor flux   126
instabilities, oscillatory   176
integrated modeling   296
integrated pyrolysis–mass spectra   276
interface
– non-planar   161
– solid–liquid   307
interface motion   146–178
internal reflections, multiple   12
internally coupled moving boundary   309
intra-layer transport   109
intrinsic charge carrier mobility   74
ion implantation   317
ionic polarization   59
$i_{ph}(U)$ measurements   18–19
iron tetrakis(N-methyl-4-pyridyl) porphyrin   216–219
iron(III) protoporphyrin IX chloride (hemin), see Hm
isotherms
– adsorption   220–221
– Frumkin   117
– Langmuir   117
isotropic Beilby layer   40

k

kinetics
– adsorption   117
– Butler–Volmer   147
– charge transfer   142–143
– Langmuir   117, 143, 164
KMC (kinetic Monte Carlo) simulation   300, 304, 307, 309
Koutecky–Levich plot   251, 255

l

Lambert–Beer's law   13–14
Langevin dynamics   301
Langmuir isotherm   117, 120, 128, 154
Langmuir kinetics   117, 143, 164
Laplacian diffusion equation   148
laser scanning
– TiO$_2$ films   19–21
– UV-   21–22
lasers
– electrochemistry applications   8–10
– thermal effects   9–10
lateral hole diffusion   24
laws and equations

– Airy's formula 9
– Butler–Volmer equation 238
– high field equation 6
– Langmuir isotherm, see Langmuir isotherm
– Langmuir kinetics 117
– Laplacian diffusion equation 148
– Master equation 302
– partial differential-algebraic equations 300–302
– Poisson equation 7
– Randles–Sevcik equation 218
– Schottky–Mott equation 31–32
– time-dependent diffusion equation 148
layer
– diffuse double 7
– diffusion boundary 28
– Helmholtz 7, 19
– isotropic Beilby 40
– nano-crystalline 42
– space charge 7, 13, 57
layer formation
– anodic passive 28–30
– electrochemical 5–7
layer thickness
– boundary, see boundary layer thickness
– hydrodynamic boundary, see hydrodynamic boundary layer thickness
– see also film thickness
leakage current 65–66, 79, 89–90
– capacitors 62
– direct 65
– Nb/Nb_2O_5 capacitors 53
– Ta/Ta_2O_5 system 93–95
leakage node 90
leakage paths 46, 65, 67–68, 70–71, 77
length scales 176–178
– multiple 293
leveling
– geometric 110, 150–152
– inhibitor-based 110, 111, 153–161
leveling power 153–155
Levich plot 251, 255
limiting current 206, 250, 264
locodynamic experiment 22
LUMO (lowest unoccupied molecular orbital) 201

m

macrocycles
– monomeric 225, 259
– ruthenation 257
– transition metal, see transition metal macrocycles
– (un)metallated 200

macrocyclic-mediated dioxygen reduction 212–219
macroscopic single crystal electrodes 29
mass conservation 118–120
mass spectrometry, pyrolysis 274–276
Master equation 304
MD (molecular dynamics) 300–301, 303–304
measurements
– capacitance 30–33
– electrochemical photocurrent 10–15
– ETR 34–35
– $i_{ph}(U)$ 18–19
– transient 33–34
memory, dynamic random access, see DRAM
MEMS (microelectromechanical systems) 107, 297–298, 306
meso-ring position 259
metal deposition
– effect of competitive adsorption 141–146
– inhibition 125–130
metal oxide semiconductor field effect transistors, see MOSFETs
metallated macrocycles 200
metalloporphyrins 203
metals
– transition metal, see transition metal
– valve, see valve metals
metrics, electrochemical 135
micro-ellipsometry, spectroscopic anisotropy, see SAME
micro-methods 4–5
micro-reflection spectroscopy, modified TiO_2 films 22–23
microelectrochemistry, photoresist 25–28
microelectrodes, photoresist, see photoresist microelectrodes
microelectromechanical systems (MEMS) 107, 297–298, 306
microelectronic applications, multiscale modeling 295–296
microfluidics 298, 306
micromanipulators 26
microscopic control 25–26
microscopic modification, TiO_2 films 19–21
microscopic sample monitoring, in situ 4
microstructure, impact of segregating adsorbates 122–125
migration, holes 19
mobility, intrinsic charge carrier 74
mobility gap energy 19
model systems
– heterogeneous electrocatalysis 244–269
– homogeneous electrocatalysis 212–219
modeling

– hierarchy 296
– integrated 296
– multiscale, see multiscale modeling
modified anode materials 73–74
modified TiO_2 films 21–25
molar paramagnetic susceptibility 212
molecular design 259, 320
molecular dynamics (MD) 300–301, 303–304
molecular nanotechnology 297
molecular orbitals 201
molecular simulation 298–304
– coarse-grained methods 303–304
– ICs 298–299
– methods 300–302
molecular speciation
– pH 209–212
– transition metal macrocycles 207–211
monolayer films 115
monomeric macrocycles 225, 259
morphological evolution 146–178
MOSFETs (metal oxide semiconductor field effect transistors) 300
– Hf/HfO_2 system 46
– Si/SiO_2 system 77–80
Mossbauer effect spectroscopy 276–277
Mott
– Cabrera–Mott interface 6–7
– Schottky–Mott, see Schottky–Mott
moving boundary, internally coupled 309
multicycle voltammetry 144
multilayer ceramic capacitors 62
multiple internal reflections 12
multiple length scales 293
multiple time scales 293
multiscale modeling 289–334
– challenges 310–315
– electrodeposition 310, 314
– rough surfaces 314
multiscale simulation 291–293, 304–309
– dynamically coupled 307–309
multiscale systems engineering 291
multistep optimization 316

n

Nafion, cyclic voltammogram 226–227
nanocrystalline layer 42
nanodielectrics 1–2, 97
nanolaminate component 46
nanoliter droplet method 25–28
nanopipette 25
nanoscale devices 291
nanoscale science 296–297
nanotechnology 77
– molecular 298

Nb/Nb_2O_5 system 49–53
nitridation, rapid thermal 90
node leakage 90
non-planar interface 161
Nordheim, Fowler–Nordheim tunneling 81
NSF Blue Ribbon Panel report 290–291
nuclei, crystallization 68

o

optical anisotropy 38, 48
optical/chemical micro-methods 5
optical/electrical micro-methods 5
optical impedance 240
optical method class, photocurrent measurements 10–15
optical/optical micro-methods 4–5
optically transparent thin layer electrode (OTTLE) 207–211
optimization, multistep 316
oscillatory instabilities 176
OTTLE (optically transparent thin layer electrode) 207–211
overpotential 139–141, 144–147, 173–175
oxidation, anodic 29
oxide collar 85
oxide formation
– anodic, see anodic oxide formation
– potentiodynamic, see potentiodynamic oxide formation
– potentiostatic, see potentiostatic oxide formation
oxide formation factor 7, 44–45, 53
oxide growth
– anodic 37
– galvanostatic 7
oxide layer formation, electrochemical 5–7
oxygen content, Ta capacitors 67–70
oxygen reduction
– heterogeneous 194
– polarization curves 271–274

p

paramagnetic susceptibility, molar 212
partial differential-algebraic equations 298–300
passivation, anodic 76
passive electronic devices 1–99
passive layer formation, anodic 28–30
PECVD (plasma-enhanced CVD) 92–93, 96
PEG (poly(ethylene glycol)) 130–146, 161, 164–166, 173–178, 313–314
– inhibition 130
permittivity of a vacuum 59
perovskite structure 1

pH 146
– aqueous electrolytes 193–199
– heterogeneous electrocatalysis 219–220
– interfacial shifts 124
– molecular speciation 212
photocorrosion 22
photocurrent measurements, electrochemical 10–15
photocurrent model, ultra-thin amorphous films 11–15
photocurrent spectra
– Ti/TiO$_2$ system 18–19
– Zr/ZrO$_2$ system 42–43
photoelectrochemistry 10–15
photoelectron spectroscopy, X-ray, see XPS
photoresist microelectrochemistry, modified TiO$_2$ films 25–28
photoresist microelectrodes
– Ti/TiO$_2$ system 28–35
– Zr/ZrO$_2$ system 43–44
phthalocyanines 191, 199
– transition metal 244–247
physical vapor deposition, see PVD
pinning 17
– Fermi level 75
pitting 36
plateau current 215, 248–251
plating, through-mask 116
Poisson equation 7
polarization 59
polarization curves
– dynamic, see dynamic polarization curves
– oxygen reduction 271–274
– steady state 198, 249
– steady state ring–disk 272–273
polycrystalline samples 16
poly(ethylene glycol), see PEG
polymers
– conductive 72–73
– foil capacitors 62
Poole–Frenkel current 89
porous coating electrodes, thin 269, 271
porphyrins 191–193, 199, 247–252
– amino functionalized 224
– Co, see Co porphyrins
– dimers 266
– face-to-face 259–268
– iron tetrakis(N-methyl-4-pyridyl)porphyrin 216–219
– metallo- 203
potential
– electrochemical 1
– formal redox 204–209
– over- 141

– standard reduction 193, 195
potential-dependent adsorption 144–145
potential-dependent depassivation 131
potential-step experiment 33
potentiodynamic oxide formation 2, 98
– Al/Al$_2$O$_3$ 53
– Nb/Nb$_2$O$_5$ 49
– Ta/Ta$_2$O$_5$ 55
– Ti/TiO$_2$ 27–28, 30–34
– Zr/ZrO$_2$ 37–38, 40–41, 44
potentiostatic oxide formation 2
– growth texture dependence 70
– Ta/Ta$_2$O$_5$ 55
– Zr/ZrO$_2$ 30–31, 33, 37, 44
Pourbaix diagram 51
precipitation 224
process simulation 298–299
production, Ta capacitors 62–65
protonation 267
Pt-based electrocatalysts 192
pulse voltammetry, differential 225
purity improvement 69
PVD (physical vapor deposition) 2, 75–76, 91, 94
pyrolysis-mass spectrometry 274–276
β-pyrrole position 255, 259–260, 264

q
QM (quantum mechanical) simulation 300, 305–306
quantum yield 18–20
quasi-in situ spectroscopy 281–283
quasi-isothermal conditions 9

r
radiation, synchrotron 230
radiotracer analysis 121
Randles–Sevcik equation 218
random access memory, dynamic, see DRAM
rapid thermal annealing 93, 299, 317–319
rapid thermal nitridation 90
rate-determining step 243
RDE (rotating disc electrode) 126, 196, 206, 214
– heterogeneous electrocatalysis 221–222
reaction rates, heterogeneous electron transfer 211–212
reactive ion etching 84–86
rectifying effect 5–6
redox active chemically modified electrodes 232–241
redox dynamics 238–241
redox groups 256–258
redox potential, formal 205–209

redox speciation 235–238
reduction
– dioxygen, see dioxygen reduction
– standard potential 193, 195
reflectance, UV–visible 233
reflection spectroscopy, micro- 22–23
reflections, multiple internal 12
relative dielectric constant 1–3, 44
resistance, corrosion 1
resonance tunneling 35
retention time 61
RIE (reactive ion etching) 84–86
ring-disk polarization curves, steady state 271–273
ring functionalization, dioxygen reduction 252–259
meso-ring position 258
ripple voltage 62
rough surfaces 109
– multiscale modeling 314
RRDE (rotating ring-disk electrode) 196–199, 213
– heterogeneous electrocatalysis 221–222
runaway, thermal 70–74
ruthenation, macrocycles 257
rutile, see TiO_2

s

SAME (spectroscopic anisotropy micro-ellipsometry) 4
– grain orientation determination 17–18
– modified TiO_2 films 23–25
– Zr/ZrO_2 system 38–42
sample monitoring, in situ microscopic 4
scanning, laser, see laser scanning
Schottky–Mott analysis 51, 56, 97
Schottky–Mott behavior 37, 50
Schottky–Mott chart/plot 30–32, 56
Schottky–Mott equation 31–32
Schottky–Mott relationship 8
seam formation 170
Secvik, Randles–Sevcik equation 218
segregation, surface 117–121
self-discharge time 61
self-healing mechanism 66
semicircular groove 152
semiconductors 74
sense amplifier 81
shallow trench isolation, see STI
shape change
– algorithms 149
– simulation 146–150, 167
SHE (standard hydrogen electrode) 16
Si grains, hemispherical, see HSG

Si oxides, as dielectric films 1–99
Si/SiO_2 system 74–96
– DRAM (dynamic random access memory) 80–82
– DRAM storage capacitor 82–90
– MOSFETs 77–80
sidewall doping 95
simulated annealing 315
simulation
– -based engineering science 290–291
– device/process 298–299
– KMC 300–304, 307, 309
– molecular 298–304
– multiscale, see multiscale simulation
– QM 305–306
– shape change 146, 150, 167
– stochastic codes 316
single-crystal electrodes, macroscopic 29
single-ring Co porphyrin 255–259
single-step dipping 72
single Ti/TiO_2 grains 18–19
– anodic passive layer formation 28–30
sinter aid 72
sinter body, anode 71–72
sinter conditions, Ta capacitors 67–70
sintered powder technology 54, 63–64
sintering 97
– Nb/Nb_2O_5 system 52
– Y- 69
site dependence, charge transfer kinetics 142–143
slow desorption 222
slow pyrolysis–mass spectra 276
smoothing mechanisms 110–113
solid–liquid interface 307
solution phase transition metal macrocycles, intrinsic properties 204–212
solution phase voltammogram 266
space charge layer 7, 13, 57
space charge polarization 59
speciation
– molecular 209–211
– redox 235–238
spectra
– electron spin resonance 203
– photocurrent, see photocurrent spectra
– pyrolysis–mass spectra 276–277
– UV–visible 210
– XPS 51
– XRD 96
spectroscopy
– electrohydrodynamic impedance 127
– in situ 281–283
– micro-reflection 22–23

- Mössbauer effect 276–277
- quasi *in situ* 281–283
- redox active electrodes 233–235
- spectroscopic anisotropy micro-ellipsometry, *see* SAME
- thermally activated macrocylces 275–281
- X-ray absorption fine structure 278–281
- XPS 51, 281
spin resonance spectra, electron 203
sponge-like structure 52, 63–64, 97
spontaneous adsorption 221–223
SPS (sulfonate-terminated disulfide) 135–148, 164–178, 313–314
- adsorption 143–145, 164
stability analysis
- catalyst-derived brightening 173–175
- geometric leveling 160–161
stabilization
- across length scales 112–113
- superconformal film growth 110–113
stack-capacitor DRAMs 91
standard hydrogen electrode, *see* SHE
standard reduction potential 193, 195
steady state polarization curves 198, 249
steady state ring-disk polarization curves 272–273
STI (shallow trench isolation) 77, 84
stochastic simulation codes 316
storage capacitor, DRAM, *see* DRAM
strongly adsorbed surfactant 118
structural properties
- modified electrodes 232–235
- thermally activated macrocylces 275–281
sub-micrometer trenches 108
substrate grain orientation, Ti 17–18
substrate texture 37
sulfonate-terminated disulfide, *see* SPS
supercapacitor 58–59, 73
superconformal film growth 107–179
- additive processes 113–146
- destabilizing influences 108–110
- interface motion 146, 178
- morphological evolution 146, 178
- stabilization 110, 114
superfilling 107–109, 157–158, 165–174
- vias 170–171
superleveling 163
surface mounted devices 65
surface segregation 117–121
surface states, number of 76
surfaces, rough 109
- rough 314
surfactant, strongly adsorbed 118
susceptibility, molar paramagnetic 212

synchrotron radiation 230
synergistic interactions 122
systems engineering, multiscale 291

t

Ta capacitors 62–66
- dislocation formation 67–70
Ta/Ta_2O_5 system 54–57
Tafel plots 34, 247
Tafel slope 248
Ta_2O_5, DT technology 92–95
Teflon-bonded electrode 279
temperature distribution, Gaussian 9
texture, Zr/ZrO_2 substrate 37
texture dependence
- HfO_2 film formation 48
- ZrO_2 film formation 40
thermal activation, transition metal macrocycles 269–283
thermal annealing 68–69
- rapid 93, 317–318
thermal conductivity, anode sinter body 71–72
thermal effects, lasers 9–10
thermal nitridation, rapid 90
thermal runaway, Ta capacitors 70–74
thermodynamic aspects, redox active electrodes 232–235
thickness
- boundary layer, *see* boundary layer thickness
- film, *see* film thickness
- hydrodynamic boundary layer, *see* hydrodynamic boundary layer thickness
thin film dielectrics, electroceramic 46
thin layer electrode, optically transparent 206–212
thin porous coating electrodes 269, 271
threshold current 160
through-mask plating 114
Ti substrate grain orientation 17–18
Ti/TiO_2 system 15–37
- DT technology 95–96
- heterogeneity 15
- single grains 18–19
time-dependent diffusion equation 148
time scales, multiple 293
TiO_2, photocurrent model 11–15
TiO_2 films
- microscopic modification 19–21
- modified 21–25
trans-$[Co([14]aneN_4)(OH_2)_2]^{3+}$ 212–216
transfer coefficient, cation 3
transient measurements 33–34
transients, chronoamperometric 137–138, 141

transistors, metal oxide semiconductor field effect, *see* MOSFETs
transition metal macrocycles   191–284
– diffusion coefficients   204–205
– electrocatalytic properties   201–204
– fundamentals   199–201
– heterogeneous electrocatalysis   219–268
– homogeneous electrocatalysis   204–219
– solution phase   204–212
– thermal activation   269–283
transition metal phthalocyanines   244–247
transition probabilities   302
transport, intra-layer   109
trenches
– deep, *see* DT
– filling   166, 168–170
– filling simulations   148
– sub-micrometer   108
tunneling
– direct   81
– direct/resonance   35
– Fowler–Nordheim   81

u

ultra-thin amorphous films, photocurrent model   11–15
unmetallated macrocycles   200
UV-laser scanning analysis   21–22
UV–visible reflectance   233
UV–visible spectra   210

v

V-doping   51
V-notch groove   151
valve metal systems
– Al/Al_2O_3 system   53–54
– electrolytic capacitor manufacturing   57–74
– Hf/HfO_2 system   46–48
– Nb/Nb_2O_5 system   49–53
– Ta/Ta_2O_5 system   54–57
– Ti/TiO_2 15–37
– Zr/ZrO_2 system   37–46
valve metals
– as dielectric films   1–99
– electrochemical oxide layer formation   5–7
– electrode $C(U)$ curves   7–8
vias   164–165, 170–172
vibrational properties, modified electrodes   229–230
void formation   170
Volmer
– Butler–Volmer equation   238
– Butler–Volmer kinetics   148
voltammetry   114, 122, 136–137, 146
– differential pulse   225, 238
– multicycle   144
voltammogram
– cyclic, *see* cyclic voltammogram
– hysteretic   136
– solution phase   266

w

wavenumber, critical   175

x

X-ray absorption fine structure spectroscopy   277–281
XANES (X-ray absorption near edge structure)   230–231, 235, 237, 275–281
XPS (X-ray photoelectron spectroscopy)   281
– spectra   51
XRD (X-ray defraction) analysis   53–54
XRD (X-ray defraction) spectra   96

y

Y-sintering   69
yield, quantum, *see* quantum yield

z

Zr/ZrO_2 system   37–46